HOW TO BUILD A BRAIN

OXFORD SERIES ON COGNITIVE MODELS AND ARCHITECTURES

Integrated Models of Cognitive Systems
Edited by Wayne D. Gray

In Order to Learn:
How the Sequence of Topics Influences Learning
Edited by Frank E. Ritter, Josef Nerb,
Erno Lehtinen, and Timothy O'Shea

How Can the Human Mind Occur
in the Physical Universe?
By John R. Anderson

Principles of Synthetic Intelligence PSI:
An Architecture of Motivated Cognition
By Joscha Bach

The Multitasking Mind
By David D. Salvucci and Niels A. Taatgen

How to Build a Brain:
A Neural Architecture for Biological Cognition
By Chris Eliasmith

HOW TO BUILD A BRAIN

A NEURAL ARCHITECTURE FOR BIOLOGICAL COGNITION

CHRIS ELIASMITH

Oxford University Press is a department of the University of Oxford.
It furthers the University's objective of excellence in research, scholarship,
and education by publishing worldwide.

Oxford New York
Auckland Cape Town Dar es Salaam Hong Kong Karachi
Kuala Lumpur Madrid Melbourne Mexico City Nairobi
New Delhi Shanghai Taipei Toronto

With offices in
Argentina Austria Brazil Chile Czech Republic France Greece
Guatemala Hungary Italy Japan Poland Portugal Singapore
South Korea Switzerland Thailand Turkey Ukraine Vietnam

Published in the United States of America by
Oxford University Press
198 Madison Avenue, New York, NY 10016

First issued as an Oxford University Press paperback, 2015.

Library of Congress Cataloging-in-Publication Data
Eliasmith, Chris.
How to build a brain : a neural architecture for biological cognition / Chris Eliasmith.
pages cm
Includes bibliographical references and index.
ISBN 978-0-19-979454-6 (hardcover : alk. paper); 978-0-19-026212-9 (paperback : alk. paper)
1. Brain. 2. Neural circuitry 3. Neural networks (Neurobiology) 4. Cognition. I. Title.
QP376.E545 2013
612.8'2—dc23
2012046211

To Jen, Alana, Alex, Celeste, Jerry, Janet, Steve, Kim, Michael, and Jessie–my family.

CONTENTS

PREFACE

I suspect that when most of us want to know "how the brain works," we are somewhat disappointed with an answer that describes how individual neurons function. Similarly, an answer that picks out the parts of the brain that have increased activity while reading words, or an answer that describes what chemicals are in less abundance when someone is depressed, is not what we have in mind. Certainly these observations are all *parts* of an answer, but none of them traces the path from perception to action, detailing the many twists and turns along the way.

Rather, I suspect that the kind of answer that we would like is the kind of answer that early pioneers in cognitive science were trying to construct. Answers like those provided by Allen Newell (1990) in his book *Unified theories of cognition*. Newell was interested in understanding how the "wheels turn" and the "gears grind" during cognitive behavior. He was interested in the details of brain function during all the variety of tasks that we might call "cognitive." Ultimately, Newell's answer, while interesting and important, was not found to be very convincing by many researchers in the behavioral sciences. Perhaps his project was too ambitious. Perhaps this is why many have since focused on characterizing the mechanisms of relatively small parts of the brain.

Regardless, I think we are now in a position to again try to provide the kind of answer Newell was after. In the grandest (and hence least plausible) sense, that is the purpose of this book. It is an attempt to provide the basic principles, and an architecture, for building the "wheels" and "gears" needed to drive cognition–in a way that integrates our knowledge of biology.[1]

[1] In an earlier paper that shares its title with this book, I speculatively outlined how I think we should proceed with this task (Eliasmith, 2007). Here I attempt to realize that outline.

The architecture I propose and develop in this book is called the "Semantic Pointer Architecture" (SPA), for reasons that will subsequently become clear. I believe a central departure of the SPA from past attempts to articulate theories of cognition is that it does not *favor* cognition. That is, it does not start with the suggestion that cognitive systems perform logic-like computations in a language of thought or that they are best described as abstract nodes connected together in a network. Rather, it gives equal time to biology. That is why it depends on a theory of how to build a *brain*. So, the architecture I propose adopts cognitively relevant representations, computations, and dynamics that are natural to implement in large-scale, biologically plausible neural networks. In short, the SPA is centrally inspired by understanding cognition as a biological process–or what I refer to as "biological cognition."

I believe the biggest benefit of this approach to specifying a cognitive architecture is that it allows for a broad unification of empirical data. If we take our models to be biologically detailed, they should be comparable to biological data (such as single cell physiology, synaptic mechanisms, neurotransmitter transport, neuroanatomy, fMRI, and so on). If we take our models to implement cognitive behavior, then they should be comparable to psychological data (such as accuracy, reaction times, individual choice behavior, population choice distributions, and so on). Building models that are able to address all such data concurrently results in models that are better constrained, more convincing, and hence more interesting than what many have come to expect from cognitive models.

I realize this is a lot to promise, and I leave it to the reader to judge how close the SPA comes to this goal. I only ask that judgement be reserved until Chapter 7, where I present a single model that is fully implemented in spiking neurons and able to perform any of eight different motor, perceptual, and cognitive tasks, without ever changing the model. These tasks include object recognition, drawing, induction over structured representations, and reinforcement learning, among others. The model is constrained by data relating to neurotransmitter usage in various brain areas, detailed neuroanatomy, single neuron response curves, and so forth, and is able to qualitatively and quantitatively reproduce human performance on these tasks. I believe this is currently the largest *functional* simulation of a brain.

There are several impressive examples of large-scale, single-cell neural simulations (de Garis et al., 2010). The best known among these include Modha's Cognitive Computation project at Almeda Labs (Ananthanarayanan & Modha, 2007), Markram's Blue Brain project at EPFL (Markram, 2006), and Izhikevich's large-scale cortical simulations (Izhikevich & Edelman, 2008). Each of these are distinctive is some way. Izhikevich's simulation is on the scale of human cortex, running approximately 100 billion neurons. Modha's work is on the scale of cat cortex (1 billion neurons) but is much faster than Izhikevich's simulations, although both share largely random connectivity in their networks. Markram's project uses a similar amount of computational resources but focuses on the cortical column in detail. It has simulated approximately 1 million neurons to date but at a level of biological detail much beyond previous work, more accurately reflecting connectivity, synaptic and neural dynamics, and so on. Critically, none of these past models have demonstrated clear perceptual, cognitive, or motor function. Random connectivity does not allow any particular function to be performed, and focus on the cortical column makes it unclear what kinds of inputs the model should have and how to interpret its output (de Garis et al., 2010). The model presented here, in contrast, has image sequences as input and motor behavior as output. It demonstrates recognition, classification, language-like induction, serial working memory, and other perceptual, cognitive, and motor behaviors. For this reason, I believe that despite its modest scale (2.5 million neurons), it has something to offer to the current state-of-the-art.

To provide the material needed to understand how this model was constructed, I begin, in Chapter 1, by providing an outline of what is expected of a cognitive theory, and then describe a systematic method for building large-scale, biologically realistic models in Chapter 2. In Chapters 3 through 6 I describe four central aspects of the SPA that allow it to capture fundamental properties of biological cognition. These include: (1) semantics, which I take to encompass the symbol grounding problem for both motor and perceptual systems; (2) syntax, which is fundamental to constructing the structured representations employed across cognitive behaviors; (3) control, which encompasses methods for both action selection and information routing; and (4) learning and memory–basic examples of adaptation. I provide example models throughout these chapters demonstrating the various methods. As mentioned, Chapter 7 integrates all of these methods into an implementation of the SPA proper.

In part two of the book, I turn to a comparison of the SPA to past work. In Chapter 8 I propose a set of criteria drawn from past suggestions for evaluating cognitive theories. I then describe current state-of-the-art approaches to cognitive modeling, evaluate them with respect to the proposed criteria, and compare and contrast them with the SPA in Chapter 9. Finally, in Chapter 10 I discuss several conceptual consequences of adopting the proposed approach to understanding cognitive function and highlight current limitations and future directions for the SPA.

I have attempted to make this book broadly accessible. As a result, the mathematical details related to various example models and theoretical discussions have been moved into appendices. As well, the first appendix is a brief overview of the mathematical background needed to understand most of the technical parts of the book.

Finally, I will note that despite my proposing a specific architecture, most elements of this book are usable independently of that architecture. There is, contrary to what you might expect, an intended *practical* reading of the title of this book–for those who wish, this book can be used to make brain building more than something you read theories about. Accordingly, I provide tutorials for constructing spiking neuron-level models at the end of each chapter. I have chosen these tutorials to highlight concepts discussed in nearby chapters, but many of these tutorials do not depend on the SPA for their relevance to cognitive or neural modeling.

The tutorials employ the freely downloadable neural simulator Nengo that was originally developed by my lab and continues to be developed in the research center of which I am a part. The first tutorial describes how to set up and run this graphical simulator, for those wishing to dive in. Many additional resources related to the simulator can be found at `http://nengo.ca/`. For those wishing to use this book as a text for a course, please visit `http://nengo.ca/build-a-brain` for additional resources and information.

ACKNOWLEDGMENTS

Attempting to reverse-engineer a system as complex as a biological brain is not a task for a single person. I am incredibly grateful for the many good friends and acquaintances who have contributed to this book. I have acknowledged particularly critical contributions to any given chapter by including co-authors as appropriate. And, I have attempted to be careful to name each person whose research project is reflected in these pages as I discuss it. These contributors include Trevor Bekolay (learning), Bruce Bobier (attention), Xuan Choo (working memory and Spaun), Travis DeWolf (motor control), Dan Rasmussen (induction), and Charlie Tang (vision and clean-up). In addition, I must give special recognition to Terry Stewart, who has been an enormous influence on the development of these ideas (especially the role of basal ganglia) and has been instrumental in demonstrating that the approach works in simulation after simulation (vision, clean-up, sequencing, Spaun, and the Tower of Hanoi).

I would also like to express my special thanks to Eric Hunsberger for his work on the vision system for Spaun, and to Matt Gingerich, who generated many figures and tables and worked hard on several tutorials and the code that supports them. A majority of the people I have already listed have done significant development work on Nengo as well, for which I also thank them. The lead developers are Bryan Tripp and Terry Stewart. Shu Wu did an excellent job on the original user interface.

During the course of writing this book, I have learned an enormous amount. Much of this has been from personal conversations. I would like to express my deepest gratitude to Charlie Anderson (NEF, learning), Cliff Clentros (learning), Doreen Fraser (physical properties), Eric Hochstein (levels),

Mark Laubach (decision making), Richard Mann (statistical inference), James Martens (statistical inference), Jeff Orchard (vision), Hans-Georg Stork (EU-funded research), Peter Strick (basal ganglia), Bryan Tripp (neural coding), and Matt van der Meer (reinforcement and decision making).

I have also had the invaluable opportunity to interact with the attendees of the Telluride Neuromorphic Engineering workshop, where we have been figuring out how to put the NEF and SPA on neuromorphic hardware and using it to control robots. Special thanks to Kwabena Boahen for a stimulating collaboration based on the Neurogrid chip and Francesco Galluppi and Sergio Davies for their dedicated hard work on the SpiNNaker hardware from Steven Furber's group. I'd also like to thank the organizers, Tobi Delbruck, Ralph Etienne-Cummings, and Timmer Horiuchi, for inviting us out and running such a unique workshop.

I am also challenged daily by the members of my lab, the Computational Neuroscience Research Group (CNRG). I mean this in the best way possible. In short, I would like to thank them, as a group, for creating such a fun, exciting, and productive environment. Similarly, I am continuously grateful to other faculty members in Waterloo's Centre for Theoretical Neuroscience for running a vibrant, inspiring, and collegial intellectual community. The many, well-expressed, and strongly reasoned views of the Centre are a constant source of inspiration.

I have also received much help directly focused on the book itself. I would like to sincerely thank Robert Amelard, Peter Blouw, Jean-Frédéric de Pasquale, and Frank Ritter for reading large sections of early drafts of the book. More dauntingly, Janet Elias (thanks, mom), Wolfgang Maas, Dimitri Pinotsis, Alex Petrov, and an anonymous reviewer read the entire text. This is an enormous amount of work, and I am deeply indebted to them. Alex's thorough and constructive contributions to early chapters deserve special mention. Paul Thagard not only read the whole book but has been supportive throughout its development. The overall shape of the book has been much improved by his contributions.

I would also like to thank a brave class of graduate students and their instructor, Mike Schoelles, at Rensselaer Polytechnic Institute for testing all of the Nengo tutorials in the book, as well as reading other portions. This kind of "battle testing" of tutorials is invaluable. Mike also served as a valuable reviewer for the entire manuscript, for which I am enormously grateful.

This research would not have been possible without the generous support of Canadian funding agencies, including NSERC, the Canadian Foundation for Innovation, the Ontario Innovation Trust, the Canada Research Chairs program, and the SharcNet computing consortium. Many months of time on Sharcnet's computers were expended running the simulations in Chapter 7. The CNRG also received enlightened, basic research funding from OpenText Corp. for some of the work described in Chapter 3.

I would like to express my genuine thanks to those who have helped guide the book through to its realization. This includes Frank Ritter, the editor of the series, and Catharine Carlin, Miles Osgood, and Joan Bossert from Oxford University Press, all of whom have been supportive, encouraging, and patient.

Finally, I cannot begin to express my appreciation to my family . . .

1. THE SCIENCE OF COGNITION

Questions are the house of answers. –Alex, age 5

1.1 ■ The Last 50 Years

"What have we actually accomplished in the last 50 years of cognitive systems research?" This was the pointed question put to a gathering of robotics, AI, psychology, and neuroscience experts from around the world. They were in Brussels, Belgium, at the headquarters of the European Union funding agency and were in the process of deciding how to divvy up about 70 million euros of funding. The room went silent. Perhaps the question was unclear. Perhaps so much had been accomplished that it was difficult to know where to start an answer. Or, perhaps even a large room full of experts in the field did not really know any generally acceptable answers to that question.

The purpose of this particular call for grant applications was to bring together large teams of researchers from disciplines as diverse as neuroscience, computer science, cognitive science, psychology, mathematics, and robotics to unravel the mysteries of how biological cognitive systems are so impressively intelligent, robust, flexible, and adaptive. This was not the first time such a call had been made. Indeed, over the course of the last 4 or 5 years, this agency has funded a large number of such projects, spending close to half a billion euros. Scientifically speaking, important discoveries have been made by the funded researchers. However, these discoveries tend not to be of the kind that tell us how to better construct integrated, truly *cognitive* systems. Rather, they are more often discoveries in the specific disciplines that are taking part in these "Integrated Projects."

For example, sophisticated new robotic platforms have been developed. One example is the iCub (Fig. 1.1), which is approximately the size of a

FIGURE 1.1 The iCub. The iCub is an example of a significant recent advance in robotics. *See* `http://www.robotcub.org/index.php/robotcub/gallery/videos` for videos. Image by Lorenzo Natale, copyright the RobotCub Consortium, reproduced with permission.

2-year-old child and has more than 56 degrees of freedom.[1] The iCub has been adopted for use by researchers in motor control, emotion recognition and synthesis, and active perception. It is clearly a wonder of high-tech robotics engineering. But, it is not a cognitive system–it is not an integrated, adaptive system able to perform a myriad of perceptual, motor, and problem-solving tasks. Rather, it has had various, typically simple, functions built into it independently.

So the pointed question still stands: "What have we actually accomplished in the last 50 years of cognitive systems research?" That is, what do we now know about how cognitive systems work that we did not know 50 years ago? Pessimistically, it might be argued that we do not know too much more than we knew 50 years ago. After all, by 1963 Newell and Simon had described in detail the program they called the General Problem Solver (GPS). This program, which was an extension of work that they had started in 1956, is

[1] In this case, a degree of freedom is one axis of independent motion. So, moving the wrist up and down is one degree of freedom, as is rotating the wrist, or moving the first joint of a finger up and down.

the first in a line of explanations of human cognitive performance that relied on production systems. Historically, production systems are by far the most influential approach to building cognitive systems. Simply put, a production system consists of a series of productions, or if-then rules, and a control structure. The job of the control structure is to match a given input to the "if" part of these productions to determine an appropriate course of action, captured by the "then" part. That course of action can change the internal state of the system and the external world, leading to a potentially new matched production. This kind of "chaining" of productions can result in relatively complex behavior.

In 1963, GPS had all of these features and it put them to good use. This "program that simulates human thought" was able to solve elementary problems in symbolic logic entirely on its own, and went through steps that often matched those reported by people solving the same problems. In short, GPS could be given a novel problem, analyze it, attempt a solution, retrace its steps if it found a dead end (i.e., self-correct), and eventually provide an appropriate solution (Newell & Simon, 1963).

The successes of GPS led to the development of several other cognitive architectures, all of which had production systems as their core. Best known among these are Soar (Newell, 1990), Executive-Process/Interactive Control (EPIC; Kieras & Meyer, 1997), and Adaptive Control of Thought (ACT; Anderson, 1983). Despite the many additional extensions to the GPS architecture that were made in these systems, they share an architecture that is based on a production system, even in their more recent hybrid forms (e.g., ACT-R; Anderson, 1996). The dominance of the representational, computational, and architectural assumptions of production systems has resulted in their being called the "classical" approach to characterizing cognition.

Although the classical approach had many early successes, the underlying architecture did not strike some as well-suited to interacting with a fast, dynamic environment. In fairness, more recent work on ACT, and its successor ACT-R , has taken dynamics more seriously, explaining reaction times across a wide variety of psychological tasks (Anderson et al., 2004). Nevertheless, explaining reaction times addresses only a small part of cognitive system dynamics in general, and the ACT-R explanations rely on assuming a 50 ms "cycle time," which itself is not explained (Anderson, 2007, p. 61; *see* Section 5.7 for further discussion). Similarly, recent work has employed Soar agents on a dynamic battlefield (Jones et al., 1999). Again, however, the time resolution of decisions by agents is about 50ms, and the short-term dynamics of perception and action is not explicitly modeled.

Tellingly, those most concerned with the fast timescale dynamics of cognition largely eschew production systems (Port & van Gelder, 1995; Schöner, 2008). If you want to build a fast cognitive system–one that directly interacts with the physics of the world–then the most salient constraints on your system are the physical dynamics of action and perception. Roboticists, as a result, seldom use production systems to control the low-level behavior of their robots. Rather, they carefully characterize the dynamics of their robot, attempt to understand how to control such a system when it interacts with the difficult-to-predict dynamics of the world, and look to perception to provide guidance for that control. If-then rules are seldom used. Differential equations, statistics, and signal processing are the methods of choice. Unfortunately, it has remained unclear how to use these same mathematical methods for characterizing high-level *cognitive* behavior–like language, complex planning, and deductive reasoning–behaviors that the classical approach has had the most success explaining.

In short, there is a gap in our understanding of real, cognitive systems: On the one hand, there are approaches centered on fast, dynamic, real-world perception and action; on the other hand, there are approaches centered on higher-level cognition. Unfortunately, these approaches are difficult to reconcile, although attempts have been made to build hybrid systems (d'Avila Garcez et al., 2008). Nevertheless, it is obvious that perception/action and high-level cognition are not two independent parts of one system. Rather, these two aspects are, in some fundamental way, integrated in cognitive animals such as ourselves. Indeed, a major theme of this book is that biological implementation underwrites this integration. But for now, I am only concerned to point out that classical architectures are not obviously appropriate for understanding all aspects of real cognitive systems. This, then, is why we cannot simply say, in answer to the question of what has been accomplished in the last 50 years, that we have identified *the* cognitive architecture.

However, this observation does not mean that work on classical architectures is without merit. On the contrary, one undeniably fruitful consequence of the volumes of work that surrounds the discussion of classical architectures is the identification of criteria for what counts as a cognitive system. That is, when proponents of the classical approach were derided for ignoring cognitive dynamics, one of their most powerful responses was to note that their critics had no sophisticated problem-solving system with which to replace classical proposals. The result was a much clearer understanding of what a cognitive system is–it is a system that integrates all aspects of sophisticated behavior. It does not put planning and reasoning ahead of motor control or vice versa. It is a system that must *simultaneously* realize a wide array of biological behaviors.

As a result, I suspect that there would be some agreement among the experts gathered in Brussels as to what has been accomplished these last 50 years. The accomplishments of cognitive systems research are perhaps not in the expected form of an obvious technology, a breakthrough method, or an agreed upon theory or architecture. Rather, the major accomplishments have come in the form of better questions about cognition. This is no mean feat. Indeed, it is even more true in science than elsewhere that, as economic

Nobel laureate Paul A. Samuelson has observed, "good questions outrank easy answers." If we truly have better questions to ask–ones that allow us to distinguish cognitive from non-cognitive systems–then we have accomplished a lot.

As I discuss in Section 1.3, I think that we do have such questions and that they have led to a set of criteria that can be used to effectively evaluate proposed cognitive architectures, even if we do not yet have a specific proposal that satisfies all of these criteria. The goal of this first chapter is to identify these cognitive criteria and articulate some questions arising from them. These appear in Sections 1.3 and 1.4, respectively. First, however, it is worth a brief side trip into the history of the behavioral sciences to situate the concepts I employ in stating these criteria.

1.2 ■ How We Got Here

To this point, I have identified the "classical" approach to understanding cognition and contrasted it with an approach that is more centrally interested in the dynamics of cognitive behavior. However, much more needs to be said about the relationship between classical and non-classical approaches to get a general lay-of-the-land of cognitive systems theorizing. Indeed, much more can be said than I will say here (*see*, e.g., Bechtel & Graham, 1999). My intent is to introduce the main approaches to (1) identify the strengths and weaknesses of these approaches, both individually and collectively; (2) state and clarify the cognitive criteria mentioned earlier; and, ultimately, (3) outline a novel approach to *biological* cognition in later chapters.

Each of the approaches I consider has given rise to specific architectures, some of which I consider in Chapter 9. To be explicit, I consider a cognitive architecture to be "a general proposal about the representations and processes that produce intelligent thought" (Thagard, 2011; *see also* Langley et al., 2009). For the time being, however, I do not focus specifically on cognitive architectures but, rather, more generally on the theoretical approaches that have spawned these architectures. In the rest of the book, I am interested in describing a new architecture that adopts a different theoretical approach.

In the last half-century, there have been three major approaches to theorizing about the nature of cognition.[2] Each approach has relied heavily on a preferred metaphor for understanding the mind/brain. Most famously, the classical symbolic approach (also known as "symbolicism" or Good Old-fashioned Artificial Intelligence [GOFAI]) relies on the "mind as computer" metaphor. Under this view, the mind is the software of the brain. Jerry Fodor, for one, has argued that the impressive theoretical power provided by

[2] Much of the discussion in the remainder of this section is from Eliasmith (2003).

this metaphor is good reason to suppose that cognitive systems have a symbolic "language of thought" that, like a computer programming language, expresses the rules that the system follows (Fodor, 1975). Fodor claims that this metaphor is essential for providing a useful account of how the mind works. Production systems, which I have already introduced, have become the preferred architectural implementation of this metaphor.

A second major approach is "connectionism" (also known as the Parallel Distributed Processing [PDP] approach). In short, connectionists explain cognitive phenomena by constructing models that consist of large networks of nodes that are connected together in various ways. Each node performs a simple input/output mapping. However, when grouped together in sufficiently large networks, the activity of these nodes is interpreted as implementing rules, analyzing patterns, or performing any of several other cognitively relevant behaviors. Connectionists, like the proponents of the symbolic approach, rely on a metaphor for providing explanations of cognitive behaviors. This metaphor, however, is much more subtle than the symbolic one; these researchers presume that the functioning of the mind is like the functioning of the brain. The subtlety of the "mind *as* brain" metaphor lies in the fact that connectionists also hold that the mind *is* the brain. However, when providing *cognitive* descriptions, it is the metaphor that matters, not the identity. In deference to the metaphor, the founders of this approach call it "brain-style" processing and claim to be discussing "abstract networks" (Rumelhart & McClelland, 1986a). In other words, their models are not supposed to be direct implementations of neural processing, and hence cannot be directly compared to many of the kinds of data we gather from real brains. This is not surprising because the computational and representational properties of the nodes in connectionist networks typically bear little resemblance to neurons in real biological neural networks.[3] This remains the case for even the most recent architectures employing the connectionist approach (e.g., O'Reilly & Munakata, 2000; Hummel & Holyoak, 2003; van der Velde & de Kamps, 2006). I discuss this work in more detail in Chapter 9.

The final major approach to cognitive systems theorizing in contemporary cognitive science is "dynamicism," and it is often closely allied with "embedded" or "embodied" approaches to cognition. Proponents of dynamicism also rely heavily on a metaphor for understanding cognitive systems. Most explicitly, van Gelder employs the Watt Governor as a metaphor for how we should characterize the mind (van Gelder, 1995). In general, dynamicist metaphors rely on comparing cognitive systems to other, continuously coupled, nonlinear, dynamical systems (e.g., the weather). Van Gelder's metaphor is useful to consider in detail because it has played a central role in his arguments for the novelty of dynamicist approaches.

[3] This is discussed in detail in chapter 10 of Bechtel & Abrahamsen (2001).

The Watt Governor is a mechanism for controlling (i.e., "governing") the speed of an engine shaft under varying loads. It is named after James Watt because he used it extensively to control steam engines, although the same mechanism was in use before Watt to control wind and water mills. Figure 1.2 depicts and explains a Watt Governor.

It is through his analysis of the best way to characterize this particular dynamical system that van Gelder argues for understanding cognitive systems as *non-representational, low-dimensional,* dynamical systems. Like the Watt Governor, van Gelder maintains, cognitive systems can only be properly understood by characterizing their state changes through time. These state changes are a function of the tightly coupled component interactions and their continuous, mutually influential connections to the environment. To proponents of dynamicism, the "mind as Watt Governor" metaphor suggests that trying to impose any kind of discreteness on system states–either

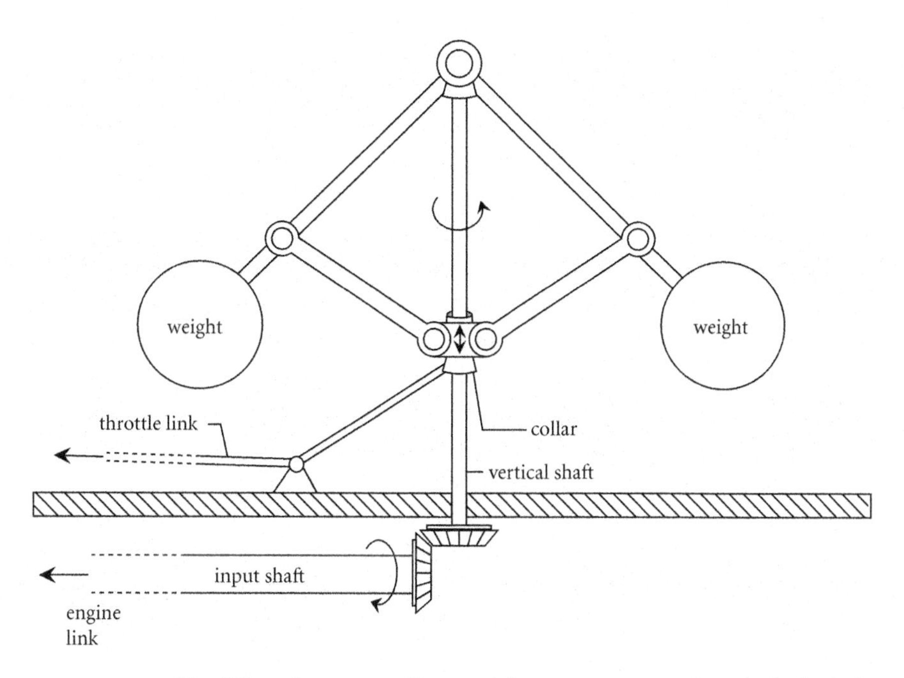

FIGURE 1.2 The Watt Governor. Two weights rotate around a vertical shaft driven by an engine. The centripetal force created by the rotation of the vertical shaft causes the weights to be pushed outward and raises a collar that is attached to the weighted arms. The collar is linked to the engine throttle, and as it rises it decreases the throttle, reducing the speed of the input shaft. The system settles into a state where the throttle is open just enough to maintain the position of the weights. Increasing or decreasing the load on the engine disturbs this equilibrium and causes the governor to readjust the throttle until the system stabilizes again.

temporal or representational–will lead to a mischaracterization of cognitive systems.

This analysis of cognitive dynamics highlights the essential *coupling* of cognitive systems to their environment (van Gelder & Port, 1995). Dynamic constraints are clearly imposed by the environment on our behavior (e.g., we must see and avoid the cheetah before it eats us). If our high-level behaviors are built on our low-level competencies, then it should not be surprising that a concern for low-level dynamics has led several researchers to emphasize that real cognitive systems are embedded within environments that affect their cognitive dynamics. Proponents of this view have convincingly argued that the nature of that environment can have significant impact on what cognitive behaviors are realized (Clark, 1997; Hutchins, 1995). Because many of the methods and assumptions of dynamicism and such "embedded" approaches to cognition are shared, in this discussion I group both under the heading of "dynamicism."

It is somewhat tempting to identify a fourth major approach to characterizing cognition–what might be called the "Bayesian" approach (*see*, e.g., Anderson, 1990; Oaksford & Chater, 1998; Tenenbaum & Griffiths, 2001; Griffiths & Tenenbaum, 2006). This work focuses on the rational analysis of human behavior using probabilistic inference. Unlike the other approaches I have mentioned, the resulting models are largely phenomenological (i.e., they capture phenomena well but not the relevant mechanisms). Proponents of this view acknowledge that the implementation of these models remains a challenge in general and specifically that "the biggest remaining obstacle is to understand how structured symbolic knowledge can be represented in neural circuits" (Tenenbaum et al., 2011, p. 1284). There are no cognitive *architectures* that I am aware of that take optimal statistical inference as being the central kind of computation performed by cognitive systems. Nevertheless, a lot has been learned about the nature of human cognition using this kind of analysis (Tenenbaum et al., 2011). Consequently, I show how the architecture proposed in this book can be given a natural probabilistic interpretation (Section 7.4). For present purposes, however, I will focus on the three approaches that have driven mechanistic cognitive modeling to date.

Crucially, the symbolic approach, connectionism, and dynamicism rely on metaphor not only for explanatory purposes, but also for developing the conceptual foundations of their preferred approach to cognitive systems. For proponents of the symbolic approach, the properties of Turing machines become shared with minds because they are the same kind of symbolic, computational system. For connectionists, the very nature of representation changes dramatically under their preferred metaphor. Mental representations are no longer unitary symbols but are taken to consist of "subsymbols" associated with each node, whereas "whole" representations are real-valued vectors in a high-dimensional property space.[4] Finally, because the Watt Governor is best described by the mathematics of dynamic systems theory, which makes no reference to computation or representation, dynamicists claim that our theories of mind need not appeal to computation or representation either (van Gelder, 1998).

I have argued elsewhere that our understanding of cognitive systems needs to move beyond such metaphors (Eliasmith, 2003). We need to move beyond metaphors because in science metaphorical thinking can sometimes unduly constrain available hypotheses. This is not to deny that metaphors are incredibly useful tools at many points during the development of scientific theory. It is only to say that sometimes metaphors only go so far. Take, for example, the development of our current theory of the nature of light. In the nineteenth century, light was understood in terms of two metaphors: light as a wave, and light as a particle. Thomas Young was the best known proponent of the first view, and Isaac Newton was the best known proponent of the second. Each used their favored metaphor to suggest new experiments and develop new predictions.[5] Thus, these metaphors played a role similar to that played by the metaphors discussed above in contemporary cognitive science. However, as we know in the case of light, both metaphors are misleading. Hence the famed "wave–particle duality" of light: sometimes it behaves like a particle, and sometimes it behaves like a wave. Neither metaphor by itself captures all the phenomena displayed by light, but both are extremely useful in characterizing some of those phenomena. So, understanding what light *is* required moving beyond the metaphors.

I believe that the same is true in the case of cognition. Each of the metaphors mentioned above has some insights to offer regarding certain phenomena displayed by cognitive systems. However, none of these metaphors is likely to lead us to all of the right answers. Thus, we ideally want a way of understanding cognitive systems that draws on the strengths of the symbolic approach, connectionism, and dynamicism but does not *depend* on the metaphors underlying any of these approaches.

In fact, I believe we are in a historically unique position to come to this kind of understanding. This is because we currently have more detailed access to the underlying biology of cognition than ever before. We can record large-scale blood flow in the brain during behavior using fMRI, we can image medium-sized networks of cells during the awake behavior of animals, we can record many individual neurons inside and outside of the functioning brain, and we

[4] See, for example, Smolensky (1988). Notably, there are also connectionist models that take activities of individual nodes to be representations. These are still often treated differently than standard symbolic representations.

[5] For a detailed description of the analogies, predictions, and experiments, *see* Eliasmith & Thagard (1997).

can even record the opening and closing of individual channels (which are about 200 nanometers in size) on the surface of a single neural cell. The time is ripe for positing theories of biological *cognition.*

Detailed information about implementation and function is exactly what we need to free ourselves of high-level, *theoretically guiding* metaphors. To be clear, I am not endorsing the claim that we should get rid of all metaphors: such a view is implausible, and I employ many metaphors throughout this book. Rather, I am advocating that we reduce our reliance on metaphors *as theoretical arbiters.* That is, we need to remove them from a role in determining what is a good explanation of cognition and what is not. If we can relate our cognitive explanations to a wide variety of specific empirical measurements, then it is those data, not a metaphor, that should be the determiner of the goodness of our explanation. If we can identify the relationship between cognitive theories and biological (including psychological) data, then we can begin to understand cognitive systems for what they are: complex, dynamic, biological systems. The three past approaches I have discussed provide few hints as to the nature of this key relationship.

Historically speaking, theories regarding cognitive systems mainly come from psychology, an offshoot of philosophy of mind in the late nineteenth century. Early in the development of psychology, very little was known about the biology of the nervous system. There was some suggestion that the brain was composed of neurons, but even this was highly contentious. There was little understanding of how cells were organized to control even our simplest behaviors, let alone our complex, cognitive ones. And so cognitive theorizing proceeded without much connection to biology.

In the early days of psychology, most such theories were generated by the method of introspection–that is, sitting and thinking very carefully, so as to discern the components of cognitive processing. Unfortunately, different people introspected different things, and so there was a crisis in "introspectionist" psychology. This crisis was resolved by "behaviorist" psychologists who simply disallowed introspection. The only relevant data for understanding cognitive systems were data that could be gleaned from the "outside"–that is, from *behavior.* Looking inside the "black box" that was the system of study was prohibited.

Interestingly, engineers of the day respected a similar constraint of only analyzing systems by looking at their input–output relationship and not worrying about what was happening inside the system itself. To understand dynamic physical systems, the central tool they employed was classical control theory. Classical control theory characterizes physical systems by identifying a "transfer function" that directly maps inputs to outputs. Adopting this method–one that does not consider the internal states of a system–turns out to impose severe constraints on analysis and synthesis of controllers. As a result, classical control theory is limited to designing nonoptimal, single-variable, static controllers and depends on approximate graphical methods and rules of thumb, while not allowing for the inclusion of noise.[6] Although the limitations of classical controllers and methods are now well known, they nevertheless allowed engineers to build systems of kinds they had not systematically built before: goal-directed systems.

This made classical control theory quite practical, especially in the 1940s when there was often a desire to blow things up (the most common goal at which such systems were directed). But, some researchers thought that control theory had more to offer. They suggested that it could provide a foundation for describing living systems as well. Most famously, the interdisciplinary movement founded in the early 1940s known as "cybernetics" was based on precisely this contention.[7] Cyberneticists claimed that living systems, like classical control systems, were essentially goal-directed systems. Thus, closed-loop control should be a good way to understand the behavior of living systems. Given the nature of classical control theory, cyberneticists focused on characterizing the input/output behavior of living systems, not their internal processes. Unfortunately for cyberneticists, in the mid-1950s there was a massive shift in how cognitive systems were viewed.

With the publication of a series of seminal papers (including Newell et al., 1958; Miller, 1956; Bruner et al., 1956; and Chomsky, 1959), the "cognitive revolution" took place. One simplistic way to characterize the move from behaviorism to "cognitivism" is that it became no longer taboo to look inside the black box. Quite the contrary–internal states, internal processes, and internal representations became standard fare when thinking about the mind. Classical control theory no longer offered the kinds of analytic tools that mapped easily onto this new way of conceiving complex biological behavior. Rather, making sense of the insides of that black box was heavily influenced by concurrent successes in building and programming computers to perform complex tasks. Thus, when they opened the lid of the box, many early cognitive scientists saw a computer.

Computers are interesting because they "show us how to connect semantical [properties] with causal properties for *symbols*" (Fodor, 1987, p. 18). Thus, computers have what it takes to be minds. Once cognitive scientists began to think of minds as computers, a number of new theoretical tools became available for characterizing cognition. For example, the computer's theoretical counterpart, the Turing machine, suggested novel philosophical theses, including "functionalism" (i.e., the notion that only the mathematical function computed by a system was relevant for its being a mind or not) and "multiple realizability" (i.e., the notion that a mind could be implemented [i.e., realized] in pretty much any substrate–water, silicon, interstellar

[6] For a succinct description of the history of control theory, see Lewis (1992).

[7] For a statement of the motivations of cybernetics, see Rosenblueth et al. (1943).

gas, etc.—as long as it computed the appropriate function). More practically, the typical architecture of computers, the von Neumann architecture, was thought by many to be relevant for understanding our cognitive architecture.

Eventually, however, adoption of the von Neumann architecture for understanding minds was seen by many as poorly motivated. Consequently, the early 1980s saw a significant increase in interest in the connectionist research program. As mentioned previously, rather than adopting the architecture of a digital computer, these researchers felt that an architecture more like that seen in the brain would provide a better model for cognition.[8] It was also demonstrated that a connectionist architecture could be as computationally powerful as any symbolic architecture. But despite the similar computational power of the approaches, the specific problems at which each approach excelled were quite different. Connectionists, unlike their symbolic counterparts, were very successful at building models that could learn and generalize over the statistical structure of their input. Thus, they could begin to explain many phenomena not easily captured by a symbolic approach, such as object recognition, reading words, concept learning, and other behaviors crucial to cognition.

By the early 1990s, however, some began to worry that the connectionists had not sufficiently escaped the influence of the "mind as computer metaphor." After all, connectionists still spoke of representations and computations and typically considered the mind a *kind* of computer. Dynamicists, in contrast, suggested that if we wanted to know which functions a system can actually perform in the real world, we must know how to characterize the system's dynamics. Because cognitive systems evolved in specific, dynamic environments, we should expect evolved control systems (i.e., brains) to be more like the Watt Governor–dynamic, continuous, coupled directly to what they control–than like a discrete-state Turing machine that computes over "disconnected" representations. As a result, dynamicists suggested that dynamic systems theory–not computational theory–was the right quantitative tool for understanding minds. They claimed that notions like "chaos," "hysteresis," "attractors," and "state-space" underwrite the conceptual tools bestsuited for describing cognitive systems. Notably absent from the list of relevant concepts are "representation" and "computation."

[8] It can be helpful to distinguish the "simulated architecture" from the "simulating architecture" in such discussions. The former is the architecture that is supposed to be isomorphic to the cognitive system, whereas the latter is the architecture of whatever machine is being used to run the simulations. For example, in connectionism the simulating architecture is a von Neumann machine (the desktop computer running the simulation), which is different from the simulated architecture (connected nodes). In the symbolic approach these are often the same. In my discussion I am comparing simulated architectures.

As we have just seen, each of these three approaches grew out of a critical evaluation of its predecessors. Connectionism was a reaction to the over-reliance on digital computer architectures for describing cognition. Dynamicism was a reaction to an under-reliance on the importance of time and our connection to the physical environment. Even the symbolic approach was a reaction to its precursor, behaviorism, which had ruled out a characterization of cognition that posited states internal to the agent. Consequently, each of the metaphors has been chosen to emphasize different aspects of cognition, and hence, each has driven researchers in these areas to employ different formalisms for describing their cognitive theories. Succinctly put, symbolic proponents use production systems, connectionists use networks of nodes, and dynamicists use sets of differential equations.

Although we can see the progression through these approaches as rejections of their predecessors, it is also important to note what was preserved. The symbolic approach preserved a commitment to providing scientific explanations of cognitive systems. Connectionism retained a commitment to the notions of representation and computation so central to the symbolic approach. And dynamicism, perhaps the most self-conscious attempt to break away from previous methods, has not truly abandoned the tradition of explaining the same set of complex behaviors. Rather, dynamicists have convincingly argued for a shift in emphasis: They have made time a non-negotiable feature of cognition.[9]

Perhaps, then, a successful break from all of these metaphors will be able to relate each of the central methods to one another in a way that preserves the central insights of each. And, just as importantly, such a departure from past approaches should allow us to see how our theories relate to the plethora of data we have about brain function.

I believe that the methods developed in this book can help us accomplish both of these goals, but I leave this argument until Section 9.4, because such arguments cannot be made with any force until much work is done. First, for example, I need to use the historical background just presented to identify what I earlier claimed was the major advance of the the last 50 years–that is, I need to identify criteria for evaluating the goodness of cognitive theories. Perhaps more importantly, I need to specify the methods that allow us to both modify and embrace all three of these perspectives–that is the objective of Chapters 2 through 7.

[9] Notably, Newell, one of the main developers of production systems, was quite concerned with the timing of cognitive behavior. He dedicated a large portion of his book, *Unified Theories of Cognition*, to the topic. Nevertheless, time is not an intrinsic feature of production systems, and so his considerations seem, in many ways, after the fact. By "intrinsic" I mean that, for example, productions of different complexity can have the same temporal properties and vice versa (for further discussion, *see* Eliasmith, 1996).

1.3 ■ Where We Are

It will strike many as strange to suggest that there is anything like agreement on criteria for identifying good cognitive theories. After all, cognitive research has been dominated by rather vigorous debate between proponents of each of the three approaches. For example, proponents of the symbolic approach have labeled connectionism "quite dreary and recidivist" (Fodor, 1995). Nevertheless, I believe that there is some agreement on what the *target* of explanation is. By this I do not mean to suggest that there is an agreed-upon definition of cognition. Rather, I want to suggest that "cognition" is to most researchers what "pornography" was to justice Potter Stewart: "I shall not today attempt further to define [pornography]; and perhaps I could never succeed in intelligibly doing so. But I know it when I see it...."

Most behavioral researchers seem to know cognition when they see it as well. There are many eloquent descriptions of various kinds of behavior in the literature, which most readers–regardless of their commitments to the symbolic approach, connectionism, or dynamicism–recognize as cognitively relevant behavior. It is not as if symbolic proponents think that constructing an analogy is a cognitive behavior and dynamicists disagree. This is why I have suggested that we may be able to identify agreed-upon criteria for evaluating cognitive theories. To be somewhat provocative, I will call these the "Core Cognitive Criteria," or CCC for short. I should note that this section is only a first pass and summary of considerations for the proposed CCC. I return to a more detailed discussion of the CCC before explicitly employing them in Chapter 8.

So what are the CCC? Let us turn to what researchers have said about what makes a system cognitive. Here are examples from proponents of each view:

- Dynamicism (van Gelder & Gelder, 1995, pp. 375–376): "[C]ognition is distinguished from other kinds of complex natural processes ... by at least two deep features: ... a dependence on knowledge; and distinctive kinds of complexity, as manifested most clearly in the structural complexity of natural languages."
- Connectionism (Rumelhart & McClelland, 1986b, p. 13): Rumelhart and McClelland explicitly identify the target of their well-known PDP research as "cognition." To address it, they feel that they must explain "motor control, perception, memory, and language."
- The symbolic approach (Newell, 1990, p. 15): Newell presents the following list of behaviors in order of their centrality to cognition: (1) problem solving, decision making, routine action; (2) memory, learning, skill; (3) perception, motor behavior; (4) language; (5) motivation, emotion; (6) imagining, dreaming, daydreaming.

These examples do not provide criteria for identifying good cognitive *theories* directly but, rather, attempt to identify which particular aspects of behavior must be explained to successfully explain cognition. Indeed, Newell (1990; 1980) provides longer lists of about a dozen functional criteria for cognitive systems. More recently, Anderson and Lebiere (2003) have expanded and discussed Newell's list in detail. To get at the motivations behind such an inventory, however, it is also helpful to distill theoretical commitments motivating the inclusion of particular functional criteria on such lists. Ultimately, the CCC I propose include both theoretical and functional criteria.

Let me begin by noting that there are several commonalities among these descriptions of cognitive behavior. First, language appears in all three. This is no surprise as language is often taken to be the pinnacle of human cognitive ability. But, there are other shared commitments evident in these descriptions. For example, each list identifies the importance of adaptability and flexibility. For van Gelder, adaptability is evident because of the dependence of cognitive behavior on knowledge. For the other two, explicit mention of memory and learning highlight a more general interest in adaptability. Additionally, although not in this specific quote from van Gelder, dynamicism is built on a commitment to the centrality of action and perception to cognition (Port & van Gelder, 1995). Perhaps surprisingly, then, it is the connectionists and symbolic proponents who clearly identify motor control and perception as important to understanding cognition. It is clear that all three agree on the important role of these more basic processes.

Although these are simple lists, I believe we can see in them the motivations for subsequent discussions explicitly aimed at identifying what it takes for a system to be cognitive. Let me briefly consider some of these more direct discussions. One of the first, and perhaps the most well known, was provided by Fodor and Pylyshyn in their 1988 paper, "Connectionism and Cognitive Architecture: A Critical Analysis." Although mainly a critique of the connectionism of the day, this paper also provides three explicit constraints on what it takes to be a cognitive system. These are productivity, systematicity, and compositionality. Productivity is the ability of a system to generate a large number of representations based on a few basic representations (a lexicon) and rules for combining them (a grammar). Systematicity refers to the fact that some sets of representations are intimately linked. For example, Fodor and Pylyshyn have argued that cognitive systems cannot represent "John loves Mary" without thereby being able to represent "Mary loves John." Finally, compositionality is the suggestion that the meaning of complex representations is a direct "composition" (i.e., adding together) of the meanings of the basic representations. Any good theory, they claim, must explain these basic features of cognition.

More recently, Jackendoff (2002) has dedicated a substantial portion of his book to the task of identifying challenges for a cognitive neuroscience of

TABLE 1.1 *Core Cognitive Criteria (CCC) for Theories of Cognition.*

1. Representational structure
   a. Systematicity
   b. Compositionality
   c. Productivity (the problem of variables)
   d. The massive binding problem (the problem of two)
2. Performance concerns
   a. Syntactic generalization
   b. Robustness
   c. Adaptability
   d. Memory
   e. Scalability
3. Scientific merit
   a. Triangulation (contact with more sources of data)
   b. Compactness

cognition. In it he suggests that there are four main challenges to address when explaining cognition. Specifically, Jackendoff's challenges are: (1) the massiveness of the binding problem (i.e., that very many basic representations must be bound to construct a complex representation); (2) the problem of two (i.e., how multiple instances of one representational token can be distinguished); (3) the problem of variables (i.e., how roles [e.g., "subject"] in a complex representation can be generically represented); and (4) how to incorporate long-term and working memory into cognition. Some of these challenges are closely related to those of Fodor and Pylyshyn and so are integrated with them as appropriate in the CCC (*see* Table 1.1).

The Fodor, Pylyshyn, and Jackendoff criteria come from a classical, symbolic perspective. In a more connectionist-oriented discussion, Don Norman has summarized several papers he wrote with Bobrow in the mid-1970s, in which they argue for the essential properties of human information processing (Norman, 1986). Based on their consideration of behavioral data, they argue that human cognition is: robust (i.e., appropriately insensitive to missing or noisy data and damage), flexible, and reliant on "content-addressable" memory. One central kind of cognitive flexibility, which is an example of the "structural complexity of natural language" noted by van Gelder, is captured by the notion of "syntactic generalization." Syntactic generalization refers to the observation that people can flexibly exploit the *structure* of language, independent of its content (*see* Section 8.2.2.1). Compared to symbolic considerations, the emphasis in these criteria moves from representational constraints to more behavioral constraints, driven by the "messiness" of psychological data.

Dynamicists can be seen to continue this trend toward complexity in their discussions of cognition. Take, for example, Schöner's (2008) discussion in his article "Dynamical systems approaches to cognition." In his opening paragraphs, he provides examples of the sophisticated action and perception that occurs during painting and playing on a playground. He concludes that "cognition takes place when organisms with bodies and sensory systems are situated in structured environments, to which they bring their individual behavioral history and to which they quickly adjust" (p. 101). Again, we see the importance of adaptability and robustness, with the addition of an emphasis on the role of the environment.

All three perspectives, importantly, take the target of their explanation to be a large, complex system. As a result, all three are suspect of "toy" models that oversimplify a given cognitive task or domain. However, practical considerations have thus far limited all theories to presenting simple models. But the ideal that any good theory must be scalable to larger, more complex problems is shared. This demand for "scalability" is an expression of the historical trend toward demanding ever more sophisticated explanations from our cognitive theories.

Before presenting a final summary of the CCC, I believe there are additional criteria that a good theory of cognition must meet. I suspect that these will not be controversial, as they are a selection of insights that philosophers of science have generated in their considerations of what constitutes a good scientific theory in general (Popper, 1959; Quine & Ullian, 1970; Kitcher, 1993; Craver, 2007). I take it as obvious that each approach assumes that we are trying to construct a scientific theory of cognitive systems. Nevertheless, these criteria may play an important role in distinguishing good cognitive theories from bad ones.

Two of the most important considerations for good scientific theories are those of unity and simplicity. Good scientific theories are typically taken to be unified: the more sources of data, and the more scientific disciplines that they are consistent with, the better the theory. One of the reasons Einstein's theory of relativity is to be preferred over Newton's theory of motion is that the former is consistent with more of our observations. I refer to this criterion as "triangulation" to emphasize the idea that a good theory contacts many distinct sources of data in a consistent and unified way.

In addition, good theories tend toward simplicity. I have called this criterion "compactness" to emphasize that good theories can be stated compactly and without *ad hoc* additions. The reason that the heliocentric theory of our solar system is preferred over a geocentric one is that, in the latter, we need to specify not only the circular paths of the planets but also the many infamous "epicycles" of each planet to explain their motions. That is there is a degree of arbitrariness that enters the geocentric theory, making it less convincing. In contrast, the heliocentric theory needs to specify one simple ellipse for each planet. Thus, the same data are explained by the heliocentric theory in a more compact way.

Though this discussion has been brief, I believe that these considerations provide a reasonably clear indication of several criteria that researchers in the behavioral sciences would agree can be used to evaluate cognitive theories–regardless of their own theoretical predispositions. As a result, Table 1.1 summarizes the CCC we can, at least on the face of it, extract from this discussion. As a reminder, I do not expect the mere identification of these criteria to be convincing. A more detailed discussion of each is presented in Chapter 8. It is, nevertheless, very useful to keep the CCC in mind as we consider a new proposal for their satisfaction in the remainder of the book.

1.4 ■ Questions and Answers

The CCC are concerned with *evaluating* a characterization of cognition, not directing the development of such a characterization. So in this section I identify four central questions that have been relatively persistent over the last 50 years. I take it that detailed answers to these questions will go a long way toward addressing most, if not all, of the CCC. My detailed answers are in Chapters 3 through 6, and my evaluation of these answers against the CCC is in Chapter 9. However, I do provide answer *sketches* here. Their purpose is to provide a high-level view of the proposed architecture and a sense of what lies ahead.

The questions are:

1. How is semantics captured in the system?
2. How is syntactic structure encoded and manipulated by the system?
3. How is the flow of information flexibly controlled in response to task demands?
4. How are memory and learning employed in the system?

A long-standing concern when constructing models of cognitive systems is how to characterize the relationship between the states inside the system, and the objects in the external world that they purportedly represent. That is how do we know what an internal state means? Of course, for any system we construct, we can simply define the meaning of a particular state, and call it a representation of that state. Unfortunately, this is very difficult to do: Consider trying to define a mapping between dogs-in-the-world and a state in your head that acts like the concept "dog." The vast psychological literature on concepts points precisely to the complexity of such a mapping. Most researchers in cognitive science are well aware of (and dread addressing) this problem, which is often called the "symbol grounding problem" (Harnad, 1990). Nevertheless, any implemented model of a cognitive system must make some assumptions about how the representations in that system get their meaning. Consequently, answering this first question will force anyone wishing to characterize cognition to, at the very least, state their assumptions about how internal states are related to external ones. Whatever the story, it will have to plausibly apply not only to the models we construct, but also to the natural systems we are attempting to explain.

In Chapter 3, I introduce the notion of a "semantic pointer" to help address this question. A semantic pointer is a neural representation that carries partial semantic content and is composable into the representational structures needed to support complex cognition. I address composability in Chapter 4, so focus on semantic content in Chapter 3. I refer to these representations as "pointers" because, as with computer science "pointers,"[10] they are compact, efficient to manipulate, and identify (or "point" to) more complex representations.

However, in computer science, pointers and their targets are arbitrarily related. In contrast, *semantic* pointers bear systematic relations to what they point to. For example, a neural representation at the top of the visual hierarchy would be a semantic pointer that points to the many lower-level visual properties of an image encoded in visual cortex. It points to these precisely because it does not explicitly contain representations of those properties, but it can be used to re-activate those lower-level representations through feedback connections.

The relation between a semantic pointer and what it points to can be usefully thought of as a kind of *compression.* Just as digital images can be compressed into small JPEG files, lower-level visual information can be compressed into semantic pointers. To identify the visual features captured by a JPEG, it must be decompressed. Nevertheless, we can manipulate the JPEG in some ways without fully decompressing it–by giving it to our friends, copying it, or perhaps flipping it upside down. However, to get back the visual details we must decompress or, in pointer parlance, "dereference" it. Dereferencing is thus part of the function of feedback connections in my earlier vision example.

Crucially, if the same process is always used for compression, then similar uncompressed input will result in similar compressed representations.[11] This is the sense in which semantic pointers are *semantic.* They can be directly compared to get an approximate sense of the similarity of the states from which

[10] In computer science, a pointer is typically a representation that has a memory address as its content. In many instances, the pointer can be manipulated without manipulating what is at that location in memory. This is efficient because the pointer is typically much smaller than the contents of memory to which it points.

[11] There are subtleties here about identifying what is "similar." Essentially, the compression algorithm will help determine the dimensions along which similarity is preserved. So input and output similarity are not independent notions by any means.

they were compressed. So, they carry with them what we might call a "surface" semantics. To get at the "deep" semantics, we need to more directly compare the uncompressed states. Of course, much of the information lost through compression resides in the process used to produce a compressed representation. For this reason, to get at deep semantics we must be able to effectively run the system "backward" to flesh out what properties likely gave rise to that pointer. This is just another description of dereferencing.

In Chapter 3, I show specific examples of how semantic pointers can be constructed from visual images in a hierarchical spiking neural model. Of course, vision is just one source of information driving internal representations, but I argue that the story is general enough to cover other modalities as well. And, crucially, I show an example of how motor control can also be understood as the process of dereferencing a semantic pointer. In particular, I show how this process can be used to control a nonlinear, 6-degree of freedom arm. These examples and the surrounding discussion describe how perceptual and motor semantics can be encoded into semantic pointers, transformed, and used to drive behavior in a biologically plausible manner.

The second question above addresses what I have already noted is identified nearly universally as a hallmark of cognitive systems: the ability to manipulate structured representations. Whether or not we think internal representations themselves are structured (or, for that matter, if they even exist), we must face the undeniable ubiquity of behavior that looks like the manipulation of language-like representations. Consequently, any characterization of cognition will have to tell a story about how syntactic structure is encoded and manipulated by a cognitive system. Answering this question will address at least the first five criteria in the CCC.

In Chapter 4 I tackle this question by showing how semantic pointers can be composed in a symbol-like manner to create complex representational structures. Specifically, I show how different semantic pointers can be bound together (e.g., bind[subject, dog]) and then collected into a group (e.g., group[bind(subject, dog), bind(relation, chased), bind(object, ball)]) to represent structured representations such as "the dog chased the ball." Importantly, I describe how to not only encode and decode such structures but also how to manipulate and learn them in spiking neural networks. There I argue that semantic pointers can scale to the sophistication of human structural representations, while respecting known anatomical and physiological constraints of the brain. I also briefly return to the issue of semantics, as many of our concepts bear structural relations to other concepts, suggesting that a full account of semantics will need to include structured representations as well. To demonstrate these methods, I describe a recent model that is able to account for human performance on the Raven's Progressive Matrices task, a test of general fluid intelligence.

The third question addresses a broadly admired feature of cognitive systems: their incredible, rapid adaptability. People faced with a new situation can quickly survey their surroundings, identify problems, and formulate plans for solving those problems–often within the space of a few seconds. Performing each of these steps demands coordinating the flow of huge amounts of information through the brain. For example, if I simply tell you that the most relevant information for performing a task is going to switch from something you are hearing to something you will be seeing, your brain can instantly reconfigure itself to take advantage of that knowledge. Somehow, your brain re-routes information such that the information you use for planning will come from the visual system instead of the auditory system. We do this effortlessly, quickly, and constantly.

In Chapter 5, I note that this kind of control depends on both generating a control signal (i.e., choosing what to do) and applying that control signal (i.e., performing the selected action). I describe the central role that the basal ganglia seems to play in action selection and develop a detailed neural model of this part of the brain. I provide several examples of how the basal ganglia can be used to select appropriate semantic pointers as actions, given others as input. I then describe how attention is an excellent example of information routing in the brain and suggest that the principles of a recent, detailed neural model of attentional routing can be employed generally across the brain. Finally, I combine these two methods to show how the basal ganglia can be used to control information routing in the brain to realize flexible control of action sequences that process structured semantic pointers. I end the chapter by describing a fully spiking neural model of the Tower of Hanoi task, which matches basal ganglia and cortical anatomy and physiology, millisecond spike data, and fMRI data, while accounting for behavioral performance on the task. I argue that this is a good example of a model of biological cognition, which informatively spans many temporal and spatial scales of analysis.

The fourth question focuses on another important source of cognitive flexibility: our ability to use past information to improve our performance on a future task. The timescale over which information might be relevant to a task ranges from seconds to years. Consequently, it is not surprising that the brain has developed mechanisms to store and exploit information over these timescales. Memory and learning are two behavioral descriptions of the impressive abilities of these mechanisms. Considerations of memory and learning directly address several of the performance concerns identified in the CCC. Consequently, any characterization of a cognitive system has to provide some explanation for how relevant information is propagated through time, and how the system can adapt using its past experience.

By Chapter 6, I have already discussed several aspects of both learning and memory. So, in that chapter I focus specifically on (1) working memory, and (2) the relationship between biologically realistic learning rules and high-level function. With regards to working memory, I describe a recent spiking neuron model of serial working memory that employs bound semantic pointers to account for a wide variety of behavioral results in the human working memory literature. I then turn to learning and describe a new learning rule that is able to account for the central biological results associated with spike-timing-dependent plasticity (STDP). I also describe how this same rule is able to learn nonlinear mappings between semantic pointers, given appropriate error information. As an example, I show that the binding operation used to construct structured representations can be learned in this manner. I also present a model that demonstrates how this rule can be incorporated into the previously introduced basal ganglia model to account for the detailed spike patterns of neurons in the ventral striatum of rats during a reinforcement learning task. I then show that this same rule can be used to account for human performance on the Wason card task, a language-based reasoning task.

Once I have presented answers to these four questions in detail, I have described what I call the "Semantic Pointer Architecture" (SPA). However, the piecemeal presentation that is necessary to explain each of these aspects of the architecture serves to obscure the utility of the architecture as a whole. So, in Chapter 7, I present a model provocatively called the Semantic Pointer Architecture Unified Network (Spaun), which integrates many of the functions considered independently (Eliasmith et al., 2012). That model performs eight distinct tasks, ranging from object recognition and motor control to learning and syntactic induction. The model is not changed from task to task (except through its own dynamics), and it is implemented in spiking neurons. It is in this chapter that I show how the mechanisms and representations employed by the architecture can be well described using probability theory. Spaun is intended as the current best exemplar of the SPA.

In the second part of the book, I evaluate the SPA using the CCC and contrast it with other current cognitive architectures. Unsurprisingly, perhaps, I suggest that the SPA has more to recommend it than competing architectures. Of course, many challenges remain, and I attempt to identify some of these in the final chapter.

It is worth emphasizing at this point that the approach offered here is *not* metaphorical in the way that past approaches have been. Clearly, I use metaphors in both my explanation and development of the ideas underlying the SPA ("semantic pointer," for one). However, the metaphors I employ "bottom out" in specific hypotheses about neural function. Neurons are physical entities that we have many means of measuring both directly and indirectly. We cannot measure symbols, connectionist "nodes," or dynamicist

TABLE 1.2 *Seven Contemporary Debates Resolved Through a Synthesis of Opposing Views by the Semantic Pointer Architecture According to Thagard (2012). Relevant Sections of This Book are Identified in Parentheses in the Third Column*

	Traditional view	*Alternative view*	*Synthesis*
1	computational	dynamical	Neural Engineering Framework (NEF) Principle 3: dynamics through neural computation (2.2.3)
2	symbolic	subsymbolic	Semantic pointers (3, 4)
3	rule-governed	statistical	Different quantitative descriptions of one implementation (7.4)
4	psychological	neural	Parallel analyses of one implementation (e.g., 4.7, 5.8, 6.6)
5	abstract	grounded	Semantic pointers (3.5, 4.8)
6	disembodied	embodied	Architectural integration of action, perception, and cognition (3.5, 3.6, 7)
7	reflective	enactive	Different timescale of analysis of the same implementation (5.7, 7)

"lumped parameters" in the brain in a similarly uncontentious manner. This, precisely, is why the SPA is not a metaphorical view in the same way that other approaches tend to be.[12]

Because the SPA is not metaphorical in this way, I think that it may be able to help unify our understanding of cognition. I have used four questions to direct my description of a cognitive architecture. Unsurprisingly, there are already many answers to these questions. Often there are multiple conflicting answers, giving rise to a variety of contemporary debates. For example, there is a well-known debate about whether cognitive representations are best understood as symbolic or subsymbolic (Fodor & Pylyshyn, 1988; Smolensky, 1988). It may thus seem natural to present a new architecture by describing which side of such debates it most directly supports.

However, Thagard (2012) has recently suggested that the SPA can be taken to resolve many of these debates by providing a synthesis of opposing views. For example, rather than claiming that the SPA is symbolic or subsymbolic, or computational or dynamical, or psychological or neural, we can claim that the SPA is all of these. The seven debates he highlights are shown in Table 1.2. I provide this table as a kind of promissory note: All of the concepts in Table 1.2 are discussed at some point in the book, and it will become clear how the SPA

[12] You may think that we need to worry about whether all theories are metaphorical to some extent or whether we ever measure anything truly directly. Regardless, I am here only proposing the less controversial claim that the SPA is *less* metaphorical and *more* directly testable than other available approaches.

relates to each. However, at this point I leave the table largely unexplained, although I return to its contents in chapter 10.

Of course, much work must be done to make a convincing case that the SPA successfully provides a resolution to any of these long-standing debates. As a result, I do not return to this claim until Chapter 9. Nevertheless, it is helpful to make clear early on that I do not take the SPA to be championing a particular cognitive approach. Rather, it more often borrows from, and hopefully unifies, what have often been depicted as competing ways of understanding cognitive function.

The general method I adopt to demonstrate such unification is to describe how a single, underlying implementation can be concurrently described as both, say, computational and dynamical, or symbolic and subsymbolic, or psychological and neural. This method is inspired by an observation of Richard Feynman, the famous physicist. Shortly before he died, Feynman wrote a few of his last, perhaps most dear, thoughts on the blackboard in his office. After his death, someone had the foresight to take a photograph for posterity (*see* `http://nengo.ca/build-a-brain/chapter1`). In large letters in the top left corner he wrote:

> *What I cannot create I do not understand.*

I believe this is an excellent motto for how best to answer our questions about cognition.[13] It suggests that *creating* a cognitive system would provide one of the most convincing demonstrations that we truly understand such a system. From such a foundation, we put ourselves in an excellent position to evaluate competing descriptions, and ultimately evaluate answers to our questions about cognition. I will not argue for this method in any great depth (although others have–e.g., Dretske, 1994). But this, in a nutshell, is why I think we should try to build a brain.

Of course, in this case, as in the case of many other complex physical systems, "creating" the system amounts to creating *simulations* of the underlying mechanisms in detail. As a result, I also introduce a tool for creating neural simulations that embodies the principles described throughout the book. This tool is a graphical simulation package called Nengo (`http://www.nengo.ca/`) that can be freely downloaded and distributed. At the end of each chapter, I provide a tutorial for creating and testing a simulation that relates to a central idea introduced in that chapter. It is possible to understand all aspects of this book without using Nengo, and hence skipping the tutorials. But, I believe a much deeper understanding is possible

[13] There are, in fact, many other mottos to choose from. For example, Herbert Simon said "The best rhetoric comes from building and testing models and running experiments. Let philosophers weave webs of words; such webs break easily" (1996, p. 272). (Disclaimer: I have nothing against good philosophers.)

if you "play" with simulations to see how complex representations can map to neural spike patterns, and how sophisticated transformations of the information in neural signals can be accomplished by realistic neuronal networks. After all, knowing how to build something and building it yourself are two very different things. In the next chapter, I begin with a description of neural mechanisms, so we can work our way towards an understanding of biological cognition.

1.5 ■ Nengo: An Introduction

Nengo (Neural ENGineering Objects) is a graphical neural simulation environment developed over the last several years by my research group and others at the Centre for Theoretical Neuroscience at the University of Waterloo.[14] All tutorials are based on Nengo 1.4, the most recent release as of this writing. In this first tutorial, I describe how to install and use Nengo and guide you through simulating a single neuron. The simulation you will build is shown in Figure 1.3. Anything you must do is placed on a single bulleted line. Surrounding text describes what you are doing in more detail. All simulations can also simply be loaded through the *File* menu. They are in the "Building a Brain" subdirectory of the standard Nengo installation.

A quick visual introduction to the simulator is available through the videos posted at `http://nengo.ca/videos`.

Installation

Nengo works on Mac, Windows, and Linux machines and can be installed by downloading the software from `http://nengo.ca/`.

- To install, unzip the downloaded file where you want to run the application from.
- In the unzipped directory, double-click "nengo" to run the program on Mac or Linux, or double-click "nengo.exe" to run the program on Windows.

An empty Nengo workspace appears. This is the main interface for graphical model construction.

[14] Nengo is free and released under MPL. It is written in Java™and fully scriptable in Python®. We encourage outside development and provide full access to the code. We also provide a model archive at `http://models.nengo.ca/` where any models written in Nengo (or other packages) can be posted.

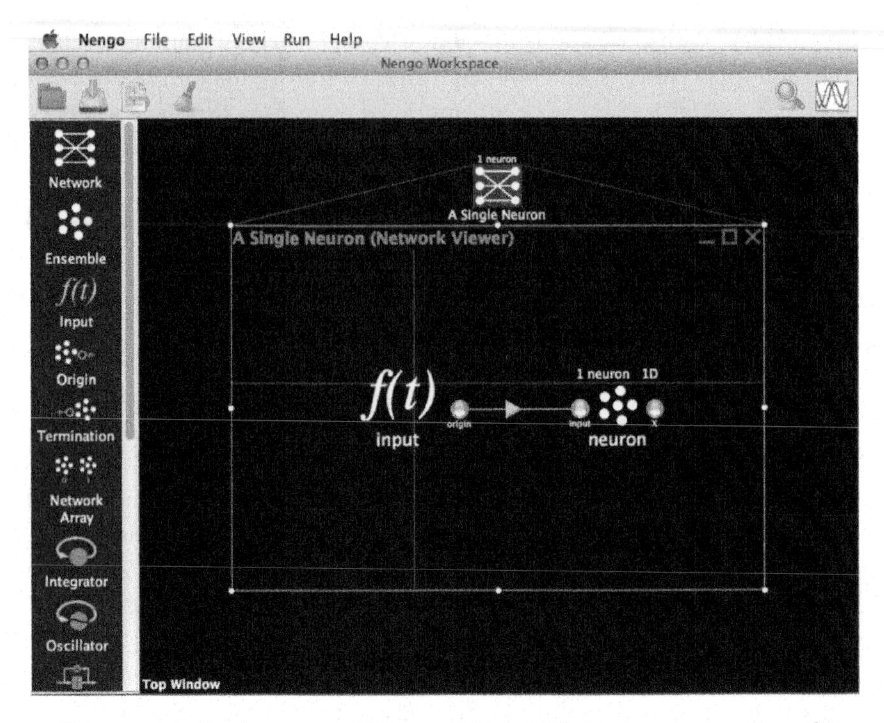

FIGURE 1.3 A single neuron Nengo model. This is a screen capture from Nengo that shows the finished model described in this section.

Building a Model

The first model we will build is very simple: just a single neuron.

- A column of icons occupies the left side of the screen. This is the *Template Bar.*
- From the template bar, click the icon above the label "Network" and drag the network into the workspace.
- Set the *Name* of the network to "A Single Neuron" and click *OK.*

All models in Nengo are built inside "networks." Inside a network you can put more networks, and you can connect networks to each other. You can also put other objects inside of networks, such as neural populations, or "ensembles" (which have individual neurons inside of them). For this model, you first need to create a neural population.

- Click the *Ensemble* icon in the template bar and drag this object into the network you created. In the dialog box that appears, you can set *Name* to "neuron," *Number of nodes* to 1, *Dimensions* to 1, *Node Factory* to "LIF Neuron," and *Radius* to 1. Click *OK.*

A single neuron is now created. This leaky integrate-and-fire (LIF) neuron is a simple, standard model of a spiking single neuron.[15] It resides inside a neural "population," although there is only one neuron. To delete anything in Nengo, right-click on it and select "Remove model." To re-adjust your view, you can zoom using the scroll wheel and drag the background to shift within the plane.

- To see the neuron you created double-click on the ensemble you just created.

This shows a single neuron called "node0." To get details about this neuron, you can right-click it and select *Inspector* and look through the parameters, although they won't mean much without a good understanding of single cell models. You can close the inspector by clicking the double arrow "≫" in the top right corner, or clicking the magnifying glass icon. To generate input to this neuron, you need to add another object to the model that generates that input.

- Close the *NEFEnsemble Viewer* (with the single neuron in it) by clicking the "X" in the upper right of that window.
- Drag the *Input* component from the template bar into the *Network Viewer* window. Set *Name* to "input" and *Output Dimensions* to 1. Click *Set Functions*. In the drop-down, select *Constant Function*. Click *Set*. Set the *Value* to 0.5. Click *OK* for all the open dialogs.

You have now created a network item that provides a constant current to injection into the neuron. It has a single output labeled "origin." To put the current into the neuron, you need to create a "termination" (i.e., input) on the neuron.

- Drag a *Termination* component from the template bar onto the "neuron" ensemble. Set *Name* to "input," *Weights Input Dim* to 1, and *PSTC* to 0.02. Click *Set Weights*, double-click the value and set it to 1. Click *OK*.

A new element will appear on the left side of the population. This is a *decoded termination* that takes an input and injects it into the ensemble. The *PSTC* specifies the "post-synaptic time constant," which I will discuss in the next chapter. It is measured in seconds, so 0.02 is equivalent to 20 milliseconds.

- Click and drag the "origin" on the input function you created to the "input" termination on the neuron population you created.

[15] An LIF neuron consists of a single differential equation that continuously sums (i.e., integrates) its input voltage while allowing a constant amount of voltage to escape per unit time (i.e., it leaks). When the summed voltage hits a pre-determined threshold value, it is reset to a resting value, and a stereotypical action potential is generated (i.e., it fires). This process then repeats indefinitely.

Congratulations, you've constructed your first neural model. Your screen should look like Figure 1.3. A few comments about organizing your screen are in order. First, the close, minimize, and maximize buttons for each window appear on the top right, and act as usual. To zoom in and out it is possible to use the scroll wheel, or right-click and drag the background. To simplify setting your view, there is a "zoom to fit" option available by right-clicking on the background of any window. As well, there is a series of icons that appear in the bottom right of the network viewer that let you save, load, and manipulate the workspace layout.

Finally, if you would rather not construct the model, we have written simple scripts to construct them for you. To run the script for this tutorial, click the folder icon in the top left of the screen, navigate to the Nengo install directory, and go to /Building a Brain/chapter1/ select the "singleneuron.py" file and click *Open*. This will load the network into the GUI and you can run it as described below. The neuron in the script is always an "on" neuron (see below), and it is also noisier, demonstrating one way to include neural variability in Nengo simulations.

Running a Model

There are two ways to run models in Nengo. The first is a "simulation mode," which lets you put "probes" into parts of the network. You then run the network for some length of time, and those probes gather data (such as when neurons are active, what their internal currents and voltages are, etc.). You can then view that information later using the built-in data viewer or export it for use with another program such as Matlab®. For future reference, to access this method of running, you can right-click on the background of the network and use the "Simulate" command under the "Run" heading.

The other way to run models is to start "Interactive Plots," which is more hands-on. I will exclusively use interactive plots in the examples in this book. An example of the output generated by interactive plots for this model is shown in Figure 1.4.

- Click on the *Interactive Plots* icon (the double sine wave in the top right corner).

This pulls up a new window with a simplified version of the model you made. It shows the "input" and "neuron" network elements that you created earlier. They are the elements with which you will interact while running the simulation. This window also has a "play" button and other controls that let you run your model. First, you need to tell the viewer which information generated by the model you would like to see.

- Right-click on the "neuron" population in the interactive plots window and select *spike raster*.

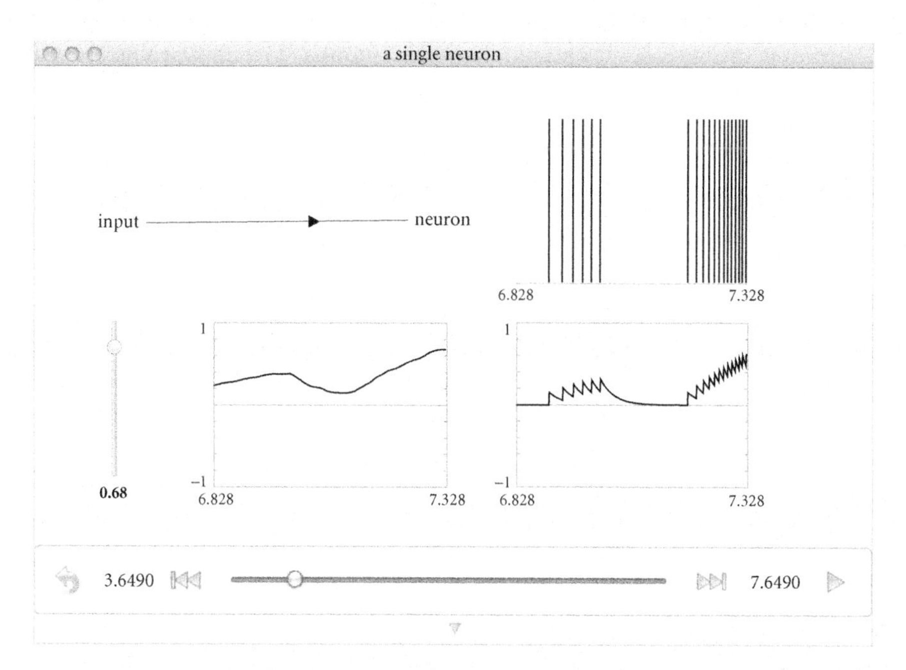

FIGURE 1.4 Running the single neuron model. This is a screen capture from the interactive plot mode of Nengo running a single neuron. The top left shows the network topology. The top right plot shows the timing of individual spikes produced by the neuron given the input (bottom middle). The bottom left is a manipulable slider that can control the input to the cell. The bottom middle plot shows the time course of the recent input, which is driving the cell. Finally, the bottom right plot shows the current that would be induced into the synapse of a cell receiving the spike train (top right). The x-axis of these plots is time in seconds.

A spike raster shows the times that the neurons in the ensemble fire an action potential spike (signifying a rapid change in its membrane voltage), which is largely how neurons communicate with one another. You can drag the background or any of the objects around within this view to arrange them in a useful manner.

- Right-click on the "neuron" population and select *value*.

This graph will show what effects this neuron will have on the current going into a neuron that receives its output spikes shown in the raster. Each small angular pulse can be thought of as a post-synaptic current (PSC) that would be induced in the receiving cell.

- Right-click on "input" and select *control*.
- Right-click on "input" and select *value*.

This shows a controller for manipulating the input as well as the value of your control input over time.

- Click the play arrow in the bottom right.

The simulation is now running. Because the neuron that you generated was randomly chosen, it may or may not be active with the given input. Either way, you should grab the slider control and move it up and down to see the effects of increasing or decreasing input. Your neuron will either fire faster with more input (an "on" neuron) or it will fire faster with less input (an "off" neuron). Figure 1.4 shows an "on" neuron with a fairly high firing threshold (just under 0.69 units of input). All neurons have an input threshold below which (or above which for "off" neurons) they will not fire any spikes.

You can test the effects of the input and see whether you have an "on" or "off" neuron and where the threshold is. You can change the neuron by pausing the simulation and returning to the main Nengo window. To randomly pick another neuron, do the following in the workspace:

- Right-click on the "neuron" population and select *Inspector.*
- Click on the gray rightward pointing arrow that is beside *i neurons (int).* It will point down and *i 1* will appear. This shows the number of neurons in the ensemble.
- Double-click on the 1 (it will highlight with blue), and hit *Enter.* This will regenerate the neuron. Click *Done.*

You can now return to the interactive plots and run your new neuron by hitting play. Different neurons have different firing thresholds. As well, some are also more responsive to the input than others. Such neurons are said to have higher sensitivity, or "gain." You can also try variations on this tutorial by using different neuron models. Simply create another population with a single neuron and choose something other than "LIF neuron" from the drop-down menu. There are a wide variety of neuron parameters that can be manipulated. In addition, it is possible to add your own neuron models to the simulator, inject background noise, and have non-spiking neurons. These topics are beyond the scope of this tutorial, but relevant information can be found at `http://nengo.ca/`.

Congratulations, you have now built and run your first biologically plausible neural simulation using Nengo. You can save and reload these simulations using the *File* menu, or the toolbar buttons.

PART I

HOW TO BUILD A BRAIN

2. AN INTRODUCTION TO BRAIN BUILDING

Before turning to my main purpose of answering the four questions regarding semantics, syntax, control, and memory and learning, in this chapter I introduce the main approach on which I rely for neural modeling. As I argued in the last chapter, I believe the best answers to such questions about cognition will be biologically based. So, in the sections that follow, I provide an overview of the relevant biology and introduce the Neural Engineering Framework (NEF), which provides methods for constructing large-scale, neurally realistic models of the brain. My purpose is to lay a foundation for what is to come: a neural architecture for biological cognition. Additional details on the NEF can be found in Eliasmith and Anderson (2003).

2.1 ■ Brain Parts

Brains are fantastic devices. For one, they are incredibly efficient. Brains consume only about 20 watts (W) of power–the equivalent of a compact fluorescent light bulb. To put this power efficiency in perspective, consider one of the world's most powerful supercomputers, "roadrunner," at Los Alamos labs in the United States. This computer, which as far as we know is unable to match the computational power of the mammalian brain, consumes 2.35 MW, which is about 100,000 times more power than the brain.

Human brains are also relatively small compared to the size of our bodies. A typical brain weighs between 1 and 2 kg and comprises only 2% of our body weight. Nevertheless, it accounts for about 25% of the energy used by the body. This is especially surprising when you consider the serious energy demands of muscles, which must do actual physical work. Of course, brains are preserved by evolution despite being such a power hog because they are doing something very important: brains control the four Fs (feeding, fleeing, fighting, and, yes, reproduction). In short, brains provide animals with behavioral flexibility that is unmatched by our most sophisticated machines. And, brains are constantly adapting to the uncertain, noisy, and rapidly changing world in which they find themselves embedded.

We can think of this incredibly efficient device as made up of a thin sheet (cortex) crammed inside our skulls, surrounding a maraca-shaped core (basal ganglia, brainstem and other subcortical structures), with a small, quarter-cabbage stuck on the back (cerebellum; *see* Fig. 2.1). The newest part of the brain is the cortex, which is equivalent in size to about four sheets of writing paper, and about 3 mm thick (Rockel et al., 1980). In almost all animals, the

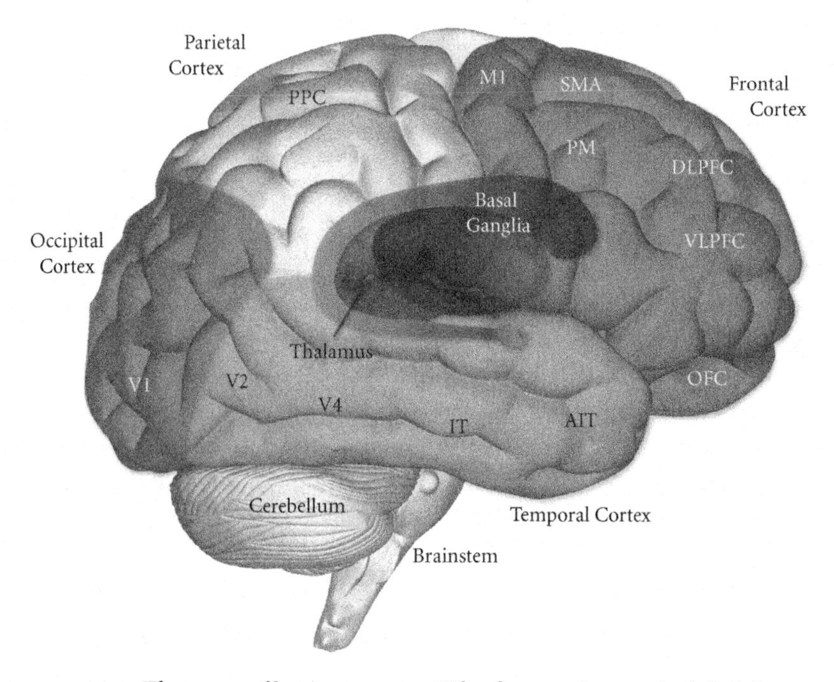

FIGURE 2.1 Elements of brain anatomy. The four major cortical divisions are shown in different shades of gray (occipital, temporal, frontal, parietal). The subcortical basal ganglia and thalamus are darker and found underneath cortex. Abbreviations identify cortical areas mentioned throughout the book. V1 - primary visual cortex; V2 - secondary visual cortex; V4 - extrastriate visual cortex; IT - inferotemporal cortex; AIT - anterior IT; OFC - orbitofrontal cortex; VLPFC - ventrolateral prefrontal cortex; DLPFC - dorsolateral prefrontal cortex; PM - premotor cortex; SMA - supplementary motor areas; M1 - primary motor cortex; PPC - posterior parietal cortex. Association cortex typically refers to all non-"primary" areas, excluding prefrontal cortex. Prefrontal cortex typically refers to frontal cortex in front of motor areas.

sheet has six distinct layers composed of three main elements: (1) the cell bodies of neurons; (2) the very long thin processes used for communication between neurons, called "axons"; and (3) glial cells, which are a very prevalent but poorly understood companion to neurons. In each square millimeter of human cortex there are crammed about 170,000 neurons.[1] So, there are about 25 billion neurons in human cortex. Overall, however, there are approximately 100 billion neurons in the human brain. The additional neurons come from "subcortical" areas, which include cerebellum, basal ganglia, thalamus, brainstem, and several other nuclei. To get a perspective on the special nature of *human* cortex, it is worth noting that monkey cortex is approximately the size of one sheet of paper, and rats have cortex the size of a Post-it note (Rockel et al., 1980).

In general, it is believed that what provides brains with their impressive computational abilities is the organization of the connections among individual neurons. These connections allow cells to collect, process, and transmit information. In fact, neurons are specialized precisely for communication. In most respects, neurons are exactly like other cells in our bodies–they have a cell membrane, a nucleus, and similar metabolic processes. What makes neurons stand out under a microscope are the many branching processes that project outward from the somewhat bulbous main cell body. These processes are there to enable short and long distance communication with other neural cells. This physical structure suggests that if we want to understand how brains work, we need to have a sense of how neurons communicate in order to compute.

Figure 2.2 outlines the main elements underlying cellular communication. The cellular projections that carry information *to* the cell body are called dendrites.[2] The cellular projection that carries information *away* from the cell body is called the axon. Dendrites carry signals to the cell body in the form of an ionic current. If sufficient current gathers in the cell body (at the axon hillock), then a series of cellular events are triggered that result in an action potential, or voltage "spike," that proceeds down the axon. Neural spikes are very brief events lasting for only about a millisecond (*see* Fig. 2.3), which travel in a wave-like fashion down the axon until they reach the end of the axon, called the bouton. When spikes reach the bouton, they cause the release of tiny packets of chemicals called neurotransmitters into the space between the bouton and the dendrite of the subsequent neuron. The neurotransmitters

[1] This estimate is based on the data in Pakkenberg and Gundersen (1997), which describes differences in density across cortical areas, as well as the effects of age and gender. This value was computed from Table 2.2 for females, using a weighted average across cortical areas.

[2] It is worth noting the following at least once: Just about everything I say about biology has exceptions. For example, information flows both ways down dendrites. That being said, I will continue with my gross oversimplifications to provide a rough sketch of biological function.

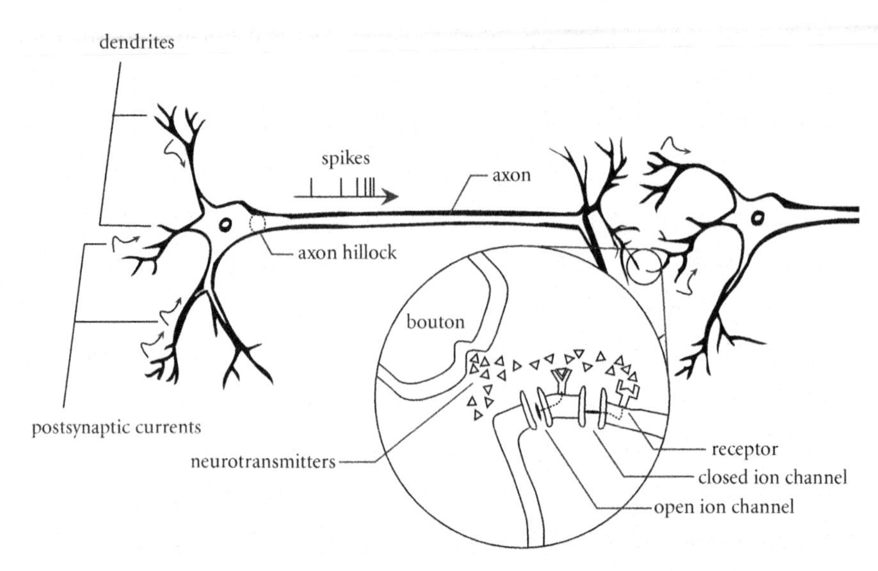

FIGURE 2.2 The main elements of neural communication. Information flows from left to right. The PSCs travel along dendrites toward the cell body of the neuron. If the current exceeds a threshold at the axon hillock, then a voltage spike is generated. Such spikes travel along the axon until they cause the release of neurotransmitters at the end of the axon. The neurotransmitters bind to matching receptors on the dendrites of connected neurons, causing ion channels to open. Open ion channels allow PSCs to be generated in the receiving cell, and the process continues.

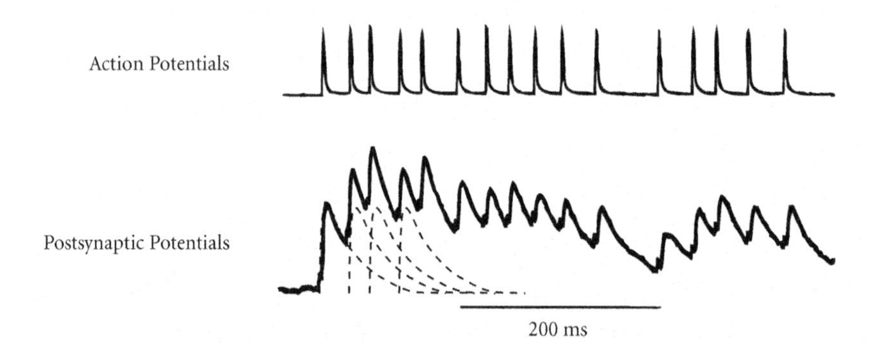

FIGURE 2.3 Electrical activity in two connected neurons. The top trace shows a series of neural action potentials (spikes) recorded intracellularly from a single rat neuron, which projects this signal to a receiving neuron. The bottom trace shows a series of postsynaptic potentials (PSPs, caused by PSCs) generated at a single synapse in the receiving neuron in response to the input spikes. The dotted lines indicate the single PSPs that would result from a spike in isolation. The PSP trace is approximately a sum of the individual PSPs. (Adapted from Holmgren et al., 2003, Figure 4, with permission.)

then bind to special proteins called "receptors" in the cell membrane of the dendrite. This binding causes small gates, or channels, in the dendrite of the next neuron to open, allowing charged ions to flow into the dendrite. These ions result in the current signal in the receiving dendrite, which is called the postsynaptic current (PSC; *see* Fig. 2.3). The process then continues as it began.

A slightly simpler description of this process, but one that retains the central relevant features, is as follows:

1. Signals flow down the dendrites of a neuron into its cell body.
2. If the overall input to the cell body from the dendrites crosses that neuron's threshold, the neuron generates a stereotypical action potential, or "spike," that travels down the axon.
3. When the spike gets to the end of the axon, it causes chemicals to be released that open channels, producing a PSC in the receiving dendrite.
4. This current then flows to the cell body of the next neuron, and the process repeats.

As simple as this story first sounds, it is made much more complex in real brains by a number of factors. First, neurons are not all the same. There are hundreds of different kinds of neurons that have been identified in mammalian brains. Neurons can range in size from 10^{-4} to 5 meters in length.[3] The number of inputs to a cell can range from about 500 or fewer (in retina) to well over 200,000 (Purkinje cells in the cerebellum). The number of outputs–that is branches of a single axon–covers a similar range. On average, cortical cells have about 10,000 inputs and 10,000 outputs (White, 1989; Abeles, 1991; Braitenburg & Shuz, 1998). Given all of these connections, it is not surprising to learn that there are approximately 72 km of fiber in the human brain. Finally, there are hundreds of different kinds of neurotransmitters and many different kinds of receptors. Different combinations of neurotransmitters and receptors can cause different kinds of currents to flow in the dendrite. As a result, a single spike transmitted down an axon can be received by many different neurons and can have different kinds of effects on each neuron, depending on the mediating neurotransmitters and receptors.

The variability found in real biological networks makes for an extremely complex system, one that at first glance seems designed to frustrate any analysis. Typically, in mathematics, homogeneous systems (those that look similar no matter where you are in the system) are much easier to understand. The brain, in contrast, is clearly highly heterogeneous–there are hundreds of kinds of neurons, many with different kinds of intrinsic dynamical properties. Even

[3] The longest such neurons are neurons in the giraffe that run from its toe to its neck. Human motor neurons are up to 1 meter in length. The famous giant squid axon is not especially long (only a few centimeters), but it is quite wide, at 1 mm thick.

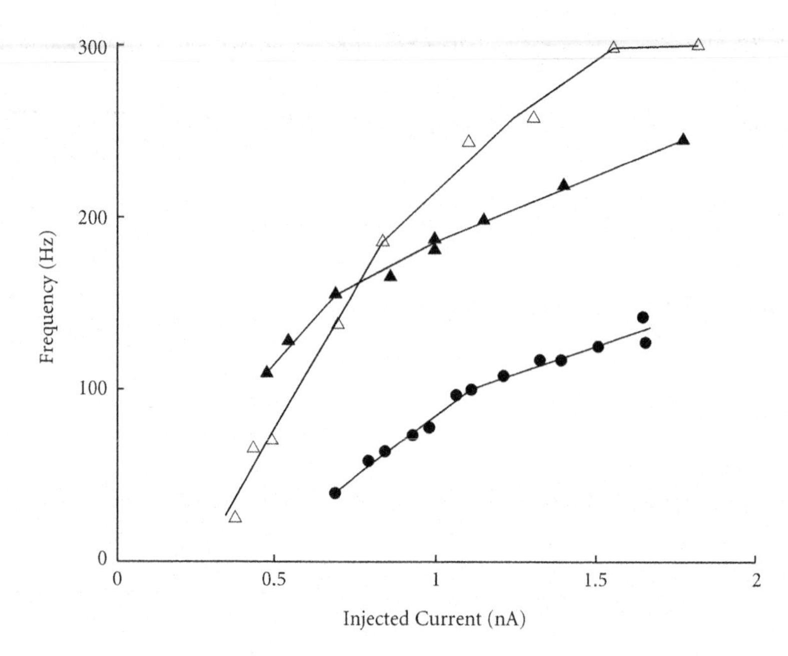

FIGURE 2.4 Response functions of regular-spiking neurons found in the mammalian neocortex. These curves demonstrate that the same kinds of neurons can have very different responses even for the same current injected directly into the soma (adapted from McCormick et al., 1985, with permission).

neurons of the same kind often respond differently–that is, generate different patterns of spikes to exactly the same current injected into their soma. And even neurons that share response properties and physiological type can still differ in the number and kinds of channels in their dendrites, meaning that the same pattern of spikes coming from preceding neurons to two otherwise similar neurons will generate different responses. For example, Figure 2.4 shows a variety of responses to directly injected current from the most common kinds of cells found in cortex. Experimental neuroscientists who record the activity of single neurons in response to perceptual stimuli will tell you that no two neurons seem to respond in the same way.

The responses of cells in such experiments are often described by their "tuning curves." These should not be confused with response functions. Tuning curves can have many different shapes, whereas response functions are more uniform. This is because tuning curves depend on where the cell is in the brain, whereas response functions depend only on intrinsic properties of the cell. Figure 2.5 shows two example tuning curves. In general, a tuning curve is a graph that shows the frequency of spiking of a neuron in response to a given input stimulus. As a result, the tuning curve helps identify what information a cell carries about a given stimulus.

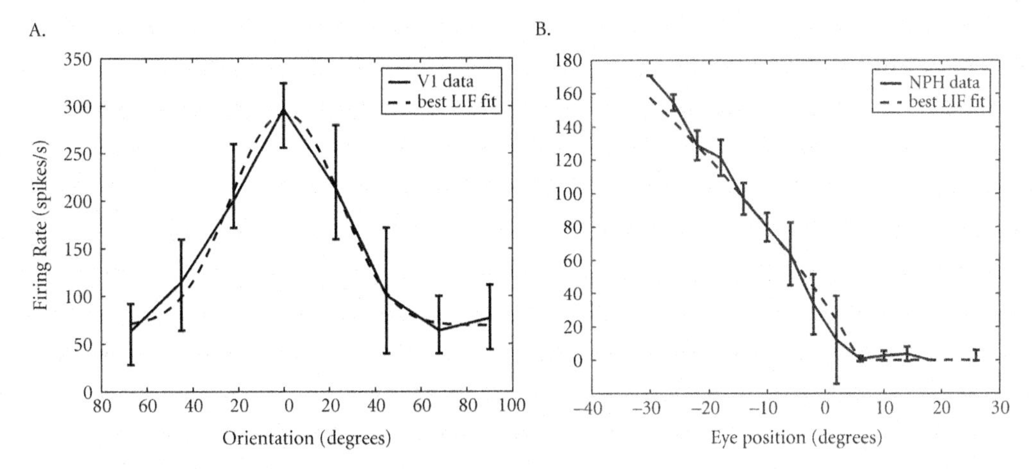

FIGURE 2.5 Example tuning curves of single cells. **A.** A cell from primary visual cortex in a macaque monkey (data provided by Dan Marcus). The x-axis indicates the angle of rotation of a bar, and the y-axis indicates the firing rate of the neuron in response to the stimulus. **B.** A cell from the nuclei prepositus hypoglossi (NPH), an area of the brain that controls eye position (data provided by Kathleen E. Cullen). The x-axis indicates the position of the eye from midline, and the y-axis indicates the firing rate of the neuron when the eye is in that position. In both graphs, error bars indicate the standard deviation of the response. The dashed line indicates the best fit of a leaky integrate-and-fire (LIF) neuron model (*see* Section 1.5) to the data. (Adapted from Eliasmith and Anderson, 2003, with permission.)

To be more precise, note that the tuning curve in Figure 2.5A is a response to an oriented bar in the cell's "receptive field." Because this is a visual cell, the receptive field indicates the part of visual space to which the neuron responds (say 1 degree in the middle of the retina). Technically, then, to completely capture the "tuning" of the cell, we should combine the receptive field and tuning curve. In the remainder of this book, I use the notion of tuning curve in this more general sense. That is, what I call a tuning curve often includes information about the receptive field of the cell in question.

Returning to the issue of heterogeneity, the reason that experimental neuroscientists would suggest that no two cells are the same is partly because their tuning curves (coupled with their receptive fields) seem to never perfectly overlap. As a result, although neurons near to the one shown in Figure 2.5A may share similar tuning properties, the peak, width, receptive field, and roll-off of the graph may be somewhat different.

We can begin to understand the heterogeneity observed in neural systems by noticing that two different sources of variability are captured by the response function and the tuning curve. As mentioned above, the response function depends on *intrinsic* cell properties, whereas the tuning curve also depends on *extrinsic* properties. Thus, intrinsic heterogeneity can be captured by variability in the models of individual cells. Extrinsic heterogeneity, in contrast, is a consequence of where a particular cell sits in the network. That is, the reason neighboring cells have different tuning curves is not merely because they are treating the same inputs differently (i.e., intrinsic heterogeneity) but also because they are receiving slightly different inputs than their neighbors. So, even if all of the cells were intrinsically homogeneous, their tuning curves might look very different depending on the processing of the cells before them. This difference between extrinsic and intrinsic heterogeneity proves important in understanding the sources of different kinds of dynamics observed in real neural networks.

Of course, this distinction does not in any way mitigate the fact of heterogeneity itself. We must still account for the observed diversity of neural systems, which has traditionally been seen as a barrier to theoretical analysis. In the next sections we will see that despite this complexity, it is possible to suggest quantified principles that do a good job of describing the functional properties of biological neural networks.

2.2 ■ A Framework for Building a Brain

In 2003, Charles H. Anderson and I wrote a book called *Neural Engineering*. In it, we presented a mathematical theory of how biological neural systems can implement a wide variety of dynamic functions. We now refer to this theory as the Neural Engineering Framework (NEF). As with most work in theoretical neuroscience, we focused on "low-level" systems, including parts of the brainstem involved in controlling stable eye position, parts of the inner ear and brainstem for controlling a vestibulo-ocular reflex (VOR), and spinal circuits in the lamprey for controlling swimming.

In more recent work, these methods have been used by us and others to propose novel models of a wider variety of neural systems, including the barn owl auditory system (Fischer, 2005; Fischer et al., 2007), parts of the rodent navigation system (Conklin & Eliasmith, 2005), escape and swimming control in zebrafish (Kuo & Eliasmith, 2005), tactile working memory in monkeys (Singh & Eliasmith, 2006), and simple decision making in humans (Litt et al., 2008) and rats (Laubach et al., 2010; Liu et al., 2011). We have also used these methods to better understand more general issues about neural function, such as how the variability of neural spike trains and the timing of individual spikes relate to information that can be extracted from spike patterns (Tripp & Eliasmith, 2007), how we can ensure that biological constraints such as Dale's Principle (the principle that a given neuron typically has either excitatory or inhibitory effects but not both) are respected by neural models (Parisien et al.,

2008), and how spike-timing can be used to learn such circuits (MacNeil & Eliasmith, 2011).

One reason the NEF has such broad application is because it does not make assumptions about what specific functions the brain performs. Rather, it is a set of three principles that can help determine *how* the brain performs some given function. For this reason, John Miller once suggested that the NEF is a kind of "neural compiler." If you have a guess about the high-level function of the brain area in which you are interested, and you know some information about how individual neurons respond in that area, then the NEF provides a way of connecting groups (or "populations" or "layers") of neurons together to realize that function. This, of course, is exactly what a compiler does for computer programming languages. The programmer specifies a program in a high-level language like Java. The Java compiler knows something about the low-level machine language implemented in a given chip, and it translates that high-level description into an appropriate low-level one.

Of course, things are not so clean in neurobiology. We do not have a perfect description of the machine language, and our high-level language may be able to define functions that we cannot actually implement in neurons. Consequently, building models with the NEF can be an iterative process: First, you gather data from the neural system and generate a hypothesis about what it does; then you build a model using the NEF and see if it behaves like the real system; then, if it does not behave consistently with the data, you alter your hypothesis or perform experiments to figure out why the two are different. Sometimes a model will behave in ways that cannot be compared to data because the data do not exist. In these lucky cases, it is possible to make a prediction and perform an experiment. Of course, this process is not unique to the NEF. Rather, it will be familiar to any modeler. What the NEF offers is a systematic method for performing these steps in the context of neurally realistic models.

It is worth emphasizing that, also like a compiler, the NEF does *not* specify what the system does. This specification is brought to the characterization of the system by the modeler. In short, the NEF is about *how* brains compute, not *what* they compute. The bulk of this book is about the "what," but those considerations do not begin until Chapter 3.

In the remainder of this section, I provide an outline of the three principles of the NEF. This discussion is admittedly truncated, and so those already familiar with theoretical neuroscience may find it somewhat lacking. There are, however, many more details elsewhere (at least a book's worth!). That being said, a few points are in order. First, we have drawn heavily on other work in theoretical neuroscience. To keep this description as brief as possible, I refer the reader to other descriptions of the methods that better place the NEF in its broader context (Eliasmith & Anderson, 2003; Eliasmith, 2005a; Tripp & Eliasmith, 2007). Second, the original *Neural Engineering* book is a useful source for far more mathematical detail than I provide here. However, the framework has been evolving, so that book should be taken as a starting point. Finally, some mathematics can be useful for interested readers, but I have placed most of it in the appendices to emphasize a more intuitive grasp of the principles. This is at least partially because the Nengo neural simulator has been designed to handle the mathematical detail, allowing the modeler to focus effort on capturing the neural data, and the hypothesis she or he wishes to test.

The following three principles form the core of the NEF:

1. Neural representations are defined by the combination of nonlinear encoding (exemplified by neuron tuning curves and neural spiking) and weighted linear decoding (over populations of neurons and over time).
2. Transformations of neural representations are functions of the variables represented by neural populations. Transformations are determined using an alternately weighted linear decoding.
3. Neural dynamics are characterized by considering neural representations as state variables of dynamic systems. Thus, the dynamics of neurobiological systems can be analyzed using control (or dynamics systems) theory.

In addition to these main principles, the following addendum is taken to be important for analyzing neural systems:

- Neural systems are subject to significant amounts of noise. Therefore, any analysis of such systems must account for the effects of noise.

The ubiquity of noise in neural systems is well documented, be it from synaptic unreliability (Stevens & Wang, 1994; Zucker, 1973), variability in the amount of neurotransmitter in each vesicle (Burger et al., 1989), or jitter introduced by axons into the timing of neural spikes (Lass & Abeles, 1975). Consequently, there are limits on how much information can be passed by neurons: it seems that neurons tend to encode approximately two to seven bits of information per spike (Bialek & Rieke, 1992; Rieke et al., 1997). Here I do not consider this addendum separately, although it is included in the formulation of each of the subsequent principles and their implementation in Nengo.

As well, it should be evident from the statement of these principles that the unit of analysis in the NEF is typically a *population* of neurons, rather than single cells. However, a population is simply a collection of single cells. Their grouping into a population is technically not important for which functions can be computed, but the grouping does often elucidate assumptions made in the model. For example, I could group visual neurons in virtue of their sharing similar sensitivity to edges or their sharing similar sensitivity to contrast. Whichever I choose will make it easier to express transformations of that property (i.e., edges or contrast). However, a given grouping does not rule out expressing transformations of any information encoded in the neurons. Choosing a specific population grouping also does not rule out interactions and overlap between the representations in different populations. Overall, grouping neurons in populations (and even populations of populations) is a convenient tool for integrating different levels of analysis, not for enforcing neuroanatomical constraints.

Let me now describe each of the three principles in turn. To make the application of these principles concrete, I adopt the example of a "controlled integrator." Qualitatively speaking, this neural circuit can be thought of as a simple memory. It can be loaded with specific data, hold the information over time, and then erase it. More quantitatively, a controlled integrator is an integrator with a tunable "leak." When the leak is set to zero, it acts as a standard integrator. It is easiest to think of a standard integrator as keeping a running sum of its input, as shown in Figure 2.6A. Consequently, the internal state of the integrator is the mathematical integral of the input. If the leak is set to a negative number, as happens at the 5-second mark in Figure 2.6B, the internal state goes exponentially toward zero. As shown, subsequent input is then only partially summed and must be maintained for the state not to go to zero. Such partial summing is an example of low-pass filtering of the input. Once there is no more input, the internal state goes to zero because of the leak.

2.2.1 Representation

A central tenet of the NEF is that we can adapt the information theoretic account of *codes* to understanding representation in neural systems. Codes are defined in terms of complimentary encoding and decoding procedures between two alphabets. Morse code, for example, is defined by the one-to-one relation between letters of the Roman alphabet, and the alphabet composed of a standard set of dashes and dots. The encoding procedure is the mapping from the Roman alphabet to the Morse code alphabet, and the decoding procedure is its inverse (i.e., mapping dots and dashes to the Roman alphabet).

To characterize representation in a neural system in this way, we must identify the relevant encoding and decoding procedures and the relevant alphabets. The encoding procedure is straightforward to identify; one typical example is the mapping of stimuli into a series of neural spikes. Encoding is what neuroscientists typically measure and is what is partly captured by the tuning curves I discussed in Section 2.1. When we show a brain an input signal (i.e., a stimulus), some neurons or other "fire" (i.e., generate spikes). This pattern of firing is often depicted by neuroscientists as a "spike raster." Spike rasters show the only response to the stimulus that is sent to other neurons. The raster in Figure 2.7 shows responses from just two neurons, whereas Figure 2.9 shows the responses from 30. Often, rasters produced during experiments show

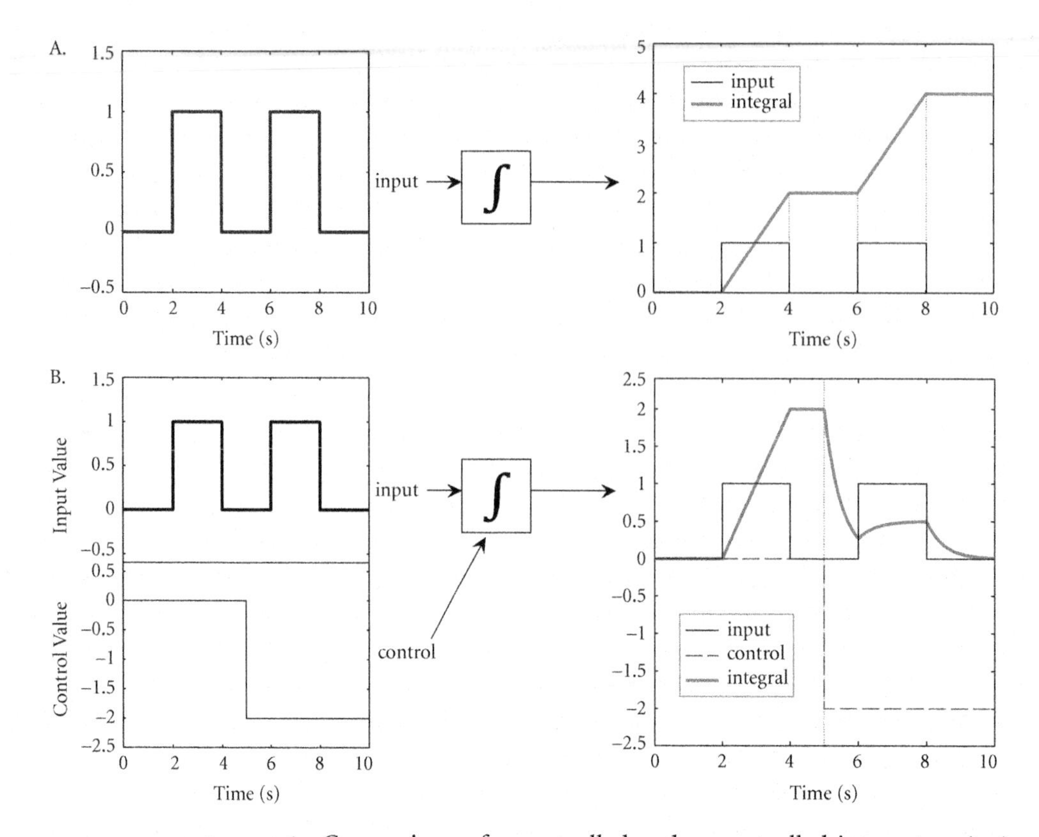

FIGURE 2.6 Comparison of a controlled and uncontrolled integrator. **A.** A standard integrator continuously adds its input to its current state. Thus it computes the mathematical integral of its input. These dynamics do not change regardless of the input signal. **B.** A controlled integrator acts like a standard integrator only if the control variable (i.e., "leak") is equal to zero. Values less than zero cause the integrator to exponentially forget its current state. The speed of forgetting is proportional to the value of the control variable. Notice that the responses of the integrators are identical for the first 5 seconds, until the control signal is changed. In a controlled integrator, the dynamics of the system is changed by its input.

the responses of the same neuron over many different trials to demonstrate the variability of the response to the same stimulus, although I will typically show the responses of many neurons on one trial.[4] The precise nature of this

[4] Because we can control the trial-to-trial variability of our single neuron models, we often keep it low to be able to better see the typical response. Nevertheless, variability is often central to characterizing neural responses, and it can be captured effectively in Nengo using the available noise parameters (*see*, e.g., Liu et al., 2011). And, as mentioned earlier, variability and noise are central to the formulation of the NEF principles.

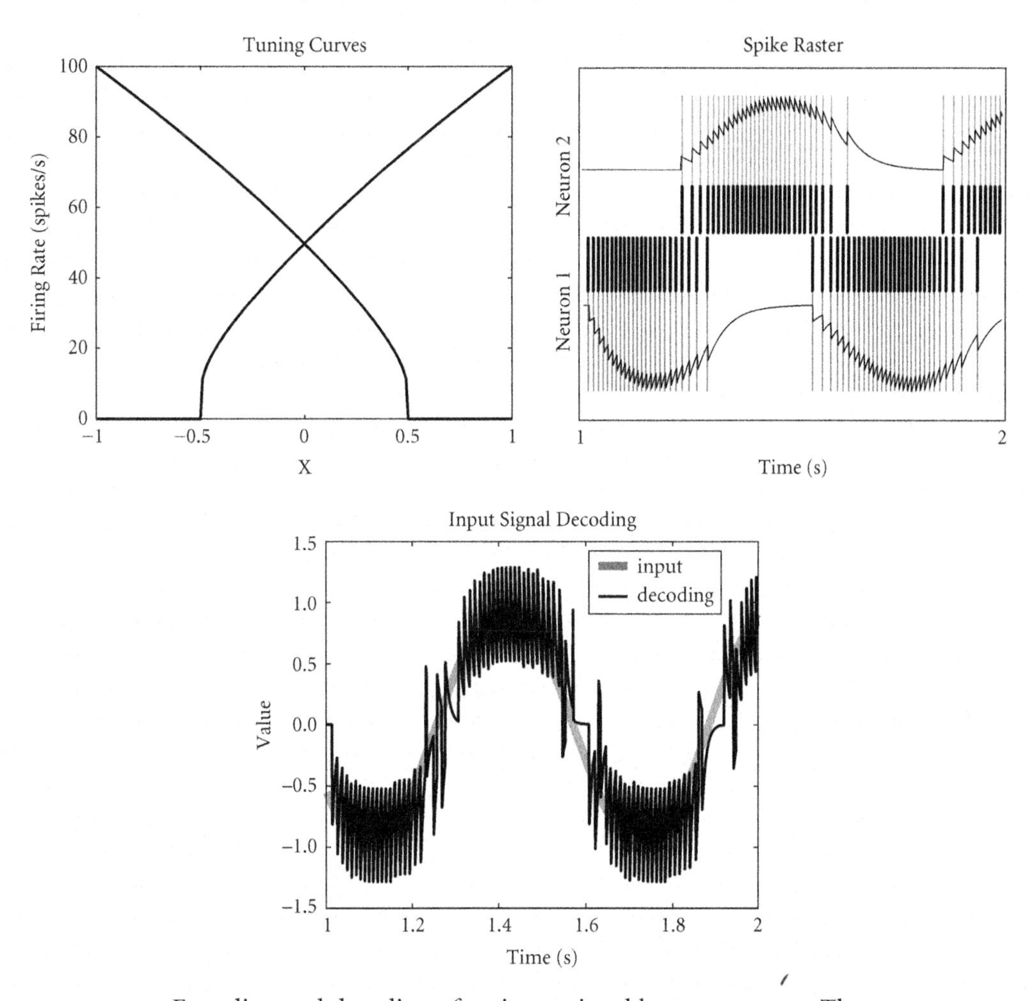

FIGURE 2.7 Encoding and decoding of an input signal by two neurons. The neurons fire at rates specified by their tuning curves, shown here to be symmetric about zero (leftmost plot). With a sinusoidally varying input, the neurons fire as shown by the black vertical lines in the spike raster. The continuous black lines in the middle plot show the PSCs that would be generated in the dendrites of a receiving neuron by the spike rasters. In the last panel, the PSCs are summed to give an estimate (black) of the original input signal (gray). The estimate is poor with two neurons but can be made much better with additional neurons (*see* Fig. 2.9).

encoding from stimulus to spikes has been explored indepth via quantitative models (*see* Appendix B.1.1).

In Figure 2.7, the spike raster is generated by injecting a sine wave into neurons that are tuned as shown on the left of that figure. As the sine wave goes below zero, the "off" neuron (sloping downward from left to right) begins to fire more spikes, and the opposite is true for the other neuron. The mapping from a sine wave to these two spike trains is the result of the encoding process.

Unfortunately, neuroscientists often stop here in their characterization of representation, but this is insufficient. We also need to identify a decoding procedure, otherwise there is no way to determine the relevance of the purported encoding for "downstream" parts of the system. For example, if no information about a stimulus can be extracted from the spikes of the encoding neurons, then it makes no sense to say that they represent the stimulus. Representations, at a minimum, must potentially be able to "stand in for" the things they represent. As a result, characterizing the decoding of neural spikes into the variable they represent is as important as characterizing the process of encoding variables into neural spikes. A simple decoding process is also shown in Figure 2.7. The top and bottom-most lines in the middle panel indicate the current that would be induced in dendrites of a neuron receiving those spike trains. Those currents are determined by summing the PSC generated from each single spike (i.e., a decaying exponential). These currents are weighted and summed to give the overall signal driving the soma of the cell as shown in the last panel. Critically, however, specifying a decoding procedure does not suggest that the brain itself must perform exactly this decoding. Nevertheless, specifying decoding is crucial to a consistent theoretical characterization of neural representation.

Quite surprisingly, despite typically nonlinear encoding (i.e., mapping a continuously varying parameter like stimulus intensity into a series of discontinuous spikes), a good linear decoding can be found.[5] To say that the decoding is "linear" means that the responses of neurons in the population are weighted by a constant (the decoder) and summed up to give the decoding. The NEF employs a specific method for determining appropriate linear decoders given the neural responses to stimuli (*see* Appendix B.1.2).

There are two aspects to the decoding of the neural response that must be considered. These are what I call the "population" and "temporal" aspects of decoding. The "population" decoding accounts for the fact that a single stimulus variable is typically encoded by many different (i.e., a population of) neurons. The "temporal" decoding accounts for the fact that neurons respond *in time* to a typically *changing* stimulus. Ultimately, these two aspects determine one combined decoding. However, it is conceptually clearer to consider them one at a time.

As depicted in Figure 2.8, population decoders are determined by finding the weighting of each neuron's tuning curve, so that their sum represents the input signal over some range (Georgopoulos et al., 1986; Salinas & Abbott,

[5] This is nicely demonstrated for pairs of neurons in Rieke et al. (1997, pp. 76-87).

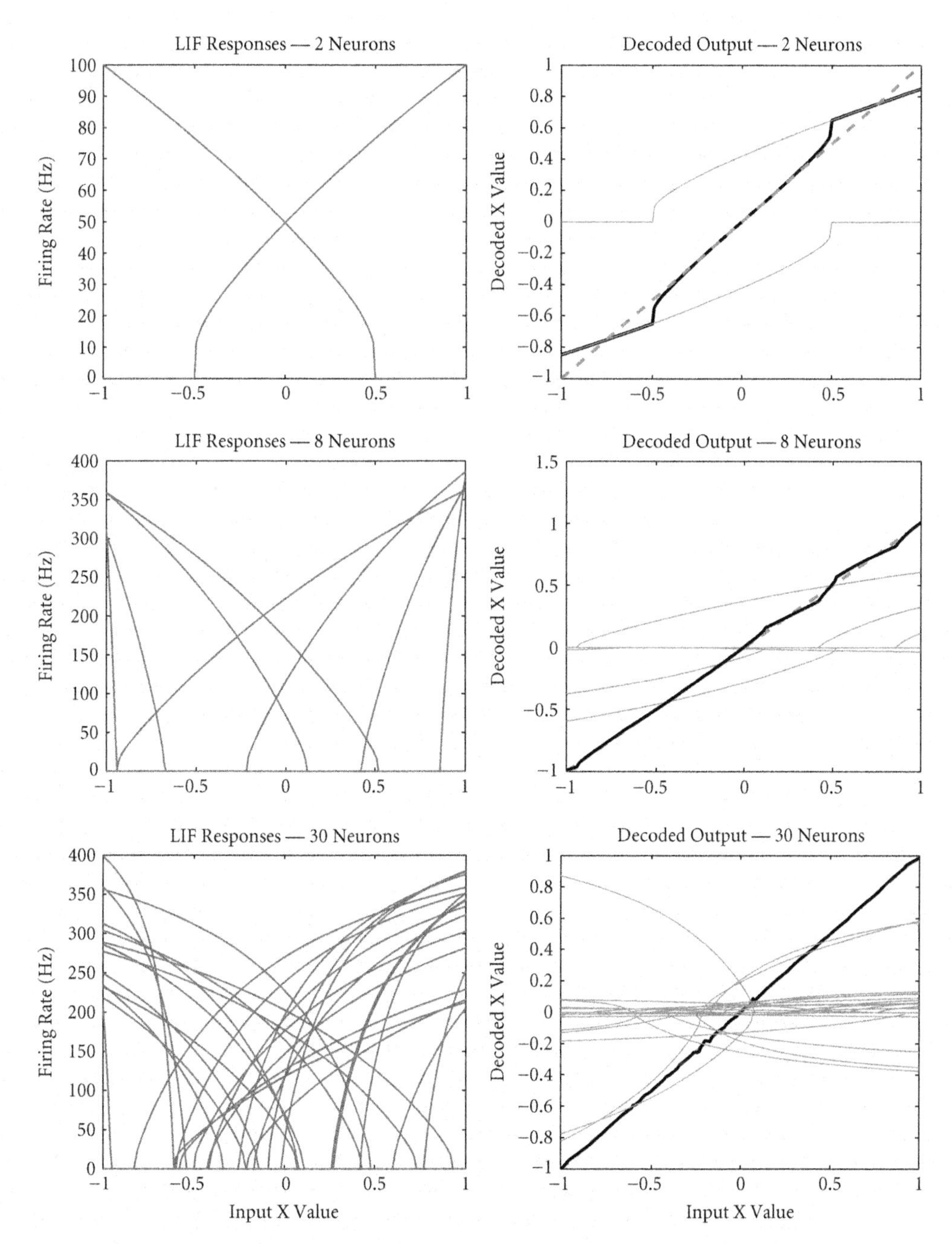

FIGURE 2.8 Finding decoders for a set of neurons. This set of figures demonstrates how neuron tuning curves (*see* Fig. 2.5) can be individually weighted by decoders to approximate an input variable, *x*, over a range of values. The left column shows the tuning curves of the neurons. The right column shows linearly weighted versions of these curves in gray and the sum of these weighted curves as a thick black line. As the number of neurons is increased, the approximation (i.e., reconstructed output) to ideal representation (i.e., the straight dashed line) improves.

1994). That is, each neuron is weighted by how useful it is for carrying information about the stimulus in the context of the whole population. If it is very useful, then it has a high weight; if not, then it has a lower weight. Finding the exact decoding weights in the NEF is accomplished by solving an optimization problem to minimize the representational error (*see* Appendix B.1.2).

As can be seen in Figure 2.8, as more neurons are added to the population, the quality of the representation improves. This is because the neuron responses are nonlinear, but ideal representation is a straight line (specifically, the estimate of the input should exactly equal the input; a representation, after all, should be able to "stand in for" [i.e., functionally replace] what it represents). Once the decoders have been found, they remain fixed for any input value. So again, it is clear from this figure that different inputs will be encoded with different accuracies (as can be inferred from the "ripples" in the thick black lines of Fig. 2.8). Typically, the input will be changing over time, so we must also determine how to decode over time (Bialek & Zee, 1990; Miller et al., 1991; Rieke et al., 1997).

Figures 2.7 and 2.9 depict temporal decoding for different numbers of neurons. For the NEF, temporal encoding and decoding are determined by the biophysics of cellular communication. In short, the PSCs discussed in Section 2.1 are used as an estimate of the input signal over time (*see* Fig. 2.7). Specifically, the spike patterns of the neurons being driven by the input are decoded by placing a PSC at each spike time and summing the result. For example, in Figure 2.7 the "on" neuron PSCs are multiplied by $+1$ and the "off" neuron PSCs are multiplied by -1 (this is a linear decoding since we are multiplying a constant function [the PSC] by a weight [± 1] and summing).

When we apply this principle to many neurons, as in Figure 2.9, the PSCs induced by each neuron need appropriate weights, which are precisely the population decoders discussed earlier. Combining these decoding weights with a standard PSC model completely defines a linear "population-temporal" decoder. In this way, many neurons can "cooperate" to give very good representations of time-varying input signals. Notice that adding additional neurons accomplishes two things. First, it allows the nonlinearities of the tuning curves to be linearized as previously discussed with respect to population decoding. Second, it allows the very "spiky" decoding in Figure 2.7 to become smoothed as the PSCs are more evenly distributed over time. These concepts are illustrated in the tutorial described in Section 2.4.

Having specified the encoding and decoding used to characterize time-varying representation, we still need to specify the relevant alphabets. Although specific cases will diverge greatly, we can describe the paradigmatic alphabets quite generally: neural responses (encoded alphabet) code physical properties (decoded alphabet). Slightly more specifically, the encoded alphabet is the set of temporally patterned neural spikes over populations

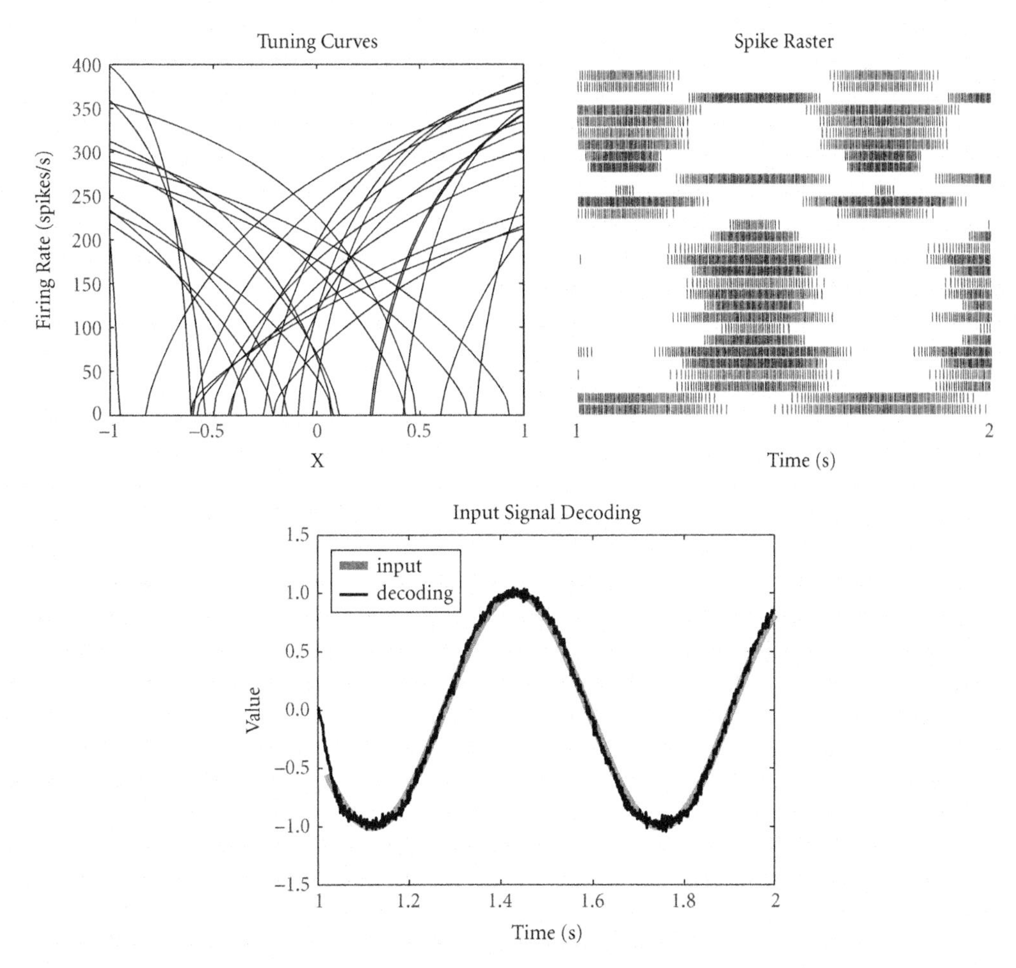

FIGURE 2.9 Encoding and decoding of an input signal by 30 neurons. As in Figure 2.7, the tuning curves of the neurons characterize their general response to input, the spike raster depicts the firing pattern of each neuron in response to a sinusoidal input signal, and the final panel shows the decoded estimate of the input signal. The temporal decoders are PSCs and the population decoders are optimal decoders (determined as depicted in Fig. 2.8).

of neurons. This is reasonably uncontroversial. However, it is much more difficult to be specific about the nature of the alphabet of physical properties.

We can begin by looking to the physical sciences for categories of physical properties that might be encoded by nervous systems. Indeed, we find that many of the properties that physicists traditionally use to describe the physical world do seem to be represented in nervous systems: there are neurons sensitive to velocity, acceleration, wavelength, temperature, and so forth. But, there are many physical properties not discussed by physicists that also seem to be encoded in nervous systems: such as hot, square, dangerous, edible, object, conspecific, and so forth. It is reasonable to begin with the hypothesis that these "higher-level" properties are inferred on the basis of representations of properties more like those that physicists talk about. In other words, encodings of "edible" depend, in some complex way, on encodings of "lower-level" physical properties like wavelength, velocity, and so forth. The NEF itself does not determine precisely what is involved in such complex relations, although I do suggest that it provides the necessary tools for describing such relations. I return to these issues–related to the meaning (or "semantics") of representations–throughout much of the book, starting in Chapter 3.

For now, we can be content with the claim that whatever is represented can be described as some kind of structure with units. A precise way to describe structure is to use mathematics. Hence, this is equivalent to saying that the decoded alphabet consists in mathematical objects (i.e., scalars, vectors, functions–including probability density functions–vector fields, etc.) with units (i.e., kilograms, meters, degrees, mates, etc.)–these, of course, describe physical properties of the world. The first principle of the NEF, then, provides a general characterization of the encoding and decoding relationship between mathematical objects with units and descriptions of patterns of spikes in populations of neurons.

To make this characterization more concrete, let us turn to considering the example of the controlled integrator (*see* Fig. 2.10) in a specific case. One of the simplest mathematical objects we might integrate (i.e., remember) is a scalar (i.e., a one-dimensional vector, which is a single number). For example, we could characterize the horizontal position of an object in the environment as a scalar, whose units are degrees from midline. There are neurons in a part of the monkey's brain called the lateral intraparietal cortex (LIP) that are sensitive to this scalar value (Andersen et al., 1985). Indeed, these parts of the brain also seem to act as a kind of memory for object location, as they are active even after an object has disappeared.[6] What is relevant for present purposes is that, as a population, neurons in this area *encode* an object's position over time.

To summarize, the representation of object position can be described as a scalar variable, whose units are degrees from midline (decoded alphabet)–that is, encoded into a series of neural spikes across a population (encoded alphabet). Using the quantitative tools mentioned earlier, we can determine the relevant decoder (*see* Appendix B.1.2). Once we have such a decoder, we can then estimate what the actual position of the object is given the neural

[6] I should be clear that I do not think a simple controlled integrator that remembers only a scalar value maps well to these specific neurons, but a similar architecture with a more complex representation has been used to model many properties of LIP activity (Eliasmith & Anderson, 2003).

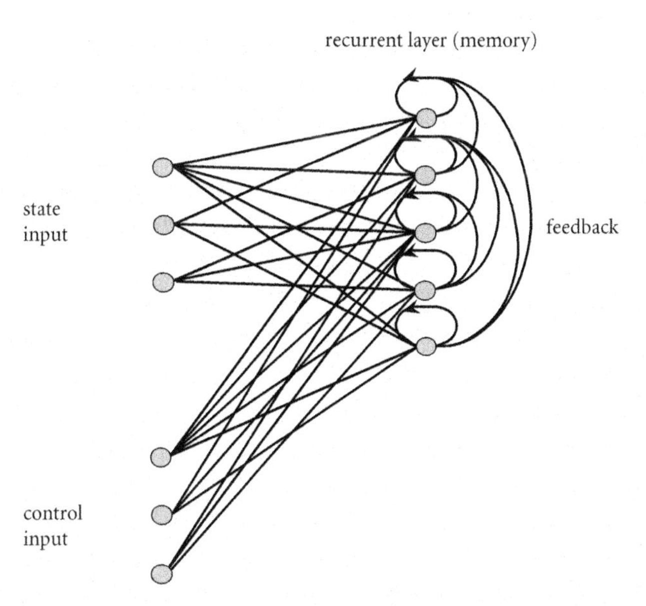

FIGURE 2.10 The architecture of a controlled integrator. This network can act like a loadable and erasable memory. Both the control input and the state input are connected to the recurrent layer of neurons, which hold a memory. The NEF allows the neural connection weights in this network of spiking neurons to be determined, once we have defined the desired representations, computations, and dynamics of the system.

spiking in this population, as in Figure 2.9. Thus we can determine precisely how well, or what aspect of, the original property (in this case the actual position) is *represented* by the neural population. We can use this characterization to understand the role that this representation plays in the system as a whole.

One crucial aspect of the NEF principle of representation is that it can be used to characterize arbitrarily complex representations. The example I have described here is the representation of a scalar variable. However, this same principle applies to representations of vectors (such as movement or motion vectors found in motor cortex, brainstem, cerebellum, and many other areas), representations of functions (such as stimulus intensity across a spatial continuum as found in auditory systems and many working memory systems), and representations of vector fields (such as the representation of a vector of intensity, color, depth, etc., at each spatial location in the visual field as found in visual cortex). See Section 2.4 for an example of 2D representation in Nengo and Section 4.9 for the use of 20D representations.

A second crucial aspect of this principle is that it distinguishes the mathematical object being represented from the neurons that are representing it. I refer to the former as the "state space" and the latter as the "neuron space." There are many reasons it is advantageous to distinguish the neuron space from the state space. Most such advantages should come clear in subsequent chapters, but perhaps most obviously, distinguishing these spaces naturally accounts for the well-known redundancy found in neural systems. Familiarity with Cartesian plots makes us think of axes in a space as being perpendicular. However, the common redundancy found in neural systems suggests that their "natural" axes are in fact not perpendicular but, rather, slanted toward one another (this is sometimes called an "overcomplete" representation). Distinguishing the state space, where the axes typically are perpendicular, from the neuron space, where they are not, captures this feature of neurobiological representation (*see* Appendix A.4 for a mathematical description of this feature).

A third crucial aspect of this principle is that it embraces the heterogeneity of neural representation. There are no assumptions about the encoding neurons being similar in any particular respect. In other work, it has been shown that such a representation is nearly as good as an optimally selected representation, in terms of both capacity and accuracy (Eliasmith & Anderson, 2003, pp. 206–217). In practice, this means that the typically heterogeneous tuning curve data from a given neural system can be well matched by a model (as the principle puts no constraints on tuning curve distributions).

The tutorial at the end of this chapter demonstrates how to build and interact with simulations of heterogeneous scalar and vector representations in Nengo. The central features of the representation principle are further highlighted there with concrete examples.

2.2.2 *Transformation*

A characterization of neural representation is not useful if it does not help us understand how the brain functions. Conveniently, the first NEF principle paves the way for a general characterization of how representations can be transformed. This is because transformations (or computations) can also be characterized using decoding. Rather than using the "representational decoder" discussed above, we can identify a "transformational decoder." A transformational decoder is a kind of *biased* decoding. That is, in determining a transformation, we extract information *other than* what the population is taken to represent. The bias, then, is away from a "pure," or representational, decoding of the encoded information.

For example, if we think that the quantity x is encoded in some neural population, then when defining the representation we determine the decoders that estimate x. However, when defining a transformation we identify decoders that estimate some function, $f(x)$, of the represented quantity. In other words, we find decoders that, rather than extracting the signal represented by a population, extract some transformed version of that signal. The same techniques used to find representational decoders are applicable in this case and result in decoders that can support both linear and nonlinear transformations (*see* Appendix B.2). Figure 2.11 demonstrates how a simple nonlinear function can be computed in this manner.

Perhaps surprisingly, this is all there is to neural transformation. Rather than decoding a representation of x from a spike train, we are decoding a function of x–that is, $f(x)$–from that same spike train. Importantly, this understanding of neural computation applies regardless of the complexity of the neural representation and thus also accounts for complex transformations. For example, it can be used to define inference relations, be they statistical (Eliasmith & Anderson, 2003, Chapter 9) or more linguistic (*see* Section 4.4). So, although linear decoding is simple, it can support the kinds of complex, nonlinear transformations needed to articulate descriptions of cognitive behavior.

For present purposes, let us again consider the specific example of a controlled integrator. Recall that the reason this integrator is "controlled" is because we can change the dynamics of the system by changing one of the inputs. Recall that a comparison of the ideal controlled and ideal noncontrolled integrators is shown in Figure 2.6. It is evident there that the dynamics of the system in the controlled integrator is a function of its input. Specifically, the "control variable" is able to make the integrator act as a standard integrator or rapidly forget its current state.

To build such a system, it is necessary to compute the product (i.e., a nonlinear function) of the current memory state and the control variable (for reasons given in Section 2.2.3). Consequently, to implement such a system in a neural network, it is necessary to transform the represented state space of the recurrent layer (*see* Fig. 2.10) in such a way that it estimates this nonlinear function. To do so, it is important to note that the state space of the recurrent layer represents both the memory and the control inputs. Hence, it contains a 2D vector representation $\mathbf{x} = [x_1, x_2]$ (*see* Appendix A.1 for mathematical background on vectors). To compute the necessary transformation of this state space, we can find transformational decoders to estimate the function $f(\mathbf{x}) = x_1 x_2$, just as we earlier found the decoders to estimate the function $f(x) = x^2$ for a 1D scalar (*see* Fig. 2.11). In general, we can characterize nonlinear functions of multiple scalars or vectors (such as $f(\mathbf{x}) = x_1 x_2$) as a linear decoding of some higher-dimensional representation in a population of neurons. The tutorial in Section 3.8 demonstrates how to construct a network that performs scalar multiplication in this manner.

Before moving on to a consideration of dynamics, it is important to reiterate that this way of characterizing representation and transformation does not demand that there are "little decoders" inside the head. That is, the NEF does not entail that the system itself needs to decode the representations it employs. In fact, according to this account, there are no directly observable counterparts to the representational or transformational decoders. Rather, as

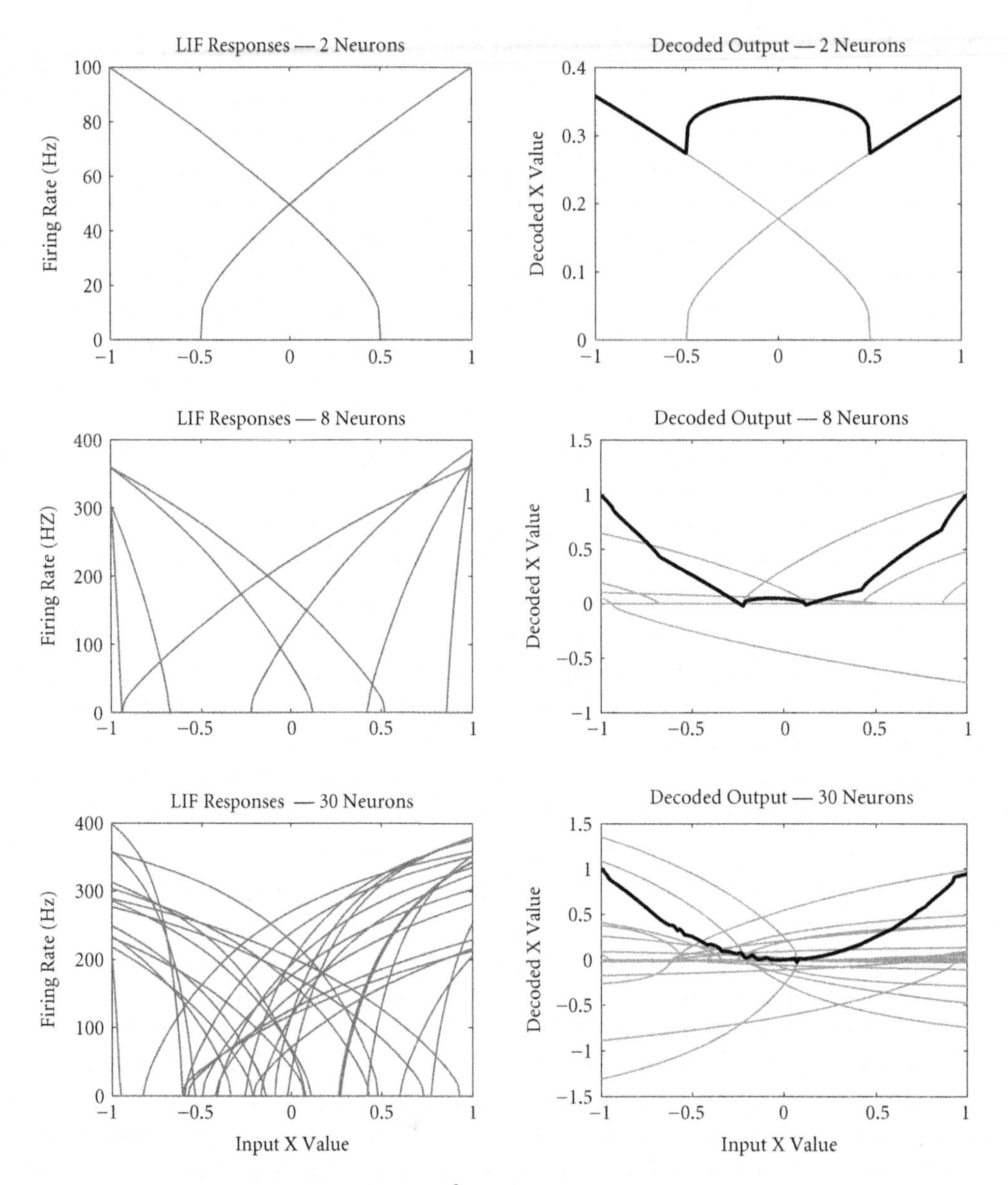

FIGURE 2.11 Computing x^2, a nonlinear function of a scalar input. The left column shows the tuning curves of the neurons, which are identical to the curves shown in Figure 2.8. The right column shows linearly weighted versions of these curves in gray and the sum of these weighted curves as a thick black line. As can be seen, the transformational decoding results in a quadratic function over the range of -1 to 1.

discussed in Section 2.2.4, they are embedded in the synaptic weights between connected neurons. That is, coupling weights of connected neurons indirectly reflect a particular population decoder, but they are not identical to the population decoder. This is because connection weights are best characterized as determined by both the decoding of the incoming signal and the encoding of the outgoing signal. Practically speaking, this means that changing a connection weight both changes the transformation being performed and the tuning curve of the receiving neuron. As is well known from both connectionism and theoretical neuroscience, this is what we should expect in such networks. In essence, the encoding/decoding distinction is not one that neurobiological systems need to explicitly respect, but it is nevertheless extremely useful in trying to understand such systems.

2.2.3 Dynamics

Throughout the history of cognitive science, computation and representation have always been central to cognitive theories, but dynamics less so. And, although it may be understandable that dynamics were initially ignored by those studying cognitive systems as purely computational, it would be strange indeed to leave dynamics out of the study of minds as physical, neurobiological systems. Even the simplest nervous systems performing the simplest functions demand temporal characterizations: moving, eating, and sensing in a constantly changing world are clearly dynamic processes. It is not surprising, then, that single neural cells have almost always been characterized by neuroscientists as essentially dynamic systems. In contemporary neuroscience, researchers often analyze neural responses in terms of "onsets," "latencies," "stimulus intervals," "steady states," "decays," and so forth–these are all terms describing temporal aspects of a neurobiological response. The fact is, the systems under study in neurobiology are dynamic systems and as such they make it very difficult to ignore time.

Notably, modern control theory was developed precisely because understanding complex dynamics is essential for building something that works in the real world.[7] Modern control theory permits both the analysis and synthesis of elaborate dynamical systems. Because of its general formulation, modern control theory applies to chemical, mechanical, electrical, digital, or analog systems. As well, it can be used to characterize nonlinear, time-varying, probabilistic, or noisy systems. As a result of this generality, modern control

[7] A nice history of classical and modern control can be found at `http://www.theorem.net/theorem/lewis1.html`. Textbooks devoted to modern control methods include Brogan (1990), Dorf and Bishop (2004), and Nise (2007). Helpful websites include `http://en.wikipedia.org/wiki/State_space_(controls)` and `http://en.wikibooks.org/wiki/Control_Systems/State-Space_Equations`.

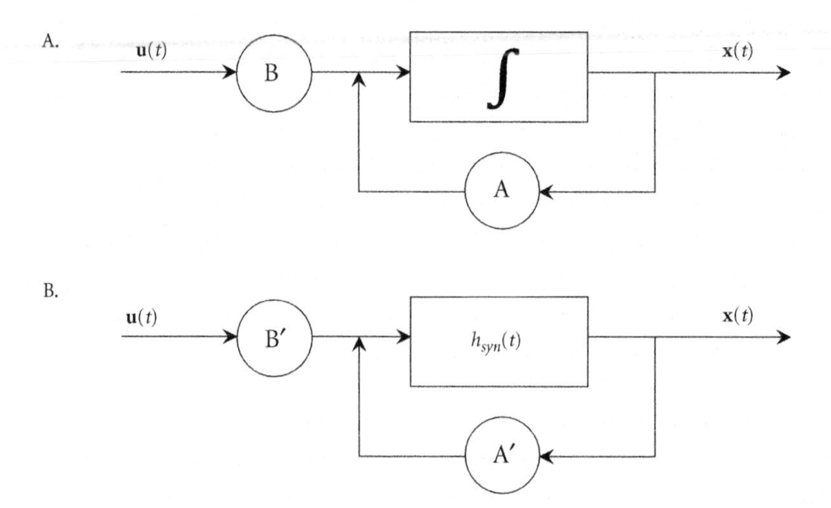

FIGURE 2.12 Standard and neural control theoretic descriptions of systems. A. The canonical diagram for any linear system in control theory. **A** determines what aspects of the current state affect the future state. **B** maps the input to the system into its state space. The transfer function is perfect integration. B. The equivalent description for a neurobiological system. The matrices A′ and B′ take into account the effects of the transfer function, $h_{syn}(t)$, which captures the intrinsic dynamics of biological networks.

theory is applied to a huge variety of control problems, including autopilot design, robotics, and chemical process control. It should not be surprising, then, that it also proves useful for characterizing the dynamics of complex neurobiological systems.

Central to employing modern control theory for understanding the dynamics of a system is the identification of the "system state variable" ($\mathbf{x}(t)$ in Fig. 2.12). It is not a coincidence that the terminology I introduced earlier has us calling the represented mathematical objects of a neural population the "state space." This is because the third principle of the NEF is precisely the suggestion that the state space representations of neural populations are the state variables of a dynamical system defined using control theory.

However, things are not quite so simple. Because neurons have intrinsic dynamics dictated by their particular physical characteristics, we must adapt standard control theory to neurobiological systems (*see* Fig. 2.12). Fortunately, this can be done without loss of generality for linear and nonlinear dynamic systems (*see* Appendix B.3). Notably, all of the computations needed to implement such systems can be implemented using transformations as defined earlier in principle 2. As a result, we can directly apply the myriad of techniques for analyzing complex dynamic systems that have been developed using modern control (and dynamic systems) theory to this quantitative characterization of neurobiological systems.

To get a sense of how representation and dynamics can be integrated, let us revisit the controlled integrator. As shown in Figure 2.6, a standard integrator constantly sums its input. If we call the input $u(t)$ and the state of the integrator $x(t)$, then we can write this relation as (*see* Appendix A.6 for mathematical background):

$$\dot{x}(t) = Ax(t) + Bu(t), \tag{2.1}$$

which is the scalar version of the system shown in Figure 2.12A. This equation says that the change in $x(t)$ (written $\dot{x}(t)$) at the next moment in time is equal to its current value plus the input $u(t)$ (times some number B). For the rest of the discussion, we can let $B = 1$.

If we also let $A = 0$, then we get a standard integrator: the value of the state $x(t)$ at the next moment in time will be equal to itself plus the change $\dot{x}(t)$. That change is equal to the input $u(t)$ (because $0 \cdot x(t) = 0$), so the value of $x(t)$ will result from constantly summing the input, just as in a standard integrator. In short,

$$x(t) = \int u(t)\,dt. \tag{2.2}$$

In many ways, this integration is like a memory. After all, if there is no input (i.e., $u(t) = 0$), then the state of the system will not change. This captures the essence of an ideal memory–its state stays constant over time with no input.

However, as noted in Figure 2.12, neuron dynamics are not well characterized by perfect integration. Rather, they have dynamics identified by the function $h_{syn}(t)$ in that figure. Specifically, this term captures the PSC dynamics of the neurons in question, several examples of which are shown in Figure 2.13. As mentioned earlier, PSC dynamics depend on the kind of neurotransmitter used. As a result, the "translation" from a control theoretic description to its neural equivalent must take this dependence into account. Specifically, that translation must tell us how to change A and B in Figure 2.12A into A' and B' in Figure 2.12B, given the change from integration ($\int$) to $h_{syn}(t)$.

For example, if we apply this translation to the neural integrator, then the feedback matrix A' is equal to 1 and the input matrix B' is equal to the time constant of the PSC. This makes sense because the neural system decays at a rate captured by the PSC time constant, so the feedback needs to "remind" the system of its past value at a similar rate. As mentioned before, this translation is defined in general for dynamical systems in Appendix B.3.

But, there is an obvious problem with this characterization of memory: a simple integrator always just *adds* new input to the current state. If you can only remember one thing, and I ask you to remember the word "cat," and

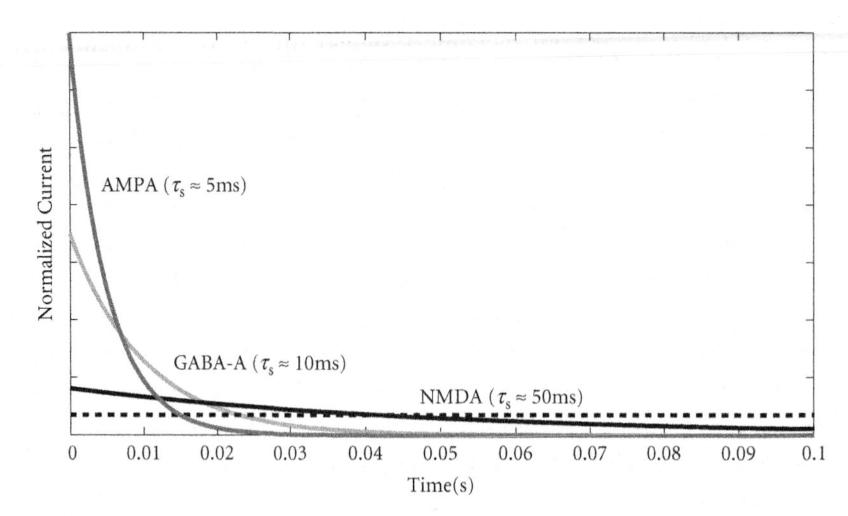

FIGURE 2.13 Example PSCs for different neurotransmitters. This figure shows a simple PSC model as a decaying exponential in response to a single neural spike. The time constant of the exponential is different for different receptors, as shown. The longer the time constant, the slower the decay. Compared to the dynamics of a perfect integrator (dashed line), PSCs lose information about their initial state over time. Each PSC here is normalized to have an area of 1 under the curve, so the peak value is not indicative of the strength of response but of the relative weighting of earlier versus later aspects of the response.

then later ask you to remember the word "dog," we would not expect you to report some kind of "sum" of "cat" and "dog" as the contents of your memory when queried later. Rather, we would expect you to *replace* your memory of "cat" with "dog." In short, there must be a way to empty the integrator before a new memory is remembered.

This is why we need real-time control of the value of A in the integrator. Consider for a moment the effect of changing the value of A in Equation 2.1 to lie anywhere between -1 and 1. Let us suppose that the current value of $x(t)$ is not 0, and there is no input. If A is positive, then all future values of $x(t)$ will move increasingly far away from 0. This is because the change in $x(t)$ will always be positive for positive $x(t)$ and negative for negative $x(t)$. In fact, this movement away from zero will be exponential because the next change is always a constant fraction of the current state (which keeps getting bigger)–adding (or subtracting) a fraction of something to itself over and over results in an exponential curve. Now suppose that A is negative. In this case, the opposite happens. The current state will move exponentially toward 0. This is because the next state will be equal to itself minus some fraction of the current state.

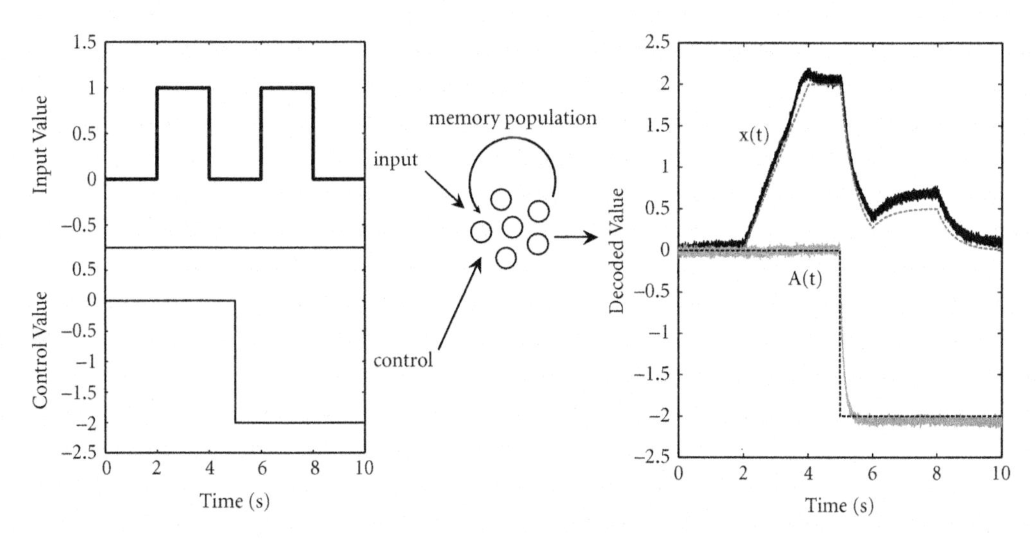

FIGURE 2.14 A spiking network implementing a controlled integrator. The memory population represents both the control signal $A(t)$ and the value of the current state $x(t)$. Decodings of these representations from the memory population are shown in the rightmost graph: the solid black line is $x(t)$, the solid gray line is $A(t)$, and the dotted lines depict the ideal values of the state and the control.

To build a controlled integrator that acts as an erasable memory, we can thus implement the dynamical system in Equation 2.1, with an additional input that controls A. We can use principle 3 to determine how to map this equation into one that accounts for neural dynamics. We can then employ principle 2 to find transformational decoders that give the product of A' and $x(t)$ (as in Section 2.2.2). And, we can employ principle 1 to represent the state variable $x(t)$, the input variable $u(t)$, and the control variable $A(t)$ in spiking neurons. The result of employing all three principles is shown in Figure 2.14. You can explore this network in Nengo by running the "controlledintegrator2.py" demo in the demos folder.

Ultimately, the principles tell us how to determine the appropriate connection weights between spiking neurons in the model (recurrent or otherwise), to give the desired dynamical behavior. Specifically, the connection weights are the product of the appropriate encoding, decoding, and dynamics parameters.

It is perhaps worth emphasizing that although the controlled integrator is simple–it computes only products and sums, it represents only scalars and a 2D vector, and it has nearly linear time-invariant dynamics–it employs each of the principles of the NEF. Further, there is nothing about these "simplicities" that is a consequence of the principles employed. NEF models can compute complicated nonlinear functions, have sophisticated high-dimensional representations, and display more interesting nonlinear dynamics.

It is perhaps also worth noting that such sophistication, although available in the NEF, is often not necessary for providing a deeper understanding of neural function. For example, Ray Singh and I used two simple (noncontrolled) neural integrators, coupled them together, and provided a novel explanation for observed spiking patterns in monkey cortex during a working memory task (Singh & Eliasmith, 2006). Specifically, the first integrator integrated a brief input stimulus to result in a constant output (i.e., a memory of the input), and the second integrator integrated the output of the first to give a "ramp" (i.e., a measure of how long since the input occurred). We demonstrated that connecting these two outputs to a population of cells representing their two-dimensional combination gave all of the observed responses in cells recorded during a working memory experiment in monkey frontal cortex (Romo et al., 1999). This included responses with surprisingly unusual dynamics. A recent analysis of a similar but larger data set verified that this kind of model (one with linear dynamics in a higher-dimensional space) captures 95% of the response variability of the over 800 observed neurons (six rather than two dimensions were used for this model; Machens et al., 2010). Another very similar model (which uses coupled *controlled* integrators) captures the population dynamics and single cell spiking patterns observed in rat medial prefrontal cortex similarly well, although on a very different task (Laubach et al., 2010). In both cases, a surprisingly good explanation of the detailed spiking patterns found in cortical areas was discovered despite the simplicity of the underlying model.

In many ways, it is the successes of these simpler models, built using the principles of the NEF, that make it reasonable to explore the available generalizations of those principles. Ultimately, I believe it is this generality that puts the NEF in a unique position to help develop and test novel hypotheses about biological cognition.

2.2.4 The Three Principles

Hopefully it is clear that this presentation of the NEF is somewhat superficial. It is not intended to satisfy those deeply familiar with theoretical neuroscience. I have side-stepped issues related to optimal decoding, nonlinear versus linear decoding, information transfer characteristics, nonlinear dendritic function, rate versus timing code considerations, and so on. Although I believe the NEF satisfactorily addresses these issues, my purpose here is to give the necessary background to make the methods plausible, usable, and not unnecessarily complex.

In that spirit, it is perhaps useful to summarize the principles diagrammatically to supplement the earlier verbal description in Section 2.2. Figure 2.15

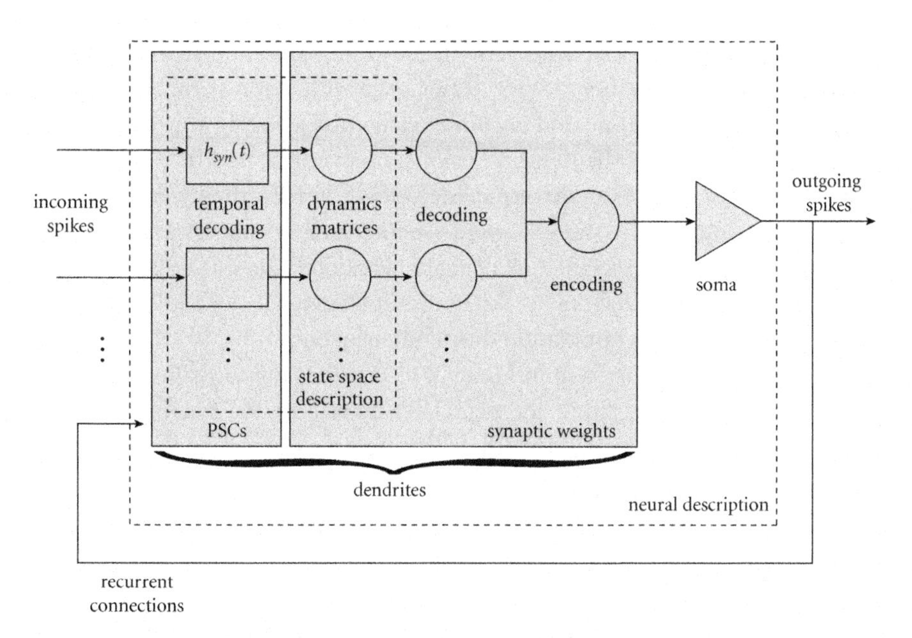

FIGURE 2.15 A generic neural subsystem. A synthesis of the preceding characterizations of representation (encoding/decoding), transformation (biased decoding), and dynamics (captured by $h_{syn}(t)$ and the dynamics matrices, A' and B'). Dotted lines distinguish neural and state space descriptions. Encoding and decoding relate these two descriptions to one another. (Adapted from Eliasmith & Anderson, 2003.)

shows a "generic" neural subsystem. The purpose of identifying such a subsystem is to note that such "blocks" can be strung together indefinitely to describe complex neural function. This subsystem is generic because it has input spikes, which generate PSCs that are then weighted and passed through a neural nonlinearity that then generates outputs spikes. This characterization is generic across almost all areas of the mammalian brain.[8]

What the NEF adds to the standard neural-level description of a population of neurons is the decomposition of synaptic weights into various elements, and a mapping of the dynamics onto biophysical properties of neurons. Again consider the neural integrator as an example of the system shown in Figure 2.12. There, the A' matrix specifies the dynamics matrix, which connects outgoing spikes of the population back to itself (the recurrent connection in Figure 2.15, equal to 1 for the integrator). The B' matrix specifies

[8] Some notable exceptions are retinal processing, some dendro-dendritic interactions, and gap junctions. An extension to the generic neural subsystem as drawn here can capture such interactions by incorporating optional elements that include these other potential sources of dendritic current. This added complexity obscures the simplicity and broad generality of the subsystem, however.

the dynamics matrix that connects the incoming spikes to this population (equal to τ_{syn} for the integrator). Associated with both of these dynamics matrices are the decoding and encoding elements, which in this case are the same and specified by the neuron responses (encoders) and the optimal linear weights for estimating the input (decoders). Finally, the dynamics of the system are captured by the PSC model being used (examples are shown in Figure 2.13). Having specified all of these elements allows us to generate the synaptic connection weights necessary to implement the defined higher-level dynamical system in a population of spiking neurons.

In this way, the subsystem in Figure 2.15 captures the contributions of the principles. That is, together the principles determine the synaptic weights. That determination depends on the contributions of each of the principles independently: the representation principle identifies encoders and representational decoders; the transformation principle identifies transformational decoders; and the dynamics principle identifies the dynamics matrices (i.e., the matrices identified in Fig. 2.12). The synaptic weights themselves are the product of these elements. We can capture the calculation of connection weights in a single equation:

$$\omega_{ij} = \mathbf{d}_i^f \mathbf{A} \mathbf{e}_j \tag{2.3}$$

which states that the connection weights between neurons in the sending population (indexed by i) and neurons in the receiving population (indexed by j) are equal to the transformational decoders $\mathbf{d}_i^f$ (for the sending population; f is the function being decoded) times the dynamics matrix $\mathbf{A}$, times the encoders $\mathbf{e}_j$ (for the receiving population).

Crucially, although this theoretical characterization of the subsystem is generic, its application is not. To determine decoders, the specific tuning curves of neurons in the circuit play a role. To determine dynamics, the kinds of neurotransmitters found in a given circuit are crucial. To determine what kind of spiking occurs, the choice of a single cell model is important. To determine encoders, both the single cell model and the tuning curves need to be known. To determine transformational decoders and dynamics, a high-level hypothesis about the system function is needed. All of these considerations can vary widely depending on which brain areas are being considered.

Before confronting the task of applying the NEF to biological cognition in the next chapter, I believe it is conceptually important to clarify a notion I will be using throughout the remainder of the book. A notion I have been using somewhat loosely to this point: the notion of "levels."

2.3 ■ Levels

Traditionally, it was widely thought that the sciences could be identified with the level of nature that they characterized. Physics was the most fundamental, followed closely by chemistry, and subsequently biology, psychology, sociology, and so on (Oppenheim & Putnam, 1958; Carnap, 1931; Hempel, 1966). The suggestion was that lower-level sciences could reductively describe the higher-level sciences. This uncomplicated view has unfortunately not withstood the test of time (Bechtel, 1988, 2008; Craver, 2007).

More recently, some have suggested that because the reduction of one science to another has failed, the sciences must be independent (Fodor, 1974; Rosenberg, 1994). Consequently, Fodor (1974) has argued that to understand cognitive systems, which lie in the domain of psychology, appeal to lower-level sciences is largely useless. The only thing such disciplines, including neuroscience, can do is to provide an implementational story that bears out whatever psychological theory has been independently developed.

However, we need not think of levels as being either reductive (Oppenheim & Putnam, 1958) or independent (Fodor, 1974). If we think of levels from an *epistemological* perspective–that is, as different *descriptions* of a single underlying system–neither view seems plausible. Sometimes we may describe a system as a "person" without much consideration of their relationship to individual atomic particles. Other times we may be concerned about a decomposition of a biological person to microscopic subsystems, leading to what looks like a kind of reductive description. The reasons for preferring one such description over another can be largely practical: in one case, we may need to predict the overall behavior of the whole; in another we may need to explain how changes in a component influence complex cellular activity.

This notion of "levels as kinds of descriptions for purposes" is flexible in a manner that acknowledges our limited intellectual capacities or simply our limited knowledge. I am suggesting that levels should be taken as pragmatically identified sets of descriptions that share assumptions. Perhaps it is useful to call this position "descriptive pragmatism" for short. This view of levels is consistent with the notion that part–whole relations, especially in mechanisms, often help pick out levels. Most systems composed of parts have parts that interact. If the parts of a system are organized in such a way that their interaction results in some regular phenomena, then they are called "mechanisms" (Machamer et al., 2000). Mechanisms are often analyzed by decomposing them into their simpler parts, which may also be mechanisms (Bechtel, 2005). So, within a given mechanism, we can see a natural correspondence among spatial scale, complexity, and levels. Note, however, that even if two distinct mechanisms can be decomposed into lower-level mechanisms, this does not suggest that the levels *within* each mechanism can be mapped *across* mechanisms (Craver, 2007). This, again, suggests that something like descriptive pragmatism better captures scientific practice than more traditional ontological views, as it allows different practical considerations to be used to identify decompositions across different mechanisms.

Further, descriptive pragmatism about levels incorporates a notion that relies on degrees of abstraction. This meaning of "levels" is perhaps best exemplified by Marr's famous three levels (Marr, 1982). To avoid confusion between these two related uses of the term, I refer to the former as levels of scale, and the latter as levels of abstraction. There are interesting relationships between these two uses of the term "levels"–for example, it is common for mechanisms at higher levels of scale to elicit descriptions at higher levels of abstraction. Nevertheless, we can clearly pull these two notions apart, as an element at one level of scale (e.g., water) can be described at varying levels of abstractness (e.g., molecular interactions, or continuous fluid flow). Consequently, I take the "level" at which we describe a system to be determined by a combination of level of scale and level of abstraction.

I should note that I in no way take this brief discussion to sufficiently specify a new characterization of levels. There are many, much deeper considerations of levels that I happily defer to others (e.g., Bechtel, 2008; Craver, 2007; Hochstein, 2011). My purpose is to clarify what general characterization of levels lies behind my use of the term throughout the book, so as to avoid confusion about what I might mean.

With this background in mind, let me return to specific consideration of the NEF. As described in Section 2.2.1 on the principle of representation, the principle applies to the representation of mathematical objects in general. Because such objects can be ordered by their complexity, we have a natural and well-defined meaning of a representational hierarchy: a hierarchy whose levels can be understood as kinds of descriptions with specific inter-relations. Table 2.1 provides the first levels of such a hierarchy.

This hierarchy maps well to notions of levels, like descriptive pragmatism, that take levels to correlate with spatial scale, complexity, part–whole relations, and abstraction. For example, if we hold the precision of the representation

TABLE 2.1 *A Representational Hierarchy*

Mathematical object	Dimension	Example
Scalar (x)	1	Light intensity at x, y
Vector ($\mathbf{x}$)	N	Light intensity, depth, color, etc., at x, y
Function ($x(\nu)$)	∞	Light intensity at all spatial positions
Vector field ($x(\mathbf{r}, \nu)$)	$N \times \infty$	Light intensity, depth, color, etc., at all spatial positions
$\vdots$	$\vdots$	$\vdots$

Each row has the same type of encoding/decoding relations with neurons. Higher-level representations (lower rows in the table) can be constructed out of linear combinations of the previous row.

constant, then the higher levels of scale in the hierarchy require more neurons and, hence, would typically be ascribed to larger areas of cortex (relatively few cells are needed to encode light intensity at a single spatial position, compared to encoding light intensity over the entire visual field). Relatedly, these higher-level representations are able to underwrite more complex behaviors (object recognition will often require detailed intensity information across large portions of the visual field). As well, the hierarchy clearly defines how the higher levels of scale are built up out of the lower levels. Hence, mechanisms trafficking in one level of representation will often be decomposable into mechanisms trafficking in lower-level representations.

Additionally, the subtle role of levels of abstraction can be understood in the context of this hierarchy. For example, a scalar might be described as being represented in spiking neurons, simpler rate neurons, or as an abstract "population" variable. These increasingly abstract characterizations are picking out slightly different relations between the scalar and its underlying implementation. Pragmatically, increasing the abstractness of our characterizations is one way to mitigate complexity as we consider progressively larger parts of a system. For example, in Nengo we can run very large models on small computers by simulating parts of the model in "direct mode," where physiological details are largely ignored (typically this is useful when designing or debugging models). However, for this book all simulations are run in "default mode" (i.e., with spiking neurons) because my purpose is to demonstrate how low-level (small spatial scale, low complexity, minimally abstract) mechanisms relate to high-level (broad spatial scale, high complexity, abstractly characterized) behavior.

Given the tight relationship between NEF principles 1 and 2, it should not be surprising that what goes for representations also goes for transformations. High-level computations tend to be carried out over larger spatial scales, can be more complex, and capture more abstractly characterized mechanisms. Further, because principles 1 and 2 are used to implement principle 3 (dynamics are defined by transformations of representations), the same can be said for dynamics. Picking which levels are appropriate for a given system depends on what explanations we are trying to provide about the system. Hence, descriptive pragmatism is a good way to understand such a hierarchy and how it is used.

The fact that all of these levels can be described quantitatively and in a standard form suggests that the NEF characterization provides a unified means of describing neurobiological systems. In addition, the NEF makes clear how we can "move between" levels and precisely how these levels are *not* independent. They are not independent because empirical data that constrain one description is about *the same system* described at a different level. So, when we move between levels, we must either explicitly assume away the relevance of lower- (or higher-) level data, or we rely on the hypothesized relationship between the levels to relate those data to the new level. Either way, the data influence our characterization of the new level (because allowable abstractions are part and parcel of identifying a level).

So, the NEF provides a consistent and general way to talk about levels in the behavioral sciences. However, only some of the interlevel relations are defined by the principles of the framework. This seems appropriate given our current state of ignorance about how best to decompose the highest-level neural systems. In the next chapter, I begin to describe a specific architecture that addresses what kinds of functions the brain may actually be computing in order to underwrite cognition. The NEF provides a method for specifying that description at many levels of detail, whereas the architecture itself helps specify additional interlevel relations not addressed by the NEF.

2.4 ■ Nengo: Neural Representation

In this tutorial, we examine two simple examples of neural representation. These examples highlight two aspects of how principle 1 in Section 2.2.1 characterizes neural representation. First, the examples make evident how the activity of neural populations can be thought of as representing a mathematical *variable*. Second, we examine a simple case of moving up the representational hierarchy to see how the principle generalizes from simple (i.e., scalar) to more complex (i.e., vector) cases.

Representing a Scalar

We begin with a network that represents a very simple mathematical object, a scalar value. As in the last tutorial, we will use the interactive plots to examine the behavior of the running model.

- In an empty Nengo workspace, drag a *Network* component from the template bar into the workspace. Set the *Name* of the network to "Scalar Representation" and click *OK*.
- Drag an *Ensemble* into the network. A configuration window will open.

Here the basic features of the ensemble can be configured.

- Set *Name* to "x," *Number of nodes* to 100, *Dimensions* to 1, *Node factory* to "LIF Neuron," and *Radius* to 1.

The name is a way of referring to the population of neurons you are creating. The number of nodes is the number of neurons you would like to have in the population. The dimension is the number of elements in an input vector you would like the population to represent (a 1-dimensional vector is a scalar). The node factory is the kind of single cell model you would like to use. The default is a simple LIF neuron. Finally, the radius determines the size of the

n-dimensional hypersphere that the neurons will be good at representing. A one-dimensional hypersphere with a radius of 1 is the range from -1 to 1 on the real number line. A two-dimensional hypersphere with the same radius is a unit circle.

- Click the *Set* button. In the panel that appears, you can leave the defaults (*tauRC* is 0.02, *tauRef* is 0.002, *Max rate* low is 100 and high is 200, *Intercept* is low -1.0 and high is 1.0).

Clicking on *Set* allows the parameters for generating neurons in the population to be configured. Briefly, *tauRC* is the time constant for the neuron membrane, usually 20 ms, *tauRef* is the absolute refractory period of the neuron (the period during which a neuron cannot spike after having spiked), *Max rate* has a high and low value (in Hertz that determines the range of firing rates neurons will have at the extent of the radius (the maximum firing rate for a specific neuron is chosen randomly from a uniform distribution between low and high), *Intercept* is the range of possible values along the represented axis where a neuron "turns off" (for a given neuron, its intercept is chosen randomly from a uniform distribution between low and high).

- Click *OK*. Click *OK* again, and the neurons will be created.

If you double-click on the population of neurons, each of the individual cells will be displayed.

- Right-click on the population of neurons, select *Plot→ Constant Rate Responses.*

The "activities" graph that is now displayed shows the "tuning curves" of all the neurons in the population. This shows that there are both "on" and "off" neurons in the population, that they have different maximum firing rates at $x = \pm 1$, and that there is a range of intercepts between $[-1, 1]$. These are the heterogeneous properties of the neurons that will be used to represent a scalar value.

- Right-click on the population of neurons, select *Plot→ Plot distortion: X.*

This plot is an overlay of two different plots. The first, in red and blue, compares the ideal representation over the range of x (red) to the representation by the population of neurons (blue). If you look closely, you can see that blue does not lie exactly on top of red, although it is close. To emphasize the difference between the two plots, the green plot is the distortion (i.e., the difference between the ideal and the neural representation). Essentially the green plot is the error in the neural representation, blown up to see its finer structure. At the top of this graph, in the title bar, the error is summarized by the mean squared error (MSE) over the range of x. Importantly, MSE decreases as the square of the number of neurons (so RMSE is proportional to 1/N), so more neurons will represent x better.

- Right-click on the population and select *Inspector*. Any of these displayed properties can be changed for the population.

There are too many properties x to discuss them all. But, for example, if you click on the arrow beside *i neurons*, and double-click the current value, you can change the number of neurons in the population. When you hit enter, the population will be regenerated with the shown number of neurons.

Let us now examine the population running in real time.

- Drag a new *Input* from the template bar into the Scalar Representation network.
- Set *Name* to "input," make sure *Output Dimensions* is 1 and click *Set Functions*. Click *OK*. (Clicking *OK* automatically sets the input to a *Constant Function* with a starting value of 0. You can edit the function by clicking *Set Functions*).

The function input will appear. You will now connect the function input to the neural population as in the previous tutorial. The output from the function will be injected into the soma of each of the neurons in the population, driving its activity. To do so:

- Drag a *Termination* component onto the "x" population. Set *Name* to "input," *Weights Input Dim* to 1, and *PSTC* to 0.02. Click *Set Weights*, double-click the value, and set it to 1. Click *OK* twice. (If you click *OK* before clicking *Set Weights*, the weight will default to the last used value.)
- Click and drag the "origin" on the input function you created to the "input" on the "x" population.

If you want to automatically organize models you generate, click on the "Feed-forward" icon that appears in the bottom right of the network viewer. This is often useful when loading a network created from a script, which may not have had a layout saved for it.

- Click the *Interactive Plots* icon (the double sine wave in the top right corner).

This should be familiar from the previous tutorial. It allows us to interactively change the input and watch various output signals generated by the neurons.

- Right-click "x" and select *value*. Right-click "x" and select *spike raster*. Right-click "input" and select *value*. Right-click "input" and select *control*.

Change the layout to something you prefer by dragging items around. If you would like the layout to be remembered in case you close and re-open these plots, click the small triangle in the middle of the bottom of the window (this expands the view), then click the disk icon under *layout* (if it is not visible, you may have to widen the window).

- Click the play button. Grab the control and move it up and down. You will see changes in the neural firing pattern and the value graphs of the input and the population.

Note that the value graph of the "x" population is the linearly decoded estimate of the input, as per the first principle of the NEF. Note also that the spike raster graph is displaying the encoding of the input signal into spikes. The spiking of only 10% of the neurons are shown by default. To increase this proportion:

- Right-click on the spike raster graph and select a larger percentage of neurons to show.

The population of neurons does a reasonably good (if slightly noisy) job of representing the input. However, neurons cannot represent arbitrary values well. To demonstrate this, do the following.

- Right-click the control and select *increase range*. Do this again. The range should now be ± 4 (to see the range, hover over the slider).
- Center the controller on zero. (To do this automatically, right-click the slider and select "zero".) The population should represent zero. Slowly move the controller up, and watch the value graph from the "x" population.

Between 0 and 1, the graph will track the controller motions well. Notice that many of the neurons are firing very quickly at this point. As you move the controller past 1, the neural representation will no longer linearly follow your movement. All the neurons will become "saturated"–that is, firing at their maximum possible rate. As you move the controller past 2 and 3, the decoded value will almost stop moving altogether.

- Move the controller back to zero. Notice that changes around zero cause relatively large changes in the neural activity compared to changes outside of the radius (which is 1).

These effects make it clear why the neurons do a much better job of representing information within the defined radius: Changes in the neural activity outside the radius no longer accurately reflect the changes of the input.

Notice that this population, which is representing a scalar value, does not in any way store that value. Rather, the activity of the cells act as a momentary representation of the current value of the incoming signal. That is, the population acts together like a variable, which can take on many values over a certain range. The particular value it takes on is represented by its activity, which is constantly changing over time. This conception of neural representation is very different from that found in many traditional connectionist networks, which assume that the activation of a neuron or a population of neurons represents the activation of a specific "concept." Here, the same population of neurons, *differently activated*, can represent different "concepts."[9] I return to this issue in Sections 9.4 and 10.1. A pre-built version of this network can be loaded from <Nengo Install Directory>/Building a Brain/chapter2/scalars.py.

Representing a Vector

A single scalar value is a simple neural representation and hence at the bottom of the representational hierarchy. Combining two or more scalars into a representation and moving up one level in the hierarchy results in a vector representation. In this tutorial, I consider the case of two-dimensional vector representation, but the ideas naturally generalize to any dimension. Many parts of cortex are best characterized as using vector representations. Most famously, Apostolos Georgopoulos and his colleagues have demonstrated vector representation in motor cortex (Georgopoulos et al., 1984, 1986, 1993).

In their experiments, a monkey moves its arm in a given direction while the activity of a neuron is recorded in motor cortex. The response of a single neuron to 40 different movements is shown in Figure 2.16. As can be seen from this figure, the neuron is most active for movements in a particular direction. This direction is called the "preferred direction" for the neuron. Georgopoulos' work has shown that over the population of motor neurons, these preferred directions, captured by unit vectors pointing in that direction, are evenly distributed around the unit circle in the plane of movement (Georgopoulos et al., 1993).

To construct a model of this kind of representation,[10] we can do exactly the same steps as for the scalar representation but with a two-dimensional representation.

[9] This is true even if we carve the populations of neurons up differently. That is, there is clearly a range of values (perhaps not the whole range in this example) over which exactly the same neurons are active, but different values are represented.

[10] Technically, Figure 2.16 only provides evidence for one-dimensional representation, as it does not include the effects of different size movements in a given direction. However, more recently such evidence has been found (Schwartz, 1994; Moran & Schwartz, 1999; Todorov, 2000), so the representation I'm considering here is consistent with the responses of motor cortex neurons.

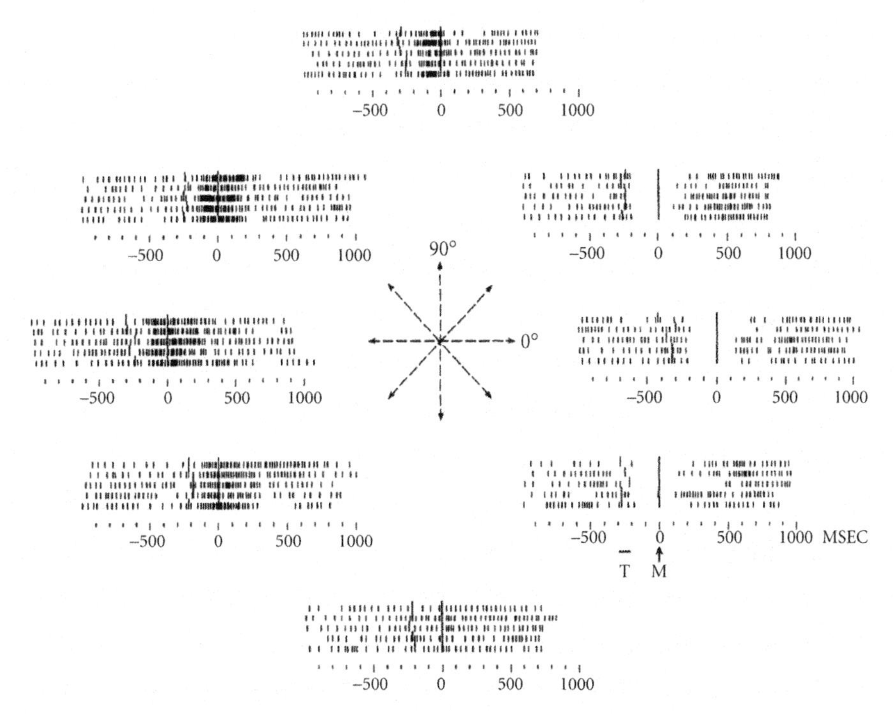

FIGURE 2.16 The response of a single neuron in motor cortex to different directions of movement. M indicates the time of movement onset. T indicates the time that the target appears. These data show five trials in each direction for this neuron. The neuron is most active when the target is between 135 and 180 degrees. The direction of the peak of this response is called the neuron's "preferred direction." In the NEF the preferred direction is represented as a unit (i.e., length one) vector pointing in that direction. (Reproduced from Georgopoulos et al. [1986] with permission.)

- In an empty Nengo workspace, drag and drop a new *Network.* (The "clear" icon in the top left makes it easy to erase the current workspace.) Set the *Name* of the network to "Vector Representation" and click *OK.*
- Create a new *Ensemble* within the network. Set *Name* to "x," *Number of nodes* to 100, *Dimensions* to 2, *Node factory* to "LIF Neuron", and *Radius* to 1.
- Click the *Set* button. In the panel that appears, you can leave the defaults (*tauRC* is 0.02, *tauRef* is 0.002, *Max rate* low is 100 and high is 200, *Intercept* is low −1.0 and high is 1.0). If you click *OK* without clicking *Set,* you will automatically use the last values entered.

These parameters mean the same as for scalar representation with two exceptions: *Max rate* and *Intercept* are defined along the preferred direction vector.

That is, it is as if each neuron tuning curve is lined up with its particular preferred direction, and then the intercept and maximum rates are chosen.

We can think of scalar representation as having only two possible "preferred directions"–positive and negative. The preferred direction vector generalizes that notion to higher-dimensional spaces. In two dimensions, preferred directions can point in any direction around the unit circle. The particular direction preferred by a given neuron is chosen randomly from a distribution you can set in Nengo.

- Click *OK* to complete the configuration of the neuron generator and return to the ensemble configuration window. Click *Advanced*. This expands the window. Ensure that the *Encoding Distribution* slider is all the way to the left (over 0.0 - *Evenly Distributed*).

The previous step gives a motor cortex-like distribution of preferred direction vectors, because the Georgopoulos data suggest that motor cortex neurons can prefer any direction around the circle with equal probability.

- Set the *Noise* parameter to 0.1.

The noise parameter sets the expected amount of noise that Nengo uses to calculate decoder values. Do *not* set this parameter to zero as this is biologically implausible and would make the ensemble extremely sensitive to noise.

- Click *OK*, and the neurons will be created.

Given these settings, preferred direction vectors will be randomly chosen from an even distribution around the unit n-dimensional hypersphere (i.e., the unit circle in two dimensions). If you plot the constant rate responses, then each neuron's response will be plotted along its preferred direction vector. Consequently, the plot is generated as if all neurons had the same preferred direction vector, or equivalently, as if all preferred direction vectors were aligned before plotting.

You can now create an input function, and run the network.

- Drag a new *Input* into the Vector Representation network.
- Set *Name* to "input," make sure *Output Dimensions* is 2, and click *Set Functions*. Ensure both are *Constant Functions*, and click *OK*.
- Add a *Termination* to the "x" ensemble. Set *Name* to "input," *Weights Input Dim* to 2, and *PSTC* to 0.02. Click *Set Weights* and set the weights to an identity matrix (the top-left and bottom-right values should be set to 1, and the remaining two values should be set to 0).

Notice that you have a matrix of weights to set. This is because the input functions can be projected to either of the dimensions represented by the neural population. For simplicity, we have told the first function to go only to the first dimension and the second function only to the second dimension.

- Click *OK* twice.
- Click and drag the "origin" on the input function you created to the "input" on the "x" population.
- Click on the *Interactive Plots* icon.

We can begin by looking at the analogous displays to the scalar representation.

- Right-click "x" and select *value.* Right-click "x" and select *spike raster.* Right-click "input" and select *value.* Right-click "input" and select *control.*
- Press play to start the simulation. Move the controls to see the effects on the spike raster (right-click the raster to show a higher percentage of neurons).

You can attempt to find a neuron's preferred direction vector, but it will be difficult because you have to visualize where in the 2D space you are because the *value* plot is over time.

- Right-click the "input" *value* plot and select *hide.*
- Right-click "input" and select *XY plot.* This shows the input plotted in a 2D space.

Now you can attempt to determine a neuron's preferred direction vector. This should be easier because you can see the position of the input vector in 2D space. There is a trail to the plot to indicate where it has recently been. The easiest way to estimate a neuron's preferred direction vector is to essentially replicate the Georgopoulos experiments.

- Move the input so both dimensions are zero, or right-click on the control and select "zero." Then move one input to its maximum and minimum values.

If a neuron does not respond (or responds minimally), then that is not its preferred direction. A neuron whose preferred direction is close to the direction you are changing will respond vigorously to the changes you make. Keep in mind that different neurons have different gains, meaning they may "ramp" up and down at different rates even with the same preferred direction.

- Right-click the "x" population and select *preferred directions* to show the neuron activity plotted along their preferred directions in the 2D space.

This plot multiplies the spike activity of the neuron with its preferred direction vector. So, the longer lines are the preferred directions of the most active neurons.

- Put one input at an extreme value, and slowly move the other input between extremes.

It should be clear that something like the average activity of the population of neurons moves with the input. If you take the mean of the input (a straight line through the middle of the blob of activity), it will give you an estimate of the input value. That estimate is something like the linear decoding for a vector space as defined in the first NEF principle (although it is not optimal, as in the principle).

- Right-click the "x" population and select *XY plot.* This shows the actual decoded value in 2D space.

Another view of the activity of the neurons can be given by looking at the neurons plotted in a pseudo-cortical sheet.

- Right-click the "x" population and select *voltage grid.*

This graph shows the subthreshold voltage of the neurons in the population in gray. Yellow boxes indicate that a spike is fired.

- Right-click on the voltage grid and select *improve layout* to organize the neurons so that neurons with similar preferred direction vectors will be near each other, as in motor cortex.
- Move the sliders and observe the change in firing pattern.

Using the same kind of exploration of inputs as before, it is reasonably evident in this view which parts of the grid have which preferred direction vectors. This network reproduces the classic view of how motor cortex is thought to encode a movement.

Recent work has challenged the idea that there is such a clean mapping between neural activity and target location in general (Churchland et al., 2010). However, the same approach to vector representation can be used to capture these recently noticed subtleties (Dewolf & Eliasmith, 2011). As a result, this network provides a useful introduction to vector representation in cortex.

It is important to keep in mind that some aspects of this tutorial on vector representation are specific to this motor cortex example. For example, using an even distribution of preferred direction vectors, using low-dimensional vectors, and thinking of preferred directions as related to actual movement direction are appropriate because we are considering motor cortex. However, many aspects are more general, including the identification of (abstract) preferred direction vectors and the use of a linear decoding to get an estimate of the population representation.

It is worth highlighting that the (2D) "represented vectors" in this characterization of neural function are *not* the same as the "activity vectors" commonly discussed in artificial neural networks. Activity vectors are typically a set of neuron firing rates. If we have three neurons in our population, and they are active at 50, 100, and 20 Hz, respectively, the 3D activity vector would be [50, 100, 20]. This vector defines a point in a space where each axis is associated with a specific neuron. In the example above, the activity vector would be in a 100D space because there are 100 neurons in the population. In contrast, the 2D space represented above has axes determined by an externally measured variable.[11]

Recall from Section 2.2.1 that the 2D space in this example is the "state space," whereas the 100D space is the "neuron space." Consequently, we can think of the 2D state space as a standard Cartesian space, where two values (x and y co-ordinates) uniquely specify a single object as compactly as possible. In contrast, the 100D vector specifies the same underlying 2D object, but it takes many more resources (i.e., values) to do so. If there was no uncertainty in any of these 100 values, then this would simply be a waste of resources. However, in the much more realistic situation where there is uncertainty (resulting from noise of receptors, noise in the channels sending the signals, etc.), this redundancy can make specifying an underlying point much more reliable. And, interestingly, it can make the system much more flexible in how well it represents different parts of that space. For example, we could use 10 of those neurons to represent the first dimension, or we could use 50 neurons to do so. The second option would give a much more accurate representation of that dimension than the first. Being able to redistribute these resources to respond to task demands is one of the foundations of learning (*see* Section 6.4).

A pre-built version of this network can be loaded from <Nengo Install Directory>/Building a Brain/chapter2/vectors.py. More extensive tutorials on neural representation are also available on the Nengo website.

[11] These axes do not need to be externally measurable, as we shall see. However, they are in many of the simplest cases of neural representation.

3. BIOLOGICAL COGNITION: SEMANTICS

C. Eliasmith, C. Tang, and T. DeWolf

In the next four chapters, I detail a cognitive architecture that I call the Semantic Pointer Architecture (SPA). Like any cognitive architecture, the SPA is a proposal about the processes and representations that result in intelligent behavior (Thagard, 2011). A particular focus of the SPA is on constraining the architecture based on biological properties of our best example cognitive systems–real brains. Consequently, models in this book are implemented in spiking neurons, have neuron response properties matched to relevant brain areas, employ neurotransmitters appropriate to the brain areas of interest, are constrained by neuroanatomical data, and so on. In the next four chapters I present many such models.

However, because the models presented in these chapters are used to demonstrate particular aspects of the architecture, they do not present the architecture as a unified whole. As a result, it is not until Chapter 7 that I am able to present an integrated model that includes all aspects of the architecture. I firmly believe that such a demonstration of integration is crucial to the claim that a cognitive architecture has been described. Perhaps the most compelling feature of biological cognitive systems is that they can perform many *different* tasks, without being "reprogrammed." The same cannot be said for most cognitive models. The model I present in Chapter 7 can perform eight very different perceptual, motor, and cognitive tasks–clearly not as impressive as real cognitive systems, but a start nonetheless.

As noted in Chapter 1, I describe the computational structures and representations employed in the SPA architecture by addressing four distinct aspects of cognition: semantics, syntax, control, and memory and learning. In this third chapter, I introduce a kind of neural representation that I call a semantic pointer. Semantic pointers are a means of relating more general neural representations to those central to cognition. As a result, I describe how semantic pointers can be generated from raw perceptual input and also how they can be used to drive motor action. I leave consideration of more sophisticated, cognitive semantics to the end of Chapter 4, as it requires a means of encoding representational structure.

To address structured representation, in Chapter 4 I provide a neurocomputational characterization of binding. Specifically, I show how to bind semantic pointers to construct structured representations that are themselves semantic pointers. I demonstrate how these representations can be transformed, decoded, and learned in a manner that scales to the level of human cognitive structures. In Chapter 5 I address features of cognitive control. There I show how semantic pointers can be embedded in an action selection system (a neural model of the basal ganglia) and used to effectively route information throughout the brain. I demonstrate the control system in a neural model of the famous Tower of Hanoi task–a cognitive task that requires planning, decision making, memory, and rule following. Finally, in Chapter 6 I describe processes related to cognitive adaptation. Specifically, I address serial working memory and learning. I show that binding of semantic pointers forms the basis of a very good model of serial working memory. As well, I demonstrate that a single biologically realistic learning rule can be used to learn binding as well as action selection strategies in changing environments. The model described in Chapter 7 includes all of the elements introduced in these earlier chapters.

Let me begin by introducing semantic pointers in more detail.

3.1 ■ The Semantic Pointer Hypothesis

Underlying the SPA is the semantic pointer hypothesis. Its purpose is to bridge the gap between the neural engineering framework (NEF), a theory that tells us how a wide variety of functions may be implemented in neural structures, and the domain of cognition (which is in need of ideas about how such neural structures give rise to complex behavior).

Here is a simple statement of the semantic pointer hypothesis:

> Higher-level cognitive functions in biological systems are made possible by semantic pointers. Semantic pointers are neural representations that carry partial semantic content and are composable into the representational structures necessary to support complex cognition.

There are two aspects of this hypothesis that must be expanded in detail. First, I must indicate how semantic information, even if partial, will be captured by the representations that we identify as semantic pointers. Second, I must give a characterization of how to construct "representational structures"–that is, account for syntax given those representations.

There have been a wide variety of proposals regarding the role that semantics and syntax should play in our theories of cognition. Traditionally, a classical cognitive system is characterized as a symbol processing system that relies on syntax to respect the semantics of the symbols and combinations of symbols found in the system (*see*, e.g., Fodor, 1975). So language-like syntax was thought to do most of the cognitive work, and semantics came along for the ride. In more connectionist approaches, semantics has been the focus, where the vector space defined by the activity of the nodes in the network was often thought of as a "semantic space" that related the meaning of different firing patterns (*see*, e.g., Churchland, 1979).

In the SPA, syntax is inspired by proponents of the symbolic approach who claim that there are syntactically structured representations in the head, and semantics is inspired by connectionists who claim that vector spaces can be used to capture important features of semantics. Both are implemented in the same biologically plausible substrate. So, with the characterization of neural representation in the previous chapter in hand, it is natural to begin with the claim that semantic pointers are well described by vectors in a high-dimensional space. In the remainder of this chapter, I address the semantics of these vector representations. In the next chapter, I discuss how representational structures can be formed out of these vectors to support syntactic manipulation.

Considering semantics for the time being, it is a well-worn claim that a high-dimensional vector space is a natural way to represent semantic relationships between representations in a cognitive system. It is difficult to visualize high-dimensional spaces, but we can get a sense of what this claim amounts to by thinking of concepts in a three-dimensional state space (a relatively low-dimensional space), as shown in Figure 3.1. I refer to such a visualization as a "conceptual golf ball." The surface of the ball represents the conceptual space, and the dimples in the surface represent concepts. Specific examples of a concept are picked out by points that lie within a given dimple. Concepts (and examples) that are semantically similar lie in relatively close proximity compared to concepts that are semantically dissimilar.

Unfortunately, things quickly get crowded in three dimensions, and so neighboring concepts are easily confused. This is especially true if there is uncertainty, or noise, in the specification of a given example. Because the presence of noise should be expected, especially in a real physical system like the brain, the number of distinct concepts that can be represented effectively may be quite small in low-dimensional spaces. With higher dimensionality, the amount of surface area available to put concepts in increases incredibly quickly, as shown in Figure 3.2. As a result, many concepts, even under noise, can be represented within a high-dimensional state space.

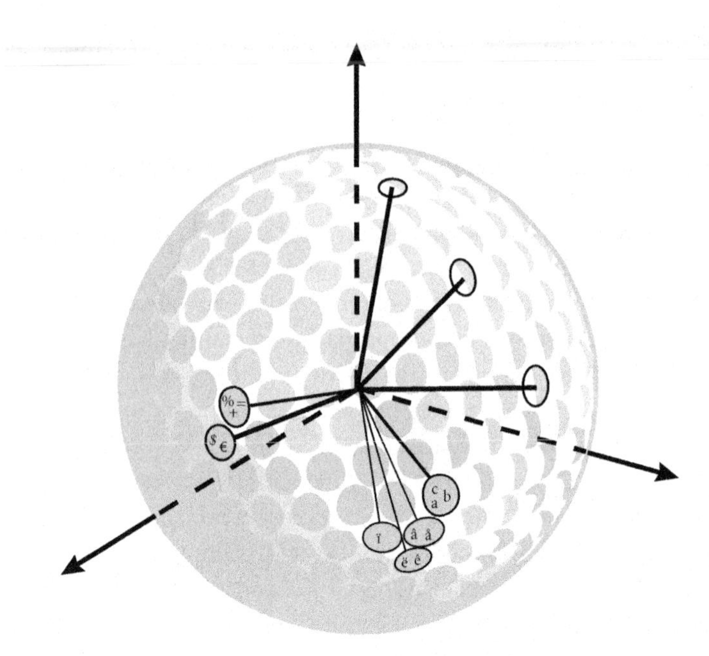

FIGURE 3.1 A "conceptual golf ball" depicting a three-dimensional semantic state space. Each dimple in the ball represents a concept, visualized here as a set of specific example symbols inside a circle. Vectors from the center of the sphere to the dimples on its surface are used to identify these concepts. In the figure, concepts for letters are clustered together while other kinds of symbols (e.g., $, %) are clustered separately; within the cluster of dimples storing letters, the letters with special accents are also grouped more closely together. The location of a concept (or specific example) on the surface of the conceptual golf ball thus carries information about its relationship to other concepts or examples.

An obvious drawback with such state spaces is that it is not immediately clear how complex representational *structures* can be placed in such a space (because the space has no obvious structure to it): I return to this issue in the next chapter. However, even without worrying about representational structure *per se*, the semantics of even simple human concepts seems to be extraordinarily rich, extending beyond a simple mapping to a single, even very high-dimensional, vector space (Barsalou, 2009, 1999). The richness of human conceptual behavior suggests that it is unlikely that all of the aspects of a concept are actively represented at the same time.

In earlier work, I have made a distinction between "conceptual" and "occurrent" representation (Eliasmith, 2000) to help clarify this issue. Non-occurrent, conceptual representations are those that have preoccupied cognitive scientists and their predecessors for centuries–representations that account for the *long-term* representational behavior of people. Occurrent

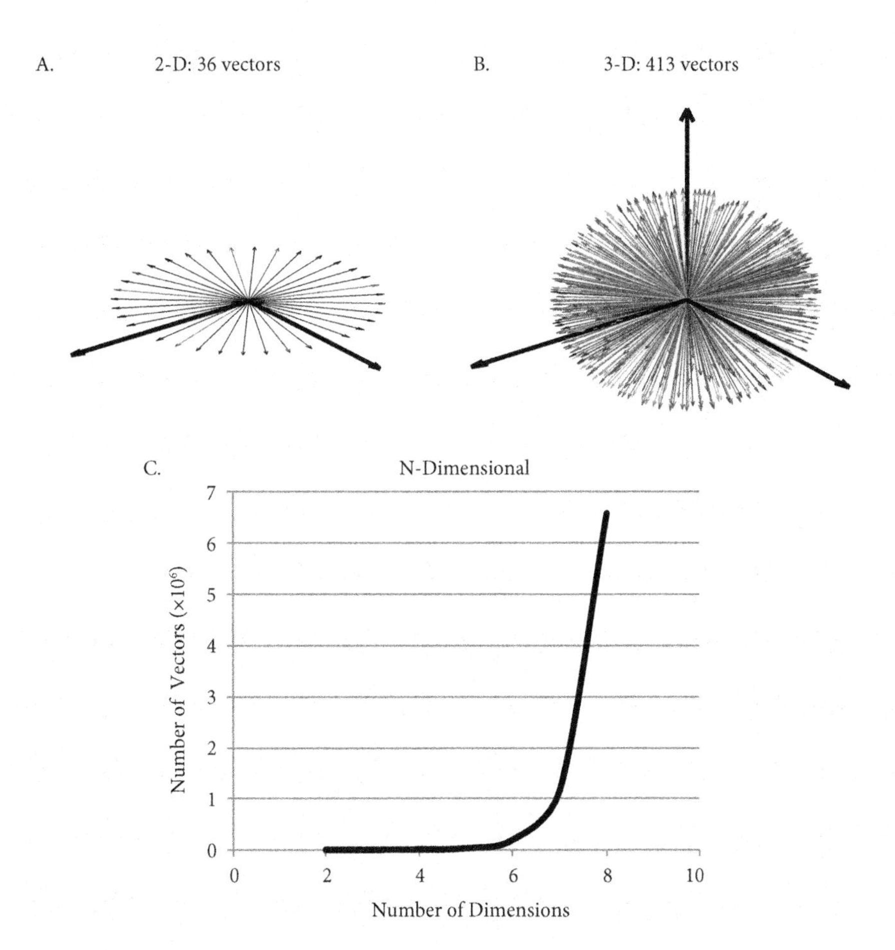

FIGURE 3.2 The scaling of the available conceptual space on the surface of a hypersphere with a given dimensionality. The figure shows the number of unit vectors that can be placed into **A**, a disk, and **B**, a ball, under the constraint that there must be a minimum angle of ten degrees between any two vectors. **C.** This plot extends the result from intuitive two- and three-dimensional shapes to higher dimensional hyperspheres.

representations are those that the *present activity* of a brain instantiates (i.e., representations as described by the NEF). Obviously the two are closely related. It is not far off to think of occurrent representations as being the currently active "parts" of complete conceptual representations.

With this distinction in hand, and acknowledging the richness of conceptual behavior, we are led to the realization that it is a mistake to assume that there are occurrent representations in the brain that carry the full semantic content of any given concept. There is simply too much information related to any particular conceptual representation to be able to actively represent and manipulate it all at the same time. This is why the SPA employs the notion of a (semantic) "pointer."

A pointer, in computer science, is a set of numbers that indicate the address of a piece of information stored somewhere in memory. What is interesting about pointers is that manipulations of pointers themselves can be performed that result in an indirect use of the information identified by the pointer, despite the fact that information itself is never explicitly accessed. Most such manipulations are quite simple. For example, if I need to pass a data structure to a function that may use it, I can simply pass the pointer, which is typically much smaller than the data structure itself. As a result, much less information needs to be moved around within the system, while still making the relevant data available for subsequent use.

One notable feature of pointers in computer science is that a pointer itself and the information contained at its address are arbitrarily related. As a result, having the pointer itself often indicates nothing about what sort of information will be found when the address to which it points is accessed. Because decisions about how to use information often depend on what the content of that information is, pointers are often "dereferenced" during program execution. Dereferencing occurs when the data at the address specified by the pointer is directly accessed. In a computer, this is a relatively cheap operation because memory is highly structured, and the pointer is easy to interpret as an address.

Given these features, pointers are reminiscent of symbols. Symbols, after all, are supposed to gain their computational utility from the arbitrary relationship they hold with their contents (Fodor, 1998). And symbols are often thought to act like labels for more sophisticated data structures (such as schemas, scripts, etc.), just as pointers act as labels for whatever happens to be at their address.

The semantic pointer hypothesis suggests that neural representations implicated in cognitive processing share central features with this traditional notion of a pointer. In short, the hypothesis suggests that the brain manipulates compact, address-like representations to take advantage of the significant efficiency and flexibility afforded by such representations. Relatedly, such neural representations are able to act like symbols in the brain.

However, the arbitrary relationship between a pointer and its contents is problematic for biological systems, because relationships between neural representations are most often learned. In contrast, the relationships between symbols in a digital computer are determined by human design decisions. It is precisely the failure of such decisions to adequately capture semantics that raises what has become known as the "symbol grounding problem" (Harnad, 1990). This, in short, is the problem of defining how symbols get their semantics–a difficult problem indeed and one over which much ink has been spilled. As a result, it would be hasty to simply import the notion of a pointer directly from computer science into our understanding of biological cognition.

Consequently, the hypothesis I am suggesting here is an extension of the standard notion of a pointer. In particular, the "semantic" in "semantic pointer" refers to the fact that the representations that play the role of a pointer contain semantic information themselves. That is, the relationship between a semantic pointer and the more sophisticated data structure to which it points is not arbitrary. Specifically, the semantic information that is contained in a semantic pointer is usefully thought of as a *compressed* version of the information contained in a more complex representation. In Section 3.5, I provide a detailed example in the visual system of how semantic pointers relate to the representations to which they point.

Bringing these preceding considerations together, semantic pointers are neural representations that are aptly described by high-dimensional vectors, are generated by compressing more sophisticated representations, and can be used to access those more sophisticated representations through decompression (a kind of dereferencing).

3.2 ■ What Is a Semantic Pointer?

In the preceding discussion I characterized semantic pointers in three different ways. First, I characterized them mathematically: Semantic pointers are described by vectors in a high-dimensional space. Second, I characterized them physically: Semantic pointers are occurrent activity in a biological neural network. And third, I characterized them functionally: Semantic pointers are compressed representations that point to fuller semantic content.

To be clear, I take these to all be characterizations of the same thing. In the previous chapter, I showed how vectors in a high-dimensional space can be used to describe neural activities in a biologically plausible network. In a way, the central purpose of the NEF's representational principle is to demonstrate how the first and second characterizations of semantic pointers are related (i.e., specified by the inter-level relation between a vector space and neural activities). The functional characterization goes beyond the NEF to specify what role semantic pointers play in the brain. In short, it suggests how those representations (implemented in neurons) can be used to underwrite biological cognition. A clear example of this function is provided in Section 3.5.

Each of these characterizations has a different explanatory role to play, but all are descriptions of the same "thing." This is a common feature of scientific concepts. For example, we can talk about an electrical current in terms of ion movements if we wish to explain the implementation of the current. We can talk about the current as being a specific number of amps (i.e., characterize it as a one-dimensional vector with units) if we want to talk about its role in Kirchoff's current law (i.e., that the sum of currents flowing into and out of a point are equal). And, we can talk about a current causing a light to illuminate if we want to explore how it is functioning in a given circuit.

I should point out that this tripartite distinction is not the same as that proposed by Marr (1982). His was a distinction to describe processes at different levels of abstraction (*see* Section 2.3). Thus, all three levels were variously detailed descriptions of the functional specification. Although it may be possible, and even useful, to talk of the SPA in this way, semantic pointers themselves are neural representations. Talking about them mathematically, physically, or functionally is not talking about them with different levels of detail. Throughout the book, I move between these ways of talking about semantic pointers. But, in every case, I take it as evident how that way of talking could be "translated" into any of the other ways of describing semantic pointers. This is clearly not the case for Marr's distinction–the translation itself is often at issue.

With the previous initial characterization of semantic pointers, and this terminological clarification in hand, we are in a position to consider recent research on semantics and examine detailed examples of how motor and perceptual semantic pointers might function in a neural architecture.

3.3 ■ Semantics: An Overview

There has been an historical tendency to start cognitive architectures with syntactic considerations and build on them by introducing more and more sophisticated semantics. However, some psychologists and linguists have more recently argued that syntactic processing can often be best explained by sophisticated semantic processing.[1] In contrast to traditional approaches, this puts semantics in the driver's seat. As a philosopher, I am used to having semantics as a primary concern. Indeed there are many well-known and influential works on the semantics of mental representations by philosophers (Dretske, 1981; Churchland & Churchland, 1981; Fodor, 1975; Millikan, 1984; Block, 1986; Cummins, 1989). As well, my own PhD dissertation is on the topic (Eliasmith, 2000).

However, it is often not obvious how such theories of meaning relate to empirical investigations of behavior. In this book I focus on an empirical approach (although what I say is consistent with the theory of "neurosemantics" I have described elsewhere [Eliasmith, 2000, 2006]). So, for present purposes, I set aside philosophical concerns and focus on more psychological ones. In psychology, interest in semantics is often expressed by asking

[1] Research in the field of "cognitive linguistics" has largely adopted this stance (e.g., Lakoff, 1987; Langacker, 1986; Fillmore, 1975), as have a number of psychologists (e.g., Johnson-Laird, 1983; Barsalou, 1999).

questions about what *kind* of semantic processing is needed to perform various cognitive tasks. In short, the question of interest becomes: For which behaviors do we need "deep" semantic processing, and which can be effectively accounted for by "shallow" semantic processing?

The distinction between deep and shallow processing can be traced back to Allan Paivio's (1986; 1971) Dual-Coding Theory. This theory suggests that perceptual and verbal information are processed in distinct channels. In Paivio's theory, linguistic processing is done using a symbolic code, and perceptual processing is done using an analog code, which retains the perceptual features of a stimulus. Paivio (1986) provided a lengthy account of the many sources of empirical evidence in support of this theory, which has been influential in much of cognitive psychology, including work on working memory, reading, and human computer interface design.

More recently, this theory has been slightly modified to include the observation that both channels are not always necessary for explaining human performance on certain tasks (Craik & Lockhart, 1972; Glaser, 1992; Simmons et al., 2008). Specifically, simple lexical decision tasks do not seem to engage the perceptual pathway. Two recent experiments have helped demonstrate this conclusion. First, Solomon and Barsalou (2004) behaviorally demonstrated that careful pairings of target words and properties can result in significant differences in response times. For example, when subjects were asked to determine whether the second word in a pair was a property of the first word, false pairings that were lexically associated took longer to process. For example, a pair like "cherry-card" resulted in 100-ms quicker responses than a pair like "banana-monkey." Second, Kan et al. (2003) observed that fMRI activation in perceptual systems was only present in the difficult cases for such tasks. Together, this work suggests that deep processing is not needed when a simple word association strategy is sufficient to complete the task.

Nevertheless, much of the semantic processing we perform on a daily basis seems to be of the "deep" type. Typical deep semantic processing occurs when we understand language in a way that would allow us to paraphrase its meaning or answer probing questions about its content. It has been shown, for example, that when professional athletes, such as hockey players, read stories about their sport, the portions of their brain that are involved in generating the motor actions associated with that sport are often active (Barsalou, 2009). This suggests that deep semantic processing may engage a kind of "simulation" of the circumstances described by the linguistic information. Similarly, when people are asked to think and reason about objects (such as a watermelon), they do not merely activate words that are associated with watermelons but seem to implicitly activate representations that are typical of watermelon backgrounds, bring up emotional associations with watermelons, and activate tactile, auditory, and visual representations of watermelons (Barsalou, 2009; McNorgan et al., 2011).

Consider the Simmons et al. (2008) experiment that was aimed at demonstrating both the timing and relative functions of deep and shallow processing. In this experiment, participants were each scanned in an fMRI machine twice. In one session, the experimenters were interested in determining the parts of the brain used during shallow semantic tasks and during deep semantic tasks. As a result, participants were given two tasks: first, "For the following word, list what other words come to mind immediately," and second, "For the following word, imagine a situation that contains what the word means and then describe it." The experimenters found that in response to the first task, language areas (such as Broca's area) were most active. In contrast, in response to the second task, participants engaged brain areas that are active during mental imagery, episodic memory, and situational context tasks (Kosslyn et al., 2000; Buckner & Wheeler, 2001; Barr, 2004). In other words, from this part of the experiment, it was evident that simple lexical association activated language areas, whereas complex meaning processing activated perceptual areas, as would be expected from the Dual-Coding Theory.

During the other part of the experiment, participants were asked to list, in their heads, answers to the question, "What characteristics are typically true of X?," where X was randomly chosen from the same set of target words as in the first session. When the two different scanning sessions were compared, the experimenters were able to deduce the timing of the activation of these two different areas. They found that the first half of the "typically true of X" task was dominated by activation in language areas, whereas the second half of the task was dominated by activation in perceptual areas. Consistent with the earlier behavioral experiments, this work shows that shallow processing is much more rapid, so it is not surprising that highly statistically related properties are listed first. Deep processing takes longer but provides for a richer characterization of the meaning of the concept.

This experiment, and many others emphasizing the importance and nature of deep semantic processing, have been carried out in Larry Barsalou's lab at Emory University. For the last two decades, he has suggested that his notion of "perceptual symbols" best characterizes the representational substrate of human cognition. He has suggested that the semantics of such symbols are captured by what he calls "simulations." Indeed, the notion of a "simulation" has often been linked to ideas of deep semantic processing (e.g., Allport, 1985; Damasio, 1989; Pulvermüller, 1999; Martin, 2007). Consequently, many researchers would no doubt agree with Barsalou's claim that deep semantic processing occurs when "the brain simulates the perceptual, motor, and mental states active during actual interactions with the word's referents" (Simmons et al., 2008, p. 107). His data and arguments are compelling.

However, the important missing component of his theory is how such simulations can be implemented in and manipulated by the brain. In a discussion of his work in 1999, one recurring critique was that his notion of "perceptual symbols" was highly underdefined. For example, Dennett and Viger pointedly noted, "If ever a theory cried out for a computational model, it is here" (1999, p. 613). More to the point, they concluded their discussion in the following manner:

> We want to stress, finally, that we think Barsalou offers some very promising sketchy ideas about how the new embodied cognition approach might begin to address the "classical" problems of propositions and concepts. In particular, he found some novel ways of exposing the tension between a neural structure's carrying specific information about the environment and its playing the sorts of functional roles that symbols play in a representational system. Resolving that tension in a working model, however, remains a job for another day.

In the conclusion to a recent review of his own and others' work on the issue of semantic processing, Barsalou stated (Barsalou, 2009, p. 1287):

> Perhaps the most pressing issue surrounding this area of work is the lack of well-specified computational accounts. Our understanding of simulators, simulations, situated conceptualizations and pattern completion inference would be much deeper if computational accounts specified the underlying mechanisms. Increasingly, grounding such accounts in neural mechanisms is obviously important.

This, of course, is one purpose of the SPA–to suggest a neurally grounded account of the computational processes underwriting cognitive function.

Given the important distinction between deep and shallow semantics, and the evidence that these are processed in different ways in the brain, a central question for the SPA is, "How can we incorporate both deep and shallow processing?" The central hypothesis of the SPA outlines the answer I pursue in the next three sections–that semantic pointers carry partial semantic information. The crucial steps to take now are to (1) describe exactly how the partial semantic information carried by semantic pointers is generated (and hence how they capture shallow semantics) and (2) describe how semantic pointers can be used to access the deep semantics to which they are related (i.e., how to dereference the pointers).

3.4 ■ Shallow Semantics

As I described earlier, the shallow semantics captured by a semantic pointer can be thought of as a kind of "compressed" representation of complex relations that underly deep semantics. Compression comes in two forms: "lossless," like the well-known zip method used to compress computer files that need to be perfectly reconstructed; and "lossy," such as the well-known

JPEG method for compressing images that loses some of the information in the original image. I take semantic pointers to be lossy compressions. To demonstrate the utility of lossy compression, and its relevance to cognition, let us consider a recently proposed class of lexical semantic representations. These representations have been developed by researchers who build algorithms to do automatic text processing. These same representations have been shown to capture many of the word-similarity effects that have been extensively studied by psychologists (Deerwester et al., 1990; Landauer & Dumais, 1997).

These representations are constructed by having a computer process a very large corpus of example texts. During this processing, the computer constructs what is called a term-document frequency matrix (*see* Fig. 3.3). The columns of this matrix index specific documents in the corpus. The rows in the matrix index words that appear in those documents. When the computer reads a particular document, it counts the number of times any word occurs in the document, and adds that value to the appropriate cell of the matrix. This way, if we look down the columns of the matrix, we can determine how many times each word appears in a given document. If we look across the rows of the matrix, we can determine how many times a given word appears in each document.

Practically speaking, there are a number of important subtleties to consider when constructing these kinds of representations to do actual textual processing. For example, such matrices tend to get very large, as many standard corpora have more than 20,000 documents and 100,000 unique words (Fishbein, 2008). Consequently, methods for reducing the size of the matrix are often employed. These methods are chosen to ensure that the most important statistical relationships captured by the original matrix are emphasized as

	Moby Dick	Hamlet	Alice in Wonderland	Sense and Sensibility
boat	330	0	0	0
enemy	5	0	0	1
mad	36	17	14	0
manners	1	1	1	34
today	1	0	1	9
tomorrow	1	0	1	15

FIGURE 3.3 Term-document frequency matrix. Each row of the matrix shows the number of occurrences of a specific word in each of the four selected texts. The row vectors of the words "today" and "tomorrow" are also noteworthy for being quite similar, suggesting some level of semantic similarity between the words. In most applications, such a matrix is compressed before it is used, and so the identity of the documents is lost.

much as possible. This is a kind of lossy compression applied to the original raw matrix.

One of the best known compression methods is singular-value decomposition, which is used, for example, in the well-known latent semantic analysis (LSA) approach (Deerwester et al., 1990). Compression can be either along the words or the documents. Let's consider the document case. After such a compression, there is still a vector for each word, but it may be only a few hundred dimensions rather than, say, 20,000. These new dimensions, however, are no longer directly associated with specific documents. Rather, each dimension is a blurred together collection of documents.

Although simple, this kind of representation has been used by researchers to capture psychological properties of language. For example, Paaß et al. (2004) have demonstrated how prototypes[2] can be extracted using such a method. The resulting representations capture standard typicality effects.[3] Additionally, this kind of representation has been used to write the Test of English as a Foreign Language (TOEFL). The computer program employing these representations scored 64.4%, which compares favorably to foreign applicants to American universities, who scored 64.5% on average (Landauer & Dumais, 1997).

However, these representations of words clearly only capture shallow semantics. After all, the word representations generated in this manner bear no relation to the actual objects that those same words pick out; rather, they model the statistics of text. Nevertheless, this work shows that semantic representations that capture basic statistics of word use can effectively support certain kinds of linguistic processing observed in human subjects. As mentioned earlier, these representations are compressed high-dimensional vectors, just like semantic pointers. That is, they capture precisely the kind of shallow semantics that semantic pointers are proposed to capture in the SPA.

In short, we have now briefly seen how vector spaces generated through compression can be used to capture some semantic phenomena. It may even be the case that the brain captures some lexical semantics by compressing word statistics into a semantic pointer. However, these text-based examples fall far short of what is needed to capture all of the semantic phenomena identified by the psychological experiments discussed in the last section. Those results suggest that there is an important and distinct role for deep semantic processing. That role is clearly not captured by the kinds of simple lexical associations used

[2] Prototypes of categories are often used to explain the nature of concepts (Smith, 1989). It has been a matter of some debate how such prototypes can be generated.

[3] Typicality effects are used to explain why subjects rate some concept instances as being more typical than others. These effects are often explained by the number of typical features that such instances have (the more typical features an instance has, the more typical it will be). Typical instances are both categorized more quickly and produced more readily by subjects. The prototype theory of concepts has been successful at capturing many of these effects.

to generate shallow representations for text processing. And shallow semantics are not sufficient for addressing the symbol grounding problem identified in Section 3.1. There is, after all, no way to get back to a "richer" representation of a word from these term-document representations. To address both deep semantics and symbol grounding, in the next section I turn to a different example of this same method for generating semantic pointers, one connected more directly to biological mechanisms.

3.5 ■ Deep Semantics for Perception

It should be clear from the preceding discussion that there are two questions of interest when considering the SPA. First, how are the semantics of the pointers generated? Second, how can the pointers be used to engage the deep semantic processing system when appropriate? The discussion in the previous section addresses only the first question. This is useful for demonstrating how shallow semantics can be captured by a semantic pointer, in the undeniably cognitive task of language processing.

In this section, I use a closely related method for generating shallow semantics that allows me to demonstrate how the resulting representation remains linked to deep semantics in a neural architecture. However, the task I consider is perceptual. In Section 4.8, I describe how these two (and other) different ways of generating and employing semantic pointers can be integrated.

Many of the most impressive results in machine vision employ *statistical modeling* methods (Hinton & Salakhutdinov, 2006a; Ranzato et al., 2007; Bengio & Lecun, 2007; Beal & Ghahramani, 2003). It is important to note that the word "model" in statistics–and in the next few paragraphs–is not used in the same way as it is throughout most of this book and in most of cognitive science. In statistics, the term "model" refers to an equation that captures relationships in a data set–there is no expectation that the elements of the equation pick out objects in the world. In contrast, "model" in nonstatistical usages typically refers to abstractions (including equations) whose parts are expected to map onto objects, processes, and states in the world. The neural models I describe throughout take their abstract parts (model neurons, brain areas, etc.) to map onto real parts of the brain. These latter models are sometimes called "mechanistic" models to distinguish them from statistical models.

In general, statistical modeling methods are centrally concerned with characterizing measured states of the world and identifying important patterns in those often noisy, measured states (Mumford, 1996; Knill & Pouget, 2004). In short, these methods have been developed to describe complex relationships given real-world data. This should sound familiar: The lexical representations described in the previous section are a kind of statistical model, attempting to describe the relationships between words in real-world text data (in which there are many spelling errors, non-words, etc.). Considered generally, describing complex real-world relationships with uncertain information is exactly the problem faced by biological systems.

For objection recognition, we can begin to formalize this problem by supposing that there is some visual data (e.g., an image), y, that is generated by the external visual world and drives neural activity. If we suppose that the purpose of perceptual systems is to construct and use a statistical model of this data, then the system must figure out a function $p(y)$ that describes how likely each state y is, so it can use that information to disambiguate future data.

Because the real world is extremely complex, the ideal statistical model will also be enormously complex (as it is the probability of all possible data at all times). As a result, the brain probably approximates this distribution by constructing what is called a *parameterized* model. Such a model identifies a small number of parameters that capture the overall shape of the ideal model. For example, if the statistics of the data y lie in the famous Bell curve (or Gaussian) distribution, we can model the data with an equation like:

$$p(y) = \frac{1}{\sigma\sqrt{2\pi}} e^{-(y-\bar{y})^2/2\sigma^2}. \tag{3.1}$$

Then, to "capture" all past data using our model, we only need to remember two parameters, $\bar{y}$ (the mean) and σ (the standard deviation), and the equation describing their relationship. This is much more efficient than remembering each value of $p(y)$ for each value of y explicitly.

To build such a model, we need to estimate the parameters (the mean and standard deviation in this case). Of course, to do any such estimating, the system needs data. As a result, a kind of bootstrapping process is necessary to construct this kind of model: We need data to estimate the parameters, then we can use our best estimate of the parameters to interpret any new data. Despite this seeming circularity, extremely powerful and general algorithms have been designed for estimating exactly these kinds of models (Dempster et al., 1977). Such methods have been extensively employed in building connectionist-type models and have been suggested to map to biological neural networks (Hinton, 2007; Friston, 2010).[4]

Note, however, that the methods for model inference do not specify the structure of the model itself (i.e., the relationships between the parameters). In the artificial neural network application of these methods, the model's structure (e.g., how many parameters there are and which parameters interact) is often "biologically inspired." One notable feature of brain structure that has proven a very useful starting point to constrain such models is its hierarchical

[4] A good place to start for state-of-the-art applications of these methods is Geoff Hinton's web page at `http://www.cs.toronto.edu/~hinton/`.

nature. Perhaps the best known example of such a structure in neuroscience is the visual hierarchy. For object recognition, this hierarchy begins with the retina, and proceeds through thalamus to visual areas V1 (primary visual cortex), V2 (secondary visual cortex), V4, and IT (inferotemporal cortex) (Felleman & Van Essen, 1991a).

In a hierarchical statistical model, each higher level in the hierarchy attempts to build a statistical model of the level below it. Taken together, the levels define a full model of the original input data (*see* Fig. 3.4). This kind of hierarchical structure naturally allows the progressive generation of more complex features at higher levels and progressively captures higher-order relationships in the data. Further, such models embody relations between hierarchical levels that are reminiscent of the variety of neural connectivity observed in cortex: feedforward, feedback, and recurrent (interlayer) connections are all essential.

The power of these methods for generating effective statistical models is impressive (Beal, 1998). They have been applied to solve a number of standard

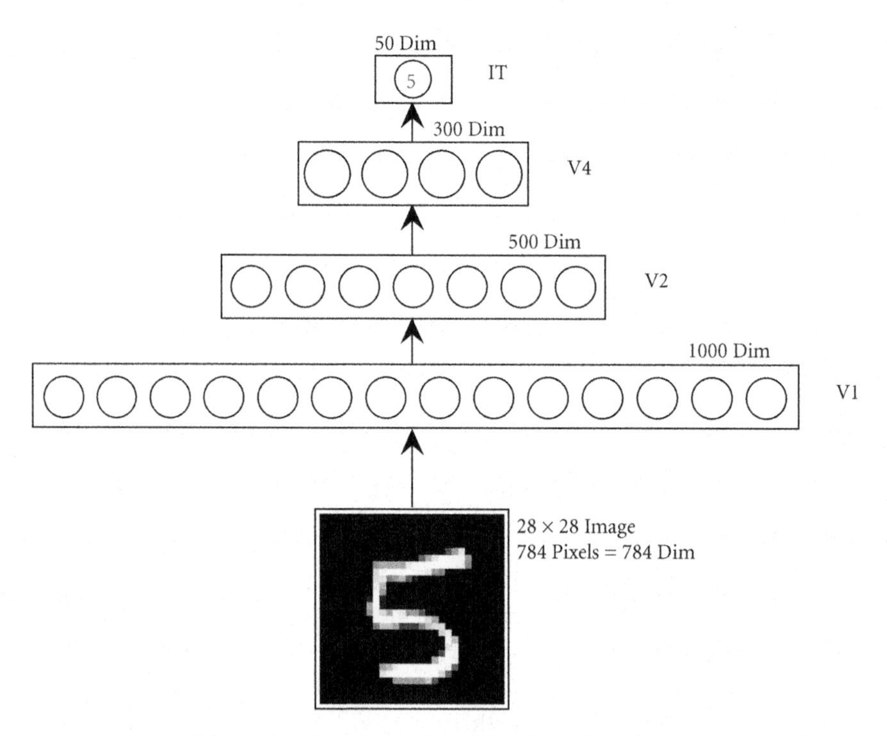

FIGURE 3.4 A hierarchical statistical model. An original image consisting of 784 dimensions is compressed into a 50-dimensional representation through a series of hierarchical statistical models. Each level can be thought of as corresponding to a level of the visual hierarchy for object recognition as indicated. The dimensionality of the state space at each level is indicated above the level.

pattern recognition problems, improving on other state-of-the-art methods (Hinton & Salakhutdinov, 2006b). Further, they have been shown to generate sensitivities to the input data that look like the tuning curves seen in visual cortex when constructing models of natural images (Lee et al., 2007). In fact, many of the most actively researched models of vision are naturally interpreted as constructing exactly these kinds of hierarchical statistical models.

To get a clearer picture of what this approach to perceptual modeling offers, how it can be mapped to a mechanistic model, and how it can be used to generate semantic pointers, let us turn to an example of such a model that was built in my lab by Charlie Tang (Tang & Eliasmith, 2010; Tang, 2010). The purpose of this model is to construct representations that support recognition of handwritten digits presented as visual input. The input is taken from the commonly used MNIST database at `http://yann.lecun.com/exdb/mnist/`. Figure 3.5 shows examples of the input. The model itself is structured as shown in Figure 3.4. The first layer of the model is initially trained on natural images to construct an input representation consistent with that found in primary visual cortex. The model is then shown the MNIST training images. Based on this training, the model tunes its parameters and is then tested on the MNIST test images.

Once we have constructed this statistical model, we can map it to a mechanistic model using the NEF. The structure of this statistical model is inspired by a specific cortical mapping (*see* Fig. 3.4), and so the NEF can respect that mapping, while determining how spiking neurons can implement the necessary underlying computations. The mathematical details of both the statistical

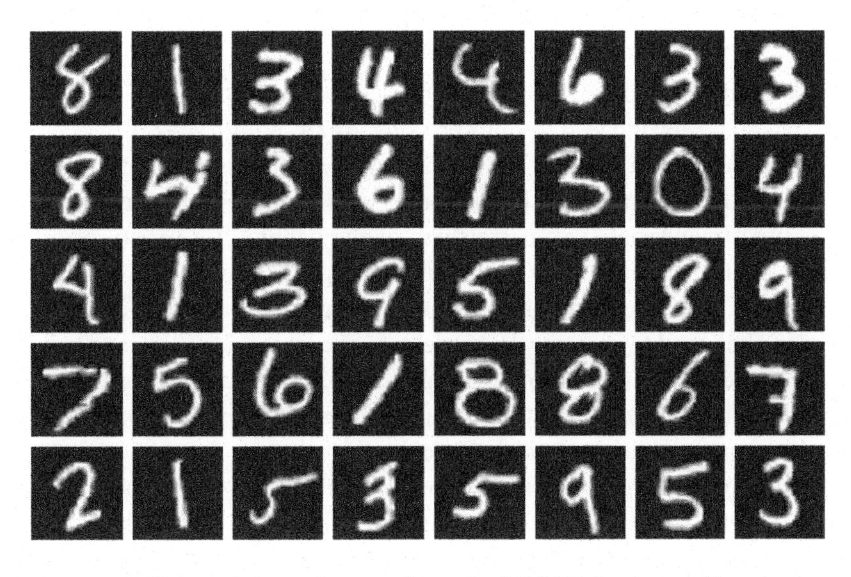

FIGURE 3.5 Example input images from the MNIST database of handwritten digits. The database consists of 60,000 training images and 10,000 test images.

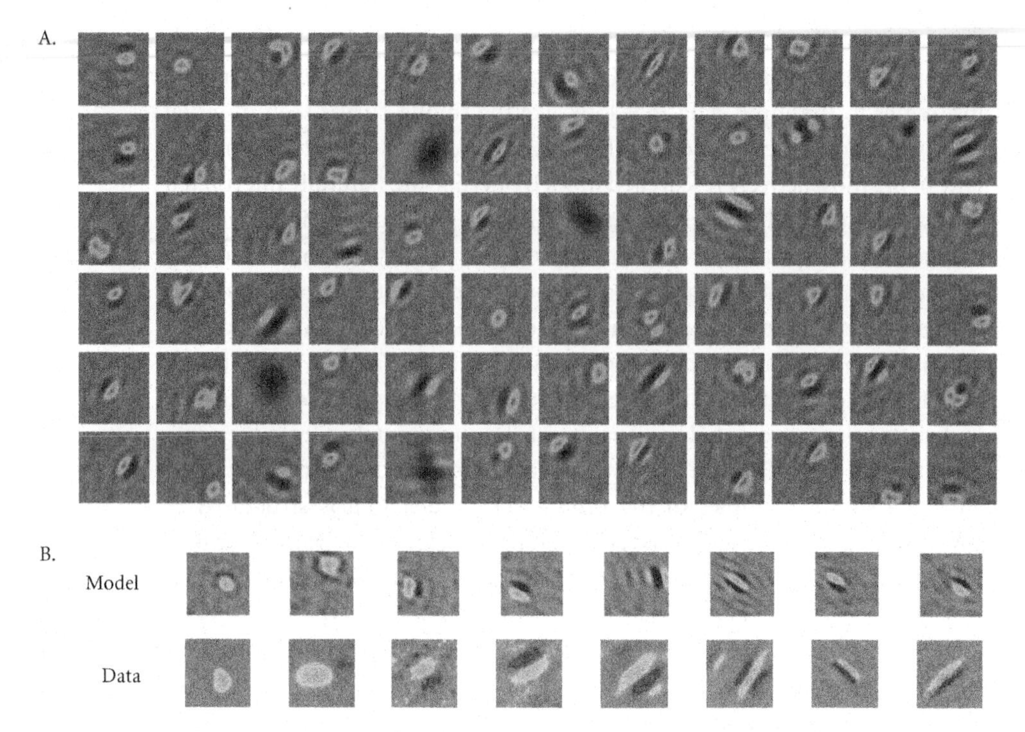

FIGURE 3.6 A sample of tuning curves of neurons in the model. **A.** These learned representations have a variety of orientations, spatial frequencies, and positions, likes those found in V1. **B.** A more direct comparison of data and model tuning curves from V1. (Data adapted from Ringach [2002] with permission.)

model, and this mapping to a cortical model are in Appendix C.1. The Nengo model can be downloaded from `http://models.nengo.ca/`.

As shown in Figure 3.6A, the tuning curves in the first layer of the cortical model capture many of the properties of V1 tuning, including a variety of spatial frequencies (i.e., narrowness of the banding); a variety of positions; variable orientations of the bands; and edge-detector and surround-like responses. Direct comparisons to biological data are shown in Figure 3.6B. The method used to generate the model tuning curves is identical to that used to generate the biological curves (Ringach, 2002). This can be done because data from both the mechanistic cortical model and the animal experiments is in the form of single neuron spike times.

This representation in V1 is used to drive the neurons in V2, and so on up the hierarchy. By the time we get to the highest level of the hierarchy, we have a much smaller dimensional (i.e., compressed) representation summarizing what has been presented to the lowest level. This compressed representation is a semantic pointer.

Specifically, the highest layer of this network represents a 50-dimensional space. Semantic pointers in this 50D space carry information about the presented digits. Clearly, however, this representation does not contain all of the information available in early visual areas. Rather, it is a summary that can be used to perform an object recognition task. This network, like most in machine vision, makes on the order of a few hundred errors on the 10,000 test digits that have not been used to train the network. In fact, recent models of this type outperform humans on this and other similar tasks (Chaaban & Scheessele, 2007).

These compressed representations (i.e., semantic pointers), like the lexical ones discussed previously, capture important information that can be used to classify the input. In both cases, shallow comparisons–that is, those directly between the semantic pointers–result in effective classification. In both cases, then, the semantic pointers themselves carry shallow semantic information sufficient to perform a useful task. However, there are two important differences between these perceptual semantic pointers and the lexical ones. First, these were generated based on raw images. This, of course, is a more biologically relevant input stream than nonvisual text: we do not have organs for directly detecting text. Consequently, the 50D pointers are grounded in natural visual input.

Second, and more importantly, these 50D pointers can be used to drive *deep* semantics. We can, in a sense, run the model "backward" to decompress (or dereference) the 50D representation. Specifically, we can clamp the semantic pointer representation at the top level of the network and then generate an input image at the lowest level.[5] This is essentially a biased sampling of the underlying statistical model in the network. Figures 3.7A and 3.7B show several examples of this process.

Figure 3.7 demonstrates that both perceptual features and category membership of an input are well captured by semantic pointers despite the compression from 784 to 50 dimensions in the state space. The good performance on recognition tasks demonstrates clearly that category membership is accessible. As well, it is clear from Figures 3.7C and 3.7D that given a category, relevant perceptual features can be generated. A natural choice for such a representation is the mean of the pointers associated with the category, which

[5] This does not suggest that the brain somehow recreates retinal images (there are no neurons that project from cortex to retina). Rather, Figure 3.7 shows the retinal images that are consistent with the unpacked cortical representations. The deep semantic information at any non-retinal level *is* accessible to the rest of the cortex. In the brain, the unpacking would stop sooner, but could still be carried out to as low a level as necessary for the task at hand. This is one reason why seeing an image is not the same as imagining one, no matter how detailed the imagining.

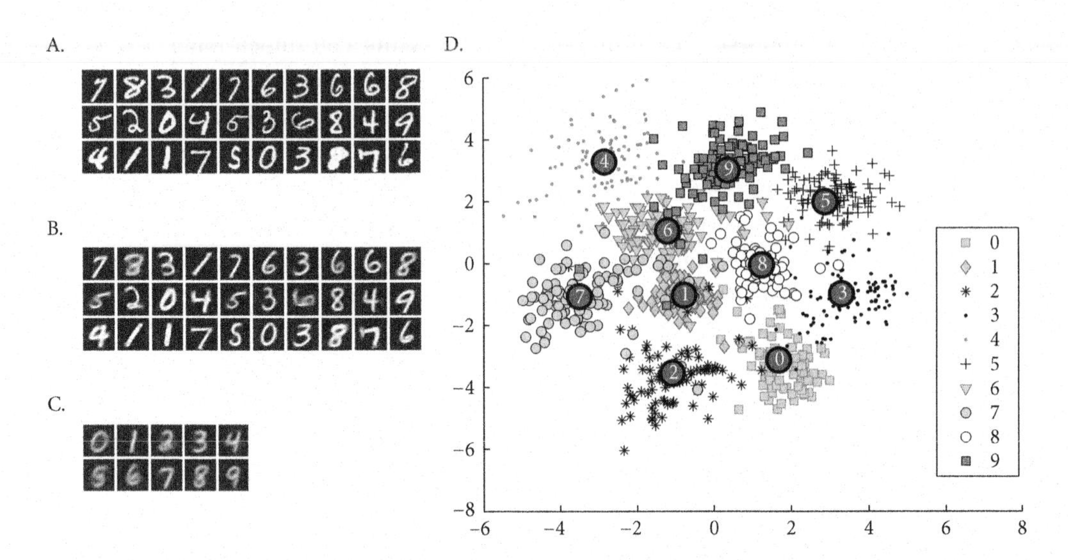

FIGURE 3.7 Dereferencing semantic pointers. **A.** The original input of several randomly chosen images. These are compressed into a 50D state space representation at the highest level of the model. **B.** The resulting images from dereferencing the 50D semantic pointers. The 50D representations generated by the network based on the input are clamped at the highest level, and the rest of the network is used to generate a guess as to what input would produce that semantic pointer. **C.** The dereferenced average semantic pointer for each category. These can be thought of as a prototype of the category. **D.** The compressed 50D representations are plotted in two dimensions using dimensionality reduction. Large dark circles indicate the mean value of the semantic pointers for each category. These can be thought of as the middle of the dimples on the conceptual golf ball (*see* Fig. 3.1).

can be thought of as a prototype of the category (*see* Figure 3.7C).[6] If other instances of the category need to be generated, then small random movements around this prototype will result in a variety of examples.

Further, the good encoding of specific perceptual features is evident from comparing Figures 3.7A and 3.7B. For example, subtleties of the way particular numbers are drawn, such as whether an "8" is slanted or not, are reconstructed from the pointer that the original image generates. Ultimately, it is

[6] Note that the "2" prototype is a combination of twos with straight bottoms and twos with loopy bottoms. This may suggest simply taking the mean is a mistake, and that there are two subcategories here. Dealing with these interesting, but additional, complexities is beyond the scope of the present discussion. However, this is accounted for in the Spaun model discussed in Chapter 7, which employs this model as its visual system.

these subtleties that capture the deep semantics of specific visual representations. The ability to regenerate such subtleties suggests that we have a good statistical model of the deep characteristics of this visual domain.

So, in general these lower dimensional representations can be dereferenced to provide detailed perceptual information, thus it is natural to think of them as "pointers" to that detailed information. Notice, however, that the dereferencing procedure depends on having the full perceptual hierarchy available, and running the representation through it. This is reminiscent of the fact that during deep semantic processing, perceptual areas are active in human subjects, as discussed earlier. It is, of course, a top-down procedure that allows these deep, grounded semantics to be regenerated. This is a procedure that only needs to be employed when the shallow semantics of the semantic pointer itself is insufficient for solving the task.

It is interesting to compare this kind of perceptual model with that employed in my earlier lexical example. Notably, the existence of a useful dereferencing procedure in the perceptual case serves to distinguish these models. This is a good example of a difference in degree becoming a difference in kind. In both cases, a statistical model of a domain is generated based on a large set of training examples. But only in the perceptual case is the detailed connection of the semantic pointer to those examples well preserved by the model. The representation of words in the linguistic context is just a co-occurrence structure, which highly underspecifies a document (e.g., word order doesn't matter). The representation of digits in the perceptual context, in contrast, captures a variety of deformations of the input relevant for visually distinguishing these digits. Indeed, the structure of the space in the second case seems to be nearly as good as that employed by people, given the similar level of performance between people and models in this classification task.[7] Lexical models are not nearly as good as humans in the analogous classification task. Consequently, only in the perceptual case would we want to claim that we have deep semantics for the domain of interest. That is, only in the case where we can define a dereferencing procedure that accounts for some of the domain's relevant structure would we claim to have captured deep semantics.

I would argue that capturing deep semantics and relating them to high-level representations solves the symbol grounding problem–*if* we can show how those high-level representations can function like symbols. Consideration of semantic pointers as being symbol-like is left for the next chapter. But, if that story is plausible, then the overall shape of the representational story underlying the SPA should be clear: semantic pointers, generated by grounded perceptual processing, can be "strippedoff" of that processing and used to

[7] I suspect the human specification is much more robust to a wider variety of noise and variability in viewing conditions (Tang & Eliasmith, 2010).

carry shallow semantics while acting like a symbol. If the deep semantics of such a representation are needed, then the semantic pointer can be used to clamp the top layer of the perceptual network that gave rise to it, and the network can regenerate the deep semantics.

Notice that in the terminology of the NEF, this story is one at the level of the state space. I have described semantic pointers and how to process them largely in terms of vector spaces. But we are concerned with *biological* cognition, so we also want to understand the functioning of this system from a biological perspective. I have given an early indication in Figure 3.6 that the model is directly comparable to neural data from early visual areas. But, of course, we can also examine the kinds of tuning curves evident in the highest level of our model, associated with processing in inferotemporal cortex (IT).

As shown in Figure 3.8, the visual model displays a wide variety of tuning curves in these cells. Some cells, such as that in Figure 3.8A, are reminiscent of the well-known Jennifer Aniston or Halle Berry neurons reported by Quiroga et al., (2005), because they are highly selective to instances of a single category. Other cells are more broadly selective for large parts of one category, as in Figure 3.8B. Such cells are also broadly reported in the neuroscientific literature (Kreiman et al., 2000). Other cells do not seem especially selective for any category, as in Figure 3.8C. Finally, some cells are simply difficult for us to classify in any systematic way, such as the cell shown in Figure 3.8D, where it seems tuned to arbitrary, specific cases from a variety of categories. These last two kinds of cells would likely not be recorded during an experiment, because they would be deemed "uninteresting." However, when data recorded from cells that are not explicitly selected by experimenters are analyzed, it has been shown that their responses are surprisingly good for determining both the category and the identity of stimuli being shown to a monkey (Hung et al., 2005). This makes sense if these cells are part of semantic pointers generated from statistical models, as shown here.

A direct consequence of observing this variety of tuning is that attempts to argue for particular kinds of "localist" representation that focus on specific hand-selected neurons (Bowers, 2009) are doomed to a degree of arbitrariness. Understanding how populations of neurons represent an entire space of perceptual input is much more relevant for constructing grounded neural representations that support cognitive function (Stewart et al., 2011). This, of course, is the purpose of introducing the notion of a semantic pointer. Semantic pointers, after all, are neural representations across *populations* of cells and identified with respect to a semantic *space*. Focusing on one kind of cell response within that context misses the more functionally important, larger picture.

In conclusion, it is worth noting that there are many limitations of models like the one I have presented. For example, although these models mimic

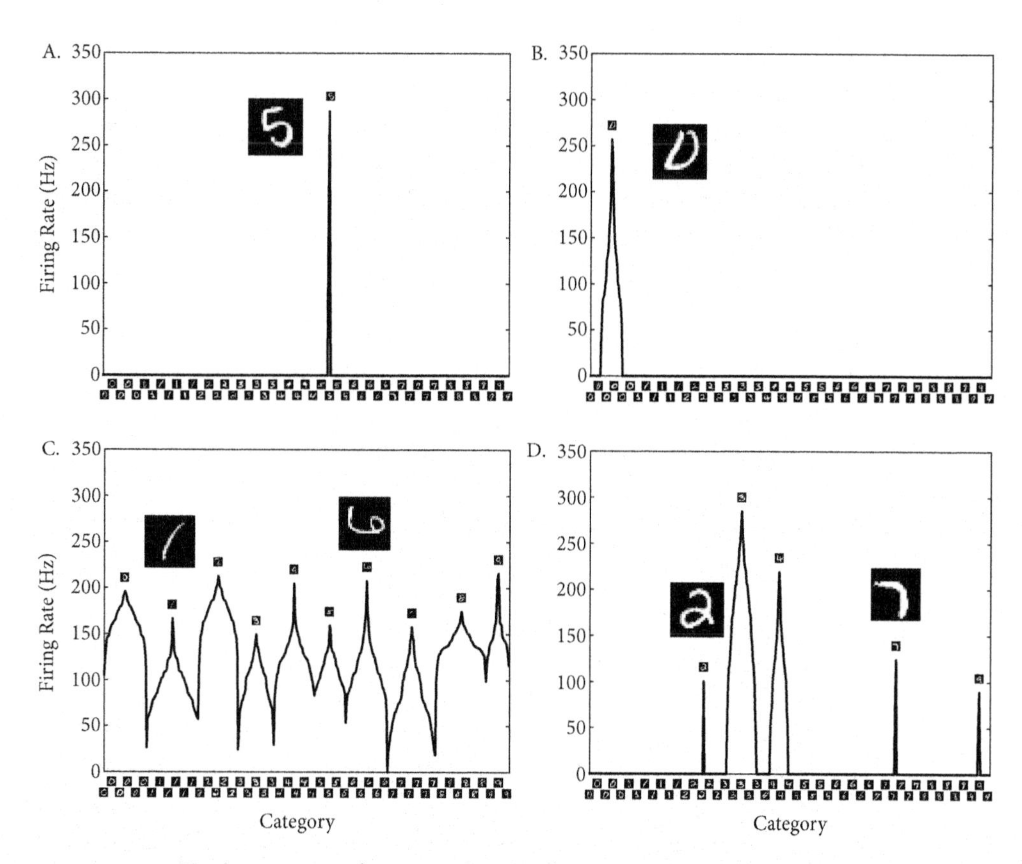

FIGURE 3.8 Tuning curves of neurons in visual area IT of this vision model. **A.** A highly specific neuron that is highly selective to an instance of a single category. **B.** A more categorical neuron that is responsive largely to a single category. **C.** A very broadly tuned neuron that does not obviously distinguish categories. **D.** A complex neuron whose response specificity is difficult to intuitively categorize. Tuning curves are generated by placing the highest firing rate image in the middle and ordering progressively outwards, by category. Sample images are shown for each category. Larger examples of the images generating peak firing are shown for clarity.

the basic hierarchical structure of perceptual systems and can approximately be mapped onto observed anatomical connections, most such models do not have connections that skip layers of the hierarchy, as observed in cortex. In addition, many other processes important for recognition, such as attention, tend to be excluded from current state-of-the-art models (c.f., Tang & Eliasmith, 2010). There also remain many questions regarding how every aspect of such models can be implemented by biological networks: Can the learning of these models be done by biologically plausible learning rules? In a full-scale system, how many dimensions should each layer of the network be taken to represent? How does the brain access information at any desired level of the perceptual network in a flexible manner? The model presented above does not address these issues in detail (although *see* Sections 4.6, 5.5, and 6.4 for hints about how to begin answering such questions).

Despite these limitations, three crucial features of this model are helpful for understanding the SPA characterization of semantics. Specifically, these are (1) the model constructs high-level representations of the perceptual input that capture statistical relationships describing the domain's structure; (2) such representations are "compressed" (i.e., lower dimensional) representations of the input; and (3) such representations can be "dereferenced" to provide detailed perceptual information.

3.6 ■ Deep Semantics for Action

Interestingly, these same three features are useful for characterizing biological representations that drive the motor system. Of course, the task for motor systems seems much different than that for perceptual systems. The motor system does not need to classify presented stimuli at all; rather, it needs to direct a nonlinear, high-dimensional system towards a desired state. But, considered generally, we might notice that perceptual systems often need to map from a high-dimensional, ambiguous state (e.g., images generated by highly nonlinear environmental processes) to a much smaller set of states (e.g., object categories), whereas motor systems often need to map from a small set of states (e.g., desired reach targets) to a high-dimensional, ambiguous state (e.g., any one of the possible configurations of muscle tensions over the body that results in reaching to a target). That is, we might notice that these tasks seem to be almost exact *opposites*.

Sometimes, opposition is a highly informative relationship. In this case, central computational features are shared by perception and motor control: both need to map low- to high-dimensional states; both need to deal with complexity and nonlinearity; both need to deal with uncertainty and ambiguity; and both need to share information between the past and future. However, the dominant direction of information flow through the two systems is in opposition. In mathematical terms, problems that share their structure in this way are called "dual problems." It is useful to identify duality between problems because if we know how to solve one such problem, then we can solve the other.

As a simple example, consider the relationship that exists between a cube and an octahedron (*see* Fig. 3.9A). Notice that there are the same number of sides in a cube as vertices in an octahedron and vice versa. Additionally, both have 12 edges, connecting the relevant vertices/faces in a structurally analogous manner. These two solids are thus duals. If we pose a problem for one

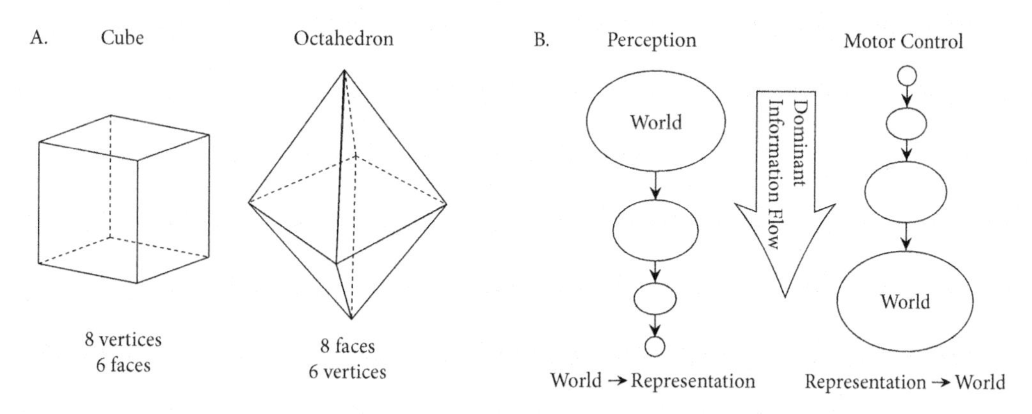

FIGURE 3.9 Dual problems. **A.** The cube and the octahedron provide a simple example. Both figures share a similar geometric description, but the number of vertices and faces of the two shapes are reversed. **B.** The perceptual and motor systems are much more complicated but can also be treated as duals. The perceptual system encodes high-dimensional, nonlinear relations using hierarchically compressed representations. The motor control system reverses this process (although with similar structure and representations) to determine high-dimensional nonlinear control signals from a low-dimensional signal.

of these solids (e.g., What is the volume of the smallest sphere that intersects all vertices [faces]?), then a solution for one of the solids provides a solution for the other (Luenberger, 1992). This is true as long as we swap analogous structural elements (i.e., faces and vertices) appropriately.

Why does this matter for understanding perception and motor control? Because this dual relationship has been suggested to exist between statistical models of perceptual processes and optimal control models of motor processes (Todorov, 2007, 2009). Figure 3.9B suggests a mapping between perceptual and motor systems that takes advantage of this duality. From an architectural point of view, this duality is very useful because it means that there is nothing different *in kind* about perceptual and motor systems. From the perspective of the SPA, in particular, this means that semantic pointers can play the same role in both perception and action. The remainder of this section describes the role of semantic pointers in the context of motor control.

To begin, like the perceptual system, the motor system is commonly taken to be hierarchical (*see* Fig. 3.10). Typically, we think of information as flowing down, rather than up, the motor hierarchy. For example, suppose you would like to move your hand toward a given target object. Once the desired goal state has been generated, it is provided to the cortical motor system. The first level of this system may then determine which direction the arm must move to

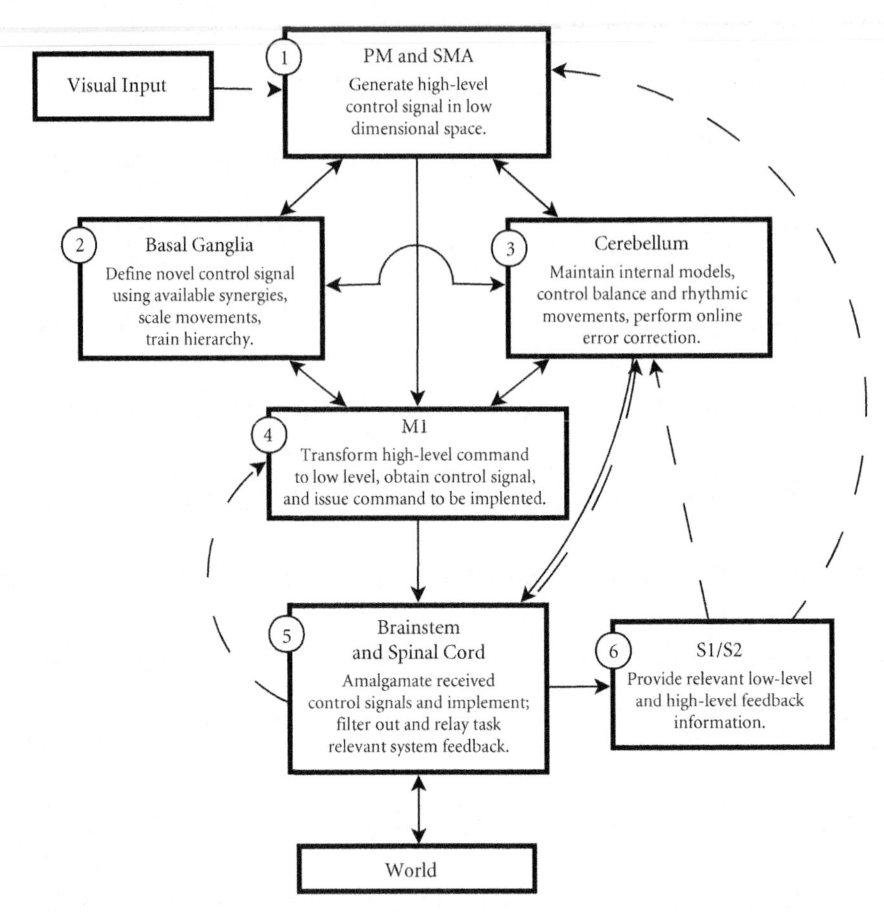

FIGURE 3.10 Neural optimal control hierarchy (NOCH). This framework maps neuroanatomy onto optimal control. The numbers indicate the approximate order of processing for a movement. Each box contains a brief description of the function of that area. Semantic pointers can be thought of as driving the top of this hierarchy, as in the perceptual system. Abbreviations: PM - premotor area; SMA - supplementary motor area; M1 - primary motor area; S1 - primary somatosensory area; S2 - secondary somatosensory area. This diagram is a simplified version of NOCH presented in Dewolf (2010).

reach that goal state.[8] Once the direction of movement has been determined, it can be used as a control signal for a controller that is lower in the hierarchy. Of course, specification of the direction of movement does not determine how

[8] This description is highly simplified and employs representations and movement decompositions that may not occur in the motor system. Identifying more realistic representations would require a much lengthier discussion but would not add to the main point of this section. For more detailed descriptions of such representations, *see* Dewolf (2010) and Dewolf and Eliasmith (2011).

torques need to be applied at the joints of the arm to realize that movement. This, then, would be the function of the next lower level in the control hierarchy. Once the specific forces needed to move the arm are determined, the specific tensions that need to be produced in the muscles to realize those forces must be determined by the next lower level in the hierarchy. Ultimately, this progression results in activity of motor neurons that cause muscles to contract and the arm to move.

Notably, just as it is inaccurate to think of information as flowing only "up" the perceptual hierarchy, so is it a mistake to think of motor control information as flowing only "down." In both systems, there are many recurrent connections as well as connections that skip hierarchical levels, so the flow of information is extremely complex. Nevertheless, the main problems to be tackled are similar. So, just as we began our characterization of perception by thinking of higher levels as constructing models of the levels below them, so we can think of lower levels in the motor hierarchy as having models of the higher levels. Lower levels can then use such models to determine an appropriate control signal to affect the behavior specified by the higher level. Notably, for perception and action, these models are bi-directional. This is why we could dereference a high-level semantic pointer.

There is good evidence for this kind of control structure in the brain (Wolpert & Kawato, 1998; Kawato, 1995; Oztop et al., 2006). That is, a control structure in which levels of the hierarchy have explicit models of the behavior of other levels. When we attempt to specify these models, a simplification in our characterization of perception becomes problematic: we assumed that the models were of static world states (i.e., images). However, there is no such thing as static movement. So, time is unavoidable. Embracing this observation actually renders the connection between perceptual and motor hierarchies even deeper–higher levels in both hierarchies must model the statistics *and* the dynamics of the levels below them.

In the previous perceptual model, the statistics at higher levels capture the regularities at lower levels. These regularities were captured by the neural tuning curves that were learned in order to represent the perceptual space. In motor control, these same kinds of regularities have been found (although they are dynamic), and are often called "synergies" (Bernstein, 1967; Lee, 1984). Synergies are useful organizations of sets of movements that are often elicited together. As before, the tuning curves of neurons reflect these synergies for the representation of the space of motor actions. These representations need to be learned based on the statistics of the space they are representing, as in the perceptual model above.

The particular dynamics of the system are thus likely to affect which synergies are most useful for controlling action. So, the representational and dynamical aspects of the system are tightly coupled. In fact, it is natural to describe this connection in a way captured by the third principle of the NEF:

the dynamical models[9] are defined over the representational state space. That is, the dynamical models of a given level of the hierarchy are defined using the synergies of that level.

There are now two main features of the motor system that mirror the perceptual system: (1) the (dynamical) model constructs representations of the domain that capture the statistical relationships that sufficiently describe the domain's structure; and (2) that such representations are "compressed" (i.e., lower dimensional) representations as we move up the hierarchy. In fact, I suspect that a dynamical hierarchy is a better model for perceptual processing than the static hierarchy considered earlier. Nevertheless, the main conceptual points relevant to the SPA remain: perception and action have dual descriptions in which semantic pointers can play the same role. Let us now consider, in more detail, how the SPA relates to motor control.

A graduate student in my lab, Travis DeWolf (2010), recently proposed a framework integrating known neuroanatomy with a hierarchical control characterization of the function of the motor system. This framework is called the Neural Optimal Control Hierarchy (NOCH), and a simplified version is shown in Figure 3.10. Models based on NOCH are able to explain the effects of various motor system perturbations, including Huntington's disease, Parkinson's disease, and cerebellar damage (Dewolf & Eliasmith, 2010). My description here is tailored to arm control to simplify the discussion. The task I consider is reaching toward a target on a flat surface (*see* Figure 3.11A).

As shown in Figure 3.11B, a NOCH model is able to generate the appropriate commands to control a highly redundant dynamical model of a physical arm.[10] Because this is a hierarchical model, it begins by generating an optimal high-level, low-dimensional (2D) control signal to move the arm to the desired target. That signal is then mapped to a mid-level, higher-dimensional (3D) signal for generating torques at the joints. There are three joints, so movement in the 2D space is redundant, consequently this is a difficult control problem. Regardless, the model is then able to map this mid-level signal to an even lower-level, higher-dimensional (6D) control signal needed to contract the six muscles appropriately to cause the desired movement. (For the mathematical derivation of such a model, *see* Appendix C.2.)

Note that this process is analogous to the dereferencing of high-level visual representations to low-level (higher dimensional) images. In this case, the

[9] For effective control, dynamical models are typically broken into two components: a forward model and an inverse model. A forward model is one that predicts the next state of the system given the current state and the control signal. An inverse model performs the opposite task, providing a control signal that can move between two given states of the system. For simplicity, I discuss both using the term "dynamical model," as this level of detail is beyond my current scope.

[10] The Matlab® code for the arm model can be downloaded at `http://nengo.ca/build-a-brain/chapter3`.

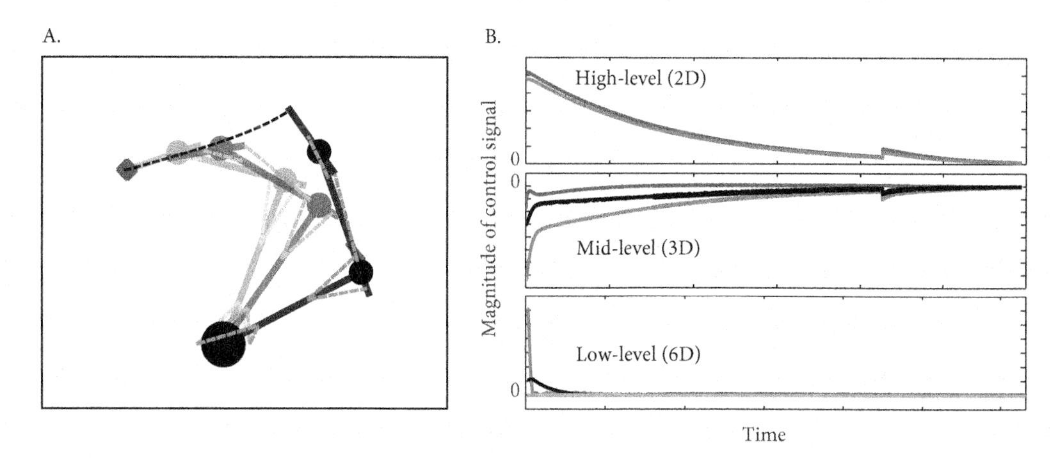

FIGURE 3.11 An example reach of an arm driven by a hierarchical controller. **A**. The movement of the arm to a target in 2D space. Black dots are joints, gray thick lines are segments, gray dotted lines are muscles. The gray square is the target and the black dotted line shows the end effector trajectory. **B**. The high-level (2D), mid-level (3D), and low-level (6D) control signals generated by the control hierarchy for the movement shown in A. Magnitude units are normalized.

intermediate representations (synergies) are not learned, but rather we have identified specific control spaces in which we can describe the dynamics of the arm. We are currently exploring methods for learning the necessary motor representations. Nevertheless, the dual relation to the perceptual hierarchy is clear: low-dimensional, high-level representations in premotor cortex (PM) are used to drive the high-dimensional, low-level spinal cord motor neurons to affect movement.

Although simple, this motor control example can be used to give a sense of how semantic pointers capture the distinction between deep and shallow semantic processing in the motor system. As in the perceptual case, a low-dimensional semantic pointer can be dereferenced by the remainder of the system into a specific, high-dimensional representation. This is demonstrated in Figure 3.12, where similar semantic pointer representations are shown to result in similar high-dimensional control signals and similar movements. Notably, however, the mapping is not a simple one. Just as in the perceptual case, complex nonlinear relationships exist between dimensions at each level.

This is consistent with a contemporary understanding of how motor cortex functions. Indeed, the work of Georgopoulos (1986) might be seen as an attempt to map some levels of this hierarchy (*see* Section 2.4, especially Fig. 2.16). Hence, it has already been shown that specific aspects of movement structure are reflected in the responses of single neurons in motor areas. More

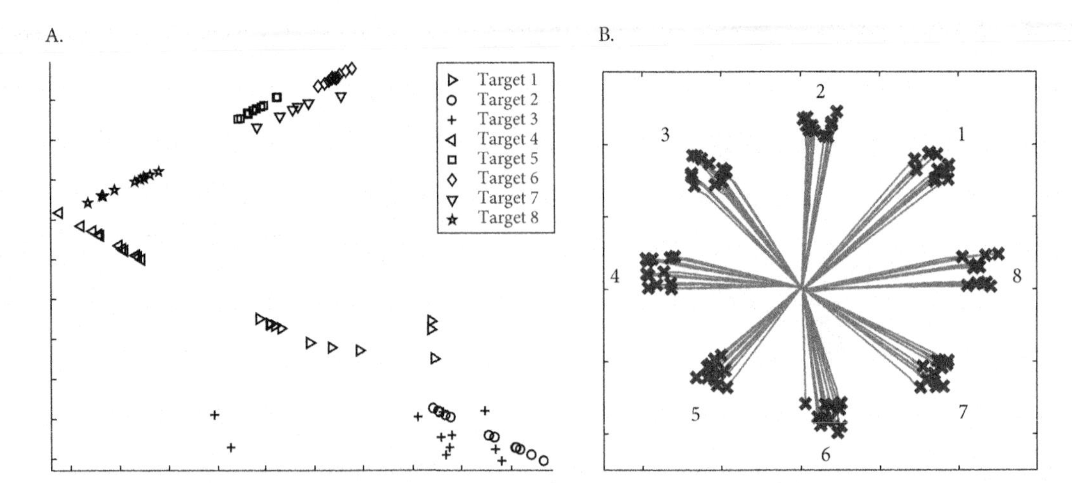

FIGURE 3.12 A semantic motor space. This figure shows how similar high-level semantic pointers can generate similar low-level movements. **A**. The clustering of 6D control signals used to drive the movements in B. **B**. The movements of the arm driven by eight different semantic pointers (the 2D coordinates of the target). Targets 1 through 8 are numbered. It is clear that there is not a simple mapping between movement in 2D and the control command in 6D space. Nevertheless, the dereferencing of the 6D pointers drive the high-dimensional system consistently in external 2D space as shown in B. Variability results from noise included in the model.

recent work in motor control has been suggesting that higher-dimensional representations are in fact the norm in motor cortex (Churchland et al., 2010). And, in our own recent work, we have shown that a spiking neuron implementation of this kind of NOCH model effectively captures a wide variety of the nonlinear temporal dynamics observed in the responses of motor neurons (Dewolf & Eliasmith, 2011). As a result, the NOCH approach maps well to the biological basis of motor control.

In cases that demand deep semantic processing in the motor system (e.g., that require estimating the precise configuration of joints), the high-level semantic pointers can be used to "run" the models in the motor system to internally generate more information about the nature of reaching movements at intermediate levels. Automatically inferring details about motor actions associated with some concepts during deep processing is a central aspect of conceptual semantics (Barsalou, 2009). In addition, there are well-known positive effects of "visualizing" motor performances before executing them (Landers & Feltz, 2007). Thus, as in the perceptual case, deep processing of motor semantics is supported by the same kind of dereferencing process of a high-level semantic pointer.

Recall that I identified two important features of semantic pointers: they capture higher-order relationships, and they are compressed representations. It should now be clear how these features are realized by the semantic pointers employed by the motor system. First, I described how this was the case in general, and then I presented a specific motor control model that embodies these features. Consequently, we have seen how motor semantic pointers capture higher-order relationships between states of the body because they can be dereferenced to coordinate those states for successful motor action. And second, we have seen that the representations at the highest level of the motor hierarchy are lower dimensional, and in many ways less specific, than those at the lowest level. As a result, they are usefully considered as compressed representations.

We have now seen how semantic pointers can play similar roles in both the perceptual and motor systems. There are, of course, important differences between the perceptual and motor models I have presented above. Most obviously, the perceptual model does not include dynamics, and the motor model does not learn the lower-level representations (i.e., synergies). There is ongoing research that attempts to address both of these challenges,[11] including work in my lab. The central theoretical features of this account, however, should remain the same as our models improve, so long as the identified duality of the sensory and motor systems remains.

3.7 ■ The Semantics of Perception and Action

My discussion of the SPA in the context of perception and action has adopted the usual approach of considering these aspects of behavior somewhat independently. However, it is hopefully clear from the above discussion that action and perception are not only tightly conceptually linked but also mutually serve to capture deep semantics in the SPA. The parallel stories provided above for action and perception are intended to lay the groundwork for more cognitive consideration of semantics (Section 4.8). For the moment, allow me to further consider the relationship between action and perception in the SPA.

Notice that both characterizations depend on a hierarchy being "run" in both directions. In the forward direction (up the hierarchy), the perceptual hierarchy allows classification of visual stimuli; in the reverse direction, it allows the generation of deep semantics. Although I did not discuss this

[11] For example, Grupen (2011), Lewis (2009), Kulic (2011), and other roboticists have been working on these issues from a motor control and reinforcement learning perspective. Hinton (2010), LeCun (2010), Bengio (2010), and others have been adopting a more statistical modeling and perceptually oriented approach.

in detail, both directions also help the system learn appropriate representations. For the motor hierarchy, the forward direction (down the hierarchy) allows for the control of a sophisticated body using simple commands, as well as the generation of deep semantics. The reverse direction allows online adaptation of the motor command to various perturbations. Again, both directions will allow the system to learn appropriate representations (i.e., synergies). So it is clear that in both cases, the semantic pointers operating at the top of the hierarchy can be thought of as "pointing to a memory," and can be used in the hierarchy to elicit details of that (motor or perceptual) memory. Of course, this use of the term "memory" is, unlike in a computer, necessarily constructive. We can make movements we have never made before, and we can imagine handwriting we have never seen before. This observation supports the notion that perceptual and motor systems are well characterized as constructing statistical models of their past input.

It is worth noting that I have been speaking of semantic pointers as if they only reside at the "top" of these hierarchies. I would like to be clear that this is a simplification to help introduce the notion. Clearly, some representations at other levels of the hierarchy are also compressed, lower-dimensional (than earlier in the hierarchy) representations. They, too, may be used by other parts of the system as pointers to more detailed semantic information. For explanatory purposes, and perhaps for reasons of efficiency, the most compressed representations are paradigmatic semantic pointers, but they need not be the only kind. So, the means of defining the semantics of the pointers and their precise location in the hierarchy are not essential to the SPA. What is central to the architecture is that the representations generated by these models act as low-dimensional summaries of high-dimensional systems engaged by the brain. This central feature highlights another defeasible assumption of the above examples: that the perceptual and motor models are generated independently.

As has been argued forcefully in a variety of disciplines,[12] it is a mistake to think of biological systems as processing perceptual information and then processing motor information. Rather, both are processed concurrently, and inform one another "all the way up" their respective hierarchies. Consequently, it is more appropriate to think of the semantics of items at the top of both hierarchies as having concurrent perceptual and motor semantics. The integrative model in Chapter 7 provides an example of this. In that case, the semantic pointer generated by the perceptual system is dereferenced by the

[12] I can by no means give comprehensive coverage of the many researchers who have made these sorts of arguments, which can be found in robotics, philosophy, neuroscience, and psychology. Some useful starting points include Brooks (1991); Churchland et al. (1994); Port and van Gelder (1995); and Regan and Noë (2001).

motor system. Thus, the deep semantics of the pointer depends on both of the models that can be used to dereference it.

The SPA thus provides a natural, but precise and quantitative, means of specifying the kinds of interaction between perception and action that have seemed to many to be central to sophisticated biological behavior. Such a characterization may remind some of the notion of an "affordance," introduced by psychologist James Gibson (1977). Affordances are "action possibilities" that are determined by the relationships between an organism and its environment. Gibson suggested that affordances are automatically picked up by animals in their natural environments, and provide a better characterization of perception (as linked to action) than traditional views.

Several robotics researchers have embraced these ideas, and found them useful starting points for building interactive perception/action systems (Scheier & Pfeifer, 1995; Ballard, 1991). Similarly, the notion of an "affordance," although somewhat vague, seems to relate to the information captured by a semantic pointer when embedded in the SPA. As just mentioned, these pointers can be used as direct links between action and perception that depend on the motor and perceptual experiences of the organism. Consequently, their semantics are directly tied to both the environment and body in which they are generated. There are, undoubtedly, many differences between semantic pointers and affordances as well (semantic pointers are not "directly perceived" in the sense championed by Gibson, for example). Nevertheless, the similarities help relate semantic pointers to concepts already familiar in psychology.

As well, affordances highlight the important interplay between perception and action. Motor control, after all, is only as good as the information it gathers from the environment. As a result, it is somewhat awkward to attempt to describe a motor controller without discussing perceptual input (notice that the NOCH in Figure 3.10 includes perceptual feedback). It is much more appropriate to conceive of the entire perception/action system as being a series of nested controllers, rather than a feed-in and feed-out hierarchy.[13]

What should be immediately evident from this structure is that the process of perceiving and controlling becomes much more dynamic, interacting at many levels of what were previously conceived of as separate hierarchies. This, of course, makes the system much more complicated, which is why it was more convenient to describe the two hierarchies separately. And, indeed, the real system is more complicated still, with connections that skip levels of the hierarchy and multiple perceptual paths interacting with a single level of the

[13] The observation that perceptual and motor cortex are both hierarchically structured and mutually interacting is hardly a new one (Fuster, 2000). What is new, I believe, is the computational specification, the biological implementation of the computations, and the integration into a cognitive hierarchy.

controller hierarchy. The question of relevance is: Does identifying this more sophisticated structure change our story about semantic pointers?

Given my suggested genesis and use of semantic pointers, I believe the answer is no. Recall that the way the highest level representations in the motor and perceptual hierarchies were generated was by characterizing statistical models that identify the relationships within the domain of interest. Whether these models are influenced solely by perceptual or motor processes, or whether they are jointly influenced by perceptual and motor processes, may change the nature of those relationships, but will not change the effective methods for characterizing those relationships. It will, then, still be the case that dereferencing a perceptual representation for deep semantic processing results in identifying finer perceptual details not available in the higher-level (semantic pointer) representation. As suggested earlier, these same pointers may also include information about relevant motor activities. So, semantic pointers will still be compressed and still capture higher-order relationships, it just may be that their contents are neither strictly perceptual nor strictly motor. So perhaps the perceptual/motor divide is one that is occasionally convenient for theorizing, even though it need not be reflected by the semantics of neural representations.

I would also like to be careful, however, not to overstate the closeness of the relationship between perception and action. Dissociation between impairment to visually guided motor control and object identification is well established (Goodale & Milner, 1992). Some patients with damage to the dorsal visual pathways can identify objects but not reach appropriately for them. Others with damage to the ventral visual pathways can reach appropriately but not classify objects. This suggests two things: (1) that motor action and visual perception can come apart; and (2) that the object classification model presented earlier only models part of the visual system (the ventral stream), at best. Again, the relevant point here is not about the completeness of the models, but that the subtle, complex character of action and perception in biological systems is consistent with the assumptions of the SPA. Capturing the appropriate degree of integration and distinctness of action and perception is a task independent on generating semantic pointers. So long as there are high-level representations with the assumed properties of semantic pointers, the SPA may be an appropriate description of the cognitive system employing those pointers.

So, semantic pointers can be compressed representations of motor/perceptual information found within a nested control structure. The story I have told so far is intended to characterize semantics, both deep and shallow, as related to perceptual and motor semantic pointers. Consequently, I would argue that this account is able to unify deep and shallow semantics along with sensory and motor semantics in a manner that can be realized in a biological cognitive system. What I must still describe, to make semantic pointers cognitively relevant, is how they can actually be used to encode complex syntactic structures. That is, how can these perceptual/motor representations be used to generate the kinds of language-like representations that underlie high-level cognition? Answering this question is the purpose of the next chapter.

3.8 ■ Nengo: Neural Computations

Characterizing neural representation as we did in the last tutorial is a useful first step to understanding biological cognition. However, to generate interesting behavior, neural representations must be transformed in various ways–that is, they must be used in neural computations. In many cases, such transformations, especially when they are nonlinear, result in novel neural representations.

In this tutorial, we examine how the second NEF principle (*see* Section 2.2.2) can be exploited in Nengo to compute transformations. This discussion builds on the previous tutorial on neural representation (Section 2.4), so it may be helpful to review that material. The following discussion is broken into two sections–the first on linear transformations and the second on nonlinear transformations. The tutorial focuses on scalar transformations, but the methods generalize readily to any level of the representational hierarchy, as I briefly discuss.

Linear Transformations

As formulated, principle 2 makes it clear that transformation is an extension of representation. In general, transformations rely on the same encoding, but exploit a different decoding than is used when representing a variable. I will introduce such transformations shortly. First, however, we can consider a transformation that is a basic property of single neurons–addition. Addition transforms two inputs into a single output, which is their sum.

- In Nengo, create a new network. Set the *Name* of the network to "Scalar Addition."
- Create a new ensemble within the "Scalar Addition" network. Name the new ensemble "A" and set *Number of Nodes* to 100, otherwise use the default settings. (*Dimensions* is 1, *Node Factory* is LIF Neuron, *Radius* is 1, *tauRC* is 0.02, *tauRef* is 0.002, *Max rate* low is 100 and high is 200, *Intercept* is low −1.0 and high is 1.0).

To ensure the ensemble uses the default settings, you can change the Nengo "Preferences." Otherwise, Nengo reuses the last set of settings used to generate an ensemble.

- Create a second ensemble within the "Scalar Addition" network. Name this ensemble "B" and use the same parameters as you did for ensemble A.

These two neural ensembles will represent the two variables being summed. As with the tutorial on neural representation, we now need to connect external inputs to these ensembles.

- Create a new function input, name it "input A," set *Output Dimensions* to 1, and click OK.
- Add a termination to ensemble A. Name the termination "input" and ensure that number of input dimensions (*Input Dim*) is set to 1. The default value of *PSTC* (0.02) is acceptable.
- Click *Set Weights* and confirm that the value in the 1-to-1 coupling matrix that appears is 1.0.
- Click *OK* to confirm the options set in each window until the new termination is created.
- Make a connection between the origin of "input A" and the new termination on ensemble A.
- Follow the same procedure to make a second constant function input and project it to ensemble B.

This process should be familiar from the previous tutorial. We have represented two scalar values in neural ensembles, and we could view the result in the *Interactive Plots* window as before. However, we wish to add the values in the two ensembles, and to do this we will need a third ensemble to represent the sum.

- Create another ensemble. Call it "Sum" and give it 100 nodes and 1 dimension. Set *Radius* to 2.0. Click *OK*.

Note that we have given the new ensemble a radius of 2.0. This means that it will be able to accurately represent values from −2.0 to 2.0, which is the maximum range of summing the input values. The "Sum" ensemble now needs to receive projections from the two input ensembles so it can perform the actual addition.

- Create a new termination on the "Sum" ensemble. Name it "input 1," set *Input Dim* to 1.
- Open the coupling matrix by clicking *Set Weights* and set the value of the connection to 1.0. Finish creating the termination by clicking *OK* as needed.
- Create a second termination on the "Sum" ensemble following the procedure outlined above, but name it "input 2."
- Form a projection between ensemble "A" and the "Sum" ensemble by connecting the origin of "A" to one of the terminations on "Sum."

- Likewise form a projection between ensemble "B" and the "Sum" ensemble by connecting its origin to the remaining termination.

You have just completed the addition network. Addition is performed simply by having two terminations for two different inputs. Recall from the discussion in Section 2.1 that when a neuron receives a signal from another neuron, postsynaptic currents are triggered in the receiving neuron, and these currents are summed in the soma. When signals are received from multiple sources, they are added in exactly the same fashion. This makes addition the default transformation of neuron inputs.

- To test the addition network, open the *Interactive Plots* viewer, display the control sliders for the inputs and the value of the "Sum" ensemble, and run the simulation.
- Play with the values of the input variables until you are satisfied that the "Sum" ensemble is representing the sum correctly. If the value graphs are difficult to see, right-click on the graph and select "auto-zoom" or zoom in by right checking and choosing *set y range.*

Another exploration to perform with this network is to set one of the inputs to zero, and change the value of the other input. You will see that the "Sum" ensemble simply re-represents that non-zero input value. This is perhaps the simplest kind of transformation and is sometimes called a "communication channel" because it simply communicates the value in the first ensemble to the second (i.e., it performs the identity transformation). It is interesting to note that the firing patterns can be very different in these two ensembles, although they are representing the same value of their respective variables. This can be best seen by viewing either the spike rasters or the voltage grid from both ensembles simultaneously. A pre-built version of this network can be loaded from <Nengo Install Directory>/Building a Brain/chapter3/addition.py.

Now that you have a working addition network, it is simple to compute any linear function of scalars. Linear functions of scalars are of the form $z = c_1x_1 + c_2x_2 + \ldots$ for any constants c_n and variables x_n. So, the next step toward implementing any linear function is to create a network that can multiply a represented variable by a constant value (the constant is often called a "gain" or "scaling factor"). Scaling by a constant value should not be confused with multiplying together two variables, which is a nonlinear operation that will be discussed below in Section 3.8.

We will create this network by editing the addition network.

- Click the *Inspector* icon (the magnifying glass in the top right corner of the window).
- Click on the input termination on the "Sum" population that is connected to "A."
- In the inspector, double-click *A transform* (at the very bottom).

This now shows the coupling weight you set earlier when creating this termination. It should be equal to 1.0.

- Double-click the 1.0 value. In the window that appears, double-click the 1.0 value again. Change this value to 2.0.
- Click *Save Changes* on the open dialog box. You should see the updated value in the *Inspector*.

Setting the weight of a termination determines what constant multiple of the input value is added to the decoded value of an ensemble. The preceding steps have set up the weights such that the value of "A" input is doubled.

- Test the network in the *Interactive Plots* window.

You should be able to confirm that the value of "Sum" is twice the value of "A," when you leave "B" set to zero. However, if you leave "A" at a high value and move "B," you will begin to saturate the "Sum" ensemble because its radius is only 2.0. The saturation (i.e., under-estimation) effect is minimal near the radius value (2.0) and becomes more evident the farther over the radius the ensemble is driven. To prevent saturation, change the radius of the "Sum" population by changing the "radii" property in the *Inspector* when the population is selected.

You can easily change the linear transformation between these ensembles by selecting any scalar value for the *A transform* of the two inputs to the "Sum" population as described above. However, you must also set an appropriate radius for the "Sum" ensemble to avoid saturation.

We can now generalize this network to any linear transformation. All linear transformations can be written in the form $\mathbf{z} = \mathbf{C}_1\mathbf{x}_1 + \mathbf{C}_2\mathbf{x}_2 + \ldots$, where $\mathbf{C}_n$ are constant matrices and $\mathbf{x}_n$ are vector variables. Computing linear transformations using vector representations rather than scalars does not introduce any new concepts, but the mapping between populations becomes slightly more complicated. We will consider only a single transformation in this network (i.e., $\mathbf{z} = \mathbf{C}\mathbf{x}$, where $\mathbf{C}$ is the coupling matrix).

- In a blank Nengo workspace, create a network called "Arbitrary Linear Transformation."
- Create a new ensemble named "x" with the default values, and give it 200 nodes and two dimensions. This will be the input.
- Create another ensemble named "z" with the default values, and give it 200 nodes and three dimensions. This will be the output.

Note that we have two vector ensembles of different dimensionality. To connect these together, we must specify the weights between each pair of dimensions in the two ensembles.

- Add a new decoded termination to ensemble "z" called "input."
- Set *Input Dim* to 2 then click *Set Weights.*

You should now see a "2 to 3 Coupling Matrix" window. Each cell in the matrix defines a weight on a dimension in ensemble "x." These weights multiply the relevent input dimension and are then summed to drive the appropriate dimension in the "z" ensemble. This works exactly like standard matrix multiplication. So, since there are 3 dimensions in the output, we have three rows. Each row weights the two input components and is then summed to give one value in the represented output vector in "z."

- Enter the matrix

$$\begin{bmatrix} 0.0 & 1.0 \\ 1.0 & 0.0 \\ 0.5 & 0.5 \end{bmatrix}$$

  into the coupling matrix and press *OK.*

This matrix swaps the first and second dimension of "x" in the output "z" and puts the average of those dimensions in the third dimension of "z." So, for example, $\mathbf{x} = [1, 2]$ becomes $\mathbf{z} = [2, 1, 1.5]$.

- Click *OK* to create the decoded termination.
- Create a projection from the origin of ensemble "x" to the termination of ensemble "z."
- Create an "input" termination on ensemble "x." Set *Input Dim* to 2 and click *Set Weights* to set the coupling matrix to an identity matrix (1.0 along the diagonal, zero elsewhere).

Creating a coupling matrix with a diagonal matrix of ones in this manner causes the input dimensions to be mapped directly to corresponding output dimensions.

- Create a new function input. Give it two dimensions and click *OK.*
- Project the new function to the termination of ensemble "x."

You've now completed a network to implement arbitrary linear transformations from a two-dimensional vector to a three-dimensional vector, represented with neurons.

- Click the Interactive Plots icon, pull up input sliders, value graphs, and anything else of interest, and test the simulation while it is running.

A simple way to check that it is performing the desired transformation is to move the first input slider to 0.5 and the second one to −0.5. The "z" value should have the black and blue lines reversed (first two dimensions), and the red line should be at zero (the third dimension, which is the average of the first two). A pre-built version of this network can be loaded from <Nengo Install Directory>/Building a Brain/chapter3/arbitrary_linear.py.

Nonlinear Transformations

In the previous section, you learned how to build a linear brain. The real world is filled with nonlinearities, however, and so dealing with it often requires nonlinear computation. In this tutorial, we compute nonlinear functions by, surprisingly, more linear decoding. In fact, nonlinear transformation can be thought of as nothing more than an "alternate" linear decoding of a represented value. In this tutorial we only consider decoding using linear combinations of *neural responses*, but, as discussed later in Section 4.3, combinations of *dendritic responses* may allow for even more efficient computation of nonlinear transformations.

- Create a new network called "Nonlinear Function."
- Inside that network create a one-dimensional ensemble with default values, 100 neurons, and name it "X."
- Create a constant function input for this ensemble and connect it to the ensemble (remember to create a decoded termination with a connection weight of 1.0 on the "X" ensemble and a *PSTC* of 0.02).

As in the preceding section of this tutorial, we are starting with a scalar ensemble for the sake of simplicity. Now we will square the value represented in "X" and represent the squared valued in another ensemble.

- Create another 100 neuron, 1D ensemble with default values named "result."
- Create a termination on the "result" ensemble (weight of 1.0, *PSTC* of 0.02).

Note that we are not using the termination to calculate the transformation. This is because the termination can only scale the decoded input by a constant factor, leading to linear transformations such as those in the first part of this tutorial. Nonlinear transformations are made with decoded *origins*.

- Drag an *Origin* from the template bar onto the "X" ensemble.
- In the dialog box that appears, name the origin "square," set *Output Dimensions* to 1, and click *Set Functions*.
- Select *User-defined Function* from the drop-down list of functions and click *Set*.

The "User-defined Function" dialog box that is now displayed allows you to enter arbitrary functions for the ensemble to approximate. The "Expression" line can contain the variables represented by the ensemble, mathematical operators, and any functions specified by the "Registered Functions" drop-down list.

- In the *Expression* field, type "x0*x0."

"x0" refers to the scalar value represented by the "X" ensemble. In an ensemble representing more than one dimension, the second variable is "x1," the third is "x2," and so forth. Returning to the preceding expression, it should be clear that it is multiplying the value in the "X" ensemble by itself, thus computing the square of the input value.

- Click *OK* to close dialogs until the decoded origin is created.
- Drag from the "square" origin you just created to the "input" termination to create a projection between the "X" and "result" ensembles.

The projection that has just been made completes the setup for computing a nonlinear transformation. The "X" and "result" ensembles have been connected together, and Nengo automatically sets connection weights appropriately to match the specified function (mathematical details can be found in Section B.2 of the appendix).

- Open the *Interactive Plots* window.
- Right-click "result" and display its value.
- Right-click your input and display its control.
- Right-click "X" and select *X→value.*
- Right-click "X" and select *square→value.*

Note that, unlike the "result" ensemble, the "X" ensemble doesn't have just one value that can be displayed. Rather, it has the representational decoding of "X," as well as any transformational decoding we have specified, in this case "square."

- Run the simulation. To demonstrate it is computing the square, move the slider as smoothly as possible between −1 and 1. You should see a straight line in the "X" value and a parabola in the "square" and "result" values.

Remember that these decoded values are not actually available to the neurons, but are ways for us to visualize their activities. However, if we only examine the relationship between the input and the spike responses in the "result" ensemble, we will see that the firing rates of the result neurons are in an approximate "squaring" relation (or its inverse for "off" neurons) to the input values. This can be seen by examining the spike raster from the "result" population while moving the slider smoothly between −1 and 1.

This tutorial has so far covered linear transformations of scalar variables, linear transformations with vectors, and nonlinear transformations of scalars. The final step is to extend nonlinear transformations to vector representations. As in the scalar case, a nonlinear transformation of a vector is implemented simply with an alternate linear decoding of a neural ensemble (as illustrated for the scalar case in Fig. 2.11 and derived mathematically in Appendix B.2). Crucially, this means that we can understand nonlinear functions of multiple scalars (e.g., $x \times y$) as a linear decoding of a higher-dimensional population. Let us examine this in practice:

- In the "Nonlinear Functions" network, create a 100 neuron, 1D ensemble named "Y" with the default values.
- Add a decoded termination to the "Y" ensemble (weight of 1, *PSTC* of 0.02).
- Add a new constant function input named "input Y" and project it to the termination on ensemble "Y."
- Now create a "default" ensemble named "2D vector" and give it two dimensions and 100 neurons.
- Create a termination named "x" on the "2D vector" ensemble. Set the dimension to 1, the weight matrix to [1, 0], and *PSTC* to 0.02.
- Create a termination named "y" on the "2D vector" ensemble. Set the dimension to 1, the weight matrix to [0, 1], and *PSTC* to 0.02.
- Connect the "x" origins of the "X" and "Y" ensembles to the "x" and "y" terminations on "2D vector," respectively.

The value from the "X" ensemble will be represented in the first component of the 2D vector and the value from the "Y" ensemble will be represented in its second component. At this point, there is no output for the network. We are going to represent the product of the two components of the "2D vector" ensemble in the "result" ensemble. To do this we cannot simply use the default "x" decoded origin on the "2D vector" ensemble. We need to create an alternate decoding of the contents of "2D vector" (as in the case of computing the square).

- Drag a new decoded origin onto the "2D vector" ensemble. Name this origin "product," give it 1 output dimension, and click *Set Functions.*
- In the *Function 0* drop-down menu, select *User-defined Function* and click *Set.*
- Set *Expression* to "x0*x1." *Input Dimensions* should be set to 2.

As mentioned earlier, x0 refers to the first component of the vector and x1 to the second. Given this expression, Nengo generates decoders that yield an approximation of the two components multiplied together.

- Click *OK* to complete all the open dialogs.

- Disconnect the "X" and "result" ensembles by dragging the "result" termination away from the ensemble. To eliminate the hanging connection entirely, right click it and select "remove."
- Create a projection from the "product" origin on "2D vector" to the termination on "result."

This completes the network. The "2D vector" population receives scalar inputs from two populations, represents these inputs separately as two components of a vector, and has its activity decoded as a nonlinear transformation in the projection to the "result" ensemble.

- Click the *Interactive Plots* icon, and show the "X", "Y", and "result" values. Show the "input" and "input y" controls. Run the simulation.

Move the sliders to convince yourself that the "result" is the product of the inputs. For example, recall that 0 times anything is 0, the product of two negative numbers is positive, the product of opposite signs is negative, and so on. You may have noticed that when computing 1 times 1, the answer is slightly too low. This is because with the two inputs set to maximum, the 2D population becomes saturated. This saturation occurs because the population represents a unit circle, which does not include the point $[1, 1]$. The farthest diagonal point it includes is $[\sqrt{2}/2, \sqrt{2}/2]$. To remove this saturation you need to increase the radius of the 2D population.

We can also examine the effects of changing the number of neurons on the quality of the computation.

- In the *Network Viewer*, select the "2D vector" ensemble, open the *Inspector*, double-click "i neurons," then change the number of neurons from 100 to 70.
- Test the network in *Interactive Plots* again.

Note that multiplication continues to work well with only 70 neurons in the vector population, but does degrade as fewer neurons are used. I appeal to this fact in later chapters when estimating the number of neurons required to compute more complex nonlinear operations. A pre-built version of this network can be loaded from <Nengo Install Directory>/Building a Brain/chapter3/nonlinear.py.

4. BIOLOGICAL COGNITION–SYNTAX

C. Eliasmith, T. Stewart, and D. Rasmussen

4.1 ■ Structured Representations

It has been forcefully argued that structured representations, like those found in natural language, are essential for explanations of a wide variety of cognitive behavior (Anderson, 2007). As I discuss in more detail in Chapter 8, researchers have often moved from this fairly uncontroversial observation to much stronger claims regarding essential properties of these structured representations (e.g., that they are compositional). For the time being, I remain largely agnostic about these stronger claims. Rather, I take the ultimate arbiter of whether our theory about such representations is a good one to be the behavioral data. The purpose of this chapter is to suggest how semantic pointers can be used in structured representations that are able to explain those data–with no additional assumptions about the essential features of structured representations used for cognition. The main behavioral test in this chapter is the Raven's Progressive Matrices, a test of general fluid intelligence that demands inferring patterns from structured representations (*see* Section 4.7). Subsequent chapters describe several additional models employing structured representations. Before turning to a specific model, however, let me consider structured representations in general.

A typical structured representation is given by natural language utterances like, "The dog chases the boy." The structure underlying this representation is the grammar of English, and the component parts are individual words. In an artificial language, like those typically used in computers, such a phrase may be represented with the structured representation `chases(dog, boy)`. Such a representation is structured because the position in which each term appears determines its grammatical role. The "verb" is first, the "agent" is second, and the "theme" is third. If the order of the elements changes, so does the meaning of the structured representation.

The fact that entire concepts, denoted by words, can be moved into different roles has historically posed a challenge for approaches that think of the representation of concepts as being patterns in a vector space (Fodor & Pylyshyn, 1988). In addition, because the kind of relation that the concept enters into changes depending on which role it is in, it is not immediately obvious how simple associative relationships can be used to capture such structures. The first of these challenges targets distributed connectionist models, and the second targets both localist and distributed connectionist models. In essence, both challenges boil down to the problem of determining how to construct a new representation that combines multiple semantic pointers playing different kinds of structural or syntactic roles. In short, how can representations be bound to roles within a structure?

As Barsalou (1999, p. 643) has commented:

> It is almost universally accepted now that representation schemes lacking conceptual relations, binding, and recursion are inadequate.

Consequently, it has been seen as a major challenge for neurally inspired architectures and theories of cognition to address the issue of how structure can be represented. For example, Barsalou's own perceptual symbol system theory, although addressing semantics well, has been criticized because (Edelman & Breen, 1999, p. 614)

> [The Theory] leaves the other critical component of any symbol system theory–the compositional ability to bind the constituents together–underspecified.

Unfortunately, Barsalou's discussion of "frames," while partly addressing the issue, does not describe how neural representations can be bound into such structures. So, despite recognizing the need for building such structures with binding in neural accounts, no precise computational suggestions have been broadly accepted. More generally, I am aware of no biologically realistic accounts that have been offered, demonstrated, and shown to scale to structures of the kind employed in human cognition.

4.2 ■ Binding Without Neurons

Connectionists, being concerned with cognitive modeling, have been working on ways of binding multiple representations since the 1990s. Binding itself has been a focus because, if we can bind vectors, then we can "tag" content vectors with structural role vectors and thereby encode structured representations. Perhaps the best known early proposal for vector binding is that offered by

Smolensky (1990). He suggested that a kind of vector multiplication called a "tensor product" could perform binding.

Despite the exotic-sounding name, tensor products are a straightforward solution to the problem of trying to multiply two vectors and then later get back one of the vectors given the other (i.e., trying to "divide" them). It is obvious how to do this if we work with scalars: $6 \times 2 = 12$ so, $12/2 = 6$ and $12/6 = 2$. Notice that to compute the unknown element given the other two (i.e., the answer and one of the multiples), we needed to use a different operation–in this case division. The question is, what happens if the mathematical objects we have are vectors? What is $(6,3) \times (2,5,4)$?

This is where tensor products come in. They are one definition of what we might mean by "$\times$" in the context of vectors, so we can have operations that mimic what we do with scalars. To distinguish this operation from regular multiplication, the "$\otimes$" sign is typically used. Specifically, tensor products are the result of multiplying *all* of the elements with one another. This is sometimes called the outer product. For example,

$$\begin{pmatrix} 6 \\ 3 \end{pmatrix} \otimes \begin{pmatrix} 2 & 5 & 4 \end{pmatrix} = \begin{pmatrix} 6\times 2 & 6\times 5 & 6\times 4 \\ 3\times 2 & 3\times 5 & 3\times 4 \end{pmatrix} = \begin{pmatrix} 12 & 30 & 24 \\ 6 & 15 & 12 \end{pmatrix}.$$

This way, if we are given the final matrix and one of the vectors, we can recover the other vector by using another operation, called the dot product (or inner product; *see* Appendix A.3):

$$\begin{pmatrix} 12 & 30 & 24 \\ 6 & 15 & 12 \end{pmatrix} \cdot \begin{pmatrix} 2 \\ 5 \\ 4 \end{pmatrix} = \begin{pmatrix} 2\times 12 + 5\times 30 + 4\times 24 \\ 2\times 6 + 5\times 15 + 4\times 12 \end{pmatrix}$$

$$= \begin{pmatrix} 270 \\ 135 \end{pmatrix} /45 = \begin{pmatrix} 6 \\ 3 \end{pmatrix}.$$

We have to divide by the length of the vectors we are querying[1] to account for originally multiplying vectors of a certain length. Often, researchers work with vectors that all have length one so they can skip this step, as I do from now on. We now have a well-defined equivalent to standard multiplication and division, but for vectors.

The relevance of tensor products to binding was evident to Smolensky: if we represent items with vectors, and structures with collections (sums) of bound vectors, then we can use tensor products to do the binding and still be able to "unbind" any of the constituent parts. Consider the previous example of `chases(dog, boy)`. To represent this structure, we need

[1] The length of each of these vectors is $\sqrt{2^2+5^2+4^2} = \sqrt{45}$ and $\sqrt{6^2+3^2} = \sqrt{45}$ so the overall scaling factor is $\sqrt{45} \times \sqrt{45} = 45$.

vectors representing each of the concepts and vectors representing each of the roles, which for this representation includes (I write vectors in bold): **dog, chase, boy, verb, agent, theme**. In the context of the SPA, each of these would be a semantic pointer.

To bind one role to one concept, we can use the tensor product:

$$\mathbf{b} = \mathbf{dog} \otimes \mathbf{agent}.$$

Now, the resulting vector **b** can be used to recover either of the original vectors given the other. So, if we want to find out what role the **dog** is playing, we can unbind it by using the dot product as before:

$$\mathbf{b} \cdot \mathbf{dog} = \mathbf{agent}.$$

Given a means of binding and unbinding vectors, we also need a means of collecting these bindings together to construct structured representations. Smolensky adopted the standard approach of simply summing to conjoin vectors. This results in a straightforward method for representing sentences. For example, a structured vector representation of the proposition "The dog chases the boy" is:

$$\mathbf{P} = \mathbf{verb} \otimes \mathbf{chase} + \mathbf{agent} \otimes \mathbf{dog} + \mathbf{theme} \otimes \mathbf{boy}.$$

In the case where all of the role vectors are orthogonal (*see* Appendix A.3),[2] decoding within this structure is similar to before–take the dot product of **P** with one of the roles. These tensor product representations are quite powerful for representing structure with vectors as elements of the structure. They can be combined recursively to define embedded structures, and they can be used to define standard LISP-like operations that form the basis of many cognitive models, as Smolensky (1990) has shown.

However, tensor products were never broadly adopted. Perhaps this is because many were swayed by Fodor and McLaughlin's contention that this was "merely" a demonstration that you could implement a standard symbolic system in a connectionist-like architecture (Fodor & McLaughlin, 1990). As such, it would be unappealing to symbolic proponents because it just seemed to make models more complicated, and it would be unappealing to connectionists because it seemed to lose the neural roots of the constituent representations. For example, Barsalou (1999, p. 643) commented:

> Early connectionist formulations implemented classic predicate calculus functions by superimposing vectors for symbolic elements in predicate calculus expressions (e.g., Pollack 1990; Smolensky 1990; van Gelder

[2] If they are not orthogonal, then either a decoding vector for each role needs to be computed (if the roles are linearly independent) or the results needs to be cleaned up, as discussed in Section 4.6.

> 1990). The psychological validity of these particular approaches, however, has never been compelling, striking many as arbitrary technical attempts to introduce predicate calculus functions into connectionist nets.

I suspect that both of these concerns have played a role.

However, there is also an important technical limitation to tensor products, one that Smolensky himself acknowledges (1990, p. 212):

> An analysis is needed of the consequences of throwing away binding units and other means of controlling the potentially prohibitive growth in their number.

He is referring here to the fact that the result of binding two vectors of lengths n and m, respectively, gives a new structure with nm elements. If we have recursive representations, then the size of our representations will quickly grow out of hand. In short, tensor product representations do not *scale* well (*see* Appendix D.5 for details). This also means that if we are constructing a network model, then we need to have some way of handling objects of different sizes within one network. We may not even know how "big" a representation will be before we have to process it. Again, it is not clear how to scale a network like this.

I believe it is the scaling issue more than any other that has resulted in sparse adoption of Smolensky's suggestion in the modeling community. However, other theorists have attacked the scaling problem head on, and now there are several methods for binding vectors that avoid the issue. For binary representations, there are Pentti Kanerva's Binary Spatter Codes or BSCs (Kanerva, 1994). For continuous representations, there are Tony Plate's Holographic Reduced Representations or HRRs (Plate, 1994, 2003). And, also for continuous representations, there are Ross Gayler's Multiply-Add-Permute, or MAP, representations (Gayler, 1998). In fact, all of these are extremely similar, with BSCs being equivalent to a binary version of HRRs and both HRRs and BSCs being a different means of compressing Smolensky's tensor product (*see* Fig. 4.1).

Noticing these similarities, Gayler suggested that the term "Vector Symbolic Architectures" (VSAs) be used to describe this class of closely related approaches to encoding structure using distributed representations (Gayler, 2003). Like Smolensky's tensor product representations, each VSA identifies a binding operation and a conjoining operation. Unlike tensor products, these newer VSAs do not change the dimensionality of representations when encoding structure.

There is a price for not changing dimensionality during binding operations: degradation of the structured representations. If we constantly encode more and more structure into a vector space that does not change in size, then we must eventually lose information about the original structure. This, of course,

$$(1)\quad \begin{pmatrix} A \\ B \\ C \\ D \end{pmatrix} \otimes \begin{pmatrix} E \\ F \\ G \\ H \end{pmatrix} = \begin{pmatrix} AE & AF & AG & AH \\ BE & BF & BG & BH \\ CE & CF & CG & CH \\ DE & DF & DG & DH \end{pmatrix}$$

$$(2)\quad \begin{pmatrix} A \\ B \\ C \\ D \end{pmatrix} \times \begin{pmatrix} E \\ F \\ G \\ H \end{pmatrix} = \begin{pmatrix} AE \\ BF \\ CG \\ DH \end{pmatrix}$$

$$(3)\quad \begin{pmatrix} 1 \\ 0 \\ 1 \\ 0 \end{pmatrix} \oplus \begin{pmatrix} 1 \\ 1 \\ 0 \\ 0 \end{pmatrix} = \begin{pmatrix} 0 \\ 1 \\ 1 \\ 0 \end{pmatrix}$$

$$(4)\quad \begin{pmatrix} A \\ B \\ C \\ D \end{pmatrix} \circledast \begin{pmatrix} E \\ F \\ G \\ H \end{pmatrix} = \begin{pmatrix} AE + BH + CG + DF \\ AF + BE + CH + DG \\ AG + BF + CE + DH \\ AH + BG + CF + DE \end{pmatrix}$$

FIGURE 4.1 Various vector binding operations. (1) Tensor products are computed as a multiplication of each element of the first vector with each element of the second vector. (2) Piecewise multiplication, used in Gayler's Multiply-Add-Permute (MAP) representation, multiplies each element of the first vector with one corresponding element in the second vector. (3) Binary Spatter Codes (BSCs) combine binary vectors using an exclusive or (XOR) function. (4) Circular convolution is the binding function used in Holographic Reduced Representations (HRRs). An overview of circular convolution and a derivation of its application to binding and unbinding can be found in Appendix D.1. Notice that (2), (3), and (4) have as many elements in the result as in either of the bound vectors, escaping the scaling issues of (1).

is a kind of compression. Representations that slowly lose information in this manner have been called "reduced" representations, to indicate that there is less information in the resulting representation than in the original structures they encode (Hinton, 1990). Thus the resulting structured representations do not explicitly include all of the bound components: if they did, there would be no information reduction and no compression. This information reduction has important consequences for any architecture, like the SPA, that employs them.

For one, it means that an architecture employing reduced representations cannot be a classical architecture. In short, the reason is that the representations in such an architecture are not perfectly compositional. This is a point I return to in Sections 8.2.1.2 and 9.2.1. Notice also that as a result of losing information during binding, any unbinding operation must return an *approximation* of the originally bound elements. This means that the results of unbinding must be "cleaned up" to a vector that is familiar to the system. Consequently, use of a VSA imposes a functional demand for a clean-up memory–that is, a memory that maps noisy versions of allowable representations onto actually allowed representations. In the previous example, all of the vectors in the structure (i.e., **boy**, **dog**, etc.) would be in the clean-up memory, so the results of unbinding could be "recognized" as one of the allowable representations. I return to this important issue in Section 4.6.

In any case, what is clear even without detailed consideration of clean-up is that there are limits on the amount of embedded structure that a reduced representation can support before noise makes the structure difficult to decode, resulting in errors. As we will see, the depth of structure that can be encoded depends on the dimensionality of the vectors being used, as well as the total number of symbols that can be distinguished.

Ultimately, however, I think it is this capacity for error that makes VSAs–and hence the SPA itself–psychologically plausible. This is because the architecture does not define an idealization of cognitive behavior like classical architectures tend to (consider, e.g., compositionality in Section 8.2.1.2) but, rather, specifies a functional, implementable system that is guaranteed to eventually fail. I take it that capturing the failures as well as the successes of cognitive systems is truly the goal of a cognitive architecture. After all, many experiments in cognitive psychology are set up to increase the cognitive load to determine what kinds of failure arise.

Now to specifics. In this book, I adopt Plate's HRR representations because they work in continuous spaces, as neurons do, and because the binding and unbinding operations are very similar (Plate, 2003). Still, most of the examples I provide do not depend on the VSA chosen. In fact much work remains to be done on the implications of this choice. In particular, it would be interesting to determine in more detail the expected behavioral differences between HRRs and MAPs. However, such considerations are beyond the scope of the present discussion. It is worth noting, however, that the structure of the SPA itself does not depend on the particular choice of a VSA. I take any method for binding that depends on a piecewise (i.e., local) nonlinearity in a vector space to be able to support the main ideas behind the SPA. Nevertheless, it is necessary to choose an appropriate VSA to provide specific examples, so I have chosen HRRs.

In the remainder of this chapter, I turn to implementing the architectural components necessary to support HRR processing in a biologically realistic setting. First, I consider binding and unbinding. I then turn to demonstrating how these operations can be used to manipulate (as well as encode and decode) structured representations, and I show how a spiking neural network can learn such manipulations. I then return to the issue of implementing a clean-up memory in neurons and discuss some capacity results that suggest that the SPA scales well. Finally, I describe an SPA model that performs structural inference, mimicking human performance in the Raven's Progressive Matrices test of general fluid intelligence.

4.3 ■ Binding With Neurons

The relevance of VSAs to what the brain does is far from obvious. As Hadley (2009) has recently noted: "It is unknown whether or how VSAs could be physically realized in the human brain" (p. 527). However, to perform binding with neurons, we can combine the above characterization of vector binding with my earlier characterizations of vector representation (Section 2.4) and transformation (Section 3.8). Notably, each of the VSA methods for binding rely on an element-wise nonlinearity (*see* Fig. 4.1), so we need to perform a nonlinear transformation. In the case of HRRs, in particular, we need to perform a linear transformation and then an element-wise nonlinearity–we need to multiply scalars, which was covered at the end of the last tutorial (Section 3.8).

Specifically, the binding operation for HRRs is called "circular convolution" and has the operator symbol "$\circledast$." The details of circular convolution can be found in Appendix D.1. What is important for our purposes is that the binding of any two vectors **A** and **B** can be computed by

$$\mathbf{C} = \mathbf{A} \circledast \mathbf{B} = \mathbf{F}^{-1}(\mathbf{FA}.\mathbf{FB})$$

where "." is used to indicate element-wise multiplication of the two vectors (i.e., $\mathbf{x}.\mathbf{y} = (x_1y_1, \ldots, x_ny_n)$). The matrix **F** is a linear transformation (specifically, a Fourier transform) that is the same for all vectors of a given dimension (*see* Appendix A.5). The resulting vector **C** has the same number of dimensions as **A** and **B**. The network that computes this binding is a standard, two-layer feedforward network (*see* Fig. 4.2).

Because we have already seen examples of both linear transformation and the multiplication of two numbers, it is straightforward to build this network in Nengo. The tutorial at the end of this chapter gives detailed instructions on building a binding network to encode structured representations (Section 4.9). Figure 4.3 shows the results of binding four sets of vectors using the circular convolution method in spiking neurons. The approximate nature of the binding is evident from the "clouds" of bindings.

To unbind vectors, the same circular convolution operation can be used. However, we need to compute the inverse of one of the bound vectors to get the other. This is analogous to computing division (e.g., 6/2) as multiplication with an inverse (e.g., $6 \times \frac{1}{2}$). Conveniently, for HRRs a good approximation to the inverse can be found by a simple linear transformation (*see* Appendix D.1). Calling this transformation **S**, we can write the unbinding of any two vectors as

$$\mathbf{A} \approx \mathbf{C} \circledast \mathbf{B}' = \mathbf{F}^{-1}(\mathbf{FC}.\mathbf{FSB})$$

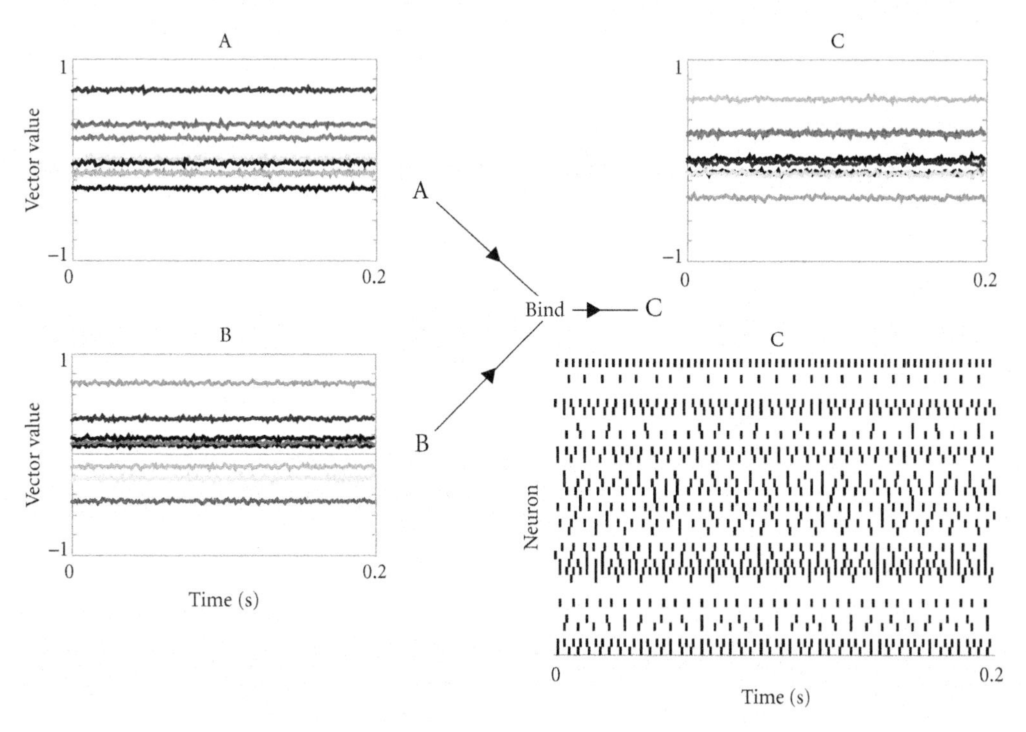

FIGURE 4.2 The network architecture of a binding network. The network is a simple two-layer, feedforward network, as there are connection weights into the Bind and C layers. In this particular network, **A**, **B**, and **C** have 150 neurons and Bind has 760 neurons. This network is binding two 8D vectors projected into the A and B neurons, whose decoded values are shown on the left side. These neurons project to the Bind layer, which forms a representation that allows the computation of the necessary nonlinearities. The spiking activity is shown for 38 randomly selected Bind neurons. This activity drives the C layer, which extracts the binding from that representation. The decoding of the C-layer vector is shown in the top right graph, as another 8D vector. The results for 200 ms of simulation time are shown.

where $'$ indicates the approximate inverse. Recall that the result of such a computation will be noisy and so must be cleaned up, as I discuss in more detail shortly (Section 4.6).

From a network construction point-of-view, we can redeploy exactly the same network as in the binding case and simply pass the second vector through the linear transformation **S** before presenting it to the network (this amounts to changing the weights between the B and Bind layers in Fig. 4.2). Figure 4.4 demonstrates, using the output of the network in Figure 4.3, that vectors can be effectively unbound in this manner.

It is perhaps unsurprising that we can implement binding and unbinding in neural networks, given the resources of the NEF. However, this does not

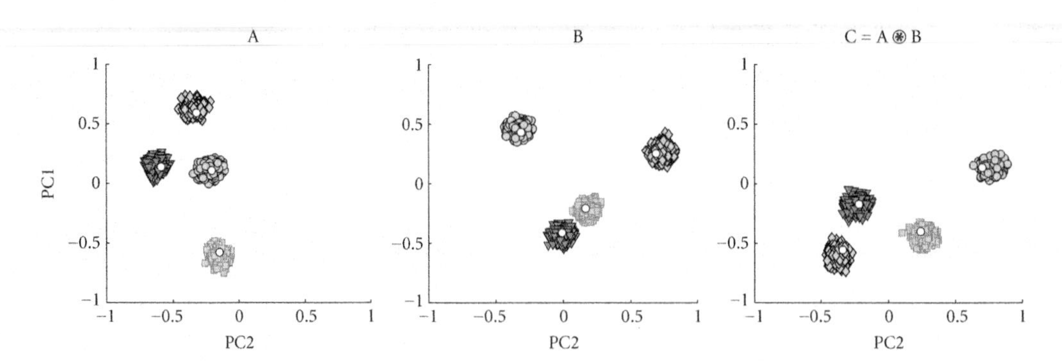

FIGURE 4.3 Binding vectors with spiking neurons. A 10D vector **A** is bound with a 10D vector **B** to produce a 10D vector **C**. This figure shows four separate instances of binding. In each instance, instantaneous decoded samples are drawn every 10 ms from a neural representation of vectors **A**, **B**, and **C** over a period of one second. This forms a collection of points that are plotted with a unique marker (e.g., the light gray circles in the first plot were bound with the light gray circles in the second plot resulting in the group of light gray circles in the third plot). These samples form a "cloud" because of the neural variability over time caused by spiking and other sources of noise. To visualize the 10D vectors used in this process, vectors are reduced to 2D points using principal component analysis (PCA) to preserve as much of the variability of the higher-dimensional data as possible. The first two principal components (PCs) define the x and y axes. Similar vectors thus map to points that are close together in these plots. Mean values of the groups of points are given by white circles.

allay concerns, like those expressed by Barsalou, that this is merely a technical trick of some kind that has little psychological plausibility. I believe the psychological plausibility of these representations needs to be addressed by looking at larger-scale models, the first of which we see later in this chapter (Section 4.7).

We might also be concerned about the *neural* plausibility of this approach. Three possible sources of concern regarding the neural plausibility of this implementation are: (1) the specific nonlinearity employed; (2) the connectivity of the network; and (3) determining how well the network scales to large lexicons. Let me consider each in turn.

The nonlinearity that is needed for the binding operation is computing the product of two scalar input signals. We have demonstrated in the tutorials how this can be done accurately using only linear dendrites, and with a small number of neurons (Section 3.8). The assumption of linear dendrites is, in fact, a limiting assumption, not a beneficial one. I make it here because it is the "textbook" assumption about dendritic processing.

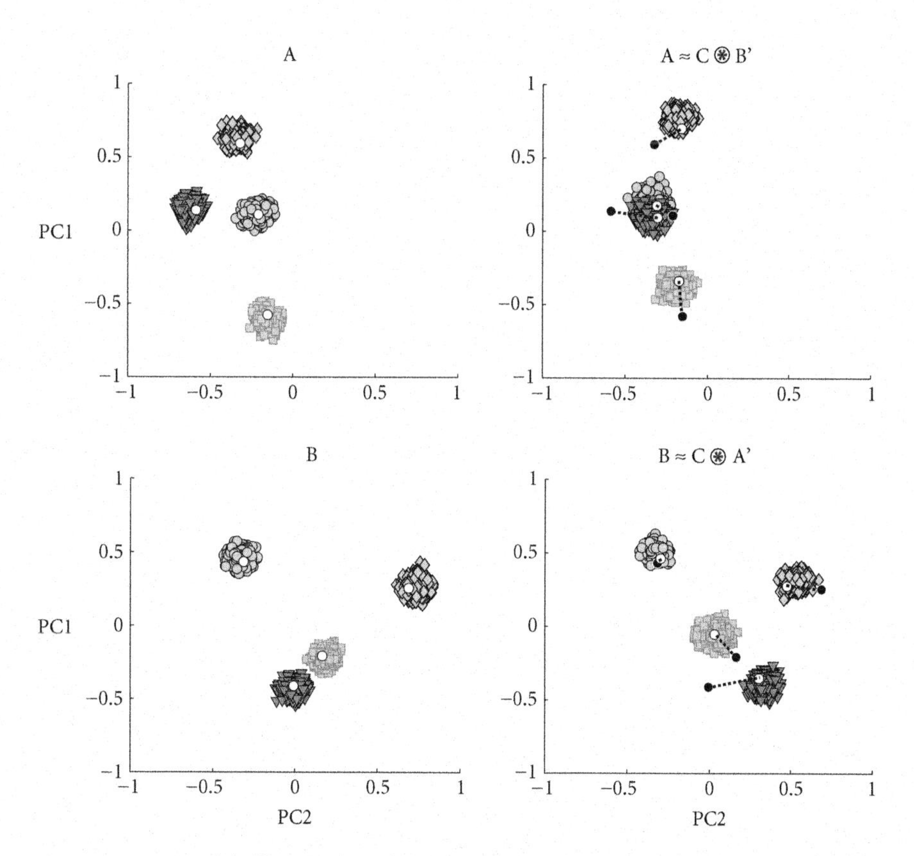

FIGURE 4.4 Unbinding vectors with neurons. The reconstructed vectors **A** and **B** were obtained by using the transformations $\mathbf{A} \approx \mathbf{C} \circledast \mathbf{B}'$ and $\mathbf{B} \approx \mathbf{C} \circledast \mathbf{A}'$, respectively. This figure uses the same initial vectors and graphing methods as Figure 4.3. The original **A** and **B** vectors have been reproduced alongside the reconstructed vectors for ease of comparison. In the graphs of the reconstructed vectors, black circles indicate the original average values of the vectors and the dotted lines indicate the distance to the new averages, which is the decoding error.

However, much recent work has strongly suggested that pyramidal cells (the most common cortical cells) have nonlinear dendritic interactions. Consequently, an alternative model of cortical pyramidal cells has been proposed based on physiological studies reporting that dendritic spikes evoked by focal synaptic stimulation often remain confined to a single dendritic branch (Schiller et al., 2000) and that within-branch stimulation from two electrodes can produce sublinear, linear, or superlinear responses, depending on the placement and timing of the stimulation (Polsky et al., 2004). This alternative model suggests that thin basal dendrites of pyramidal cells, where the majority of excitatory synapses are located, act as processing subunits, where the activity of each branch is determined by its own sigmoidal nonlinearity (Poirazi et al., 2003). A second stage of processing *then* occurs at the soma, similar to the standard linear model, where the input is a weighted summation of the subunit activities that determines the neuron's activity. This two-stage process has been observed in a study of subthreshold synaptic integration in basal dendrites of neocortical cells, where stimulation within a single terminal dendritic branch evoked a sigmoidal subunit nonlinearity, but the activity summed linearly when different terminal branches were stimulated (Schiller et al., 2000). In effect, this provides evidence that a single neuron may have the processing power of a two-layer network (although the connectivity is more restricted).

If our binding networks employed such nonlinearities, then we could eliminate the middle layer currently used to compute the nonlinearity. Consequently, the binding networks would use significantly fewer neurons than in the examples I consider here. So, the examples I present are worst case scenarios for the number of neurons that are needed to generate structured representations. In Section 5.5 I consider a model that uses nonlinear dendrites.

The second concern regarding neural plausibility is connectivity. That is, we know that in general cortical circuits are fairly locally and sparsely connected (Song et al., 2005; Hellwig, 2000; Lund et al., 1993). Do these binding networks respect this constraint? To begin, we need to be more accurate about what we mean by locally connected. In their anatomical work that involves microinjections of a tracer across several areas of monkey cortex, the Levitt group has found that small injections of up to 400 μm (i.e., 0.4 mm) spread to areas of approximately 3 mm by 3 mm (i.e., 9 mm^2) or larger (Lund et al., 1993). There is an interesting structure to this projection pattern, in which there are patches of about 300–400 μm in diameter separated by spaces of a similar size. These patches correspond well to the sizes of dendritic fields of the pyramidal cells in these areas. Given that there are 170,000 neurons per mm^2 of cortex, the injections affect about 25,000 neurons and spread to about one million others. In addition, a single neuron is typically thought to have 5,000–50,000 input and output connections (with a mean of 10,000). So "local" connectivity seems to be local within about 9 mm^2. and to about 5% of the neurons in that region. This pattern of connectivity has been found throughout visual, motor, somatosensory, and prefrontal cortex, and across monkeys, cats, and rodents.

The plausibility of connectivity is closely tied to the third issue of scaling. To determine whether a binding network can be plausibly fit within this cortical area, it is important to know what the dimensionality of the bound vectors is going to be. It is also important to know how the size of the binding network changes with dimensionality. I discuss in detail the dimensionality of the bound vectors in Section 4.6, because the necessary dimensionality is determined by how well we can clean up vectors after binding. There I show that we need vectors of about 500 dimensions to capture large structures constructed with an adult-sized vocabulary. It is also clear that the binding network itself scales linearly with the number of dimensions because the only nonlinearity needed is the element-wise product. Thus, one additional dimension means computing one additional product. Recall from Section 3.8 that 70 neurons per multiplication results in good-quality estimates of the product, and four multiplications are needed per dimension. Taken together, these considerations suggest that about 140,000 neurons are needed to bind two 500D vectors. Similarly using 70 neurons per dimension to represent the input will provide good representation and use 35,000 neuron per input. The Fourier transform requires all-to-all connectivity, but this dense connectivity between two populations of 35,000 is well within the maximum of 50,000 connections per cell. After the transform, the binding itself requires only highly local connectivity into the binding layer as there are only interactions between at most two dimensions. In sum, binding networks of the appropriate size to capture human-like structured representations can be constructed in this manner.

Recalling that there are about 170,000 neurons per mm^2 of cortex, suggesting that we need about 1 mm^2 of cortex to perform the necessary binding. The projection data suggest that local projections cover at least 9 mm^2 of cortex, so the binding layer can fit comfortably within this area. If we want to include the two input populations, the binding layer, and the output population with similar assumptions, then we would need at most 2 mm^2 of cortex. Recall that the unbinding network requires the same resources. Consequently, these networks are consistent with the kind of connectivity observed in cortex. As well, the architecture scales linearly so this conclusion is not highly sensitive to the assumptions made regarding the dimensionality of the representations. It is worth recalling that if we allow dendritic nonlinearities, the binding layer is not needed, so the entire network would require less cortex still.

Together, these considerations suggest that the vector binding underlying the SPA can be performed on the scale necessary to support human cognition in a neurally plausible architecture.

4.4 ■ Manipulating Structured Representations

To this point, my discussion of syntax in the SPA has focused on how structures can be encoded and decoded using binding. However, to make such representations cognitively useful, they must be *manipulated* to support reasoning. An interesting feature of VSA representations is that they can often be manipulated without explicitly decoding the elements. This has sometimes been called "holistic" transformation, because a vector representing a sentence can be manipulated in a way that affects the entire sentence at once without individually manipulating the entities it encodes.

Several researchers have demonstrated transformations of structured representations using distributed representations, although experiments have been limited to fairly simple transformations, and the transformations tend to be hand-coded.[3] Here, I begin by considering simple, hand-coded transformations as well but demonstrate in the next section how a broad class of transformations can be learned.

From the discussion in the previous section, I take it as evident that the binding of two semantic pointers results in a semantic pointer as well. Convolutional binding, after all, is a kind of compression, although the result of binding lies in the same vector space as the bound elements. Conveniently, performing transformations of semantic pointers that encode structure in this way relies on binding as well. Consequently, the same network architecture shown in Figure 4.2 can be used to perform syntactic manipulation. To see how these manipulations work, let us return to the `chases(dog, boy)` example, where this sentence is encoded into the semantic pointer **P** as

$$\mathbf{P} = \mathbf{verb} \circledast \mathbf{chase} + \mathbf{agent} \circledast \mathbf{dog} + \mathbf{theme} \circledast \mathbf{boy}.$$

Recall that given this encoding, we can decode elements by binding **P** with the appropriate inverses. For example, if we multiply **P** by the approximate inverse of **chase** (i.e., **chase**′) we get:

$$\begin{aligned}\mathbf{chase}' \circledast \mathbf{P} &= \mathbf{chase}' \circledast (\mathbf{verb} \circledast \mathbf{chase} + \mathbf{agent} \circledast \mathbf{dog} + \mathbf{theme} \circledast \mathbf{boy}) \\ &= \mathbf{chase}' \circledast \mathbf{verb} \circledast \mathbf{chase} + \mathbf{chase}' \circledast \mathbf{agent} \circledast \mathbf{dog} \\ &\quad + \mathbf{chase}' \circledast \mathbf{theme} \circledast \mathbf{boy} \\ &= \mathbf{verb} + \mathbf{noise} + \mathbf{noise} \\ &\approx \mathbf{verb}.\end{aligned}$$

So we can conclude that the "chase" plays the role of a verb in **P**.

A few comments are in order. First, this is an algebraic demonstration of the effects of binding **chase**′ to **P** and does not require the decoding of any of the other sentence elements (i.e., we don't have to "remove" the agent and theme). Second, this demonstration takes advantage of the fact that $\circledast$ can be treated like regular multiplication as far as allowable operations are concerned (i.e., it is communicative, distributive, and associative). Third, I have used the ′ sign to indicate the pseudo-inverse and distinguish it from a true inverse. This pseudo-inverse is computed by a simple linear operation (a shift), as mentioned earlier (Section 4.3). Fourth, **chase**′ $\circledast$ **verb** $\circledast$ **chase** is replaced

[3] Niklasson and van Gelder (1994) demonstrated how to transform a logical implication into a disjunction, as did Plate (1994); Pollack (1990) transformed reduced representations of the form *loved(x,y)* to the form *loved(y,x)*; Legendre et al. (1994) showed how to transform representations for active sentences into representations for passive sentences (and vice versa).

by **verb** because a vector times its inverse is equal to one. Or, more accurately, a vector convolved with its pseudo-inverse is approximately equal to one, and any difference is absorbed by the **noise** term.

Finally, the last two terms are treated as noise because circular convolution has the important property that it maps the input vectors to an approximately orthogonal result (*see* Appendix A.3). That is, the result $\mathbf{w} = \mathbf{x} \circledast \mathbf{y}$ is unlike either **x** or **y**. By "unlike" I mean that the dot product (a standard measure of similarity) between **w** and **x** or **y** is close to zero. Consequently, when we add in these new but unfamiliar items from the last two terms, they will slightly blur the correct answer, but not make it unrecognizable. In fact, the reason we really think of these as noise is because they will not be similar to any known lexical item (because they are approximately orthogonal to all known items; Plate, 2003) and hence during the clean-up operation will be ignored. Once again, it is clear that clean-up, although not an explicit part of most VSAs, plays a central role in determining what can be effectively represented and manipulated. This is why it is a central part of my discussion of the SPA (*see* Section 4.6).

We can think of this simple decoding procedure as a kind of manipulation of the structure encoded in **P**. More interestingly, we can perform non-decoding manipulations, such as switching roles and changing the relation. Consider the hand-constructed semantic pointer **T**:

$$\mathbf{T} = \mathbf{agent}' \circledast \mathbf{theme} + \mathbf{chase}' \circledast \mathbf{hug} + \mathbf{theme}' \circledast \mathbf{agent}.$$

Convolving our original representation of the sentence **P** with **T** produces:

$$\mathbf{T} \circledast \mathbf{P} = \mathbf{verb} \circledast \mathbf{hug} + \mathbf{agent} \circledast \mathbf{boy} + \mathbf{theme} \circledast \mathbf{dog} + \mathbf{noise}.$$

That is, we now have a representation of `hug(boy, dog)`. This occurred because the first term in **T** converts whatever was the **agent** into the **theme** by removing the **agent** vector (since $\mathbf{agent}' \circledast \mathbf{agent} \approx 1$) and binding the result with **theme**. The last term works in a similar manner. The second term replaces **chase** with **hug**, again in a similar manner. The **noise** term is a collection of the noise generated by each of the other terms. Crucially, the original semantic pointer **P** did not have to be decoded into its constituents to affect this manipulation.

For a demonstration of the usefulness of such manipulations, we can consider a simple task that requires syntactic manipulation: question answering. Figures 4.5 and 4.6 show the same network of 10,000 neurons answering four different questions about different sentences. For each sentence and question, a different input is provided to the same network for 0.25 seconds.[4] In Figure 4.6, the results of clean-up are shown over time, so the similarity

[4] The length of time is not intended to be reflective of timing in a task with human subjects. For this purpose, such an operation must be embedded in an architecture that accounts

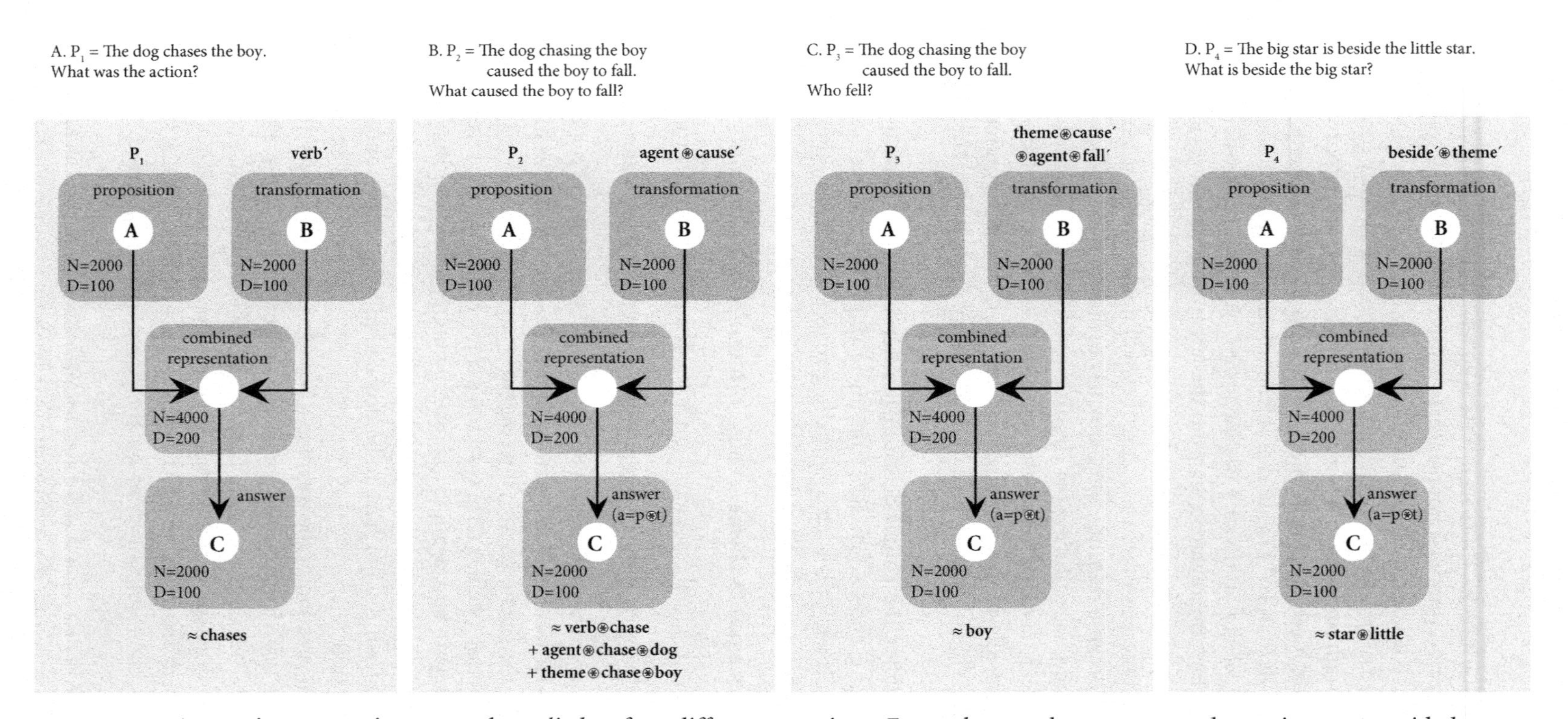

FIGURE 4.5 A question-answering network applied to four different questions. For each case, the sentence and question are provided as separate inputs (**A** and **B**). Notably, the network is in no way modified during this task, so this one network is capable of answering any question of a similar form using these lexical items. Behavior of the network over time is shown in Figure 4.6.

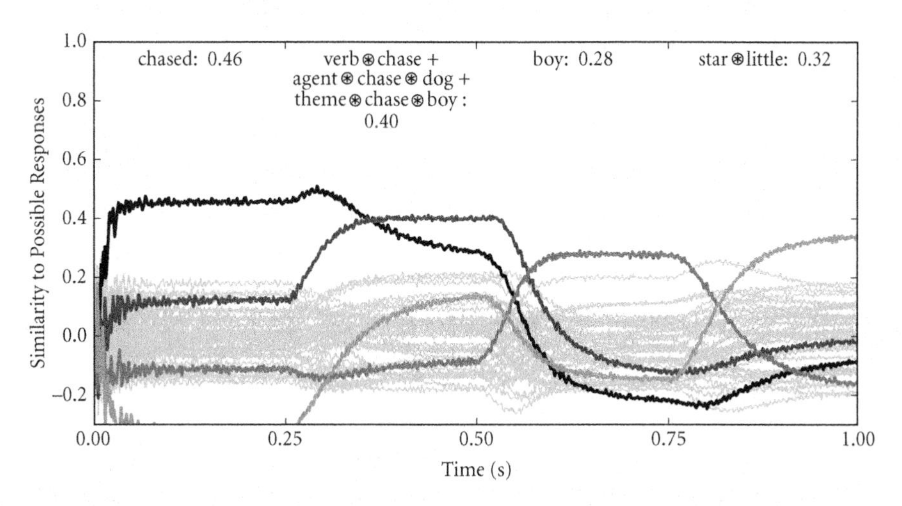

FIGURE 4.6 The network answering four questions over time. Each question is presented for 0.25 seconds by changing the input to the neural groups representing vectors **A** and **B** in Figure 4.5. The resulting spiking behavior of neural group **C** is interpreted as a semantic pointer and compared to the possible responses. This plot shows the similarity of the vector **C** to possible responses (i.e., the top line in the plot represents the output of a clean-up memory). Dark gray through medium gray lines indicate the similarity to the four correct responses. The lightest gray lines are the similarities of the output with 40 randomly chosen other semantic pointers in the lexicon, representing other possible responses. Because the output is closer to the correct response than any other answer in each case, this network successfully answers all of the questions.

between the output of the network and possible responses is plotted. The system is considered to have given a correct response if its output is more similar to the correct response than any other possible response.

To understand these results, let us consider each question in more detail. The first case is nearly identical to the first transformation we considered above. Rather than determining the role of **chase**, the network determines what the action of the sentence is.

Turning to the second question, we see that the sentence has a more complex, embedded structure. To capture such structures, we need a way to map natural language sentences onto semantic pointers. For the second case in Figure 4.5, the roles "agent" and "theme" are multiply instantiated. Consequently, roles need to be tagged with the verb they are roles for, so the items they are bound to do not get confused. For example, if the roles were not

for response times. The Spaun model presented in Chapter 7 provides an example of question answering in this kind of context.

tagged, it would be difficult to distinguish whether **dog** was the agent of **chase** or of **fall**. This kind of role tagging can be accomplished through binding as follows:

$$\begin{aligned}\mathbf{P}_2 \quad = \quad & \mathbf{verb} \circledast \mathbf{cause} + \mathbf{agent} \circledast \mathbf{cause} \circledast (\mathbf{verb} \circledast \mathbf{chase} + \\ & \mathbf{agent} \circledast \mathbf{chase} \circledast \mathbf{dog} + \mathbf{theme} \circledast \mathbf{chase} \circledast \mathbf{boy}) + \\ & \mathbf{theme} \circledast \mathbf{cause} \circledast (\mathbf{verb} \circledast \mathbf{fall} + \mathbf{agent} \circledast \mathbf{fall} \circledast \mathbf{boy}).\end{aligned}$$

This additional tagging of elements by their verb does not affect previous results (i.e., I could have tagged all roles with the verb but did not for clarity). Here, as in other more sophisticated linguistic structures, it is essential to remove this simplification and distinguish different roles to preserve the appropriate structure. Using this encoding, we can now answer the question "What caused the boy to fall?" by convolving $\mathbf{P}_2$ with $\mathbf{agent} \circledast \mathbf{cause}'$ and the correct answer is provided.

The third case presents a more challenging kind of question that queries the contents of subphrases that make up the sentence–that is, "Who fell?" This question is asked in the third case by determining the $\mathbf{agent} \circledast \mathbf{fall}$ of the $\mathbf{theme} \circledast \mathbf{cause}$ all at once. That is,

$$\begin{aligned}& \mathbf{P}_2 \circledast (\mathbf{theme} \circledast \mathbf{cause} \circledast \mathbf{agent} \circledast \mathbf{fall})' \\ & \approx \mathbf{boy}.\end{aligned}$$

The same answer would result from applying the two transformations in series.

Finally, we can consider the fourth case from Figure 4.5. Here the system answers the question "What is beside the big star?" for the sentence "The big star is beside the little star." This example was chosen to demonstrate that there is no inherent "problem of two" for this approach (*see* Section 1.3). This is not surprising because representation of the same object twice does not result in representations that cannot be distinguished. This case is structurally identical to the first case, with the difference that there are modifiers ("big" and "little") bound to the nouns.

It is perhaps useful at this point to reiterate that the representations being manipulated in these examples are not at all like classical symbols. Each of the semantic pointers discussed above should be taken as distributed, semantic representations, just like those found in the visual model presented earlier (Section 3.5). Although it is helpful for me to refer to them by names like "dog" and "boy," the representations themselves can be dereferenced to extract relevant lexical, perceptual, and so forth, features. As demonstrated in detail in the context of a fuller model in Chapter 7, the same semantic pointers are used in such structured representations as are used for encoding low-level perceptual features.

To conclude this section, it is worth noting two important challenges that are raised by considering the manipulation of language-like representations. First, there is the issue of mapping language onto a VSA representation like that used by the SPA. There are important decisions regarding how we should represent complex structure that are insufficiently addressed here. I have employed a representation that has proven useful for several models (including Eliasmith, 2005b; Stewart et al., 2011; Stewart & Eliasmith, 2011b; Rasmussen & Eliasmith, 2011a) but does not have a solid linguistic justification.

The second closely related challenge is how natural language questions should be mapped to transformations of the encoded representational structures. There are multiple possible ways to affect such transformations. For example, in this chapter I have been considering transformations that are bound to the representations in one step. However, it is quite likely that in many circumstances, the way questions are answered is partly determined by how representations are moved through the system. If the question demands deep semantic analysis, for example, then the employed pointers need to be dereferenced. A simple syntactic manipulation will not suffice. Consequently, the next chapter, on control structures in the SPA, is relevant for understanding how such question answering is, in fact, performed in a biological system. In short, I want to be clear that the treatment provided here is self-consciously superficial. I do not want to suggest that all reasoning can be accomplished by syntactic manipulation of the kind presented here, although I do think such manipulation is likely important to much structure-based cognition.

4.5 ■ Learning Structural Manipulations

Unlike most connectionist approaches, the NEF does not rely on a learning process to construct model networks (although it can, *see* Chapter 6). This is demonstrated, for example, by the binding networks derived above. However, learning clearly plays a large role in cognition, and is important for explaining central features of cognition such as fluid intelligence and syntactic generalization (which I consider in more detail in Sections 4.7 and 6.6, respectively).

In my previous example of structural manipulation, as in most past work on syntactic transformations, the transformation vectors **T** are hand-picked based on previous knowledge of the vectors and the ways in which they are combined. This is generally undesirable from a cognitive modeling perspective, because it requires significant design decisions on the part of the modeler, and these tend to be driven by particular examples. It is even more undesirable from a psychological perspective because one of the distinctive features of cognition is the ability to generate new rules given past experience. This, of course, is called *induction*.

Inductive reasoning is the kind of reasoning that proceeds from several specific examples to a general rule. Unlike deductive reasoning, which forms the foundation of classical logic and guarantees the truth of any correctly derived conclusion, inductive reasoning can be correctly used to generate rules that may well be false. Nevertheless, induction is perhaps the most common kind of reasoning used both in science, and in our everyday interactions with the world.

In the language of the SPA, performing induction is equivalent to deriving a transformation that accounts for the mapping between several example structure transformations. For example, if I tell you that "the dog chases the boy" maps to "the boy chases the dog," and that "John loves Mary" maps to "Mary loves John" and then provide you with the structure "the bird eats the worm," then you would probably tell me that the same map should give "the worm eats the bird" (semantic problems notwithstanding). In solving this problem, you use induction over the syntactic structure of the first two examples to determine a general structural manipulation that you then apply to a new input. We would like our architecture to do the same.

Intuitively, to solve the above example we are doing something like generating a potential transformation given information only about the pre- and post-transformation structures. Neumann (2001) presented an early investigation into the possibility of learning such structural transformations using a VSA. She demonstrated that with a set of example transformations available, a simple error minimization would allow extraction of a transformation vector. In short, she showed that if we had examples of the transformation, we could infer what the transformation vector is by incrementally eliminating the difference between the examples and the transformations we calculated with our estimate of **T** (*see* Appendix D.2).

A simple extension to her original rule that allows the transformation to be calculated online[5] is the following (Eliasmith, 2004):

$$\mathbf{T}_{i+1} = \mathbf{T}_i - w_i \left[\mathbf{T}_i - \left(\mathbf{A}_i' \circledast \mathbf{B}_i\right)\right]$$

where i indexes the examples, **T** is the transformation vector we are trying to learn, w is a weight that determines how important we take the current example to be compared to past examples, and **A** and **B** are the pre- and post-transformation structured vectors provided as input. This rule essentially estimates the transformation vector that would take us from **A** to **B** and uses this to update our current guess of the transformation. Specifically, the

[5] Online learning is often distinguished from batch learning. The former uses one example at a time to learn, whereas the latter needs many examples at the same time to be employed. Online learning is typically taken to be more psychologically plausible.

rule computes the convolution between the inverse of the pre-transformation vector **A** and the post-transformation vector **B**. That is, it assumes that $\mathbf{B} = \mathbf{T} \circledast \mathbf{A}$, and hence $\mathbf{T} = \mathbf{A}' \circledast \mathbf{B}$. The result of that computation is subtracted from the current estimate of the transformation vector $\mathbf{T}_i$ to give an error. If that error is zero, then the transformation has been learned. If it is not zero, then the rule uses that difference to update the current transformation to something different (specifically, to a running average of inferred transformations given the provided inputs and outputs).

Although simple, this rule is surprisingly powerful, as I show in Section 4.7. Before doing this, however, I need to return to the issue of clean-up memory, which is also used in that model.

4.6 ■ Clean-Up Memory and Scaling

The result of applying transformations to structured semantic pointers is usually noisy. This is true for any compressed representation employed in VSAs. In the case of HRRs, the noisy results are in a high-dimensional continuous vector space, and they must be "cleaned up" to the nearest allowable representation. Most typically, we can think of "allowable" representations as those vectors that are associated with words in a lexicon. However, they might also be sentences, subphrases, or other structured representations that are pertinent to the current context. Ultimately, what makes a representation allowable is its inclusion in a clean-up memory.

As is clear from the examples in Section 4.4, the more complex the manipulation and the more elements in a structure, the more noise there is in the results of a manipulation. Consequently, to demonstrate that the SPA is feasible in its details, it is crucial to show that a scalable, effective clean-up memory can be constructed out of biologically plausible neurons. Two researchers in my lab, Terry Stewart and Charlie Tang, have worked out the details of such a memory (Stewart et al., 2010). Here I describe some of the highlights of that work. I begin by describing the clean-up memory itself and then turn to issues of scaling.

The particular implementation I discuss here is relevant for non-SPA architectures as well. This is because mapping a noisy or partial vector representation to an allowable representation plays a central role in several other neurally inspired cognitive models (Sun, 2006; Anderson & Lebiere, 1998; Pollack, 1988). Indeed, there have been several suggestions as to how such mappings can be done, including Hopfield networks, multilayer perceptrons, or any other prototype-based classifier: in short, any type of auto-associator can fulfill this role.

In the SPA, an ideal clean-up memory would take a given structured semantic pointer **A** and be able to correctly map any vector extracted from it to a valid lexical item. For present purposes, we can suppose that **A** is the sum of some number of "role-filler" pairs. For example, there are three such pairs in the previously discussed encoding of "The dog chases the boy."

Algorithmically, decoding any such structure would consist of:

1. Convolving input **A** with a probe vector **p**;
2. Measuring the similarity of the result with all allowable items; and
3. Picking the item with the highest similarity over some threshold and returning that as the cleaned-up result.

Perhaps the most difficult computation in this algorithm is measuring the similarity with all allowable items. This is because the many-to-one mapping between noisy input and clean output can be quite complicated.

So, unfortunately, many simple auto-associators, including linear associators and multilayer perceptrons, do not perform well (Stewart et al., 2010). These are simple because they are feedforward: the noisy vector is presented on the input, and after one pass through the network the cleaned up version is on the output. Better associators are often more complicated. The Hopfield network, for example, is a dynamical system that constantly feeds its output back to the input. Over time, the system settles to an output vector that can be considered the cleaned-up version of an input vector. Unfortunately, such recurrent networks often require several iterations before the results are available, slowing down overall system response. Ideally, we would like a clean-up memory that is both fast and accurate.

To build such a memory, we can exploit two of the intrinsic properties of neurons described in Section 2.1 and demonstrated in the tutorial on neural representation (Section 2.4). The first is that the current in a neuron is the dot product of an input vector with the neuron's preferred direction in the vector space. The dot product is a standard way of measuring the similarity of two vectors. So, a similarity measure is a natural neural computation. The second property is that neurons have a nonlinear response, so they do not respond to currents below some threshold. This means they can be used to compute nonlinear functions of their input. Combining a similarity measure and a nonlinear computation turns out to be a good way to build a clean-up memory.

Specifically, for each item in the clean-up memory, we can set a small number of neurons to have a preferred direction vector that is the same as that item. Then, if that item is presented to the memory, those neurons will be highly active. And, for inputs near that item, these neurons will also be somewhat active. How near items must be to activate the neurons will depend on the firing thresholds of the neurons. Setting these thresholds to be slightly positive in the direction of the preferred direction vector makes the neuron insensitive to inputs that are only slightly similar. In effect, the inherent properties of the neurons are being used to clean up the input.

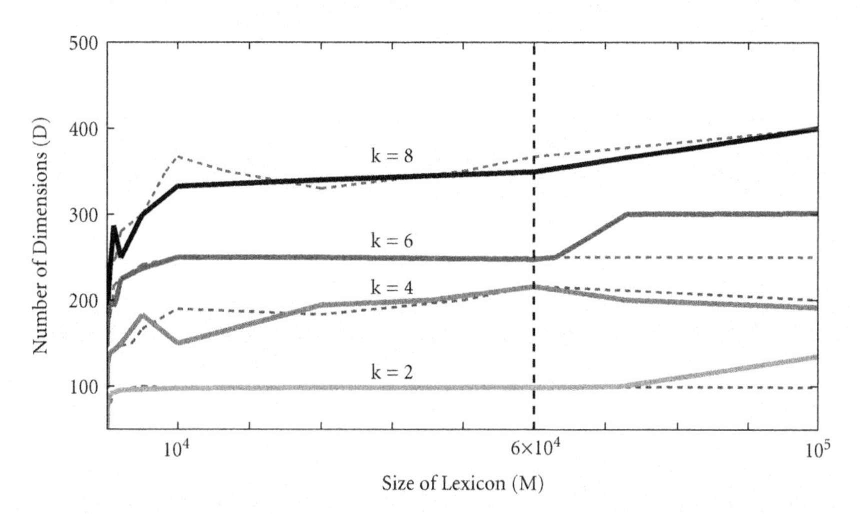

FIGURE 4.7 Scaling properties of a neural clean-up memory. A structured semantic pointer is formed by binding k pairs of random lexical semantic pointers together and then adding the bound pairs. The clean-up memory must recover one of the lexical semantic pointers that formed the input by convolving the input with a random probe from that set. This figure plots the minimum number of dimensions required to recover a lexical item from the input 99% of the time. Data were collected using average results from 200 simulations for each combination of k, M, and D values. The vertical dashed line indicates the approximate size of an adult lexicon (Crystal, 2003). The horizontal dashed lines show the performance of a non-neural clean-up directly implementing the three-step algorithm. The neural model performs well compared to the purely computational implementation of clean up.

Figure 4.7 shows how this neural clean-up memory scales. The parameters affecting how good the clean-up is are: k, the number of role-filler items in the input representation; M, the number of valid lexical items to compare to; and D, the dimensionality of the representation. In the simulations shown here, performance for values up to $k = 8$ are shown because Miller's (1956) classic limits of seven plus or minus two on working memory fall in about the same range. The graphs show M up to a value of 100,000 as this is well over the size of an adult lexicon (Crystal, 2003, suggests 60,000 terms). The plot tracks the dimensionality of the vectors that allow for 99% clean-up accuracy over these ranges of k and M. In sum, this clean-up memory is very accurate for adult-sized vocabularies and approximately 400 dimensions.

Notably, this clean-up memory is also very fast. The network is purely feed-forward, and so clean-up occurs on the timescale of the neurotransmitter used in the network. For excitatory connections in cortex, this is on the order of about 5 ms (i.e., for AMPA receptors).

So, what conclusions can we draw about the SPA with such a model? Although it would be difficult to argue that we are in a position to definitively set a "maximum" number of dimensions for semantic pointers in the SPA, these simulations provide good evidence that the architecture is scalable to within the right neighborhood. Because this model is constructed with 10 neurons per lexical item, we need approximately 600,000 neurons to implement the clean-up memory model described here for an adult-sized lexicon. This works out to about 3.5 mm^2 of cortex, which fits well within the connectivity patterns discussed in Section 4.3. In addition, the connectivity matrix needed to implement this memory is of the same dimensionality as in the binding network, and so it respects the anatomical constraints discussed there as well.

When coupled with the previous discussion regarding the neural plausibility of the binding network (Section 4.3), there is a strong case to be made that with about 500 dimensions, structure representation and processing of human-like complexity can be accomplished by the SPA. That is, we have good reason to suppose that the SPA encompasses methods for constructing models able to perform sophisticated structure processing in a biologically plausible manner. I present an example of such a model in the next section.

Before doing so, however, it is worth considering a few issues related to clean-up memory in the SPA. In general, the SPA is not very specific with respect to the anatomical location or number of clean-up memories we should expect to find in the brain (*see* Section 10.5). The calculations above suggest that they may be ubiquitous throughout cortex, but perhaps many structural manipulations can be performed before clean-up is needed, which would be more efficient. As well, construction of clean-up memories may be a slow process, or a rapid one, or (I suspect most likely) both.

That is, it makes sense to have long-term clean-up memories that account for semantic memory, as well as clean-up memories constructed on-the-fly that are able to track the current context. The construction of long-term distributed associative memories has been an active area of research for many years (Hinton & Anderson, 1981; Barto et al., 1981; Miikkulainen, 1991; Knoblauch, 2011). Although the clean-up memory described here is sparse, and implemented in a spiking network, it is not deeply different in kind from those considered in past work. Similarly, it is reminiscent of our current understanding of the hippocampus, an area of the brain known for rapid learning of sparse distributed representations (Hasselmo, 2011). This suggests that hippocampus might play a central role in constructing on-the-fly clean-up memories. The considerations in Chapter 6 speak to the general issue of learning, so I consider it no further here.

Before turning to a specific model employing clean-up, I would like to address a concern that has been raised about any attempt to choose a fixed number of dimensions for semantic pointers and, hence, clean-up memories. The concern is that we have no good reason to expect that all representations will be of the same dimension. Perhaps some simple representations may not be of sufficiently high dimension or perhaps perceptual and motor systems work with semantic pointers that are of very different dimensions. Or, perhaps, if we choose a "maximum" number of dimensions, then the system simply stops working if we go over that number.

There are two kinds of responses to such concerns: theoretical and practical. On the theoretical side, we can note that any vectors with lower than the maximum number of dimensions can be perfectly represented in the higher-dimensional space. In mathematical terms, note that vector spaces with smaller numbers of dimensions are subspaces of vector spaces with more dimensions. Hence, there is nothing problematic about embedding a lower-dimensional space in a higher-dimensional one.

In the opposite case, if there are more dimensions in a representation than in the vector space, then we can simply choose a random mapping from the higher-dimensional representation to the more limited vector space (simple truncation often works well). There are more sophisticated ways of "fitting" higher-dimensional spaces into lower-dimensional ones (such as performing singular-value decomposition), but it has been shown that for high-dimensional spaces, simple random mappings into a lower-dimensional space preserve the structure of the higher-dimensional space very well (Bingham & Mannila, 2001). In short, all or most of the structure of "other dimensional" spaces can be preserved in a high-dimensional space like we have chosen for the SPA. Vector spaces are theoretically quite robust to changes in dimension.

On the more practical side, we can run explicit simulations in which we damage the simulation (thereby not representing the vector space well, or losing dimensions). Or we can add significant amounts of noise, which distorts the vector space and effectively eliminates smaller dimensions (as measured by the singular values). Performing these manipulations on the binding networks described earlier demonstrates significant robustness to changes in the vector space.

For example, after randomly removing neurons from the binding population, accurate performance is evident even with an average of 1,221 out of 3,168 neurons removed (*see* Fig. 6.20). As well, the discussion surrounding Figure 6.20 shows that the network is accurate when up to 42% Gaussian noise is used to randomly vary the connection weights. Random variation of the weights can be thought to reflect a combination of imprecision in weight maintenance in the synapse, as well as random jitter in incoming spikes. Consequently, such simulations speak both to concerns about choosing a specific number of dimensions and to concerns about the robustness of the system to expected neural variability.

Like other components of the SPA, binding and clean-up networks allow for graceful degradation of performance. This should not be too surprising. After all, previous considerations of neural representation make it clear that there is a smooth relationships between the number of neurons and the quality of the representation (*see* Section 2.4). As well, we have just seen that there are a smooth relationships between the quality of a clean-up memory, the number of dimensions, and the complexity of the representation (Fig. 4.7). There are similarly smooth relationships between dimensionality and complexity in ideal VSAs (Plate, 2003). Because the SPA combines VSAs with neural representation, it is unsurprising that it inherits the kinds of graceful degradation characteristic of these approaches.

In sum, I have described how we can implement a fast and accurate clean-up memory and how such a memory scales. These scaling considerations suggest that 500 dimensional semantic pointers should be appropriate for human cognitive simulations. And, I have argued that clean-up, and the methods of the SPA generally, are robust to variable numbers of dimensions, noise, and physical damage.

4.7 ■ Example: Fluid Intelligence

In cognitive psychology, "fluid intelligence" is distinguished from "crystallized intelligence" to mark the difference between sophisticated cognitive behavior that depends on currently available information and cognitive behavior that depends on potentially distant past experiences. Fluid intelligence is exemplified by solving difficult planning problems, whereas crystallized intelligence is exemplified by recalling important facts about the world to solve a given problem. Most tests that have been designed to measure general intelligence focus on fluid intelligence. One of the most widely used and respected tools for this purpose is the Raven's Progressive Matrices (RPM) (Raven, 1962; Raven et al., 2004).

In the RPM subjects are presented with a 3×3 matrix, where each cell in the matrix contains various geometrical figures, with the exception of the final cell, which is blank (Fig. 4.8). The task is to determine which of eight possible answers most appropriately belongs in the blank cell. To solve this task, subjects must examine the contents of all of the cells in the matrix to determine what kind of pattern is evident and then use that pattern to pick the best answer. This, of course, is an excellent example of induction, which I mentioned earlier in Section 4.5.

The RPM is an extremely widely used clinical test, and many experiments have been run using the test. Nevertheless, our understanding of the processes underlying human performance on this task is minimal. To the best of my knowledge, there have been no neural models of the advanced RPM

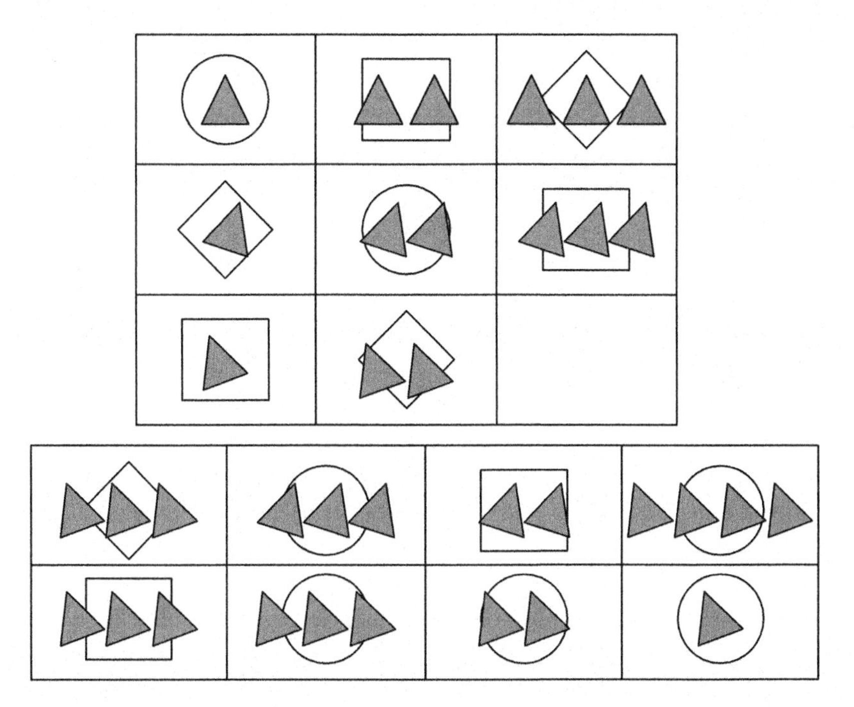

FIGURE 4.8 Example Raven's Progressive Matrix (RPM). A subject must examine the first two rows (or columns) of the matrix to determine a pattern. The task is pick one of the eight possible answers shown along the bottom to complete that pattern for the third row (or column). Matrices used in the RPM cannot be published, so this is an illustrative example only. (Figure from Rasmussen & Eliasmith, 2011a.)

that include the inductive process of rule generation. The best-known model of RPM includes all of the possible rules that the system may need to solve the task (Carpenter et al., 1990). In solving the task, the model chooses from the available rules and applies them to the given matrix. However, this treats the RPM like a problem that employs crystallized intelligence, which contradicts its acceptance as a test of fluid intelligence (Prabhakaran et al., 1997; Gray et al., 2003; Perfetti et al., 2009).

Recently, however, a graduate student in my lab, Daniel Rasmussen, proposed a model of the RPM that is implemented in a spiking neural network, using the methods of the SPA (Rasmussen & Eliasmith, 2011a). To understand how his model works, we can consider the example Raven's-style matrix shown in Figure 4.8. To correctly solve this problem, a subject needs to extract information from the filled-in cells that identifies three rules: (1) the number of triangles increases by one across the row; (2) the orientation of the triangles stays the same across the row; and (3) each cell also contains a shape in the background, which, across the row, includes each of a circle, a square, and a diamond. This combination of rules identifies the correct response (i.e., three triangles tilted to the left on top of a circle), which can be determined by applying those rules to the last row. Although not all subjects will explicitly formulate these rules, they must somehow extract equivalent information (or get lucky) to get this matrix correct.

In Rasmussen's graduate work, he argues that there are three classes of rule needed to solve the entire set of problems on the RPM (Rasmussen, 2010). The first class is induction rules, which generalize across neighboring cells of the network (e.g., increase the number of triangles by one). The second class are set completion rules, which account for set completion across an entire row (e.g., each of a circle, a square, and a diamond are represented within a row). The third class are visually based same/different rules, which account for the presence and absence of particular visual features across a row (these are not demonstrated by the above example). In addition, the choice and coordination of the application of these rules is accounted for by an executive system. In recent work he has shown that a full model using 500-dimensional semantic pointers gets 22.3/36 ($\sigma = 1.2$) correct on average, whereas college undergraduates get a reported 22/36 correct on average (Bors & Stokes, 1998). However, for simplicity, here I only focus on the induction rules (for discussion of other aspects of the model and many other simulations *see* Rasmussen [2010]; Rasmussen & Eliasmith [2011b]).

In Section 4.5, I described how we can perform induction to infer a transformation based on input/output examples. To make this characterization more concrete, consider the simplified Raven's-style matrix shown in Figure 4.9A. Each cell in this matrix is represented by a semantic pointer that consists of role-value pairs. For example, the top right cell could be represented as:[6]

$$\mathbf{cell} = \mathbf{shape} \circledast \mathbf{circle} + \mathbf{number} \circledast \mathbf{three}.$$

Once all of the cells are represented, the previously described induction rule is provided the pairwise examples in the order they tend to be read by subjects (i.e., across rows, and down columns; Carpenter et al., 1990). As the examples are presented, the inferred transformation is built up out of the average of each of the pairwise extracted transformations. That inferred transformation is then applied to the second last cell in the matrix. This results in a vector that

[6] Surprisingly few decisions need to be made about which role-filler pairs to include in the representations of cells. As long as the information necessary to identify the required transformation is available, the induction rule is generally able to extract it. Thus, including additional information like $\mathbf{color} \circledast \mathbf{black} + \mathbf{orientation} \circledast \mathbf{horizontal} + \mathbf{shading} \circledast \mathbf{solid}$, and so forth, in this instance does not affect performance. Although this model does not include visual processing to generate such representations, the Spaun model in Chapter 7 performs exactly the same kind of induction from raw image input.

A.

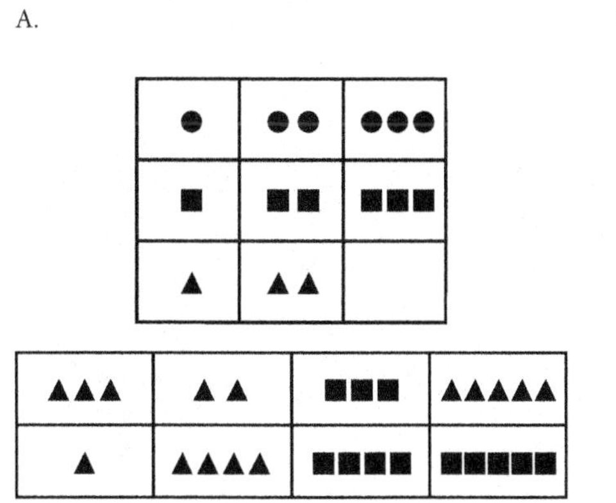

B.

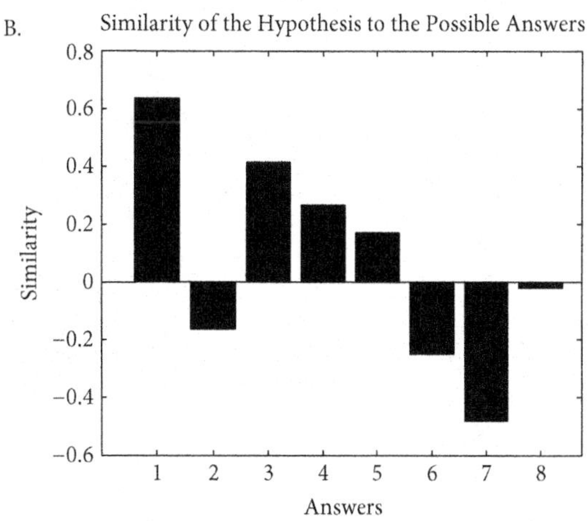

FIGURE 4.9 A simple induction RPM example. **A.** A simple RPM matrix in which the number of items increases by one across each row. The model is presented with consecutive pairs of cells and infers a transformation. That transformation is applied to the last cell, and the similarity to each answer is computed. **B.** The similarity at the end of the run, indicating that the model correctly identifies the first answer (i.e., three triangles) as the correct one. (Figure from Rasmussen & Eliasmith, 2011a.)

is compared to each of the possible answers, and the most similar answer is chosen as the correct one by the model, as shown in Figure 4.9B.

Comparing a fully spiking neural model across many runs to average human performance on all the matrices that include an induction rule (13 of 36 examples), humans score about 76% on average (Forbes, 1964), and the model scores 71% (σ =3.6%) on average (chance is 13%). So the model is performing induction at a level similar to human subjects.

More generally, the model can account for several other interesting experimental observations regarding the RPM. For example, subjects improve with practice if given the RPM multiple times (Bors, 2003) and also show learning within the span of a single test (Verguts & De Boeck, 2002). As well, a given subject's performance is not deterministic; given the same test multiple times, subjects will get previously correct answers wrong and vice versa (Bors, 2003). In addition, there are both qualitative and quantitative differences in individual ability; there is variability in "processing power" (variously attributed to working memory, attention, learning ability, or executive functions), but there are also consistent differences in high-level problem-solving strategies between low-scoring and high-scoring individuals (Vigneau et al., 2006). This is not an exhaustive list, but it represents some of the features that best characterize human performance, and all of these features (and others) are captured by the full RPM model (Rasmussen, 2010; Rasmussen & Eliasmith, 2011b).

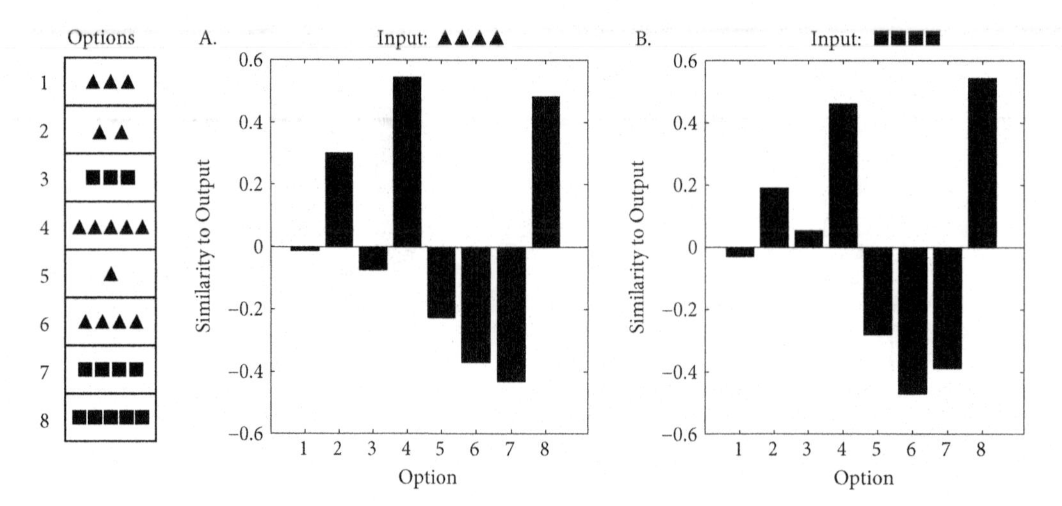

FIGURE 4.10 Broader application of the rule induced from Figure 4.9. **A**. Applied to four triangles, the model chooses five triangles as the next item in the sequence. This is correct because the item type is retained (triangles) and the number is incremented by one (to five), as in the rows of the example matrix. **B**. Applied to four squares, the model chooses five squares as the next item in the sequence. This is correct for the same reason as in A. Notably, for inference run down columns, the expected responses would differ.

That being said, the theoretically most important feature of the model for my purposes is its ability to successfully perform induction and extract rules. However, simply choosing the best of the eight possible answers might not convince us that it has, in fact, found a structural regularity. Much more convincing is determining how the learned transformation applies in circumstances never encountered by the model during training. Figure 4.10 shows two examples of this kind of application. In the first instance, the model is asked to apply the learned transform to four triangles. As is evident from Figure 4.10A, the model's preferred answer corresponds to five triangles. This suggests that the inferred transformation encodes the correct "increase by one" aspect of the rule, at least in the context of triangles. Figure 4.10B shows that, in fact, the "increase by one" is generic across objects. In this case, the rule is applied to four squares, and the preferred result is five squares. In short, the model has learned the appropriate counting rule and generalized this rule across objects.

This is a simple, but clear, example of what is sometimes called "syntactic generalization." Syntactic generalization is content insensitive generalization–that is, generalization driven by the syntactic structure of the examples (Gentner, 1983). It is important to note that precisely this kind of generalization has often been claimed to distinguish cognitive from noncognitive systems (Fodor & Pylyshyn, 1988; Jackendoff, 2002; Hummel &

Holyoak, 2003). In this case, it is clear that the generalization can be relatively content insensitive because "increase by one" is applied for squares, triangles, or what have you. Note that even if we may suspect that "pure" syntactic generalization does not occur, it is still critical to explain such inductive performance. I provide more examples of syntactic generalization in Section 6.6.

It might not be too surprising that to perform cognitively interesting induction, one has to perform syntactic generalization. What is more interesting is that once we have an appropriate representational system (semantic pointers and binding), the method for performing such syntactic induction is quite simple (i.e., averaging over examples) and can be built into a biologically realistic network. In addition, the same basic architecture is able to capture the other two kinds of reasoning usually identified in cognitive science: abduction and deduction. In previous work, I have shown that the approach underlying the SPA can be used to capture central features of human deductive reasoning (Eliasmith, 2004); I describe this work in more detail in Section 6.6. Thagard and Litt (2008) have also shown that a version of this model can be used to capture abductive reasoning (i.e., inference to the best explanation).

This unification of kinds of reasoning in a simple neural mechanism is appealing because it brings together a range of what might otherwise seem very different cognitive strategies. Further, having a consistent architecture in which to explore these mechanisms in a biologically constrained manner provides a useful way to begin constructing models of biologically based reasoning. As I noted earlier, central issues regarding the control, contextual effects, competing strategies, and so forth are not accounted for by this mechanism alone. Nevertheless, such a mechanism provides fertile ground for building more sophisticated models, especially in the broader context of the SPA.

I take it that it is no accident that past models of the advanced RPM have not had mechanisms for inferring rules. Such mechanisms are not in those models because there is no obvious way to infer the appropriate cognitive rules given the example transformations in a symbolic representation. There is little sense to be made of "averaging" a series of symbolic structures. And, there is no simple replacement for "averaging" with another operation. With the SPA, however, averaging past examples of transformations becomes both very natural, and very powerful. It is powerful because semantic pointers encode changes across examples as a kind of "noise," while allowing whatever elements of the representation that are consistent to be reinforced. Crucially, this is true for both the content and the syntactic structure of the representation.

One final note, which strikingly distinguishes this SPA implementation from standard connectionist approaches, is that there are no connection weight changes in this model. This is true despite the fact that the model learns based on past experience and is able to successfully generalize. I return to this observation in my discussion of learning in Chapter 6.

4.8 ■ Deep Semantics for Cognition

In the last chapter, I distinguished deep and shallow semantics for both perception and action, but I did not address the issue of deep semantics for more cognitive representations. This is because I could not address sophisticated, structured representations without discussing syntax. As I commented in the previous chapter, I also did not describe a means of integrating the variety of possible sources of semantics, including the different modalities, motor systems, lexical relationships, and so on. Now, however, we have seen a method for building structures out of sets of semantic pointers, and so I can return to these issues.

It is a natural extension of the previous discussion on semantics to suggest that the complex kind of semantic structures found in human cognition can be constructed by binding and combining semantic pointers from different sources. So, for example, the perceptual features of a "robin" might be represented as:

$$\mathbf{robinPercept} = \mathbf{visual} \circledast \mathbf{robinVis} + \mathbf{auditory} \circledast \mathbf{robinAud} + \mathbf{tactile} \circledast \mathbf{robinTouch} + \ldots.$$

The pointers **robinVis**, **robinAud**, and so forth are those at the top of the relevant hierarchy (analogous to the visual semantic pointers generated in the previous chapter). Those perceptual features can be integrated into a more lexical representation of the concept "robin" in a similar manner:

$$\mathbf{robin} = \mathbf{perceptual} \circledast \mathbf{robinPercept} + \mathbf{isA} \circledast \mathbf{bird} + \mathbf{indicatorOf} \circledast \mathbf{spring} + \ldots.$$

Importantly, the resulting representations **robin** and **robinPercept** are also semantic pointers. After all, circular convolution is a compression operation, so the results of binding two vectors is a compressed representation of the original content (as for any semantic pointer). In this case, what is being represented is not only perceptual features but also conceptual structure. And, just as with perceptual semantic pointers, structured semantic pointers must also be dereferenced by effectively "running the model backward." In the case of structure, "running the model backward" means unbinding and cleaning up the results of that operation. In the case of perceptual semantic pointers, this means decompressing the representation by seeding the model that had been constructed based on perceptual experience. The only difference, it may seem, is that the perceptual model is hierarchical and clean-up memory is not. But this, we will see shortly, may not be a difference after all.

In any case, we now have, theoretically at least, a means of bringing together shallow and deep semantics by binding them into a single semantic pointer. But, we can go further. One additional, interesting feature of the preceding characterization of the SPA is that transformations of semantic pointers are themselves semantic pointers. For example, when the model solving the Raven's matrices induced a transformation to explain the changes in the structures it had been presented with, the result was a transformation vector in the same vector space as the structures it transformed. In fact, these vectors can be decoded to determine what transformation structure they represent (Rasmussen & Eliasmith, 2011b).

Because semantic pointers can act as transformations, and conceptual structure can be built up out of bound semantic pointers, our conceptual representations can include characterizations of transformations in them. That is, not only static properties but also the kinds of changes associated with objects can be encoded in (or pointed to by) the lexical representation of that object. More precisely, a compressed description of an object's possible transformations can be encoded in the lexical representation of that object. As with any "part" of such a representation, it would need to be dereferenced, likely by another area of the brain, to decode its semantics. I believe this is an intriguing and natural way to include an element of time in our understanding of lexical conceptual representations, an element that is often missing in discussions of concepts. That semantic pointers can be used to represent temporal transformations should not be too surprising. This is, after all, precisely what occurs in the motor model described in Section 3.6.

To sum up, I have suggested that SPA conceptual structures can be very complex, including the sum of bound representations of properties, perceptual features, lexical relationships, dynamics, and so on. But, there seems to be an important problem lurking here–namely, that in my previous description of a clean-up memory, I noted that we could accurately decode about eight role-filler pairs in a network that was anatomically plausible. Certainly, more than eight roles will be needed to include all of these kinds of information in a single semantic pointer. How then are we going to be able to encode sophisticated concepts in a biologically realistic manner using the resources of the SPA?

I would like to propose a solution to this problem that allows the architecture to scale well, even for complex structures. Specifically, rather than constraining the model to a single clean-up memory for decoding, we can allow the chaining of memories together. In short, I am suggesting that the model perform successive decodings in different clean-up memories. This provides a chain of decoding, where initial decoding is performed in one memory, the results of that decoding are then further decoded in a second memory, and so on. This process can be repeated several times. At each stage of such a chain, the semantic pointer to be decoded comes from the clean-up of the previous stage, ensuring that the full capacity of the current stage is available for further decoding of the semantic pointer. As a result, the effective capacity scales as the power of the number of stages in such a chain. Let us consider a concrete example.

Consider an example of how a "dog" semantic pointer might be constructed. It could include role-filler pairs for a wide variety of information, as described earlier. Suppose for this simple example that there are only three roles:

$$\mathbf{dog} = \mathbf{isA} \circledast \mathbf{mammal} + \mathbf{hasProperty} \circledast \mathbf{friendly} + \mathbf{likes} \circledast \mathbf{bones}.$$

Suppose that in a 100-dimensional space, we can decode any elements in a three-role representation with 99.9% accuracy. So, if we query our "dog" concept, wanting to know what it is (i.e., determining the filler in the "isA" role) we discover that a dog is a mammal. That is, the result of this operation will be a clean version of the "mammal" semantic pointer.

We might then decode this new semantic pointer with a second memory in our chain (Collins & Quillian, 1969). Suppose that this semantic pointer is the sum of three other role-filler pairs, perhaps

$$\mathbf{mammal} = \mathbf{isA} \circledast \mathbf{animal} + \mathbf{hasProperty} \circledast \mathbf{hair} + \mathbf{produces} \circledast \mathbf{milk}.$$

We can then decode any of the values in those roles with 99.9% accuracy. That is, we could determine that a mammal is an animal, for example. And so the chaining might continue with the results of that decoding. What is interesting to note here is that in the second memory we could have decoded any three role-filler pairs from any of the three role-filler pairs in the memory earlier in the chain. So, effectively, our original concept can include nine ($= 3^2$) role-filler pairs that we would be able to decode with near 100% accuracy (the expected accuracy is 99.9%$\times$99.9% = 99.8%).

In short, we can increase the effective number of role-filler pairs by chaining the spaces. As Figure 4.11 shows, the scaling of chained spaces significantly outperforms an unchained space. It is clear from this figure that even with the same total number of dimensions, an unchained space cannot effectively decode very many pairs (the number of pairs is approximately linear in the number of dimensions). In contrast, the number of decodable pairs in the chained space is equal to the product of the number of decodable pairs in each space separately. A comparison of Figures 4.11C and D, demonstrates that we can effectively decode several hundred rather than about 20 role-filler pairs with the same number of dimensions.

We can extrapolate the expected accuracy for much larger sets of pairs by turning to our earlier consideration of clean-up memory. For example, we know that in a 500-dimensional space we can clean up about 8 pairs from a vocabulary of 60,000 terms with 99% accuracy using the number of neu-

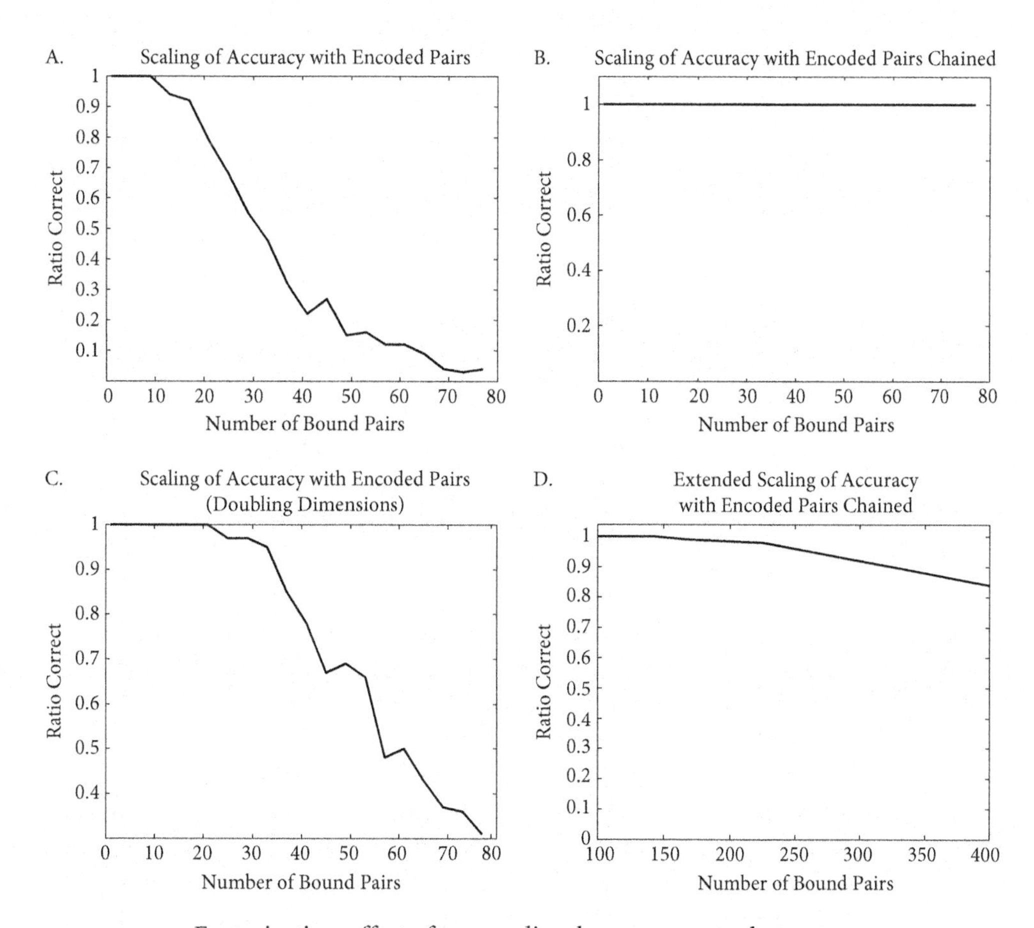

FIGURE 4.11 Factorization effects for encoding large conceptual structures. **A.** The accuracy versus number of encoded role-filler pairs rapidly decreases in a single memory. This is a 500-dimensional space with 60,000 possible terms. The accuracy falls to 3% for 80 pairs. **B.** The accuracy of a pair of memories that are chained. Each space is 500-dimensional and has 60,000 possible terms. The decoding accuracy is 100% up to the limits of the graph. **C.** The accuracy of a single memory with the same total number of dimensions as the chained space. This is a 1000-dimensional space with 60,000 possible terms. The decoding accuracy is 31% for 80 pairs. **D.** The extended scaling of the chained space between 100 and 400 pairs. The scaling begins to decline smoothly around 150 pairs but is still 84% accurate for 400 pairs. All points are averages of 100 runs.

rons consistent with observed connectivity (*see* Fig. 4.7). If we chain four such memories together, then we would be able to decode $8^4 = 4096$ pairs with $0.99^4 = 0.96$, or 96%, accuracy, although the pairs cannot be completely arbitrary. This kind of scaling makes earlier concerns about encoding sophisticated concepts much less pressing.

Why do we get such good scaling? In short, it is because we have introduced a nonlinearity (the clean up) into our decoding process that is not available in a single memory. That nonlinearity acts to align the output of one memory with the known values in the next semantic space in the chain. It may seem that this enforces a kind of hierarchy on our conceptual structure, but it does not. The reason I use the word "chain" rather than "hierarchy" is because there are many kinds of chains that could be implemented in this way, only some of which are hierarchies. For example, the second memory in a two-level chain could be the same as the first (a recursive chain). Or, each possible role could have its own dedicated memory for decoding items of that role-type (a strict hierarchy). It is unclear which of these, or other, kinds of structure will be most appropriate for characterizing human-like semantic spaces.[7] It is intriguing, however, that a hierarchy is a very natural kind of chain to implement and that there is some evidence for this kind of structure to human concepts (Collins & Quillian, 1969). As well, parsimony makes it tempting to assume that because of the hierarchical nature of perceptual and motor systems (also used for capturing deep semantics), cognition will have a similar hierarchical structure. However, mapping any particular structure carefully to data on human concepts is beyond the scope of this book (*see*, e.g., Rogers & McClelland [2004] for interesting work along these lines).

Regardless of how a specific mapping turns out, in general I believe chained semantic spaces provide a crucial demonstration of how syntax and semantics can be used together to represent very large, intricate conceptual spaces in ways that can be plausibly implemented in the brain. These semantic spaces are tied directly to perceptual and motor experiences as described in the previous chapter and can also be used to encode temporal transformations, language-like structures, and lexical relationships. Further, we can transform these representations in structure-sensitive ways and even learn to extract such transformations from past examples. The stage is now set for considering how to build a system that itself uses, transforms, and generates semantic pointer representations in the next chapter.

4.9 ■ Nengo: Structured Representations in Neurons

This chapter introduced a method for constructing structured representations using semantic pointers. Semantic pointers themselves are high-dimensional vector representations. When we think of semantic pointers as acting like symbols, we can name them to keep track of different semantic pointers in

[7] Matt Gingerich and Eric Crawford in my lab have recently encoded all 117,000 terms from WordNet into a recursively chained clean up. They have shown that the very deep WordNet structures can be recovered and that the same representations are usable, and decodable, in sentence structures (unpublished).

the same high-dimensional space. The collection of named vectors in a space forms a kind of "vocabulary" of known semantic pointers in that space. Structured representations can then be built out of this vocabulary. In this tutorial, I discuss how to run simulations in Nengo that create and use semantic pointer vocabularies.

- Open a blank Nengo workspace and create a network named "Structured Representation."
- Create a default ensemble in the network, name it "A," give it 300 neurons, and make it 20 dimensions.

This ensemble is representing a vector space just as in the simpler low-dimensional (i.e., 2D and 3D) cases we considered earlier. To use semantic pointers, we need to work in higher-dimensional spaces.

- Open the network with *Interactive Plots*. Right-click the "A" population and click "value." Right-click the population again and click "semantic pointer." You should now have two graphs.
- Run the simulation for a moment.

Displaying the "value" graph in *Interactive Plots* shows the value of individual components of this vector.[8] The "semantic pointer" graph compares the vector represented by the ensemble to all of the elements of the vocabulary and displays their similarity. Initially, the vocabulary contains no vectors and thus nothing is plotted in the semantic pointer graph. As well, because there is no input to the population, the value graph shows noise centered around zero for all components.

- Right-click the semantic pointer graph and select "set value." Enter "a" into the dialog that appears and press *OK*.
- Run the simulation again.

Using the "set value" option of the semantic pointer graph does two things: (1) if there is a named vector in the set value dialog that is not part of the vocabulary, then it adds a new randomly chosen vector to the vocabulary of the network and associates the given name (e.g., "a") with it; and (2) it makes the represented value of the ensemble match the named vector. The result is that the ensemble essentially acts as a constant function that outputs the named vocabulary vector. As a result, when the simulation is run, the semantic pointer graph plots a 1, because the representation in the ensemble is exactly

[8] By default, the plot shows five components–to change this you can right-click the "value" graph and choose "select all," but be warned that plotting all 20 components may cause the simulation to be slow. Individual components can be toggled from the display by checking or unchecking the items labeled "v[0]" to "v[19]" in the right-click menu.

similar (i.e., equal) to the "a" vector. Similarity is measured using the dot product. The value graph shows the values of several dimensions of the randomly chosen "a" vector.

- Right-click the "semantic pointer" graph and select "set value." Enter "b" into the "Set semantic pointer" dialog and press *OK*.
- Run the simulation.
- Switch between setting "a" and setting "b" on the semantic pointer graph while the simulation is running.

Setting the value of the semantic pointer to "b" adds a second, random 20-dimensional vector to the vocabulary. The "value" plot reflects this by showing that the neural ensemble is driven to a new vector. Switching between the "a" and "b" vectors, it should be clear that although the vectors were randomly chosen initially, each vector is fixed once it enters the vocabulary. The semantic pointer graph changes to reflect which vocabulary item is most similar to the current representation in the ensemble.

The named vectors are the vocabulary items that can be combined to form structured representations. To experiment with the binding (i.e., convolution) and conjoin (i.e., sum) operations, we require additional ensembles.

- In the "Structured Representation" network in the Nengo workspace, add a default ensemble named "B" with 300 neurons and 20 dimensions.
- Add a third ensemble named "Sum" with the same parameters.
- Add a decoded termination to the "Sum" ensemble. Name the termination "A" and set the number of input dimensions to 20, and *PSTC* to 0.02.
- The coupling matrix of the termination should be set to an identity matrix by default (a grid of zeros except for a series of ones along the diagonal). Click *Set Weights* to make sure, and keep this default. Click *OK* twice.
- Add a second decoded termination to the "Sum" ensemble. Name the termination "B" and set all other parameters as for the "A" termination.
- Add projections from ensembles "A" and "B" to the "Sum" ensemble.
- Add a fourth ensemble to the network named "C" with the same parameters as all of the previous ensembles.
- Drag the icon for the "Binding" template from the bar on the left side of the screen into the network viewer for the "Structured Representation" network.
- In the dialog box that is opened, enter the name "Bind," use "C" as the name of the output ensemble, and set 70 neurons per dimension. The two inversion options should not be checked.
- Project the "A" and "B" ensembles to the "Bind" network that was created.

The "Sum" ensemble should be familiar from the previous tutorial about transformations (Section 3.8). However, the "Bind" network uses a Nengo template to construct a subnetwork, which we have not done before. Nengo can create subnetworks by placing one network inside another and exposing selected terminations and origins of the inner network to the outer network. This is a useful technique for organizing a model, and it is used here to group the ensembles required to compute a binding operation within a single subnetwork element.

In general, the template library allows common network components to be created quickly and with adjustable parameters. The templates call script files that advanced Nengo users can write to aid rapid prototyping of networks (scripting is the topic of the Section 7.5 tutorial). Instructions for creating templates can be found at `http://nengo.ca/docs/html/advanced/dragndrop.html`.

The "Bind" network is a component that computes the circular convolution of its two inputs using nonlinear transformations as discussed in the preceding tutorial. To look at the subnetwork elements, you can double-click it. You cannot see the connections into and out of the subnetwork when you open it, but only the terminations and origins–the connections are visible only in the parent network. Let us examine the behavior of this binding network.

- Open the *Interactive Plots* viewer. Right-click the background and select the "A," "B," "C," "Bind," and "Sum" nodes if they aren't already displayed.
- Show the semantic pointer graphs for the "A," "B," "C," and "Sum" nodes by right-clicking the nodes and selecting "semantic pointer."
- Set the value of the "A" ensemble to "a" and the value of the "B" ensemble to "b" by right-clicking the relevant semantic pointer graphs.
- Right-click the semantic pointer graph of ensemble "C" and select "show pairs."
- Right-click the semantic pointer graph of ensemble "C" and select "a*b."
- Repeat the previous two steps for the "Sum" ensemble, so that both "show pairs" and "a*b" are checked.
- Run the simulation for about 1 second, and then pause it.

The "C" population represents the output of a neurally computed circular convolution (i.e., binding) of the "A" and "B" input vectors, whereas the "Sum" population represents the sum of the same two input vectors. The label above each semantic pointer graph displays the name of the vocabulary vectors that are most similar to the vector represented by that neural ensemble. The number preceding the vector name is the value of the normalized dot product between the two vectors (i.e., the similarity of the vectors).

In this simulation, the "most similar" vocabulary vector for the "C" ensemble is "a*b." The "a*b" vector is the analytically calculated circular convolution of the "a" and "b" vocabulary vectors. This result is expected, of course. Also of note is that the similarity of the "a" and "b" vectors alone is significantly lower. Both of the original input vectors should have a low degree of similarity to the result of the binding operation. Toggling the "show pairs" option of the graph makes the difference even clearer. The "show pairs" option controls whether bound pairs of vocabulary vectors are included in the graph.

In contrast, the "a" and "b" vectors have a relatively high (and roughly equal) similarity to the vector stored in the "Sum" ensemble. The sum operation preserves features of both original vectors, so the sum is similar to both. Clearly, the conjoin and bind operations have quite different properties.

We have now seen how we can implement the two functions needed to create structured representations in a spiking network. However, to process this information, we must also be able to extract the information stored in these conjoined and bound representations. This is possible through use of an inverse operation.

- Set the value of the "A" semantic pointer to "a*c+b*d."
- Set the value of the "B" semantic pointer to "~d."
- Hide the "Sum" and "C" semantic pointer graphs.
- Display the "C" semantic pointer graph again by right-clicking "C" and selecting "semantic pointer" (hiding and displaying the graph allows it to adjust to the newly added "c" and "d" vocabulary items).
- Right-click the graph of "C" and select "select all" (the letters "a" through "d" should be checked).
- Run the simulation for about 1 second.

In this simulation, we first set the input to a known, structured semantic pointer. The value "a*c+b*d" is the semantic pointer that results from binding "a" with "c" and "b" with "d" then summing the resulting vectors. This vector is calculated analytically as an input, but it could be computed with neurons if need be. The second input, "~d", represents the pseudo-inverse of "d" (*see* Section 4.3). The pseudo-inverse vector is a shifted version of the original vector that approximately reverses the binding operation. Binding the pseudo-inverse of "d" to "a*c+b*d" yields a vector similar to "b" because the "b*d" operation is inverted whereas the "a*c" component is bound with "d" to form a new vector that is dissimilar to anything in the vocabulary and so is ignored as noise.

- Experiment with different inverse values as inputs to the "B" ensemble ("~a," "~b," or "~c"). Note the results at the "C" ensemble for different inputs.
- You can also try changing the input "statement" given to the "A" ensemble by binding different pairs of vectors together and summing all the pairs.

- Binding more pairs of vectors together will degrade the performance of the unbinding operation. If "B" is set to "~a," then setting "A" to "a*b+c*d+e*f+g*h" will not produce as clean an estimate of "b" as when "A" is set only to "a*b."

Rather than naming the vectors "a," "b," and so on, you can name them "dog," "subject," and so on. Doing so makes it clear how this processing can be mapped naturally to the various kinds of structure processing considered earlier in this chapter. Note that the name "I" is reserved for the identity vector. The result of binding any vector with the identity vector is the vector itself. The lowercase "i" does not refer to the identity vector.

There are two competing constraints that determine the accuracy of information processing in this network. First, the number of dimensions. With a small number of dimensions, there is a danger that randomly chosen vectors will share some similarity because of the small size of the vector space (*see* Fig. 3.2 in Chapter 3). So it would be desirable to have very high-dimensional spaces. However, the second constraint is the number of neurons. A larger number of dimensions requires a larger number of neurons to accurately represent and manipulate vectors. There are, of course, only a limited number of neurons in the brain. So, there is an unsurprising trade-off between structure processing power and the number of neurons, as I explored in Section 4.6. The network presented in this tutorial is using a rather low number of dimensions and few neurons for its semantic pointer representations for the sake of computational performance. A pre-built version of this network can be loaded from <Nengo Install Directory>/Building a Brain/chapter4/structure.py.

5. BIOLOGICAL COGNITION–CONTROL

C. Eliasmith, B. Bobier, and T. Stewart

5.1 ■ The Flow of Information

To this point in the book, I have discussed two concepts central to cognitive science: semantics and syntax. Both notions are primarily concerned with aspects of representation that provide insight into the roles representations play in a cognitive architecture. A third, crucial but often overlooked, aspect of any cognitive architecture is control: the control of what course of action to pursue next, the control of which representations to employ, the control of how to manipulate current representations given the current context, and so on. All of these kinds of control allow representations to be effectively exploited, and so all are central to making an architecture cognitive. In short, understanding how the flow of information is controlled throughout the brain is paramount to understanding the flexible, robust, and adaptive nature of biological cognition.

I believe that control is underexamined in contemporary behavioral science because it does not become absolutely indispensable until we focus on cognitive flexibility. Consequently, many models–those that focus on specific cognitive phenomena or tasks–do not really address the general problem of control. Such models are restricted to a single, often tightly constrained, domain–be it associating pictures and words, matching a sample, remembering a target, copying a drawing, making a simple decision, or what have you. Biological cognitive systems, in contrast, need to be able to perform all of these tasks, switching between them as appropriate. Consequently, any prospective cognitive architecture needs to specify how information can be routed to different areas of the brain to support a given task, how that same information can be interpreted differently depending on the context, how the system can determine what an appropriate strategy is, and so on.

In general, the process of control can be usefully broken down into two parts: (1) determining what an appropriate control signal is and (2) applying that control signal to affect the state of the system. The first of these is a kind of decision making. That is, determining what the next course of action should be based on currently available information. The second is more of an implementational issue: How can we build a system that can flexibly gate information between different parts of the system to realize the chosen course of action?

In this chapter, I describe the aspects of the Semantic Pointer Architecture (SPA) most important for control. I begin by presenting the work of a member of my lab, Terry Stewart, on the role of the basal ganglia in action selection, with a focus on cognitive actions. This discussion largely revolves around determining the appropriate control signal. I then describe work pursued by Bruce Bobier in my lab regarding how actions that demand the routing of information can be implemented in a neurally realistic circuit; this addresses the biologically plausible application of a control signal. Bruce's work focuses on routing information through the visual cortex to explain aspects of visual attention. However, I describe how lessons learned from this characterization of attention can be exploited throughout the SPA. The chapter concludes with an example model that combines all aspects of the SPA we have seen so far and solves the well-known Tower of Hanoi problem in a psychologically and neurally plausible manner.

5.2 ■ The Basal Ganglia

The basal ganglia are a highly interconnected cluster of brain areas found underneath the neocortex and near the thalamus (*see* Fig. 5.1A). For well over 20 years, the basal ganglia have been centrally implicated in the ability to choose between alternative courses of action. This process is unsurprisingly called "action selection." Damage to the basal ganglia is known to occur in several diseases of motor control, including Parkinson's and Huntington's diseases. Interestingly, these "motor" diseases have also been shown to result in significant cognitive defects (Frank & O'Reilly, 2006). Consequently, both neuroscientists (e.g., Redgrave et al., 1999) and cognitive scientists (e.g., Anderson et al., 2004) have come to understand the basal ganglia as being responsible for "action" selection broadly construed (c.f. Turner & Desmurget, 2010)–that is, including both motor and cognitive actions (Lieberman, 2006, 2007).

The basal ganglia circuit shown in Figure 5.1B highlights the connectivity accounted for by the classic Albin/Delong model of basal ganglia function

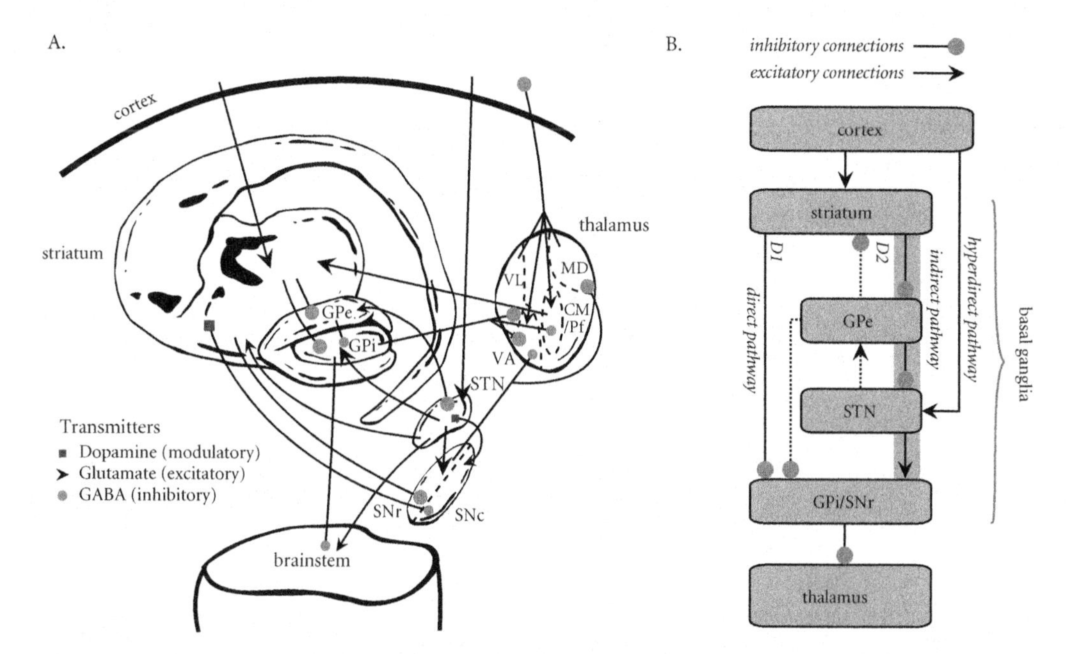

FIGURE 5.1 The basal ganglia. **A.** Major connections of the cortico-basal ganglia-thalamo-cortical system. Markers are roughly proportional to the density of projections. Abbreviations used are: centromedian nucleus of the thalamus (CM); parafascicular nucleus (Pf); ventral anterior nucleus (VA); ventral lateral nucleus (VL); medio-dorsal nucleus (MD); substantia niagra pars compacta (SNc); external globus pallidus (GPe); internal globus pallidus (GPi); substantia niagra pars reticulata (SNr); and subthalamic nucleus (STN). **B.** A schematic diagram of the basal ganglia showing the standard direct/indirect pathway and the hyperdirect pathway. Other major connections that have been recently discovered are shown with dotted lines. D1 and D2 refer to two classes of dopamine-modulated cells in striatum that have different kinds of dopamine receptors.

(DeLong, 1990; Albin et al., 1989). This model is able to qualitatively account for many of the symptoms of Parkinson's and Huntington's diseases. In this model, there is a "direct pathway," where excitatory inputs from cortex to the D1 dopamine cells in the striatum inhibit corresponding areas in GPi and SNr, which then inhibit areas in the thalamus. And, there is an "indirect pathway" from the D2 dopamine cells in the striatum to GPe, STN, and then GPi/SNr. However, more recent evidence has shown other major connections, including a "hyperdirect" excitatory pathway straight from cortex to STN (Nambu et al., 2002), and other feedback connections, shown as dotted lines.

A more recent model of the basal ganglia has suggested that these extra connections can effectively underly action selection (Gurney et al., 2001). This newer model also takes advantage of the great deal of topological structure in

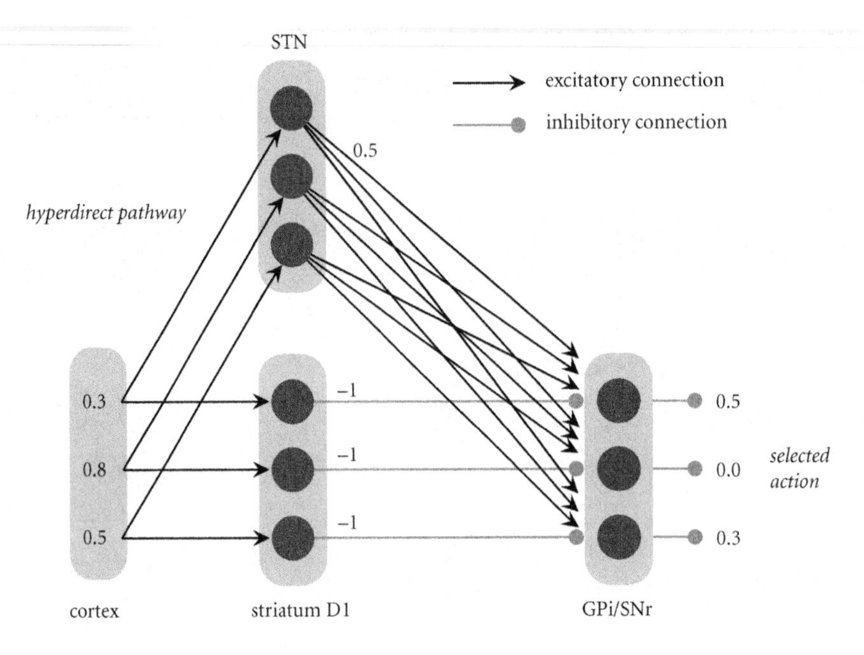

FIGURE 5.2 Action selection via the striatum D1 cells and the subthalamic nucleus (STN). Cortical activity provides input identifying three possible actions with different levels of activity (i.e., 0.3, 0.8, and 0.5). As described in the main text, the input with the highest utility (0.8) causes the corresponding output in the GPi/SNr to drop to zero, indicating that the action is selected. (Adapted from Stewart et al., 2010.)

the inhibitory connections in basal ganglia. Neurons in the striatum project to a relatively localized area in the GPi, GPe, and SNr, whereas the excitatory connections from STN are very broad (Mink, 1996). The Gurney et al. model, which forms the basis of our model, makes use of this structure to explain basic action selection.

To see how, consider the striatum, STN, and GPi/SNr circuit shown in Figure 5.2. This circuit includes the direct and hyperdirect pathways. In this figure, there are three possible actions that have different "desirabilities" (or "utilities") indicated by the inputs to the striatum and STN from cortex (i.e., 0.3, 0.8, and 0.5). It is important to note that "selecting" an action in the basal ganglia corresponds to *inhibiting* the corresponding output.[1] That is, the GPi output should be low for the selected action, but high for all other actions. The equally weighted inhibitory connections in the direct pathway cause the

[1] This is because the output of GPi/SNr itself is inhibitory, so inhibiting cells in GPi/SNr corresponds to releasing cells they connect to in thalamus from inhibition. Consequently, in thalamus cells corresponding to selected actions are more active.

most active input to most greatly suppress the action to be selected. However, if not for the additional excitation provided by the STN, all of the actions might be concurrently suppressed (meaning more than one action had been selected). The very broad STN connections thus take the input from the hyperdirect pathway and combine it to provide a level of background excitation that allows only the most inhibited action to be selected.

It should be evident that the example shown in Figure 5.2 has carefully selected input utilities from cortex. In fact, if there are many actions with large utilities, or all actions have low utilities, then this circuit will not function appropriately. For this reason, a control system is needed to modulate the behavior of these neural groups. Gurney et al. (2001) have argued that the external globus pallidus (GPe) is ideally suited for this task, as its only outputs are back to the other areas of the basal ganglia, and it receives similar inputs from the striatum and the STN, as does the internal globus pallidus (GPi). In their model, the GPe forms a circuit identical to that in Figure 5.2, but its outputs project back to the STN and the GPi (*see* Fig. 5.3). This regulates the action selection system, allowing it to function across a wide range of utility values.

The model described so far is capable of performing action selection and reproducing a variety of single-cell recording results from electrostimulation and lesion studies. However, it does so with rate neurons–that is, the neurons do not spike and instead continually output a value based on their recent

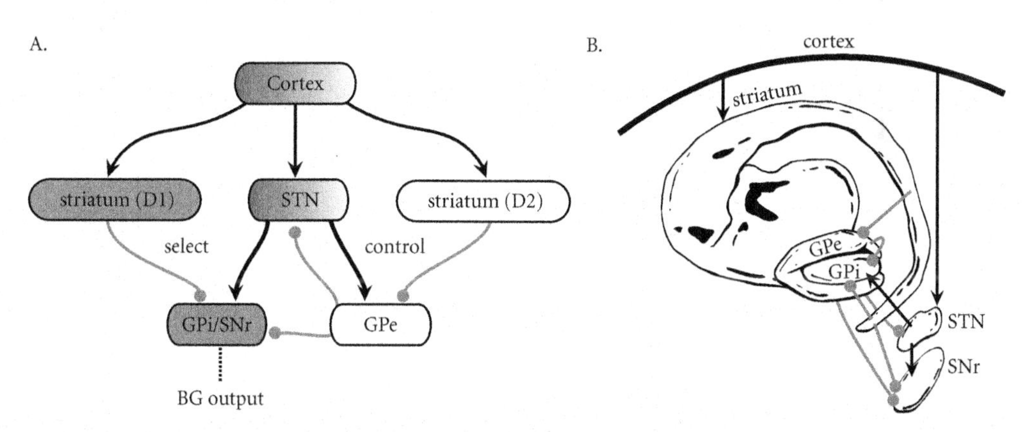

FIGURE 5.3 The full model of action selection in the basal ganglia presented by Gurney et al. (2001). **A.** The schematic circuit building on that in Figure 5.2. The circuit in the original figure is shown in gray, the structurally similar circuit encompassing cortex, STN, the striatal D2 cells, and GPe are shown in white (adapted from Gurney et al. [2001], Fig. 5). Gurney et al. demonstrated that the D2 circuit allows action selection to function across the full range of utility values. **B.** The same circuit shown in the neuroanatomy, capturing all major projections shown in Figure 5.1.

input. This makes it difficult to make precise timing predictions. Further, the model has no redundancy, since exactly one neuron is used per area of the basal ganglia to represent each action. The original version of the model shown in Figure 5.3 uses a total of nine neurons to represent three possible actions, and if any one of those neurons is removed, then the model will fail.

However, given the resources of the NEF and the SPA, these shortcomings can be rectified. In particular, the SPA suggests that rather than a single neuron representing potential actions, a mapping from a high-dimensional semantic pointer into a redundant group of neurons should be used. Additionally, the NEF provides a means of representing these high-dimensional semantic pointers and their utilities in spiking neurons and provides a method for reproducing the transformations suggested by the original model. I return to the role of semantic pointers in Section 5.3. For the time being, allow me to consider a simpler model that represents actions and utilities as scalar values. Figure 5.4 demonstrates the ability of this new model to select actions. There it

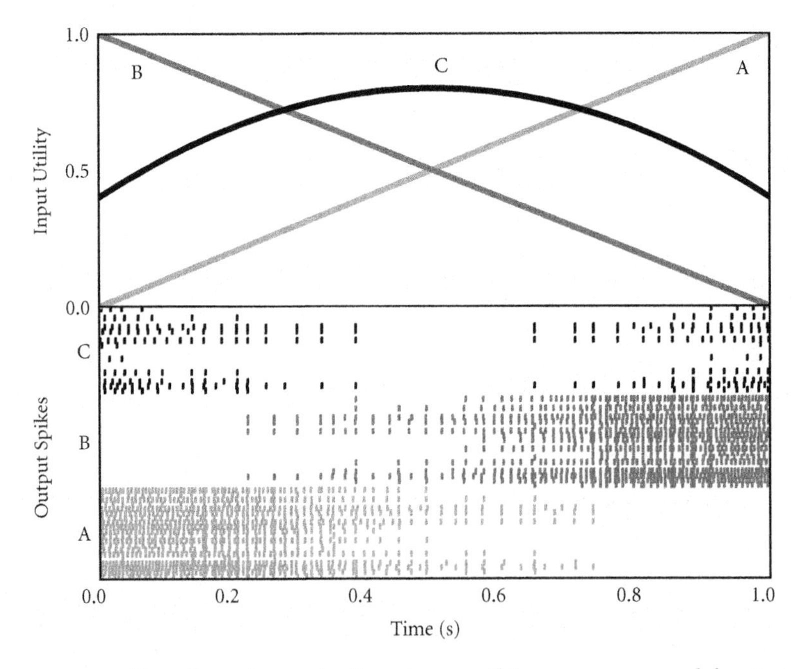

FIGURE 5.4 Simple action selection in a spiking neuron model. Output spikes from GPi/SNr produced for three possible actions (A, B, and C), as their utility changes (top graph). The highest utility action is selected, as demonstrated by a suppression of the spiking of the group of neurons corresponding to the related action. In order, the highest utility action is B then C then A. (Figure from Stewart et al. [2010].)

can be seen that the highest utility action (B then C then A) is always selected (i.e., inhibited, and so related neurons stop firing).

Crucially, this extension of the Gurney et al. model allows us to introduce additional neural constraints into the model that could not previously be included. In particular, the types of neurotransmitters employed in the excitatory and inhibitory connections of the model have known effects on the timing of a receiving neuron's response. All of the inhibitory connections involve GABA receptors (with time constants between about 6 ms and 10 ms; Gupta et al., 2000), whereas the excitatory connections involve fast AMPA-type glutamate receptors (with time constants of about 2 ms to 5 ms; Spruston et al. 1995). The time constants of these neurotransmitters have a crucial impact, on the temporal behavior of the model. I discuss these temporal properties and their related predictions in Section 5.7.

5.3 ■ Basal Ganglia, Cortex, and Thalamus

We have now seen how a spiking neuron model of the basal ganglia can be used to select simple actions. However, the purpose of the SPA is to provide a framework for building large-scale cognitive models. These, of course, require selection of complex, sequenced actions, driven by sophisticated perceptual input. In fact, there is evidence that suggests that although single actions can be selected without basal ganglia involvement, chains of actions seem to require the basal ganglia (Aldridge et al., 1993).

Previous chapters demonstrated how the SPA can support representations sufficiently rich to capture the syntax and semantics of complex representations. Here, I consider how these same representations can be used by the basal ganglia to control cognitive behavior. I begin, in this section, with an overview of the central functions and connections of cortex, basal ganglia, and thalamus. In the rest of the chapter, I demonstrate first how this circuit can select fixed sequences of actions and subsequently demonstrate how it can also select flexible sequences of actions.

To construct such models, I rely on the well-known cortex-basal ganglia-thalamus loop through the brain (*see* Fig. 5.5). Roughly speaking, the SPA assumes that cortex provides, stores, and manipulates representations; the basal ganglia map current brain states to courses of action; and the thalamus applies routing signals to cortical pathways.

Let me begin with a brief consideration of cortex. It is clear that in general, cortex is able to perform intricate tasks. In the SPA, most such tasks are built out of combinations of four basic functions: temporal integration (for working memory; *see* Section 6.2), multiplication (for control, compression, and syntactic manipulation; *see* Section 4.9), the dot product (for clean-up memory and other linear transformations; *see* Sections 3.8 and 4.6),

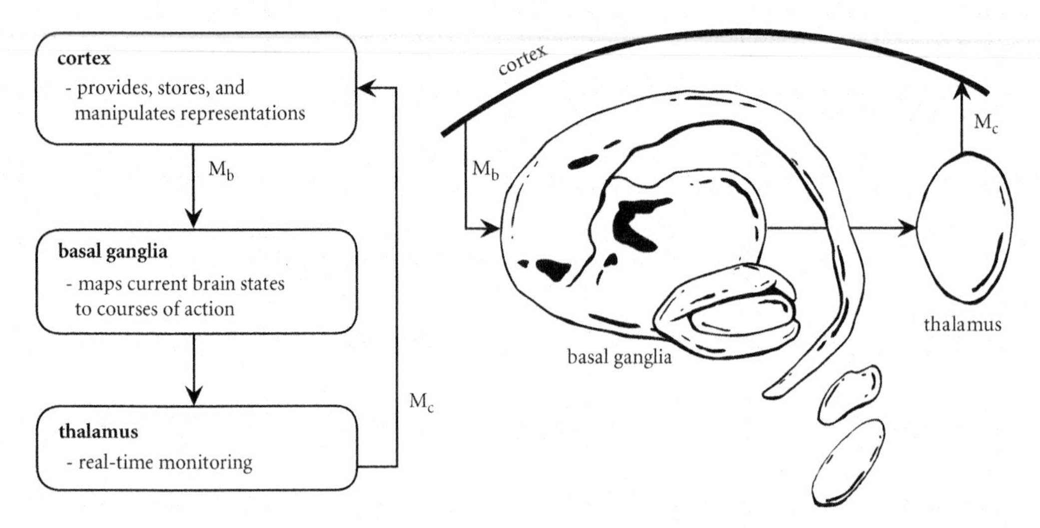

FIGURE 5.5 The cortex-basal ganglia-thalamus loop. Arrows indicate connections between the three areas. At a functional level, brain states from the cortex are mapped through the $\mathbf{M}_b$ matrix to the basal ganglia. Each row in such a matrix specifies a known context for which the basal ganglia will choose an appropriate action. The product of the current cortical state and $\mathbf{M}_b$ provides a measure of how similar the current state is to each of the known contexts. The output of the basal ganglia disinhibits the appropriate areas of thalamus. Thalamus, in turn, is mapped through the matrix $\mathbf{M}_c$ back to the cortex. Each column of this matrix specifies an appropriate cortical state that is the consequence of the selected action. The relevant anatomical structures are pictured on the right based on a simplified version of Figure 5.1.

and superposition, which is a kind of default operation. I have discussed each of these functions in detail in past sections, so I will not describe them any further here. The SPA does not carry strong commitments about the specific organization of these operations, and this is clearly an avenue for future work. Determining the most appropriate combination of these (and perhaps other) functions to mirror all cortical function is far beyond the scope of this book. Additionally, learning clearly plays a large role in establishing and tuning these basic functions. I leave further consideration of learning until later (Section 6.5). In sum, the SPA takes cortex to be an information processing resource that is dedicated to manipulating, remembering, binding, and so forth representations of states of the world and body. I take this claim to be general enough to be both correct and largely uninteresting. Far more interesting are the specific proposals for how such resources are organized, as captured in the many example models throughout the book.

As I was at pains to point out in the introduction to this chapter, for these cortical operations to be flexibly exploited, it is essential to control the flow of information between them. Interestingly, all areas of neocortex, with the exception of the primary visual and auditory cortices, project to the basal ganglia. Thus, controlling the flow of information between cortical areas is a natural role for the basal ganglia. Although there is evidence that cortex can perform "default" control without much basal ganglia influence (Kim et al., 1997), the basal ganglia make the control more flexible, fluid, and rapid. In short, in the SPA, as in other cognitive architectures, the basal ganglia control the exploitation of cortical resources by selecting appropriate motor and cognitive actions, based on representations available in cortex itself.

Finally, because basal ganglia and cortex are in many ways diffuse, performing many functions over a wide area, and because they are built on top of older, more basic control systems (e.g., brainstem and thalamus), their outputs often need to be integrated with signals coming from elsewhere in the brain–it would not do for an animal to be stuck "cognizing" about an auditory input while in imminent danger. The thalamus is structured in a way ideally suited for playing the role of a coordinator and monitor of various brain systems. For example, all output from the basal ganglia goes through the thalamus before returning to cortex, to which thalamus projects broadly. As well, the thalamus receives direct projections from every sense (except smell) and from all cortical areas. In the thalamus, the cortical projections are organized by cortical region, providing a somewhat topographical map of cortex.

There are some exceptions, however. Interestingly, the reticular nucleus of the thalamus forms a kind of shell around most of the thalamus and thus communicates with and regulates the states of many other thalamic nuclei. Consequently, the thalamus is ideally structured to allow it to monitor a "summary" of the massive amounts of information moving through cortex and from basal ganglia to cortex. Not surprisingly, thalamus is known to play a central role in major shifts in system function (e.g., from wakefulness to sleep) and participates in controlling the general level of arousal of the system. In the models presented here, thalamus acts much like a basic relay (as it was long thought to be), because the cognitive coordination of the system can be accounted for by basal ganglia for the considered tasks. I suspect, however, that thalamus has a much more important role to play in the overall system, though this role is largely untapped by the examples I consider. Nevertheless, its contribution to timing effects are important, so the thalamus is included in the models I discuss.

Understanding the elements of the cortex-basal ganglia-thalamus loop in this manner allows the SPA to assign specific transformations to the projections between these areas. As depicted in Figure 5.5, communication between cortex and basal ganglia can be mathematically characterized as mapping the representations of current cortical states to the striatum and STN (the basal ganglia input nuclei) through a matrix, $\mathbf{M}_b$. This mapping is a means of determining the utilities that drive the basal ganglia model in Section 5.2. One natural and simple interpretation of the rows of this matrix is that they specify the "if" portion of a rule (i.e., the antecedent).[2] Consider the rule "if there is an A in working memory, then set working memory to B." By multiplying $\mathbf{M}_b$ with the semantic pointer currently encoded in working memory, we get a list of similarities between its rows and the current state of working memory (i.e., $\mathbf{s} = \mathbf{M}_b\mathbf{w}$, where $\mathbf{s}$ are the similarities [i.e., dot product] between each of the rows of $\mathbf{M}_b$ and the semantic pointer $\mathbf{w}$, the input from working memory). That vector of similarities then acts as input to basal ganglia, which select the highest similarity (utility) from that input, as described in Section 5.2.

The output from basal ganglia results in a release from inhibition of the connected thalamic neurons, which are then mapped back to cortex through another matrix, $\mathbf{M}_c$. This matrix can be thought of as specifying the "then" part of a rule (i.e., the consequent), with one consequent in each column of the matrix. The output from basal ganglia selects particular actions in thalamus, whose output can be thought of as a vector, $\mathbf{a}$, with all zeros except for the selected action, which is one. Multiplying this matrix and vector will result in output equal to the column of $\mathbf{M}_c$ corresponding to the selected action– for example, setting working memory to a new state, B. More generally, the $\mathbf{M}_c$ matrix can be used to specify any consequent *control* state given a current cortical state. That is, $\mathbf{M}_c$ could specify how to route information differently through the brain given the current cortical state (not just update information as specified in this example). This loop from cortex through basal ganglia and thalamus and back to cortex forms the basic control structure of the SPA.

Two points are in order regarding this rough sketch of the cortex-basal ganglia-thalamus loop. First, as discussed in Section 7.4, the $\mathbf{M}_b$ mapping is more general than the implementation of simple if-then rules. It provides for statistical inference as well. Second, this additional inferential power is available partly because all of the representations being manipulated in this loop (i.e., of cortical states, control states, etc.) are semantic pointers. Complex representations in the SPA allow for complex control. Let me now turn to consideration of a simple control structure of this form, to illustrate its function.

5.4 ■ Example: Fixed Sequences of Actions

A simple but familiar example of a fixed sequence of actions is rehearsal of the alphabet. This is an arbitrary sequence, with no systematic rule connecting one state to its successor (unlike counting, for example). For simplicity, we are not going to consider hierarchical representation or chunking of the type observed in human alphabet encoding (Klahr et al., 1983). Consequently, we

[2] Ultimately, however, the rules implemented in this model are more subtle and powerful than standard logical "if-then" rules (*see* Section 7.4).

need to use 25 rules to traverse the entire sequence. All such rules can be of the form:

IF working memory contains **letter** + **A**
THEN set working memory to **letter** + **B**,

where bold indicates that the item is a semantic pointer. For present purposes, these pointers are randomly generated in a 250-dimensional space. The inclusion of **letter** in each of the pointers provides minimal semantic structure that can be exploited to trigger actions appropriate for any letter (*see* Section 5.6).

To implement the IF portion of the rules, an $\mathbf{M}_b$ matrix with rows consisting of all the letter representations is constructed and embedded in the connection weights between cortical working memory and the basal ganglia input (using the methods detailed in Section 3.8). This allows the basal ganglia to determine what rule is most applicable given the current state of working memory.

The THEN portion of the rule is implemented by the $\mathbf{M}_c$ matrix in a similar manner, where only one column is activated by disinhibition (as determined by the basal ganglia), sending a given letter representation to working memory. As working memory is being constantly monitored by the basal ganglia, this new state will drive subsequent action selection, and the system will progress through the alphabet.

To run the model, it is initialized by setting the working memory neurons to represent the **letter** + **A** semantic pointer. After this, all subsequent activity results from the interconnections between neurons. Figure 5.6 shows the model correctly following the alphabet sequence. From the spiking pattern we see that the correct action for each condition is successfully chosen by turning off the appropriate inhibitory neurons in the GPi. The top plot is generated by comparing the semantic pointer of each of the 26 possible letters to the current semantic pointer in working memory (decoded from spiking activity) by using a dot product, and plotting the top eight results.

This model demonstrates that a well-learned set of actions with a specific representation as an outcome can be implemented in an SPA model. However, this model is not particularly flexible. For example, we might assume that our working memory is being driven by perceptual input–say, from vision. If so, we would have a connection between our visual system and the working memory system that is driving action selection. However, if this is the case, then changing the visual input during the fixed action sequence will cause the sequence to shift suddenly to the new input. In fact, just leaving the visual input "on" would prevent the model from proceeding through the sequence, as the working memory would be constantly driven to the visually presented input despite the actions of the basal ganglia, as shown in Figure 5.7.

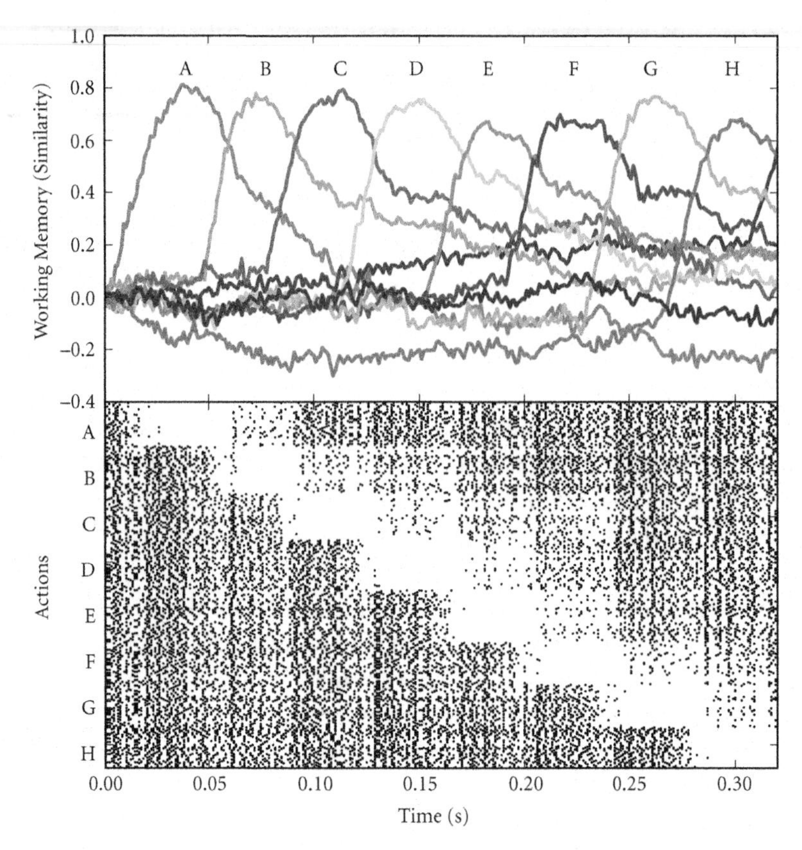

FIGURE 5.6 Rehearsal of the alphabet. **(Top)** Contents of working memory generated by taking the dot product of all possible semantic pointers with the decoded contents of working memory (top eight values are shown). **(Bottom)** The spiking output from GPi indicating the action to perform. The changes in activity demonstrate that the population encoding the currently relevant IF statement stops firing, disinhibiting thalamus and allowing the THEN statement to be loaded into working memory.

In short, the current model is not sufficiently flexible to allow the determined action to be one that actually changes the control state of the system. That is, the chosen action does not yet gate the flow of information between brain areas using the output of the basal ganglia. Rather, the chosen action simply updates the representational state of the system. To fix this problem, we need to load visual input into working memory only when appropriate. In sum, we would like basal ganglia to determine how information is *routed* in cortex, as well as being able to introduce new representations in cortex. Routing information flexibly through the brain is a fundamental neural process, one sometimes called "attention."

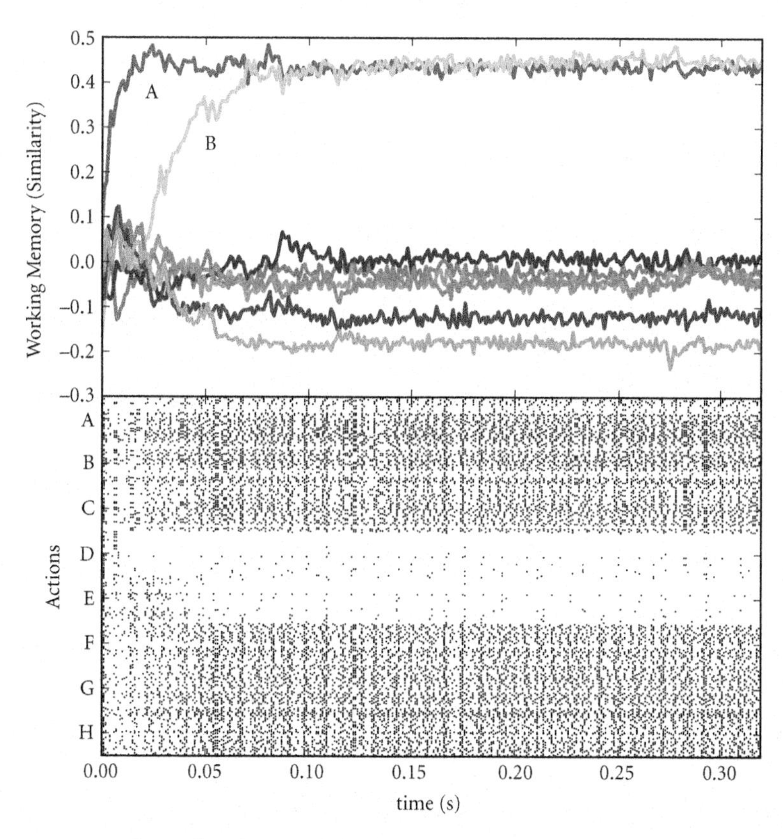

FIGURE 5.7 The inflexibility of the initial control structure is demonstrated by connecting a visual system to working memory. The perceptual system (**letter** + **A**) continually drives working memory and prevents it from properly moving to **letter** + **B** and subsequent states as shown in the top plot. The spiking out from GPi reflects this same "frozen" state in the bottom plot.

5.5 ■ Attention and the Routing of Information

Attention has been quite thoroughly studied in the context of vision. And, although visual attention can take on many forms (e.g., attention to color, shape, etc.), the spatial properties of visual attention are often of special interest. Traditionally, there are two main functions discussed when it comes to understanding visuo-spatial attention: selection and routing (notably, analogous to the two aspects of any control problem). Selection, perhaps the more studied of the two, deals with the problem of identifying what the appropriate target of the attentional system is given current task demands and perceptual features. Routing, in contrast, deals with how, once a target has been selected, a neural system can take the selected information and direct additional resources toward it. Figure 5.8 identifies many of the brain areas thought

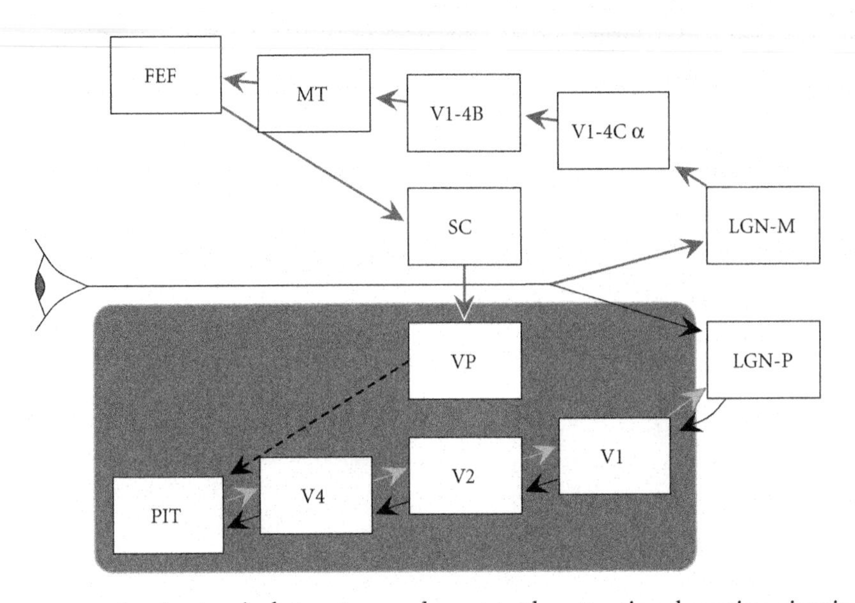

FIGURE 5.8 Anatomical structures relevant to the attentional routing circuit (ARC). The lateral geniculate nucleus (LGN) is a structure within the thalamus that receives information directly from the optic nerve. The LGN contains magnocellular and parvocellular cells that form the LGN-M and LGN-P layers, respectively. The other structures shown are the posterior inferior temporal cortex (PIT), visual cortical areas (V1, V1-4B, V1-4C α, V2, V4), frontal eye fields (FEF), the middle temporal area (MT), superior colliculus (SC), and ventral pulvinar (VP). Feedforward visual information projects through the ventral stream areas (black lines). The pathway from LGN-M leading to FEF (dark gray lines) is one possible route through which the focus of attention (FOA) may be rapidly determined. Areas within the rounded gray box are included in the ARC model. VP projects a coarse signal (dashed black line) indicating the location of the FOA to control neurons in PIT. Based on the VP signal, local control signals are computed in PIT. The results are then relayed to the next lower level of the ventral stream (light gray lines), where relevant local control signals are computed and again relayed.

to be involved in both selection and routing in vision. In this section, I only discuss the routing problem, which has not received as much consideration as selection (although *see* Gisiger & Boukadoum [2011] for a recent summary of potential neural mechanisms for routing).

In the last 15 years there have been several proposed models of attentional routing (Olshausen et al., 1993; Salinas & Abbott, 1997; Reynolds et al., 1999; Wolfrum & von der Malsburg, 2007; Womelsdorf et al., 2008). However, none of these use biophysically plausible spiking neurons, most are purely mathematical models, and few characterize attentional effects beyond a single neuron or small cortical area. Consequently, most–if not all–have been criticized as being unscalable (e.g., there are not enough neurons in the relevant brain structures to support the required computations). Given the importance of both scalability and biological plausibility to the SPA, Bruce Bobier (from my lab) has developed a model of attentional routing called the "Attentional Routing Circuit" (ARC), which addresses these issues.

The ARC draws on past attentional models in several ways. For example, it relies on nonlinearities to perform attentional routing, as many past proposals do. It also incorporates connectivity constraints and general architectural considerations, and shares a concern for explaining behavioral and psychophysical data with several of these models. It is most directly a descendant of the "shifter circuit" model (Olshausen et al., 1993). Nevertheless, it improves upon the shifter circuit by being significantly more biologically plausible, scalable, and hence able to account for detailed neurophysiological findings not addressed by that model. Consequently, the ARC provides an especially good account of routing, and one that can be generalized to other parts of the SPA (*see* Section 5.6). Let me consider the ARC in more detail.

As shown in Figure 5.8, the ARC consists of a hierarchy of visual areas (from V1 to PIT), which receive a control signal from a part of the thalamus called the ventral pulvinar (VP). A variety of anatomical and physiological evidence suggests that pulvinar is responsible for providing a coarse, top-down attentional control signal to the highest level of the visual hierarchy (Petersen et al., 1985, 1987; Stepniewska, 2004).

A more detailed picture of the model's connectivity between these visual areas is provided by Figure 5.9. In that figure, an example focus of attention that picks out approximately the middle third of the network is shown. To realize this effective routing (i.e., of the middle third of V1 up to PIT), VP provides a control signal to PIT indicating the position and size of the current focus of attention in V1. Control neurons in PIT use this signal to determine what connections to "open" between V4 and PIT and then send their signal to control neurons in V4. The signal that is sent to the next lower level of the hierarchy (V4) is similarly interpreted by that level to determine what gating is appropriate and then passed on (i.e., to V2). Again, the gating allows the flow of information only from those parts of the next lower level that fall within the focus of attention. And so this process of computing and applying the appropriate routing signal continues to V1.

To "open" a connection essentially means to multiply it by a non-zero factor. Consider a simple example. Suppose we have a random signal going from population A to population B in a communication channel (*see* Section 3.8). At the inputs to population B, we might insert another signal that can be 0 or 1, coming from our control population, C. If we multiply the inputs to B by C, then we cause the representation in B to be either 0 (if C = 0) or the

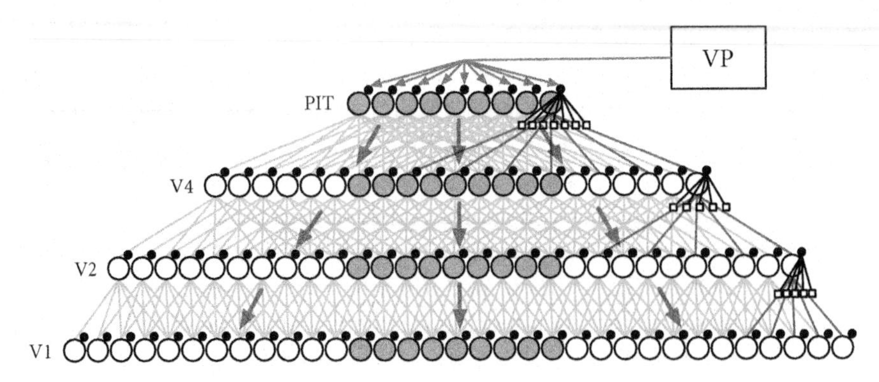

FIGURE 5.9 The architecture of the attentional routing circuit (ARC) for visual attention. Each level has a columnar and retinotopic organization, where columns are composed of visually responsive pyramidal cells (white circles) and control neurons (black dots). Filled gray circles indicate columns representing an example focus of attention. Neurons in each column receive feed-forward visual signals (gray lines) and a local attentional control signal from control neurons (black lines). These signals interact nonlinearly in the terminal dendrites of pyramidal cells (small white squares). Coarse attentional signals from ventral pulvinar (VP) are relayed through each level of the hierarchy downward to control neurons in lower levels (large gray arrows). Control connectivity is highlighted for the right-most columns only, although other columns in each level have similar connectivity.

current value of A (if C = 1). Population C thus acts as a gate for the information flowing between A and B. The control neurons in ARC are performing a similar function.

In essence, the control neurons determine which lower-level (e.g., V4) neurons are allowed to project their information to the higher level (e.g., PIT). In the ARC, this gating is realized by the nonlinear dendritic neuron model mentioned in Section 4.3. This kind of neuron essentially multiplies parts of its input in dendrites, and sums the result to drive spiking in the soma. Consequently, these neurons are ideal for acting as gates, and fewer such neurons are needed than would be in a two-layer network performing the same function.

The mapping of the computational steps employed in the ARC to cells in specific cortical layers of these visual areas is shown in Figure 5.10. As shown here, neurons in layer V receive the top-down control signal specifying the size and position of the desired routing. These then project to layer VI neurons that determine an appropriate sampling and shift consistent with the desired routing. The results of this computation are sent to layer IV neurons that act to gate the feed-forward signals into layer II/III neurons, which project to higher levels of the visual hierarchy. A more detailed discussion of the anatomy and physiology underlying this mapping can be found in Bobier (2011).

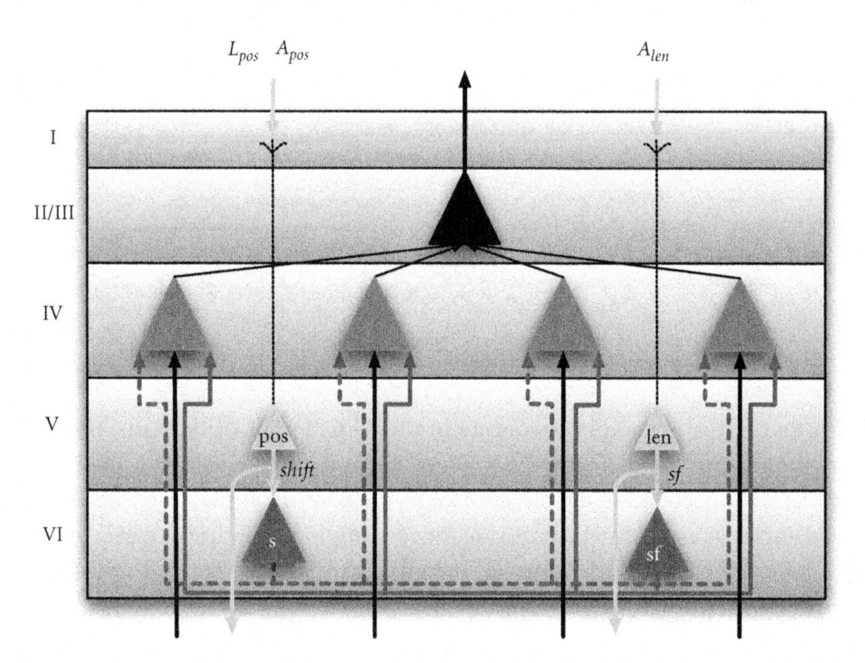

FIGURE 5.10 Proposed laminar arrangement of neurons in ARC for a single column. Global attention signals (length A_{len} and position A_{pos} of selected region) and a local signal indicating the position of the target in the current level (L_{pos}) are fed back from the next higher cortical level to layer I of the current level, where they connect to apical dendrites of layer-V cells. The layer-V neurons relay this signal to the next lower level with collaterals projecting to control neurons in layer VI that compute the sampling frequency (*sf*, solid lines) and shift (*s*, dotted lines). These signals, along with feed-forward visual signals from retina (solid black lines from the bottom) are received by layer-IV pyramidal cells, where the routing occurs (i.e., layer-VI signals gate the feed-forward signals). Cells in layer II/III (black) pool the activity of multiple layer-IV neurons and project the gated signal to the next higher visual level.

This organization and the computations it underwrites repeated throughout the model (consistent with most contemporary accounts of cortical organization). Consequently, the model, although detailed, is reasonably straightforward as its central features are simply repeated over and over throughout the width and depth of the hierarchy shown in Figure 5.9.

Figure 5.11 shows an example of the shifting and resizing of information presented to V1 made possible by this mechanism. As can be seen in that figure, only the right side of the signal in V1 is sent to the top levels of the network's hierarchy. The model preferentially copies data from the nodes that are the focus of attention to higher levels, resulting in the observable scaling and shifting effects. It should be evident that the circuit is thus effectively taking the information on the lowest level of the hierarchy and "routing" it to

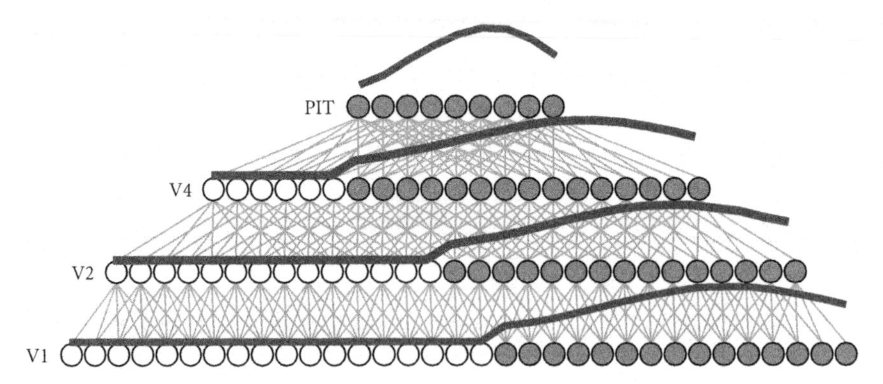

FIGURE 5.11 An example of routing in the ARC. Levels of the visual hierarchy are laid out as in Figure 5.9. Above each of these levels, the thick line indicates the decoded signal represented by the neurons at that level. The darkened circles in each level indicate the focus of attention in the network. In this example, the focus of attention has the effect of both shifting and scaling the original data from V1 to fit within PIT.

always fit within the available resources at the highest level. There is ample psychophysical evidence that precisely this kind of processing occurs in the visual system (Van Essen et al., 1991). Additionally, this model is able to account for several detailed observations about the neural mechanisms of attention.

For example, it accounts for the known increase in size of the receptive fields of neurons farther up the hierarchy, is consistent with the patchy connectivity in the visual hierarchy (Felleman et al., 1997), and captures the topographic organization of receptive fields in these areas (Tanaka, 1993). The ARC also accounts for the observation that the timing of attentional modulation of neural activity begins at the top of the hierarchy and proceeds down it (Mehta, 2000; Buffalo et al., 2010). Many other anatomical, psychophysical, and physiological properties of attention are captured by the ARC (Bobier, 2011). More importantly, I want to focus on whether the ARC is plausible in its neural details.

To demonstrate the biological plausibility of ARC, let me consider just one set of experiments in detail. In a set of challenging experiments reported in Womelsdorf et al. (2006), recordings were taken from monkeys performing a spatial attention task (*see* Fig. 5.12). Specifically, the animals foveated on a fixation point, after which a cue stimulus (S1) was presented indicating where the animal should covertly attend. Following a delay, three stimuli were presented–one at the target location and two distractor stimuli (S2 and S3–one inside and one outside of the recorded cell's receptive field). The animal had to indicate when it saw a brief change in the stimulus at the cued location, some random interval after the three stimuli were presented. During the interval, the animal was taken to have sustained spatial attention to S1, and

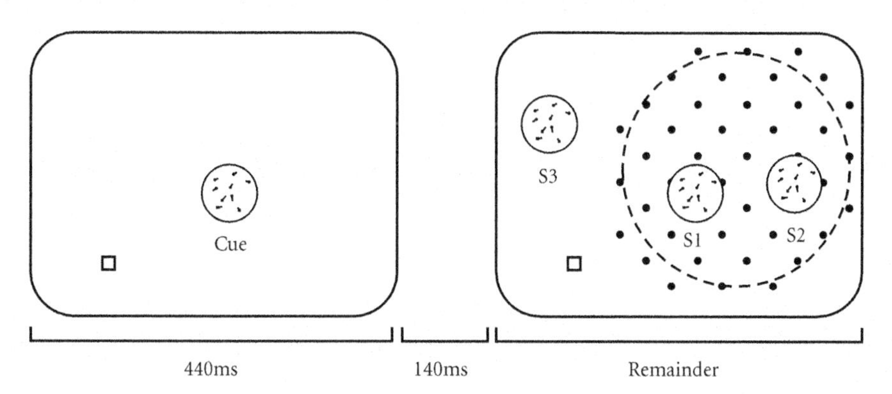

FIGURE 5.12 The Womelsdorf task (Womelsdorf et al., 2006, 2008). A trial begins when a monkey foveates in a small square area near the fixation point (small square). Then a stationary random dot pattern appears for 440 ms (the cue). After a brief delay period, three moving random dot patterns are shown. Two of these (S1 and S2) were placed within the neuron's receptive field (dotted circle). The third (S3) was placed outside the receptive field. The receptive field of the neuron was mapped during the delay between the cue and stimulus change with high contrast probes that were randomly placed on a grid (black dots). The experiment allowed systematic probing of changes in receptive fields for different attentional targets.

the receptive field of the cell was mapped. This experimental design allowed Womelsdorf et al. to map the receptive field during sustained states of selective attention to the stimuli inside the receptive field or to the stimulus outside the receptive field. The experimenters wanted to understand how the receptive fields of individual neurons changed, given that they were within the focus of attention or not.

To compare the model to these experimental results, we can run analogous experiments on the model. We can also use the same methods for analyzing the spiking data to determine neuron receptive fields as in the experiment. However, a central difference between the model and the monkey experiments is that in the model all spikes could be collected from all neurons. In addition, we could run the experiment on 100 different realizations of the model. Each version has the same general ARC architecture, but the neurons themselves are randomly chosen from a distribution of parameters that matches the known properties of cells in the relevant parts of cortex. Consequently, rather than having just over a hundred neurons from two animals as in the experiment, we have thousands of neurons from hundreds of model-animals. This means that we have a much better characterization of the overall distribution of neuron responses in the model than is available from the experimental data for the animals, and we also have a distribution over animals, which is not available from the experimental data because there are only two monkeys.

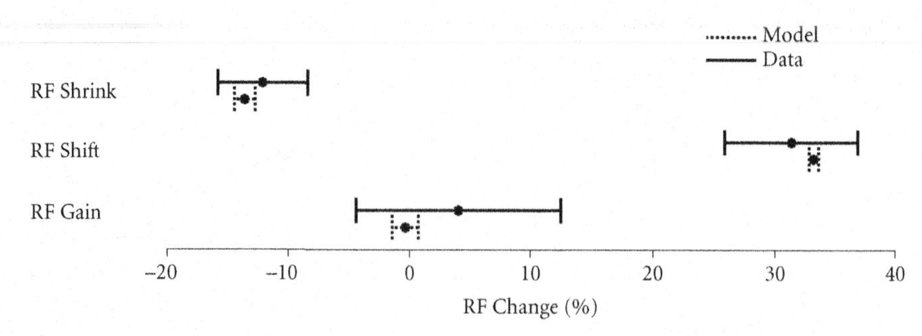

FIGURE 5.13 Attentional effects (mean and 95% confidence intervals) for 100 model-animals (dashed lines) and data from Womelsdorf et al. (2008) (solid lines). Because we can record from many more spiking neurons in the model, the distribution is much better characterized, and so the confidence intervals are tighter. These results demonstrate that the ARC quantitatively accounts for the receptive field (RF) shrink, shift, and gain observed experimentally.

To compare the model and data, we can consider the three main effects described in the experimental work. These effects were seen by comparing attention being directed at a stimulus inside the receptive field to attention being directed at a stimulus outside the receptive field. The effects were: (1) a change in the peak firing rate (i.e., gain); (2) a shift of the receptive field center (i.e., a shift); and (3) a change in the receptive field size (i.e., a shrink). In each case, several statistics were calculated to compare model and experimental data (*see* Bobier, 2011, for details). Overall, the model and data match extremely well on each of these effects, as shown in Figure 5.13.

Although this is just a single experiment, the fact that the model is anatomically, physiologically, and functionally constrained and that the data analysis methods are identical suggests that the underlying mechanisms are likely similar. Further, a comparison with this same level of detail to neurophysiological experiments from Treue and Martinez-Trujillo (1999) and Lee and Maunsell (2010) is performed in Bobier (2011). The first of these experiments demonstrated that the width of tuning to visual features is not affected by attention, although the gain of the neuron responses is affected. The second experiment demonstrated that neuron gain responses are best explained as response gain and not contrast gain. Both sets of experimental results are well captured by the exact same ARC model.

One final note about this model is that there are only two free parameters. One governs the width of the routing function and the other the width of feed-forward weights. Both of these parameters are set to match the known receptive field sizes in visual cortex independently of any particular experiment. That is, they were in no way tuned to ensure the ability of the model to predict any specific experiment. The remaining parameters in the model are randomly chosen between different model-animals from distributions known to match general physiological characteristics of cells in visual cortex. This means that the model does not simply capture the mean of the population (of cells or animals) but actually models the *distribution* of responses, which is more difficult. Nevertheless, the model is able to provide a good characterization of changes in several subtle aspects of individual cell activity across a variety of experiments from different labs. Consequently, the ARC provides a breadth of experimental consistency not found in other attention models.

Overall, these results suggest that the detailed mechanism embodied by the ARC for routing information through cortex is plausible in its details. That is, the ARC provides a biologically plausible method for *applying* control, which was missing from our model of fixed sequences of actions.

5.6 ■ Example: Flexible Sequences of Actions

With the ARC characterization of routing in hand, we can return to the model of Section 5.4 and repair its shortcomings. The model as presented was unable to appropriately ignore visual input (*see* Fig. 5.7). However, if we allow thalamus to generate gating signals that control the flow of information between visual input and working memory, then we can prevent such errors.

As mentioned, the model of visual attention presented in the preceding section provides a biologically realistic mechanism for exactly this function. Although it is clearly geared to a different part of cortex, it demonstrates how thalamic gating signals (in that case, from pulvinar) can be used to control the flow of information through a cortical circuit. As well, there is nothing physiologically unusual about visual cortex, so it would be unsurprising to see similar functions computed in similar ways in other parts of cortex. In short, this means that computing a nonlinearity between a thalamic input and cortical signals can be considered a plausible and effective means of routing information in many areas of cortex. So, we can introduce a similar circuit (essentially just one level of the ARC) into our action-selection model and provide for much more flexible control.

The addition of this mechanism means that the model can selectively ignore or be driven by the visual input (*see* Fig. 5.14). Notably, in this network a much more general kind of rule is employed than we have seen before. Specifically, there is now a mapping of the form

> IF visual cortex contains **letter**+?
> THEN copy visual cortex to working memory.

The "?" stands for any valid letter. So when this rule is activated, the contents of visual cortex will be loaded into working memory, which is monitored by the same rules as before. Now, only when this state is available in visual cortex will it have any effect on the activity of the model.

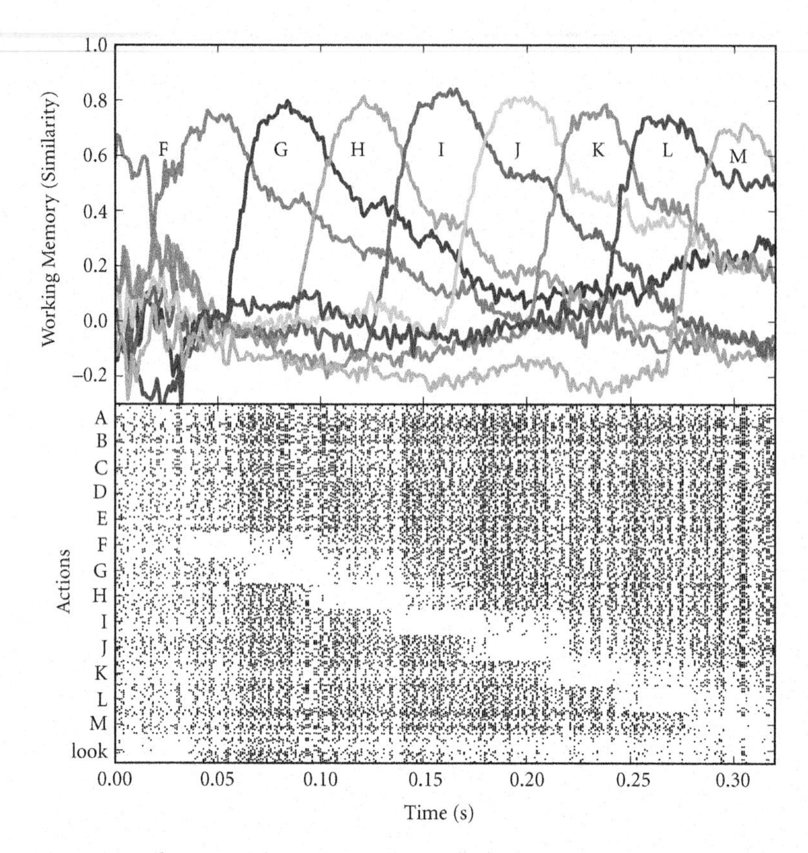

FIGURE 5.14 The consequences of routing information in a simple sequence of actions. The contents of working memory are shown on top. The similarity of vocabulary vectors is found with respect to the decoded value of the working memory ensemble. The lower half of the graph shows spiking output from GPi indicating the action to perform. The **letter**+? action takes information from visual cortex and routes it to working memory (in this case **letter** + **F**). This representation of **F** stays constantly on the input throughout this run (as in Fig. 5.7). In this case, however, it does not interfere with the sequence because of the routing. This circuit is constructed in the tutorial in Section 7.5.

Crucially, this "copy visual cortex" command is the specification of a control state that consists of gating the information between visual inputs and working memory (unlike rules that only update representational states in the previous model). This demonstrates a qualitatively new kind of flexibility that is available once we allow actions to set control states. In particular, it shows that not only the content of cortical areas but also the communication between such areas can be controlled by our cognitive actions.

There is a second interesting consequence of being able to specify control states. Notice that the specified rule applies to every letter, not just the one that happens to be in the visual input at the moment. This allows for rules to be defined at a more general category level than in the previous model. This demonstrates an improvement in the flexibility of the system, in so far as it can employ instance-specific or category-specific rules.

In fact, routing can provide yet more flexibility. That is, it can do more than simply gate information flow between different cortical areas. In the simple alphabet model above, the routing action was essentially "on" or "off" and hence gate-like. However, we can use the same neural structure described in ARC to actually *process* the signals flowing between areas. Recall that the SPA method of binding semantic pointers is to use circular convolution, which is a linear transformation followed by multiplication. As described in the section on attention, gating can also be accomplished by multiplication (and linear transformations will not interfere with the multiplication). If we allow our gating signal to take on values *between* 0 and 1, then we can use the same gating circuits to bind and unbind semantic pointers, thus not only routing but also *processing* signals. Essentially, we can give the gating signals useful content.

Introducing this simple extension means that the same network structure as above can be used to perform syntactic processing. So, for example, we can implement a dynamic, controlled version of the question-answering network described in Section 4.4. In this network, we define semantic pointers that allow us to present simple language-like statements and then subsequently ask questions about those statements. So, for example, we might present the statement

$$\mathbf{statement} + \mathbf{blue} \circledast \mathbf{circle} + \mathbf{red} \circledast \mathbf{square}$$

to indicate that a blue circle and red square are in the visual field. We might then ask a question in the form

$$\mathbf{question} + \mathbf{red},$$

which would be asking: "What is red?"

To process this input, we can define the following rules

IF the visual cortex contains **statement**+?
THEN copy visual cortex to working memory,

which simply gates the visual information to working memory as before. We can also define a rule that performs syntactic processing while gating

IF visual cortex contains **question**+?
THEN apply visual cortex to the contents of working memory.

Here, "apply" indicates that the contents of visual cortex are to be convolved with the contents of working memory, and the result is stored in the network's output. More precisely, the contents of visual cortex are moved to a

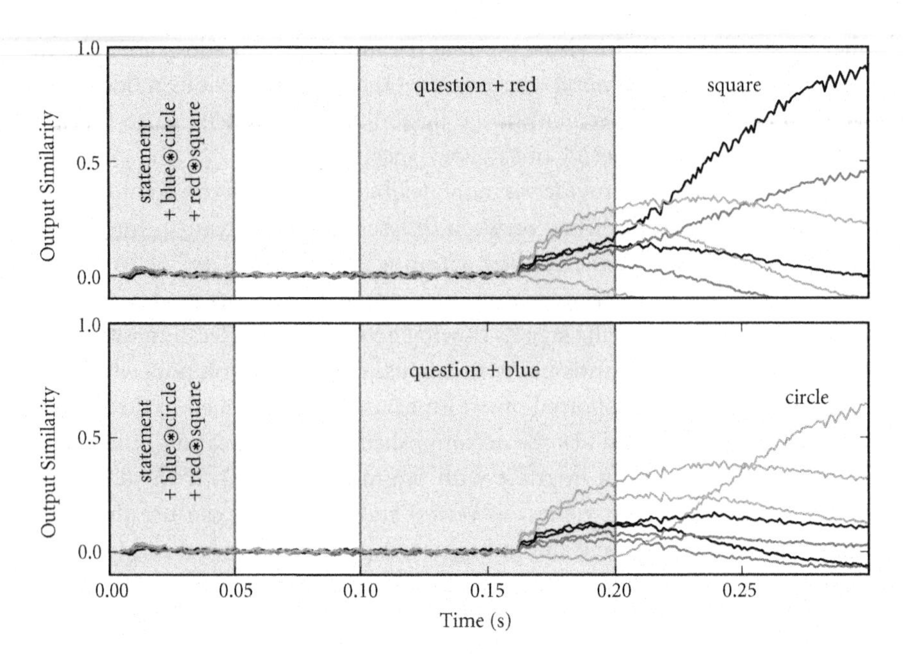

FIGURE 5.15 Answering two different questions starting from the same statement. Gray areas indicate the period during which the stimuli were presented. The similarity between the contents of network's output and the top seven possible answers is shown. The correct answer is chosen in both cases after about 50 ms. (Figure from Stewart et al. [2010].)

visual working memory store (to allow changes in the stimulus during question answering, as above), and the approximate inverse (a linear operation) of visual working memory is convolved with working memory to determine what is bound to the question. This result is then stored in an output working memory to allow it to drive a response.

The results of this model answering two different questions from the same remembered statement are given in Figure 5.15. The circuit implementing these two generic rules can store any statements and answer any questions provided in this format. This circuit is constructed in the tutorial provided in Section 5.9. Notably, this same model architecture can reproduce all of the sequencing examples presented to this point. This means that the introduction of these more flexible control structures does not adversely impact any aspects of the simpler models' performance as described above.

In addition, we can be confident that this control circuit will not adversely affect the scaling of the SPA. Only about 100 neurons need to be added for each additional rule in the basal ganglia. Of those, about 50 need to be added to striatum, which contains about 55 million neurons (Beckmann & Lauer, 1997), 95% of which are input (medium spiny) neurons. So, about one million rules can be encoded into a scaled-up version of this model. There is currently no reason to think that this is an unreasonable constraint. Consequently, in combination with the reasonable scaling of the representational aspects of the SPA (*see* Section 4.6), I believe it is clear that the SPA as a whole scales well.

In sum, we have seen how the detailed neural mechanisms related to attentional routing can be adapted to provide an account of informational routing in the SPA in general. I have shown that this approach to routing can be usefully integrated into a functional model to perform syntactic processing and flexible sequencing. So, unlike past neural mechanisms that have been suggested to account for attention (for a review see Gisiger & Boukadoum, 2011), the mechanism adopted by the SPA is simultaneously biologically detailed and functionally effective.

5.7 ■ Timing and Control

Because the model of the basal ganglia that I have presented is constructed using a variety of biological constraints, it is possible to ask questions of this model that have not been addressed adequately in past models. Specifically, because we know the kinds of neurons, their spiking properties, and the temporal properties of the neurotransmitters in basal ganglia, we can make specific predictions about the timing of action selection (Stewart et al., 2010).

For example, Ryan and Clark (1991) showed that in the rat basal ganglia the output neurons stop firing 14 ms to 17 ms after a rapid increase in the utility of one of the possible actions. It is a simple matter to run such an experiment on the basal ganglia model. To keep things straightforward, we include two possible actions, A and B, and rapidly change the utility by directly changing the input to striatum (i.e., by changing the cortical firing pattern to match that of one action). An example run of this model is shown in Figure 5.16A, which results in a cessation of firing after approximately 15 ms, consistent with the observations in rats.

More interestingly, we can systematically explore the effects of the difference in utility between the two actions on the length of time such firing persists. Figure 5.16B presents these more informative results, and it can be seen that the latency can increase to about 38 ms for actions that are only slightly different in utility after the change. For the largest utility differences, the latency drops to about 14 ms, the same lower bound as the Ryan and Clark experiment. As far as I am aware, this latency profile as a function of utility differences remains an untested but strong prediction of the model.

We can also look at the role basal ganglia plays in the timing of the entire cortex-basal ganglia-thalamus loop. This timing is implicitly demonstrated by Figure 5.6. There it can be seen that in the fixed action selection case, it takes about 40 ms for the system to switch from one action to another. This is more fully characterized in Figure 5.17, where the mean timing is shown

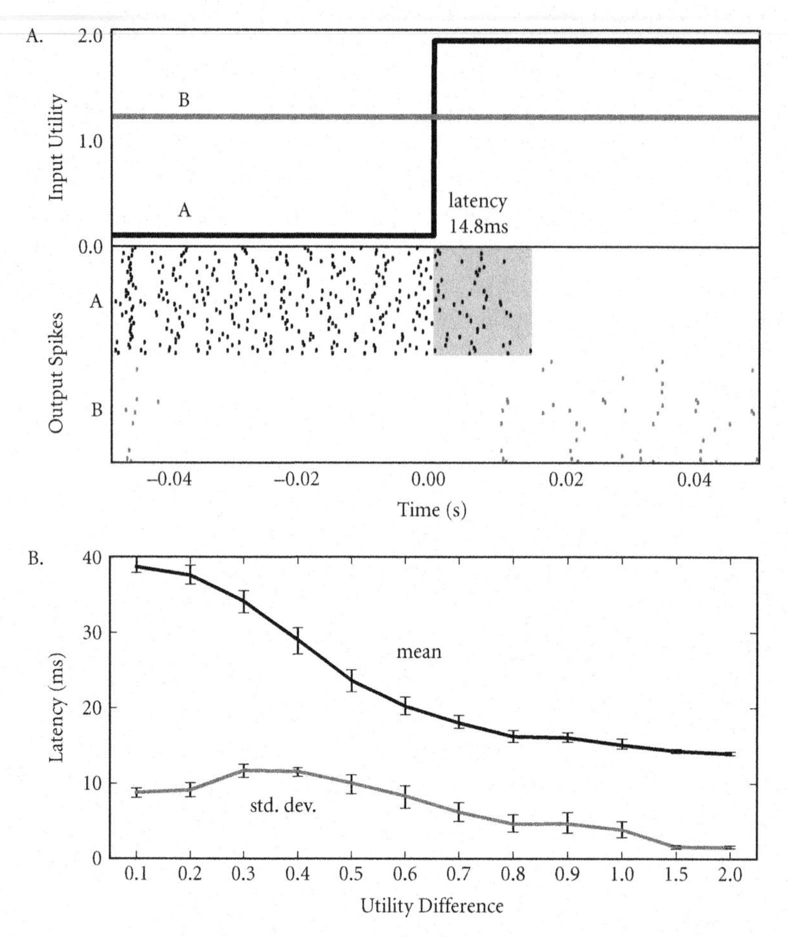

FIGURE 5.16 Timing predictions for changes in action utility. **A.** An example of the delay time between a rapid change in utility (top) and the change in spiking activity choosing a new action (bottom). Here, firing for action A stops 15.1 ms after the change in utility, matching the 14 ms to 17 ms delay observed by Ryan and Clark (1991). **B.** Averages and standard deviations of such changes over many utility differences. Error bars are 95% confidence intervals over 200 runs. (Figure from Stewart et al. [2010].)

over a range of time constants of the neurotransmitter gamma-aminobutyric acid (GABA). GABA is the main neurotransmitter of the basal ganglia and has been reported to have a decay time constant of between 6 ms and 11 ms (Gupta et al., 2000). We have identified this range on the graphs with a gray bar. On this same graph, we have drawn a horizontal line at 50 ms because this is the standard value assumed in most cognitive models for the length of time it takes to perform a single cognitive action (Anderson et al., 1995; Anderson, 2007).

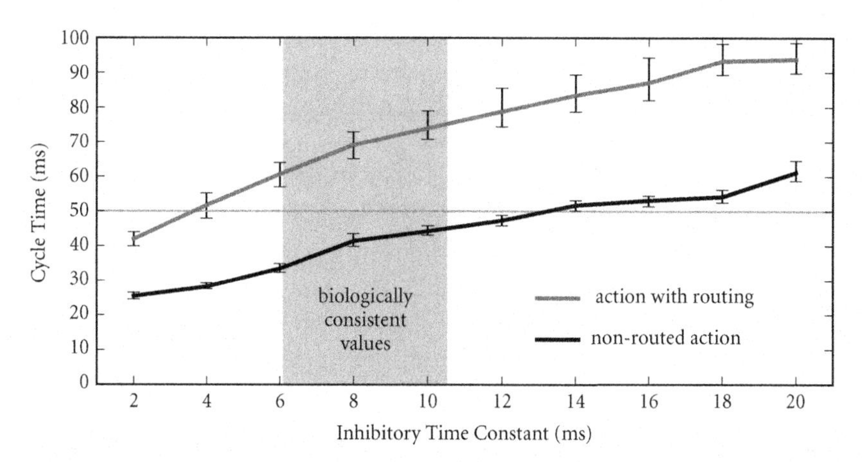

FIGURE 5.17 The effects of neurotransmitter time constant on the timing of cognitive actions. This plot shows the effect that increasing the time constant of the GABA neurotransmitter has on cycle time and compares a simple, non-routed action to a more complex action involving routing. Biologically plausible values for the time constant of the neurotransmitter are highlighted with the gray bar. The gray horizontal line indicates a 50-ms cycle time, which is usually assumed by cognitive models in their matches to behavioral data (Anderson, 2007). (Figure from Stewart et al. [2010].)

Interestingly, the cycle time of this loop depends on the complexity of the action being performed. Specifically, the black line in Figure 5.17 shows the simplest fixed action cycle time, whereas the gray line shows the more complex flexible action cycle time, discussed in Section 5.6. The main computational difference between these two types of action is that the latter has a control step that can re-route information through cortex. The cycle time increases from about 30 ms to 45 ms in the simple case to about 60 ms to 75 ms in the more complex case. These two instances clearly bracket the standard 50-ms value. Notably, this original value was arrived at through fitting of behavioral data. If the tasks used to infer this value include a mix of simple and complex decisions, then arriving at a mean between the two values described here would be expected.

There is some evidence in the ACT-R literature that this has indeed happened. In a recent model of the effects of fatigue on driving, the cycle time needed to be reduced to 40 ms (Gunzelmann et al., 2011). Similarly, in a model and experiment on attention by Wang et al. (2004), simple attentive elements (orienting and alerting) of a task were found to be quicker than an element that required comparison (that they called "executive control"). To model this, they had to use a reduced cycle time of 40 ms to capture these differences. This modeling result has been replicated (Hussain, 2010), consistently finding a best-fit of 40 ms for the simpler steps. However, these authors further assumed that the "executive control" part of the task required two cycles. The SPA suggests an alternative–that the simpler tasks may be of a different kind than the slower comparison task (i.e., non-routing versus routing, respectively). This strikes me as a more parsimonious explanation than suggesting that the comparison task required two "steps" and that rule firing was unusually fast in this attentional setting for an unspecified reason.

In general, the timing results from this basal ganglia model help to highlight some of the unique properties of the SPA. For example, the control structures of the SPA provide an explanation of the genesis of what has been taken to be a "cognitive constant"–that is, the 50 ms cycle time. This is especially helpful in cases where the constant is not constant. Notably, the SPA explanation is not available to more traditional cognitive modeling approaches, such as ACT-R, because the relevant biological constraints (e.g., neurotransmitter time constants, anatomy and physiology of the basal ganglia, etc.) play no role in these approaches.

Similarly, the SPA directly addresses available neural data about the impact of utility on selection speed, making detailed predictions about this relationship. Such explanations and predictions cannot be provided by a basal ganglia model implemented in rate neurons, because such models do not include biological parameters affecting the relevant timing information (e.g., neurotransmitter time constants, cellular neurophysiology parameters).

In general, the dynamical properties of the SPA do not merely help us match and explain more data, they also suggest behavioral and neurobiological experiments to run to test the architecture. Behaviorally, the SPA suggests that it should be possible to design experiments that distinguish the kinds of cognitive action that take longer (or shorter) to execute with reference to their complexity. Physiologically, the SPA suggests that it should be possible to design experiments that manipulate the length of time neurons in basal ganglia fire after actions are switched. Results from such experiments should provide a more detailed understanding of the neural underpinnings of cognitive control.

The central theme of this chapter, "control," is tightly tied to neural dynamics. Although I have discussed control more in terms of flexibility of routing information because of my interest in presenting an architecture that can manipulate complex representations, the dynamical properties underlying such control are ever-present and unavoidable. In many ways, measuring these dynamical properties is simpler than measuring informational properties, because *when* an event occurs (such as a spike or a decision) is more amenable to quantification (e.g., using a clock) than determining *what* such an event represents (e.g., determining what content the observed spikes carry).

But, of course, the "when" and the "what" *together* determine behavior. As a result, architectures that can be constrained by empirical data relating to both dynamical and informational properties are subject to stricter constraints.

Being able to simultaneously meet such constraints is a critical test for any proposed architecture. This is why the ability of the SPA to relate to temporal constraints, while performing interesting information processing, is an important strength of the approach.

There are, of course, many more kinds of temporal constraints than those I have discussed in this chapter. In the next chapter, I consider temporal constraints related to learning and memory. Although the behavior under consideration is different, the theme is the same: the resources of the SPA can provide a biologically based account of the detailed dynamical and information processing processes underlying cognition. In support of that theme, let me consider one more example–one that combines flexible sequences of actions, routing, and binding to solve a classic task in cognitive science.

5.8 ■ Example: The Tower of Hanoi

The Tower of Hanoi task involves three pegs and a fixed number of disks of different sizes with holes in them such that they can be placed on the pegs. Given a starting position that stacks disks from largest to smallest on the first peg, the goal is to move all of the disks to the third peg, subject to the constraints that only one disk can be moved at a time and a larger disk cannot be placed on top of a smaller disk (*see* Fig. 5.18). The Tower of Hanoi task has a rich history in cognitive science as a good example of human problem solving (e.g., Simon, 1975), and as a result, there are a variety of symbolic cognitive models that match expert human behavior well (e.g., Altmann & Trafton, 2002).

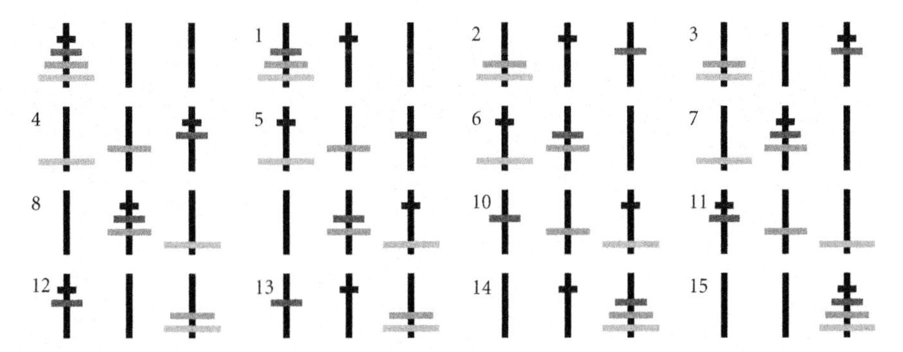

FIGURE 5.18 The Tower of Hanoi task for four disks. Each move made by an expert solver is shown. Simon's (1975) original "Sophisticated Perceptual Strategy" results in these moves and is the basis for the neural model presented here. (From Stewart & Eliasmith, 2011b.)

There are also several neural network models that solve the task, but these typically treat the problem as a pure optimization problem. Consequently, they only show that an abstract network architecture (e.g., a Hopfield network or a three-layer network) can solve a problem that can be mapped to the task (e.g., Kaplan & Gzelis, 2001; Parks & Cardoso, 1997). They do not show any match to behavioral timing or neural data, so there is little reason to think such models solve the task in the same way that humans do. To the best of my knowledge, there are no neural architectures that have been shown to capture the details of human performance. This is likely because a model must implement rule following, planning, and goal recall to solve the task in a human-like manner. Of course, the point of the SPA is to provide a means of constructing just such models that conform to the known anatomy, connectivity, and neural properties of the basal ganglia, thalamus, and cortex. In my lab, Terry Stewart has recently exploited these properties of the SPA to construct just such a model (Stewart & Eliasmith, 2011b).

There are many algorithms that can be used to produce the series of moves shown in Figure 5.18. As a result, one focus of cognitive research on this task has been determining which algorithm(s) people are using. This is done by examining factors such as the time taken between steps and the types of errors produced. In this SPA model, the basic algorithm used is the "Sophisticated Perceptual Strategy" suggested by Simon (1975).

Let us suppose for a moment that we are a subject employing this algorithm. To do so, we begin by examining the largest disk that is not in the correct location. We then examine the next smaller disk. If it blocks the move we want to make, then our new goal is to move that disk to the one peg where it will not be in the way. We iterate these steps, going back to previous goals once we have accomplished the current one.

To realize this algorithm, we need to keep track of three things: the disk we are currently attending to, the disk we are trying to move, and the location we are trying to move it to. To keep track of these varying states, we can introduce three cortical working memory circuits that I will label ATTEND, WHAT, and WHERE for ease of reference (*see* Fig. 5.19).

In addition, the algorithm requires the storage and recall of old goals regarding which disk to place where. Storing a single goal such as "disk 4 on peg C" would be easy–we could simply add the vectors together (**disk4** + **pegC**) and store the result. However, multiple goals cannot be stored in this manner, as (**disk4** + **pegC**) + (**disk3** + **pegB**) cannot be distinguished from (**disk4** + **pegB**) + (**disk3** + **pegC**). This is another instance of the binding problem discussed in Section 4.2. So, to store a set of goals, we compute the sum of the bound vectors (**disk4** ⊛ **pegC** + **disk3** ⊛ **pegB**). Then, to recall where we wanted to place a particular disk (e.g., **disk3**), we can unbind that disk from the overall goal state.

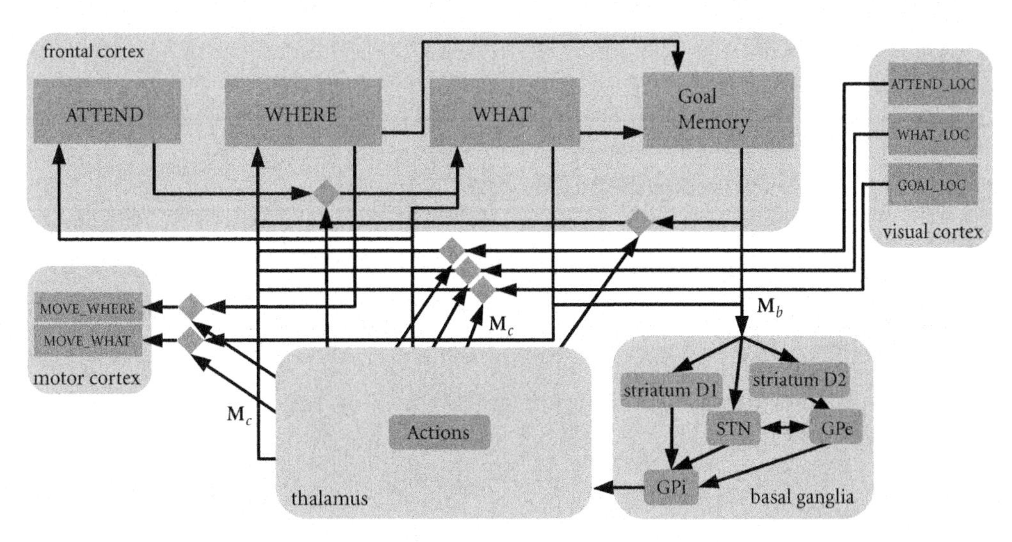

FIGURE 5.19 The architecture used for the Tower of Hanoi. As described in more detail in the text and Appendix E.1, elements in frontal cortex act as memories, with the goal memory including the ability to bind items to keep track of multiple goals. Visual and motor cortices are treated as inputs and outputs to the model, respectively. The basal ganglia and thalamus are as described in Figure 5.3. Note that diamonds indicate gates that are used for routing information throughout cortex, and are largely the targets of selected actions. $\mathbf{M}_b$ is defined through the projections to the basal ganglia, and $\mathbf{M}_c$ is defined through the multiple projections from thalamus to cortex (only two of which are labeled with the matrix for clarity). The model has a total of about 150,000 neurons. (Adapted from Stewart & Eliasmith, 2011b.)

Finally, we need to consider what input and output is needed for the model. Here, the focus is on cognitive operations, so I will not consider motor or perceptual systems in detail. As a result, I will assume that perceptual systems are able to supply the location of the currently attended object (ATTEND_LOC), the location of the object we are trying to move (WHAT_LOC), and the location of the goal we are trying to move the object to (GOAL_LOC). Similarly, the two output motor areas in the model will have the results of deliberation that indicate what disk should be moved (MOVE_WHAT) and where it should be moved to (MOVE_WHERE), without worrying about how those moves will be physically executed.

We have now specified the essential architecture of the model and can turn to characterizing the set of internal actions the model needs to perform, and the conditions under which it should perform each action. These rules define the $\mathbf{M}_b$ (IF) and $\mathbf{M}_c$ (THEN) matrices from Figure 5.5. For each action, we determine what state cortex should be in for that action to occur and connect the cortical neurons to the basal ganglia using the resulting $\mathbf{M}_b$. We then determine what new cortical state should result from being in a given cortical state and connect thalamus to cortex using the resulting $\mathbf{M}_c$. A full listing of the 16 rules used in the final model is in Appendix E.1.

Let us consider a few steps in this algorithm for illustration. First, the model will ATTEND to the largest disk (placing the **disk4** vector into the ATTEND memory). Next, the model forms a goal to place **disk4** in its final location, by routing ATTEND to WHAT and GOAL_LOC to WHERE. The model now has **disk4** in ATTEND and WHAT and **pegC** in WHERE. Next, it checks whether the object it is trying to move is in its target location. If it is (i.e., WHERE=WHAT_LOC), then it has already finished with this disk and needs to go on to the next smallest disk (loading **disk3** into WHAT and routing GOAL_LOC to WHERE).

If the disk in WHAT is not where it is trying to move it (i.e., WHERE is not equal to WHAT_LOC), then it needs to try to move it. First, it looks at the next smaller disk by sending **disk3** to ATTEND. If it is attending to a disk that is not the one it is trying to move (ATTEND is not WHAT) and if it is not in the way (ATTEND_LOC is not WHAT_LOC or WHERE), then it attends to the next smaller disk. If it is in the way (ATTEND_LOC=WHAT_LOC or ATTEND_LOC=WHERE), then it needs to be moved out of the way.

To do this, the model sets a goal of moving the disk to the one peg where it will not be in the way. The peg that is out of the way can be determined by sending the value $\mathbf{pegA} + \mathbf{pegB} + \mathbf{pegC}$ to WHAT and at the same time sending the values from WHAT_LOC (the peg the disk it is trying to move is on) and ATTEND_LOC (the peg the disk it is looking at is on) to WHAT as well, but multiplied by -1. The result will be $\mathbf{pegA} + \mathbf{pegB} + \mathbf{pegC} - \mathit{WHATLOC} - \mathit{ATTENDLOC}$, which is the third peg in this example.

This algorithm, with the addition of a special case for attending the smallest disk (it is always possible to move **disk1**, because nothing is ever in its way), is sufficient for solving Tower of Hanoi. However, this algorithm does not make use of the memory system. That is, it must rebuild its plans from the beginning after every move. The considerable data on human expert performance on this task indicates that people do keep track of their previous goals.

To address this, we can route the WHAT and WHERE information to the memory, and add a rule to do nothing if RECALL is not the same as WHERE. This occurs if the model has set a new goal, but there has not been enough time for the memory to hold it. Next, we can add rules for the state where the model has just finished moving a disk. Rather than starting over from the beginning, the model sends the next largest disk to ATTEND and WHAT and routes the value from RECALL to WHERE. This recalls the goal location for the next largest disk and continues the algorithm. With these final additions, the model does a good job of matching human performance.

To fully implement this algorithm, including better use of the memory system, requires specifying 16 actions in the two **M** matrices (*see* Table E.2). The behavior of the model is captured by Figure 5.20. The model is able to successfully solve the Tower of Hanoi, given any valid starting position and any valid target position. It does occasionally make errors and often recovers from them. Figure 5.20A shows several examples of successful action selection, where the decoded spiking output from the thalamus is shown. As can be seen, different actions are selected as the task proceeds. Unfortunately, it is difficult to directly compare detailed neural responses with human data, because such data do not exist for this particular task.

However, we can compare the model to human performance in other ways. For example, Anderson, Kushmerick, and Lebiere (1993) have provided a variety of measures of human performance on this task. Figure 5.20B compares the time taken between each move in the case where no mistakes are made. To generate this comparison, there are only two free parameters for the model: the amount of time needed to move a disk (1.8 seconds) and the memory weighting factor that determines how quickly new information is encoded in memory (0.08). All other timing parameters are taken from the neurophysiology of the various brain regions (e.g., time constants of the relevant neurotransmitter types, membrane time constants, etc.). As a result, the timing behavior of the system is not the result of a parameter fit to the human data but, rather, falls naturally out of the dynamics intrinsic to the SPA.

As can be seen, in both the human and model data, steps 1, 9, and 13 show longer pauses as a new set of goals are established. As well, shorter pauses on steps 3, 7, and 11 are also consistent across the model and data. The only point of disagreement is step 5, where the model is taking longer than humans. We suspect this is because humans are employing a heuristic shortcut that we have not included in the set of rules used in the model (analogous to the use of the memory system not originally included). Nevertheless, the overall agreement between the model and human data is quite good, suggesting that it is reasonable to claim that we have constructed the first spiking, neurally plausible model to capture the behavioral features of this task.

However, one main purpose of constructing a biological model is to make better contact with neural data, not just to explain behavioral data. As I have mentioned, the low-level spiking data are not available, but high-level fMRI data are available. As shown in Figure 5.21, the SPA model matches the fMRI activity recorded in a variety of cortical areas during most portions of a closely related task. Although a symbolic model of this data (ACT-R; Anderson et al., 2005) has been shown to have a similar fit, there are crucial differences between how the SPA generates these results and how ACT-R does.

First, like most cognitive models, ACT-R will provide the same prediction for each run of the model unless noise is injected into the system. As a result,

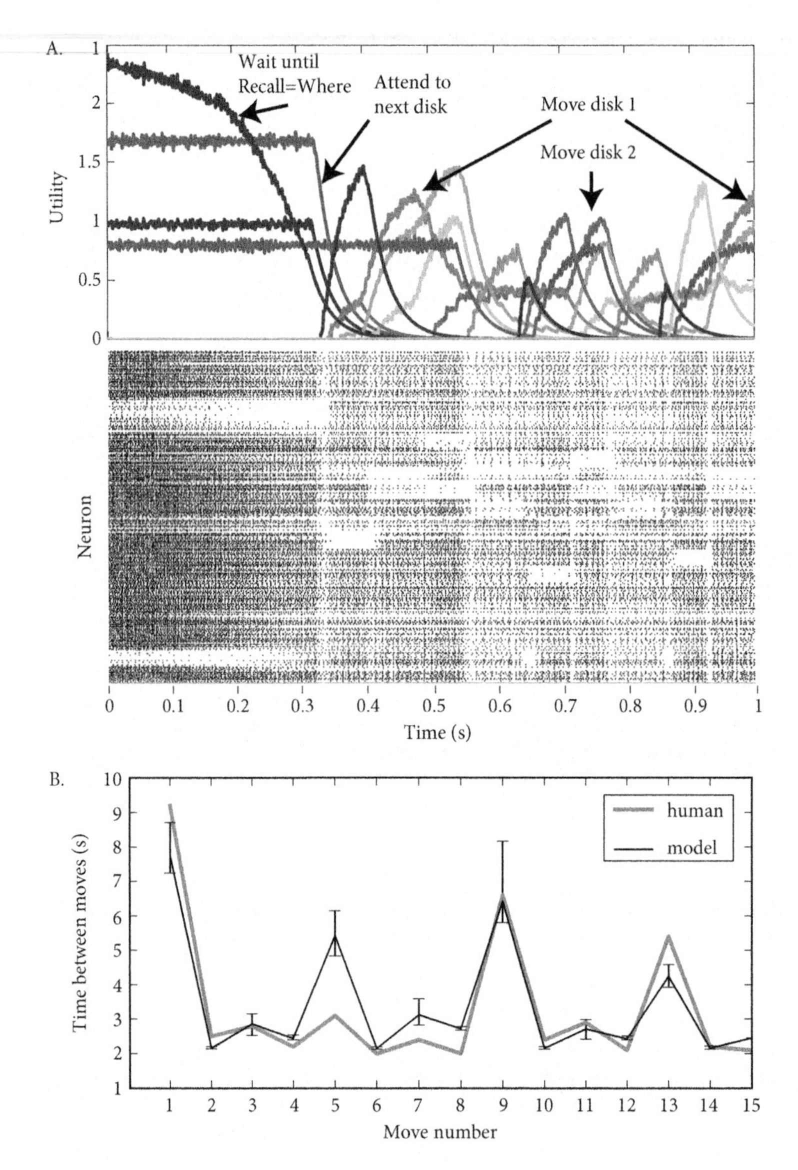

FIGURE 5.20 Behavior of the Tower of Hanoi model. **A**. The decoded spiking activity from thalamus to cortex (top), and the raw spiking of neurons in GPi (bottom). This activity shows which actions are selected as the task progresses. Different shades of gray indicate the activity associated with different actions. Some actions are labeled as examples. GPi neurons stop spiking when associated actions are selected. (Adapted from Stewart & Eliasmith, 2011b.) **B**. Behavioral performance of the model compared to human subjects (only means are available for human subjects). Move times are determined by using basal ganglia processing times (e.g., in A) and adding 1.8 s every time an actual move is made to account for motor processing. This method of comparing the model to behavioral data is borrowed from standard ACT-R models of this behavior (Altmann & Trafton, 2002; Anderson et al., 1993).

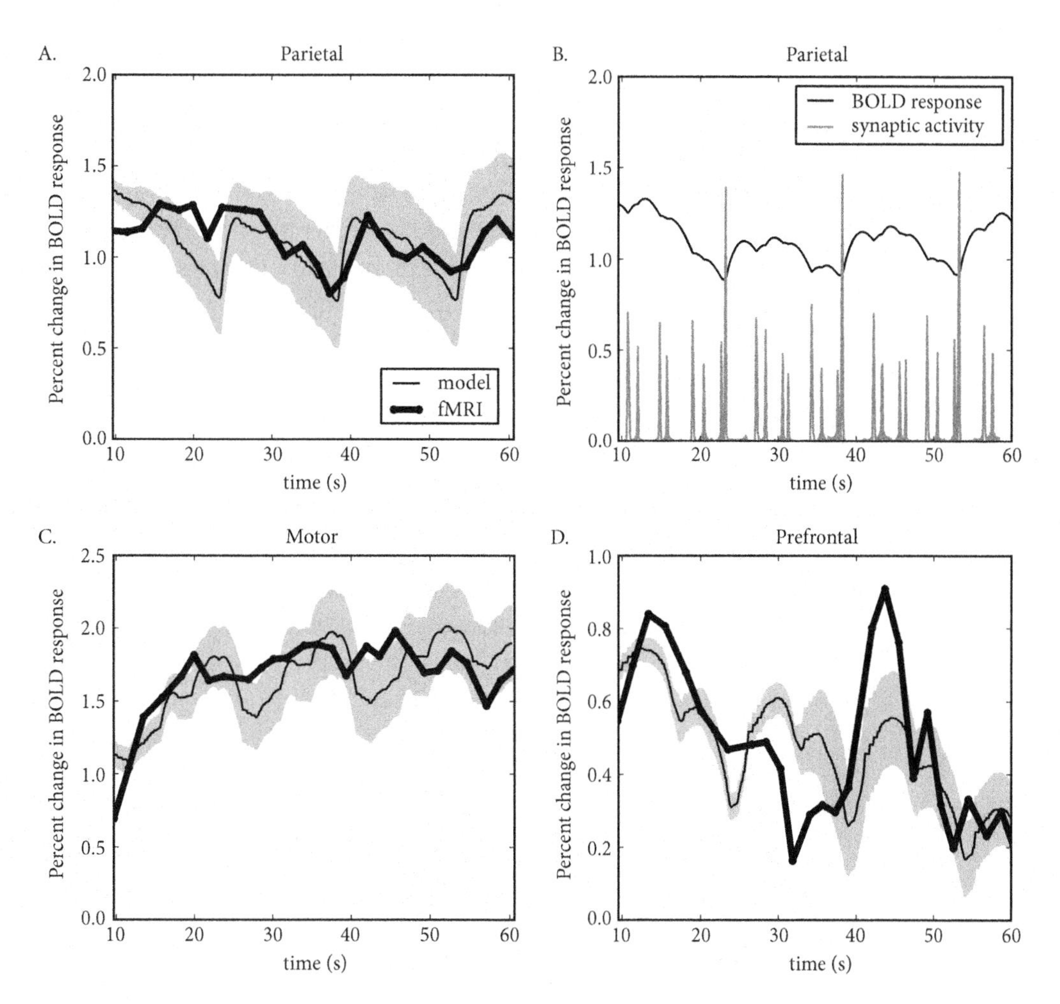

FIGURE 5.21 fMRI results comparing human data to the SPA model during a Tower of Hanoi-like task. Human fMRI data are from Anderson et al. (2005). Scans were taken when subjects pressed a button indicating a move. The first button press was at 10 seconds. Thick black lines are the average human fMRI data. Thin black lines are the average model data. **A.** Parietal cortex activity. One standard deviation is shown in gray. **B.** The parietal activity from a single run. Gray lines indicate the synaptic activity (neurotransmitter usage) that generates the fMRI prediction, which is shown in black. **C.** Motor cortex activity. **D.** Prefrontal cortex activity.

the model mostly predicts mean performance, with any distributions highly dependent on the chosen statistical model of the runtime noise. In contrast, the SPA model can predict a distribution of results (as shown by the standard deviation plotted in Fig. 5.21) based on individual implementation variability. That is, the SPA model has different neurons for each individual, not just different random noise used during runtime. In short, the SPA includes individual as well as runtime variability, whereas ACT-R includes only runtime variability. Thus the SPA model is more amenable to modeling individual differences found in populations of subjects, because people exhibit both kinds of variability.

Second, and more importantly, the SPA generates these predictions based on an explicit biophysical property of the model: neurotransmitter usage (*see* Fig. 5.21B). There is strong evidence that dendritic processing, driven by neurotransmitter usage, underwrites the BOLD signal (Logothetis & Wandell, 2004). It is that BOLD signal that is actually measured by MRI machines. Consequently, MRI measures neurotransmitter usage, *not* neural activity (i.e., spiking of neurons) in a brain area. Neurotransmitter usage is actually determined by neurons that project into the measured area, not the neurons in the area itself. Here we predict fMRI data by filtering neurotransmitter usage in the model. This filter is determined by a fixed model of how neurotransmitter fluctuations demand energy and give rise to the blood-usage-based BOLD signal.

In contrast, the ACT-R model relies on the proportion of time that a particular cognitive module is "active." There is thus no specification of a physiological process that underwrites the BOLD signal in ACT-R predictions. This also has the consequence that for the ACT-R model to fit the fMRI data, the BOLD filter is characterized differently for each cortical area, which is difficult to justify physiologically. In short, unlike ACT-R, the SPA is able to make physiologically based predictions of physiological measurements. This is more informative about neural mechanisms than making area-by-area timing-based predictions of physiological measurements.

Overall, the Tower of Hanoi model provides a concrete example of how the SPA bridges the gap between the biological substrate and high-level cognitive behavior. Although the model is anatomically and physiologically constrained and runs at the level of individual spiking neurons, it demonstrates rule following, planning, and goal recall. This allows the model to be simultaneously compared to both neural data from fMRI and more traditional behavioral reaction time data.

5.9 ■ Nengo: Question Answering

This tutorial brings together methods used in each of the previous tutorials, and introduces a basic control structure. This will allow us to construct a moderately complex model that relies on SPA methods for characterizing representation, transformation, and control. The specific goal of this tutorial is to build a network that will output answers to questions based on supplied statements. The statements and questions are supplied to the network as semantic

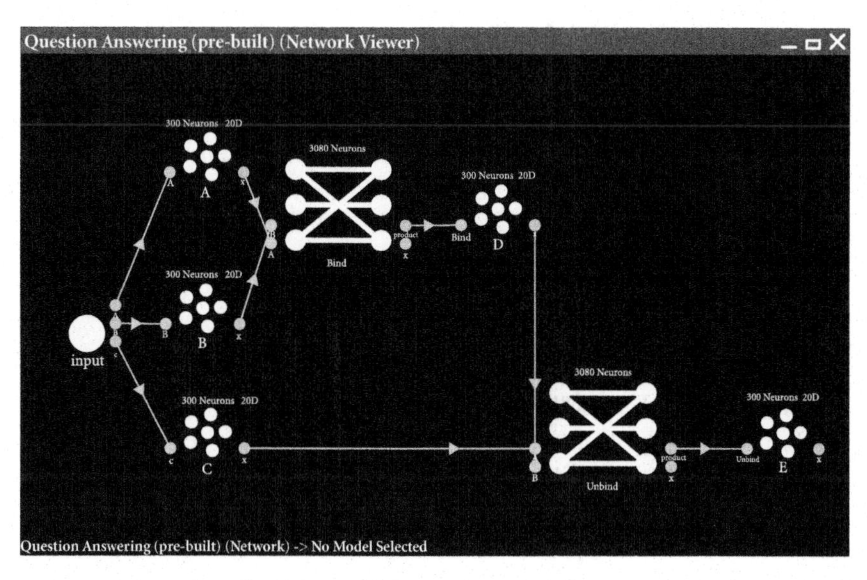

FIGURE 5.22 The initial question-answering network constructed in Nengo. It is similar to the network discussed in Section 4.4. Later Sections of this tutorial add memory and control to this basic network. Figures like this one can be generated using the *Save View to PDF* command in the *File* menu.

pointers placed in a "visual cortical area" and the output answers are semantic pointers generated by the model and sent to a "motor cortex."

The final question-answering network is somewhat complex, so I introduce it in three stages. The first network that we make binds two semantic pointers together and then binds the result with the pseudo-inverse of a third semantic pointer (*see* Fig. 5.22). Note that this is very similar to the network in the preceding tutorial (section 4.9). However, in this network, unbinding is accomplished by calculating the relevant inverse as part of the operation, rather than having an inverted semantic pointer supplied.

- Open a blank Nengo workspace and create a new network named "Question Answering."
- Create five new default ensembles with the names "A" ,"B," "C," "D," and "E." Use 300 neurons and 20 dimensions per ensemble.

If ensemble creation is slow, once the first ensemble is made it can be copied and pasted to form the other four ensembles. Select "A" and use the *File* menu, right-click menu, or shortcut keys to copy the ensemble. Then paste (and rename) the ensemble. The ensemble will paste where your mouse pointer is. This will result in the same neurons in each ensemble, rather than each being randomly generated, but that will not significantly affect the operation of this network.

- Drag the "Binding" template into the network. Set *Name* to "Bind" and *Name of output ensemble* to "D." Use 70 neurons per dimension and don't invert either input.
- Project from the "A" and "B" ensembles to the "A" and "B" terminations on the "Bind" network.
- Drag another "Binding" template into the network. Set *Name* to "Unbind" and *Name of output ensemble* to "E." Use 70 neurons per dimension and check the box to invert input A but not B.
- Project from the "C" ensemble to the "A" termination on the "Unbind" network and from the "D" ensemble to the "B" termination of the "Unbind" network.

To select multiple network elements, hold *Shift* while dragging the marquee around them. Then release *Shift* to manipulate the selected items.

- Open *Interactive Plots* and ensure the semantic pointer graphs for "A," "B," "C," "D," and "E" are shown.
- Use *Set value* on the semantic pointer graph of "A" and set it to "RED." Similarly, set the value of "B" to "CIRCLE," and the value of "C" to "RED."
- Run the simulation for about 1 second.

Depending on your computer, this may run slowly and interactive plots may only refresh occasionally. To speed up the simulation, you can decrease the "recording time" (click the gray arrow in the middle bottom of the window). You can also use fewer neurons in the ensembles, but the simulations may not work well. However, if you run the ensembles in "direct" mode (accessible by right-clicking on a network or ensemble), the number of neurons will not matter, but there will be no neural activity. Direct mode is an approximation and cannot be reliably assumed to perform the same as actual neurons.

As seen in the previous tutorial, binding the "A" and "B" inputs together results in a semantic pointer in "D" that does not share a strong similarity to any of the vocabulary vectors. However, the added "Unbind" network takes the generated semantic pointer in "D" and convolves it with the inverse of "C," which is the same as one of the inputs, resulting in an estimate of the other bound item in "E."

- Alternate setting the value of "A" to "RED" and "BLUE," the value of "B" to "CIRCLE" and "SQUARE," and the value of "C" to any one of these four labels. Run and pause the simulation between changes (pausing is not necessary if the simulation runs quickly).

The inputs to the "A" and "B" ensembles can be thought of as forming the statement "D." The statements in this case could be interpreted as temporary associations, such as "squares are blue" and "circles are red." The input to the "C" ensemble then poses a question about the statement. Thus an input of "SQUARE" asks "what is associated with squares" and the expected answer is "blue." A pre-built version of this network can be loaded from <Nengo Install Directory>/Building a Brain/chapter5/question.py. In the pre-built version, some inputs have been defined for you so you can see the network working. If you delete the "input" node in the network viewer, you can easily control the input as described earlier in this tutorial.

Question Answering With Memory

This network is not a very impressive demonstration of cognitive ability because answers to questions are always provided at the same time as the questions themselves. However, the network is a good start because it forms a bound representation from its input and then successfully extracts the originally bound elements.

The usefulness of this operation becomes apparent when memory is introduced to the network, so we will now add a memory buffer.

- Remove the projection from "D" to "Unbind" by clicking the "B" termination on "Unbind" and dragging it off the "Unbind" network.
- Drag the "Integrator" template from the template bar into the "Question Answering" network.
- In the template dialog, set *Name* to "Memory," *Number of neurons* to 1000, *Number of dimensions* to 20, *Feedback PSTC* to 0.4, *Input PSTC* to 0.01, and *Scaling factor* to 1.0. Click *OK*.
- Connect the projection from "D" to the "input" termination on "Memory."
- Project from the "X" origin of "Memory" to the "B" termination on "Unbind."

The integrator network is quite simple, but behaviorally very important. This is because it can act to prolong the availability of information in the system, and hence act as a memory. With no input, the present state of the ensemble is projected back to itself, allowing it to retain that state. If input is present, then it moves the system to a new state until the input ceases, at which point the new state is retained. So, without input the integrator acts as a memory, and with input it acts to update its current state based on recent history. In either case, it helps to track changes in the environment. Crucially, the decay of the memory is much slower than (although related to) the decay rate of the feedback.

In previous tutorials, we have included default time constants for all synapses (by defining *PSTC*), but the models have been largely feed-forward. As a result, the dynamics have been relatively straightforward. However, the integrator has a feedback connection, and so dynamics become more subtle (*see* the discussion of the third NEF principle in Section 2.2.3). In this model we have set the feedback time constant to be artificially high (400 ms), which gives a relatively stable memory (several seconds). However, a more physiologically realistic time constant of 100 ms (a time constant associated with NMDA receptors) would require us to significantly increase the number of neurons for the same memory performance and would significantly slow the simulation. So, this setting is a useful simplification for current purposes, but should be changed if we want to make the simulation more biologically plausible.

Before proceeding, I will note that it is possible that the following simulation gets some of the expected answers wrong. This is because I have attempted to use as few neurons as possible to allow the network to run more quickly. Because the vocabulary and neuron tunings are generated randomly, some networks may not work perfectly. Increasing the number of neurons and the number of dimensions will alleviate any problems but will also slow the simulation.

- Open *Interactive Plots* and show the same graphs as before if they are not visible by default. (To see the "Memory" network, you have to close and reopen the *Interactive Plots*, but be sure to save the layout before doing so.)
- Right-click the semantic pointer graph for "*E*" and choose "select all."
- Set "A" to "RED," "B" to "CIRCLE," and "C" to "RED."
- Run the simulation for 0.5 seconds.

This results in the same behavior as before, where the result in "E" is "CIRCLE."

- With the simulation paused, set "A" to "BLUE" and "B" to "SQUARE."
- Run the simulation for another half second and pause.

During the first half second of the simulation, the network receives inputs from the "A" and "B" ensembles and binds them (i.e., resulting in *RED* $\circledast$ *CIRCLE*). This is stored in the "Memory" ensemble. During the next half second, a second binding (i.e., *BLUE* $\circledast$ *SQUARE*) is presented. It too is added to the "Memory" so the value stored in the integrator memory after 1 second is approximately *RED* $\circledast$ *CIRCLE* + *BLUE* $\circledast$ *SQUARE*. As a result, the output "E" remains as "CIRCLE" because the "C" input has stayed constantly equal to "RED" and so unbinding will still result in "CIRCLE," which is bound to "RED" in the memory. If you open the semantic pointer graph for the "Memory" population, be sure to select *show pairs*.

- Right-click the semantic pointer graphs for "A" and "B" and select "release value." Run the simulation for 0.5 seconds.

You will notice that the result in "E" is still "CIRCLE" because the memory has not changed, and it already encoded the inputs.

- Change "C" to a different vocabulary vector ("BLUE," "CIRCLE," "SQUARE").
- Run the simulation for 0.5 seconds and pause.
- Repeat the previous steps for each vocabulary vector.

For each of these cases, the memory is encoding the past state of the world, and only the question is changing in "C." The result in "E" changes because it continues to show the result of unbinding the semantic pointer in memory with the question input. The network is now performing question answering, because it can remember pairs of items and successfully recall which items are associated with each other. So it has internally represented a simple structure and can answer queries about that structure. A pre-built version of this network can be loaded from <Nengo Install Directory>/Building a Brain/chapter5/question-memory.py.

Question Answering With Control

We have not yet accomplished our original goal, which is to have statements and questions supplied through a single "visual input" channel and produce replies in a "motor output." The final stage of building the question-answering network introduces the control structures needed to shift between a state of accepting new information to store in memory and a state of replying to questions about those memories.

- Clear the "Question Answering" network from the workspace.
- Create a new network named "Question Answering with Control."
- Drag the "Network Array" template from the template bar into the network. Set *Name* to "Visual," *Neurons per dimension* to 30, *Number of dimensions* to 100, *Radius* to 1.0, *Intercept (low/high)* to −1 and 1, respectively, *Max rate (low/high)* to 100 and 300, *Encoding sign* to "Unconstrained," and *Quick mode* to enabled.
- Repeat the previous step to create two additional network arrays named "Channel" and "Motor," with the same parameters as the "Visual" network.
- Drag the "Integrator" template into the main network. Set the name of the integrator to "Memory" and give it 3000 neurons, 100 dimensions, a

feedback time constant of 0.4, an input time constant of 0.05, and a scaling factor of 1.0.

The network array template creates a subnetwork of many low-dimensional ensembles that together represent a high-dimensional vector. In this case, each internal ensemble represents only one dimension (the number of dimensions per ensemble can be changed through scripting). In general, neural ensembles can represent many dimensions. However, the network array can be constructed much more quickly for computational reasons.[3] Because we have increased our dimensionality to 100, performance will degrade significantly if we do not use network arrays (unless you have a very fast computer). Additionally, we have employed these arrays in a way that should not affect the performance of the network. This increase in dimensionality is needed because it allows for more complex structures to be created while lowering the odds that semantic pointers will be similar through coincidence (*see* Section 4.6).

The integrator template automatically makes use of network arrays to create integrators with eight or more dimensions. This is both for computational and stability reasons. High-dimensional stable integrators are much more difficult to make than low-dimensional ones. So, the template creates a high-dimensional stable integrator by employing many low-dimensional ones for integrators larger than 8 dimensions.

We can now begin to connect the circuit:

- Add a decoded termination to the "Channel" network called "visual." Set *PSTC* to 0.02, *Input Dim* to 100, and use the default coupling matrix (the identity matrix).
- Create a projection from the "Visual" network array to the "visual" termination.
- Create a projection from the "Channel" network to the "input" termination on the "Memory" network.
- Drag the "Binding" template into the "Question Answering with Control" network. Set *Name* to "Unbind," *Name of output ensemble* to "Motor," and *Number of neurons per dimension* to 70. Enable the *Invert input A* option.
- Project from the origin of the "Memory" network to the "B" termination of the "Unbind" network.
- Project from the origin of the "Visual" network to the "A" termination of the "Unbind" network.

[3] With a network array, many low-dimensional ensembles can be created, rather than one large ensemble. Creating ensembles requires performing SVD on an $N \times N$ matrix, which is an $O(N^3)$ operation, so it is much slower for large N than small N.

At this stage, we have roughly recreated the basic "Question Answering" network, albeit with much larger populations representing a higher-dimensional space. The remaining elements of the network will introduce action selection and information routing.

- Drag the "Basal Ganglia" template into the network. Set the name to "Basal Ganglia," the number of actions to 2, and the *PSTC* to 0.01. Click *OK.*

This creates a subnetwork that is structured like the basal ganglia model described earlier (Section 5.2). To see the elements of this model, double-click the subnetwork.

- Drag the "BG Rule" template onto the new "Basal Ganglia" network. Set *Rule Index* to 0, *Semantic Pointer* to "STATEMENT," *Dimensionality* to 100, *PSTC* to 0.01, and enable *Use Single Input.* Click *OK.*

Choosing *Use Single Input* sends any input to basal ganglia to all three input elements of the basal ganglia (i.e., striatal D1 and D2 receptors, and STN).

- Drag another "BG Rule" template onto the basal ganglia network. Set *Rule Index* to 1, *Semantic Pointer* to "QUESTION," *Dimensionality* to 100, *Input PSTC* to 0.01, and enable *Use Single Input.*
- Project from the "Visual" network to both the "rule_00" and "rule_01" terminations on the "Basal Ganglia" network.

The basal ganglia component receives projections from ensembles representing semantic pointers, assesses the similarity of the input to the set of "rule antecedents," which are also vocabulary vectors, and computes a winner-take-all-like function. In this model, the basal ganglia receives projections from the "Visual" network, then outputs the vector (0, 1) if the input is similar to the semantic pointer designated "STATEMENT" or the vector (1, 0) if the input is similar to the semantic pointer "QUESTION" (recall that the output of basal ganglia inhibits the selected action).

- Drag a "Thalamus" template into the main network. Set *Name* to "Thalamus," *Neurons per dimension* to 30, and *Dimensions* to 2.
- Add a termination to the "Thalamus" network. Name it "bg," set *Input Dim* to 2, and *PSTC* to 0.01. Click *Set Weights* and set the diagonal values of the matrix (the two that are 1.0 by default) to −3.0.
- Create a projection from the output origin of "Basal Ganglia" to the "bg" termination of "Thalamus."

The large negative weights between basal ganglia and thalamus reflect the inhibitory input of basal ganglia. Thus, when an action is selected, its activity goes to zero, which will be the only non-inhibited channel into the thalamus, releasing the associated action consequent in thalamus, allowing the related neurons to spike. We can now define action consequents.

- Drag a "Gate" template into the main network. Name the gate "Gate1," set *Name of gated ensemble* to "Channel," *Number of neurons* to 100, and *Gating PSTC* to 0.01.
- Add a decoded termination to the "Gate1" ensemble. Set *Name* to "thalamus," *Input Dim* to 2, and click *Set Weights.* Set the weights to [1 0]. Set the *PSTC* to 0.01.
- Create a projection from the "xBiased" origin on the thalamus network to the termination on "Gate1."

In this model, each consequent of the thalamus results in ungating information flow in the "cortex." Essentially, the non-selected outputs of the basal ganglia inhibit the thalamus, whereas the selected outputs stop inhibiting connected parts of thalamus, allowing those parts to fire. This thalamic firing in turn inhibits a cortical gate, which then stops preventing a flow of information into the network it controls. So far, we have set up the circuit that will allow "Visual" information to flow into the "Channel" when "STATEMENT" is presented to the basal ganglia.

- Drag a second "Gate" template into the main network. Name the gate "Gate2," set *Name of gated ensemble* to "Motor," *Number of neurons* to 100, and *Gating PSTC* to 0.01.
- Add a decoded termination to the "Gate2" ensemble. Set *Name* to "thalamus," *Input Dim* to 2, and click *Set Weights.* Set the weights to [0 1]. Set the *PSTC* to 0.01.
- Create a projection from the "xBiased" origin on the thalamus network to the termination on "Gate2."

We have now created a second action consequent, which allows information to flow to the "Motor" cortex from the "Unbind" network when "QUESTION" is presented to basal ganglia. We can now run the model.

- Open *Interactive Plots* and display only the "Visual" and "Motor" nodes (hide other nodes by right-clicking on them).
- Turn on the semantic pointer graphs for the "Visual" and "Motor" ensembles by right-clicking these nodes and selecting "semantic pointer."
- Set the value of the "Visual" semantic pointer to "STATEMENT+RED*CIRCLE" by right-clicking the semantic pointer graph.
- Run the simulation for 0.5 seconds of simulation time and pause.
- Set the value of the "Visual" semantic pointer to "STATEMENT+BLUE*SQUARE." Run for another half-second and pause.

- Set the value of the "Visual" semantic pointer to "QUESTION+BLUE."
- Run the simulation until the semantic pointer graph from the "Motor" population is settled (this should take under half a second).
- Set the value of the "Visual" semantic pointer to "QUESTION+CIRCLE." You can substitute "RED," "SQUARE," or "BLUE" for "CIRCLE."
- Run the simulation and repeat the previous step for each possible question.

If the question-answering model has been properly constructed, it will successfully store bound pairs of vectors given inputs that have been summed with the semantic pointer designating statements. If you are unable to get the network to function, then be sure to check all of the weight settings. If all else fails, a pre-built version of this network can be loaded from <Nengo Install Directory>/Building a Brain/chapter5/question-control.py. If your computer runs the interactive plots in a choppy manner, let it run for about 3 simulation seconds and hit the pause button.

Once it pauses, you can drag the slider at the bottom of interactive plots to smoothly scroll through all of the data as it was originally generated. If you drag the slider to the start of the simulation and hit the play button, it will smoothly display the data as if it were being generated on-the-fly.

There are several ways you might go beyond this tutorial. For example, you could increase the vocabulary of the network to determine how many items it can store in memory before being unable to successfully recall items. You could add additional rules to the basal ganglia to incorporate other actions. Or, you could determine what happens when there are multiple answers to a question, and so on.

The completed question-answering network contains about 20,000 neurons and demonstrates rudimentary examples of the main SPA elements (except learning). These same elements are used to construct the advanced models in this book (e.g., Tower of Hanoi, Raven's Matrices, Spaun, etc.).

6. BIOLOGICAL COGNITION–MEMORY AND LEARNING

C. Eliasmith, X. Choo, and T. Bekolay

6.1 ■ Extending Cognition Through Time

Both memory and learning allow neural systems to adapt to the intricacies of their environment based on experience. It is perhaps not unfair to think of learning as the accumulation of memories. For example, the memory of recent or long past events can be very important for determining an appropriate behavior in the present. Those memories can be naturally characterized as having been learned based on the experience of the relevant past events.

Such a general characterization does not capture the diverse kinds of learning and memory that researchers have identified, however. For example, there is unsupervised, supervised, reinforcement, and associative learning. Similarly, there is short-term, long-term, semantic, procedural, and implicit memory. Most of these (and many other) kinds of memory and learning have been characterized behaviorally. In developing the SPA, I am largely interested in the neural mechanisms underlying this diversity.

Unfortunately, these different kinds of learning and memory are difficult to cleanly map to specific neural mechanisms. Traditionally, in the neural network literature, when researchers speak of "learning," they are referring to changing the connection weights between neurons in the network. The ubiquitous "learning rules" proposed by these researchers are a means of specifying how to change connection weights given the activity of neurons in the network. However, notice that the example of learning that we have seen earlier in the book (Section 4.7) *looks* like a standard case of learning, but depends only on updating the neural activity of a memory system. That is, no connection weights are changed in the network although the system learns a general rule that allows past experience to change what behaviors are chosen in the present.

Similarly, the traditional contrast between working memory (which is often taken to be activity in a recurrent network) and long-term memory (which is often taken to be postsynaptic connection weight changes) is difficult to sustain in the face of neural evidence. For example, there is wide range of timescales of adaptive processes influencing neural firing in cortex (Ulanovsky et al., 2004). Some changes on the timescale of working memory (seconds to minutes) are related to postsynaptic weight changes (Whitlock et al., 2006), and others are not (Varela et al., 1997; Romo et al., 1999). Consequently, some "working" memories might be stored in the system in virtue of connection weight changes, and others might be stored in virtue of presynaptic processes, or stable network dynamics. The timescales we use to identify kinds of memory behaviorally seem to pick out more than one neural mechanism.

Because of such complexities in the mapping between kinds of memory and neural mechanisms, I am not going to provide a comprehensive description of memory and learning. Rather, I provide specific examples of how activity-based and connection weight-based mechanisms for adaptation can be employed in the SPA. These examples help demonstrate how several of the kinds of memory and learning identified at the behavioral level fall naturally out of the SPA.

More specifically, this chapter addresses two challenges. The first is related to activity-based explanations of working memory, and the second is related to learning connection weight changes. The first challenge is to extend current neural models of working memory to make them more cognitively relevant. There are many neurally plausible models of working memory that rely on recurrently connected networks (Amit, 1989; Zipser et al., 1993; Eliasmith & Anderson, 2001; Koulakov et al., 2002; Miller et al., 2003). These models are able to explain the sustained firing rates found in many parts of cortex during the delay period of memory experiments. Some are able to explain more subtle dynamics, such as the ramping up and down of single cell activity seen during these periods (Singh & Eliasmith, 2006; Machens et al., 2010). However, none of these models address complex working memory tasks that are essential for cognition, such as serial working memory (i.e., remembering an ordered list of items). This is largely because methods for controlling the loading of memories, clearing the network of past memories, and constructing sophisticated representations have not been devised. In the next two sections, I describe a model that addresses many of these limitations. The tutorial at the end of the previous chapter employed a memory that has some of these features (*see* Section 5.9).

The second challenge is to introduce a biologically realistic mechanism for connection weight changes that can learn manipulations of complex representations. In the last three sections of this chapter, I describe a spike-timing-dependent plasticity (STDP) rule that is able to learn linear and nonlinear transformations of the representations used in the SPA. I demonstrate the rule by showing that it can be used to learn the binding operation employed in the SPA (i.e., circular convolution), that it can be used to learn action selection in the model of the basal ganglia (Section 5.2), and that it can be used to learn different reasoning strategies in different contexts to solve a language processing task. I argue that this rule is unique in its combination of biological realism and functional utility, so it is a natural fit for the SPA. The tutorial at the end of this chapter demonstrates how to use this rule in Nengo.

6.2 ■ Working Memory

The working memory system is typically characterized as an actively engaged system used to store information that is relevant to the current behavioral situation. Typical tests of working memory in monkeys consist of having a monkey fixate at a central location and presenting a target stimulus somewhere in the periphery of the monkey's visual field. The stimulus is removed, a delay of about 3 to 6 seconds long is then initiated, and finally the fixation target changes to indicate to the monkey that it should respond by saccading or pointing to the remembered location of the stimulus. Many experiments like these have been used to characterize the activity of single cells during working memory tasks (e.g., Pesaran et al., 2002). Additionally, many models have been built that explain a wide variety of the properties of single cell tuning curves observed during such tasks (Zipser et al., 1993; Amit et al., 1997; Koulakov et al., 2002; Miller et al., 2003; Singh & Eliasmith, 2006; Machens et al., 2010).

However, much greater demands are put on working memory during cognitive tasks. Rather than encoding and recalling a single item, many situations require encoding the order in which several items occur (e.g., a telephone number). Cognitive psychologists have studied these more demanding tasks since the 1960s (Jahnke, 1968; Rundus, 1971; Baddeley, 1998). The ability to store and recall items in *order* is called serial working memory. Unlike the simple single target case, serial working memory is seldom studied in animal single cell studies (Warden & Miller, 2007), and there are no spiking single cell models of which I am aware (although there are some neurally inspired models, e.g., Beiser & Houk [1998]; Botvinick & Plaut [2006]). It is thus both crucial to have serial working memory in a cognitive architecture and unclear how such a function can be implemented in a biologically realistic network.

In studying serial working memory, two fundamental regularities have been observed: primacy and recency. Both primacy and recency can be understood by looking at an example of the results of a serial recall task, like those shown in Figure 6.1. In this task, subjects are shown lists of items of various lengths and asked to recall the items in their original order. Primacy is

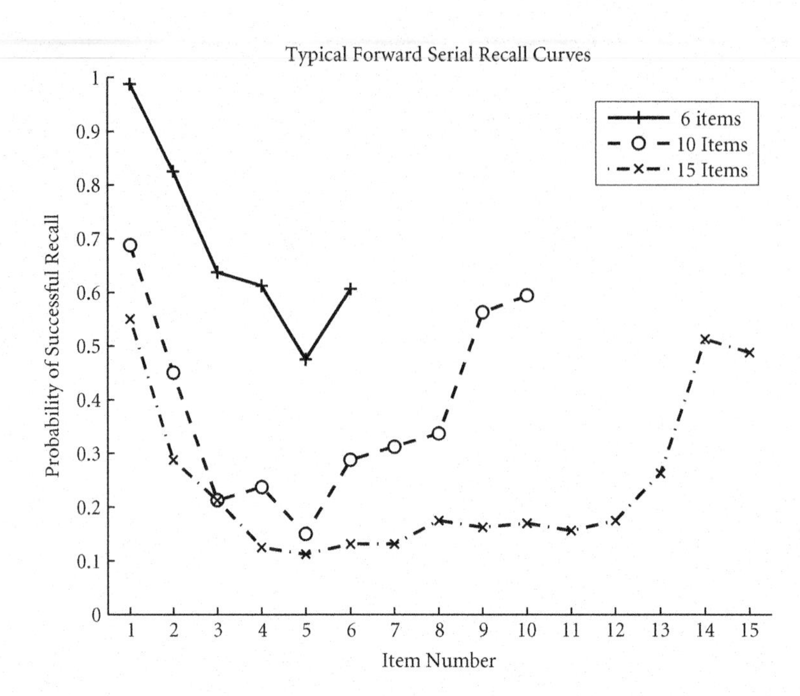

FIGURE 6.1 Recall accuracy from human behavioral studies showing primacy and recency effects (data from Jahnke, 1968). Participants were asked to recall ordered lists of varying length.

identified with the observation that items appearing earlier in the list have a greater chance of being recalled accurately, regardless of the length of the list. Recency is identified with the observation that the items most recently presented to subjects have an increased chance of being recalled as well. Together, primacy and recency account for the typical U-shaped response probability curve seen in serial working memory tasks.

Interestingly, this same Ushape is seen in free recall tasks (where order information is irrelevant; *see* Fig. 6.6). So it seems likely that the same mechanisms are involved in both free recall and serial recall. In fact, as I discuss in more detail in the next section, it seems likely that all working memory behavior can be accounted for by a system that has serial working memory. In contrast, models of working memory that can account for the behavior of monkeys on single item tasks cannot account for serial working memory behavior. Consequently, serial working memory is more fundamental when considering human cognition.

Recently, Xuan (pronounced "Shawn") Choo in my lab has implemented a spiking neural network model of serial working memory using the resources of the SPA. Based on a consideration of several models from mathematical and cognitive psychology (e.g., Liepa, 1977; Murdock, 1983, 1993; Henson, 1998; Page & Norris, 1998), Xuan has proposed the ordinal serial encoding

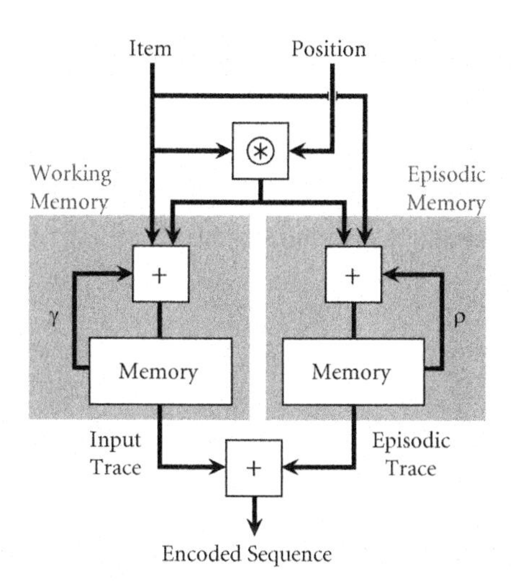

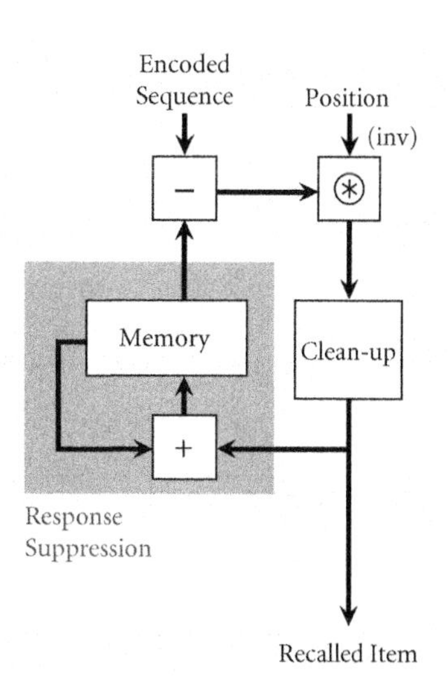

FIGURE 6.2 Network-level diagrams of the OSE encoding network (left) and the OSE item recall network (right). The encoding network receives an item and position semantic pointer as input. These are bound and sent to both memory systems, along with the item vector itself. The inputs are added to the current memory state in each memory. Each memory thus stores the information by integrating it over time, although the decay rates (γ and ρ) are different for the two memories. The decoding network begins with the originally encoded memory and decodes it by unbinding the appropriate position. The result is cleaned up and placed in a memory that tracks already recalled items. The contents of that memory are subtracted from the originally encoded sequence before additional items are recalled to suppress already recalled items.

(OSE) model of working memory (Choo, 2010). In the next section, I present several examples of how this model can account for human psychological data on serial and free recall tasks. First, however, let us consider the component parts and the basic functions of the model.

Figure 6.2 depicts the process of encoding and decoding items to be remembered by the OSE model. As can be seen there, the model consists of two main components that I have called a working memory and an episodic memory. I take these to map onto cortical and hippocampal memory systems, respectively. Both memories are currently modeled as neural integrators (*see* Section 2.2.3). A more sophisticated hippocampal model (e.g., Becker, 2005) would improve the plausibility of the system, although we expect that there would only be minor changes to the resulting performance. Notably, modeling these memory systems as neural integrators amounts to including past models of single item working memory as a component of this model (such as the model proposed in Singh & Eliasmith [2006]).

The input to the model consists of a semantic pointer representing the item and another semantic pointer representing its position in a list. These two representations are bound, using a convolution network (Section 4.2), and fed into the two memory systems. Simultaneously, the item vector alone is fed into the two memories to help enforce the item's semantics for free recall. Each memory system adds the new representation–constructed by binding the item vector to a position vector–to the list that is currently in memory, and the overall representation of the sequence is the sum of the output of the two memories.

Decoding of such a memory trace consists of subtracting already recalled items from the memory trace, convolving the memory trace with the inverse of the position vector, and passing the result through a clean-up memory (Section 4.6). In short, the decoding consists of unbinding the desired position from the total memory trace and cleaning up the results. Concurrently, items that have already been generated are subtracted so as to not be generated again. In the case of free recall (where position information is not considered relevant), the unbinding step is simply skipped (this can be done by setting the position vector to be an identity vector). The equations describing both the encoding and decoding processes can be found in Appendix D.3.

As can be seen from Figure 6.2, this model has two free parameters, ρ and γ. These govern the dynamic properties of the two different memory systems captured by the model. Specifically, ρ captures the effect of rehearsing items early in the list during the encoding process and is associated with hippocampal function. The value of this parameter was set by fitting it to an experiment conducted by Rundus (1971) in which the average number of rehearsals was found as a function of the item number in a serial list task. The other system, influenced by γ, characterizes a generic cortical working memory process. The value of this parameter was set by reproducing the simple working memory experiment described in Reitman (1974) and choosing the value that allowed the accuracy of the model to match that of humans. Reitman's experiment showed that after a 15-second delay (with no rehearsal) human subjects could recall, on average, 65% of the items that they recalled immediately after list presentation.

It is important to note that the two free parameters of this model were set by considering experiments not being used to test the model. That is, we do not use either the Reitman or Rundus experiments to suggest that the OSE model is a good model. Rather, all of the test experiments use the same parameters determined from fitting the Reitman and Rundus experiments. Consequently, the performance of the model on these new experiments in no way depends on tuning parameters to them. As such, the results on the new experiments are independent predictions of the model.

Finally, it is worth highlighting that the preceding description of this model is reasonably compact. This is because we have encountered many of the underlying elements–the necessary representations, the basic transformations, and the dynamical circuits–before in the SPA. So, in many ways this working memory system is not a new addition to the SPA but, rather, a recombination of functions already identified as central to the architecture (e.g., integration, binding, and clean-up). This is encouraging because it means our basic architecture does not need to get more complicated to explain additional, even reasonably sophisticated, functions.

6.3 ■ Example: Serial List Memory

Figure 6.3 shows the OSE model capturing the basic recency and primacy effects observed in the human data, shown earlier in Figure 6.1. This demonstrates not only that the model captures basic features of serial memory but that setting the parameters of the model using different experiments did not limit the system's ability to explain this behavior.

As is clear from the fact that the model, and the subjects, do not have perfect recall accuracy, mistakes are made in recalling serial lists. These mistakes have been analyzed in several experiments (Henson et al., 1996). Interestingly, these same experiments were taken to show that recurrent network models would not be able to reproduce the human data (Botvinick & Plaut, 2006). However, as made clear in Botvinick and Plaut (2006), this is not the case. The model presented here supports that conclusion and extends it to show that recurrent models can address more dynamical aspects of serial memory, such as delayed recall, and a wide variety of recall demands, such as reverse and free recall, not considered previously (Choo, 2010). Returning

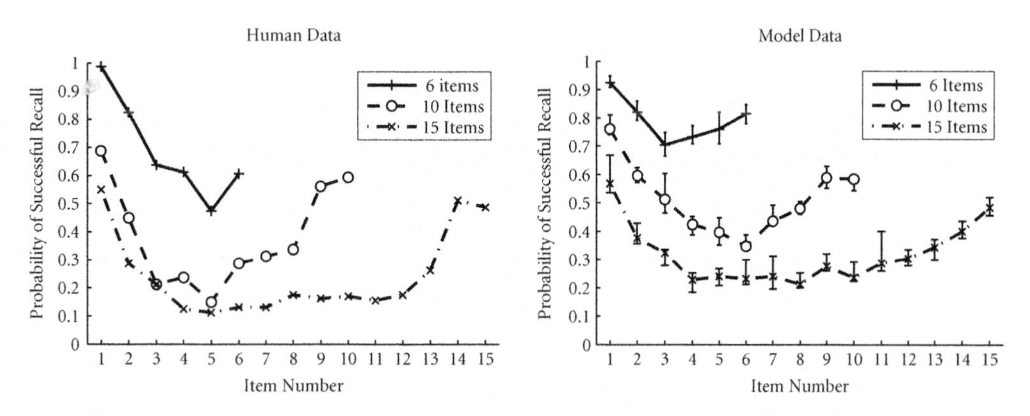

FIGURE 6.3 A comparison of human and model data on serial list recall tasks. The model data are presented with 95% confidence intervals, whereas for human data only averages were reported. The human data are from Jahnke (1968) and reproduced from Figure 6.1.

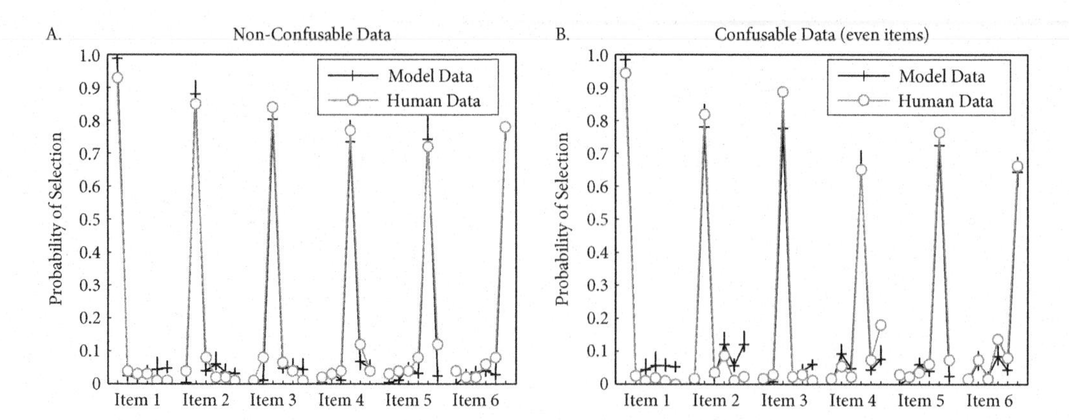

FIGURE 6.4 Transposition gradients for serial recall tasks. Given the task of recalling six items in order, these plots show the probability of selecting each item for each of the six possible positions in the list. The transposition gradients thus show the probabilities of errors involving recalling an item outside its proper position. **A.** The six items in the list are all easy to distinguish. **B.** Even numbered items in the list are similar and more easily confused. Human data are from Henson et al. (1996).

to the empirical work that characterizes error, Figure 6.4A shows a comparison of the transposition gradients found in the model with those measured from human subjects. The transposition gradient measures the probability of recalling an item outside of its correct position in a list. Both the model and the human data show that errors are most likely to occur in positions near the original item position, as might be expected.

As you might also expect, the similarity between items in a list can affect how well items can be recalled. Henson et al. (1996) designed an experiment in which they presented subjects with lists containing confusable and non-confusable letters. Because the stimuli were heard, confusable letters were those that rhymed (e.g., "B," "D," and "G"), whereas non-confusable letters did not (e.g., "H," "K," and "M"). The experimenters presented four different kinds of lists to the subjects: lists containing all confusable items, those containing confusable items at odd positions, those containing confusable items at even positions, and those containing no confusable items. In the model, items were randomly chosen semantic pointers where those with similarity (as measured by the dot product) below 0.25 being considered non-confusable, and those with similarity between 0.25 and 0.5 being considered confusable. The probability of successful recall for these lists for both the human subjects and the model are shown in Figure 6.5; a comparison of the transposition gradients from the model and human subjects on a recall task with confusable items at even positions is given in Figure 6.4B.

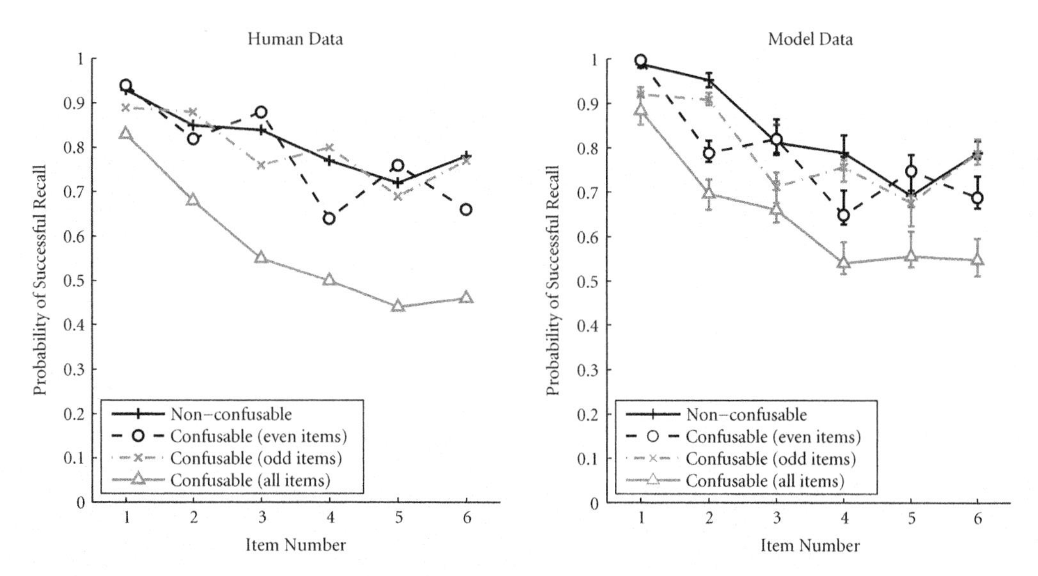

FIGURE 6.5 Serial recall of confusable lists. These graphs plot human and model performance recalling lists containing four different patterns of easily confusable versus non-confusable items. Model data includes 95% confidence intervals, whereas for human data only averages were reported. The human data are from Henson et al. (1996), experiment 1.

Again, this example shows that the model does a good job of capturing behavioral data, both the likelihood of successful recall and the pattern of errors that are observed in humans. The same model has also been tested on several other serial list tasks, including delayed recall tasks, backward recall tasks, and combinations of these (Choo, 2010), although I do not consider such tasks here.

More important for present purposes, the same model can also explain the results of free recall experiments. As shown in Figure 6.6, the accuracy of recall for the model is very similar to that of humans for a wide variety of list lengths. In these tasks, there is no constraint on the order in which items are recalled from memory. Nevertheless, both the model and human data show the typical U-shaped response probability curves.

Taken together, these examples of the OSE model performance on a wide variety of memory tasks (none of which were used to tune the model) make us reasonably confident that some of the principles behind human working memory are captured by the model. Because it uses the same kinds of representations as the rest of the SPA, we can be confident that it will integrate easily into larger-scale models as well. I take advantage of this in the next chapter.

However, before leaving consideration of this model I want to highlight what I think is perhaps its most theoretically interesting feature–namely, that this model only works if it is implemented in neurons. As demonstrated by

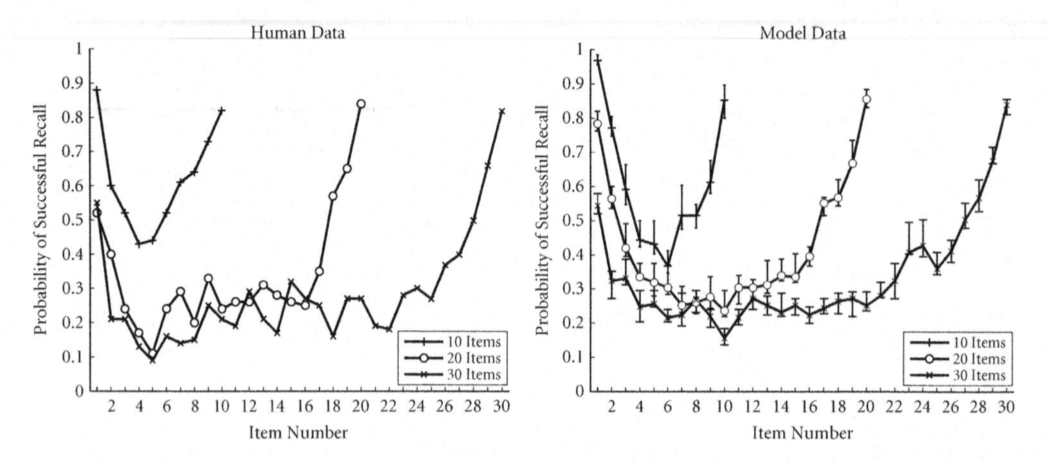

FIGURE 6.6 Human and model data for a free recall task. Unlike the serial recall task of Figure 6.3, subjects can recall the contents of the lists in any order. The free recall results continue to demonstrate primacy and recency effects with U-shaped curves. Human data are from Postman and Phillips (1965).

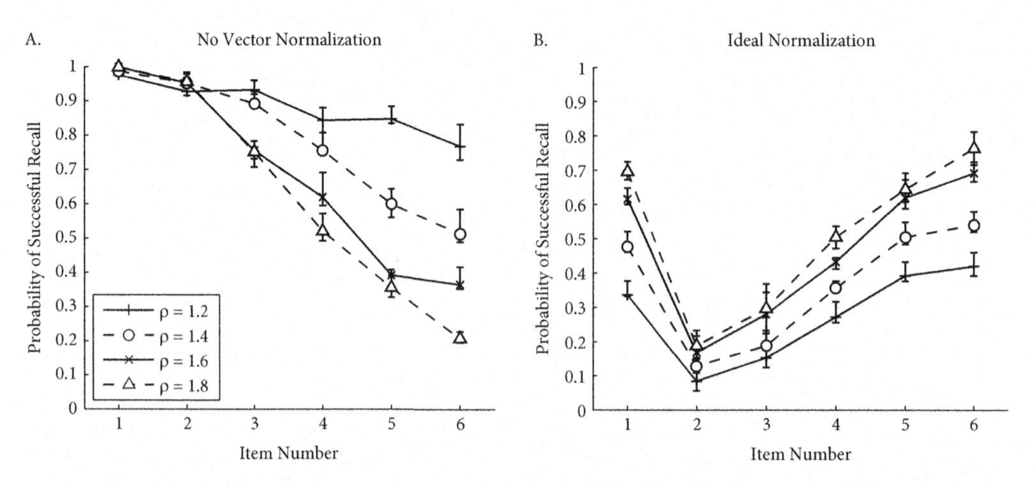

FIGURE 6.7 Non-neural implementations of the model. **A.** The model is implemented by directly evaluating the equations in Appendix D.3. Vectors are not normalized and can grow arbitrarily long. **B.** The model is again implemented without spiking neurons, but vector lengths are held constant using ideal vector normalization. Varying the parameters does not help. The effects of changing ρ are shown.

Figure 6.7A, if we directly simulate the equations that describe this model (*see* Appendix D.3), then it is unable to accurately reproduce the recency and primacy effects observed in the human data. Initially, it seemed that this failure might have been caused by the semantic pointer vectors in the system becoming arbitrarily long as additional items were added to the memory trace. Consequently, we also implemented the model using standard vector normalization, which guarantees that the vectors always have a constant length. But again, as seen in Figure 6.7B, the model is unable to capture the human data (i.e., neither case has the appropriate U-shaped curve, although normalization is closer).

Consequently, we realized that one of the main reasons that this model is able to capture the human data as it does is that the individual neurons themselves saturate when participating in the representation of large vectors. This saturation serves as a kind of "soft" normalization, which is neither ideal mathematical normalization nor a complete lack of normalization. Rather, it is a more subtle kind of constraint placed on the representation of vectors in virtue of neuron response properties. And, crucially, this constraint is directly evident in the behavioral data (i.e., it enables reconstructing the correct U-shaped curve).

This observation is theoretically interesting because it provides an unambiguous example of the importance of constructing a neural implementation for explaining high-level psychological behavior.[1] All too often, researchers consider psychological and neural-level explanations to be independent–a view famously and vigorously championed by Jerry Fodor (Fodor, 1974). But in this case, the dependence is clear. Without constructing the neural model, we would have considered the mathematical characterization a failure and moved on to other, likely more complex, models. However, it is now obvious that we would have been doing so unnecessarily. And, unnecessary complexity is the bane of intelligible explanations.

6.4 ■ Biological Learning

I mentioned earlier that the NEF methods are useful partly because they do not demand that learning be central to designing models (Section 2.2.4). But, of course, this does not mean that learning is not an important aspect of our cognitive theories–I have emphasized this point, too (Section 1.4). So, in this section I consider not only how learning can be included in SPA-based models but how our understanding of learning can be enhanced by adopting a method that allows for the direct construction of some networks. As well, I have been at pains to argue for the biological plausibility of the methods I adopt throughout the book and will continue this trend in my consideration of learning.

It is difficult to specify what counts as a "biologically plausible" learning rule, as there are many mechanisms of synaptic modification in the brain

[1] Another example of the importance of building neural implementations to match psychological data can be found in Hurwitz (2010).

(Feldman, 2009; Caporale & Dan, 2008). However, the vast majority of characterizations of synaptic plasticity that claim to be biologically plausible still adhere to Hebb's (1949) suggestion: "When an axon of cell A is near enough to excite cell B and repeatedly or persistently takes part in firing it, some growth process or metabolic change takes place in one or both cells such that A's efficiency, as one of the cells firing B, is increased" (p. 62). Or, more pithily: "Neurons that fire together, wire together." Most importantly, this means that any modification of synaptic connection weights must be based on information directly available to the cell whose weight is changed (i.e., it must be based on presynaptic and postsynaptic activity alone). Unfortunately, most learning rules stop here. That is, they introduce variables that standin for pre- and postsynaptic activity (usually these are "firing rates"), but they typically do not work with actual neural activity (i.e., temporal spiking patterns).

This is problematic from a plausibility perspective because there is increasingly strong evidence that the modification of synaptic strength can be highly dependent on the *timing* of the pre- and postsynaptic spikes (Markram et al., 1997; Bi & Poo, 1998). However, rules that employ firing rates do not capture relative spike timing. Such spike-driven learning has become known as STDP (spike-timing-dependent plasticity; Fig. 6.8). Spike-timing-dependent plasticity refers to the observation that under certain circumstances, if a presynaptic spike occurs before a postsynaptic spike, there is an increase in the likelihood of the next presynaptic spike causing a postsynaptic spike. However, if the postsynaptic spike precedes the presynaptic spike, then there is a decrease in the likelihood of the next presynaptic spike causing a postsynaptic spike. Or, more succinctly, a presynaptic spike followed closely in time by a postsynaptic spike will potentiate a synapse, and the reverse timing depresses it. Ultimately, this can be thought of as a more precise statement of "fire together, wire together."

The learning rules proposed to explain STDP computationally focus, as the original experiments did, on comparing two spikes at a time. However, to explain more recent plasticity experiments, it has become necessary to consider triplets of spikes (Pfister & Gerstner, 2006). Specifically, Pfister and Gerstner have shown that by adding an additional depression with a pre-post-pre triplet, and additional potentiation with post-pre-post triplets, a better fit to the experimental data can be achieved. In particular, triplets allow the rule to capture the effects of different input spike frequencies evident in the neural data.

Interestingly, there is also recent evidence that it is not specifically postsynaptic *spikes* themselves that drive this plasticity. Rather, other indicators of postsynaptic activity like synaptic potentials and dendritic calcium flux are much more important for generating synaptic plasticity (Hardie & Spruston, 2009). Consequently, non-spiking activity of the cell may be a more appropriate measure of postsynaptic activity for learning rules.

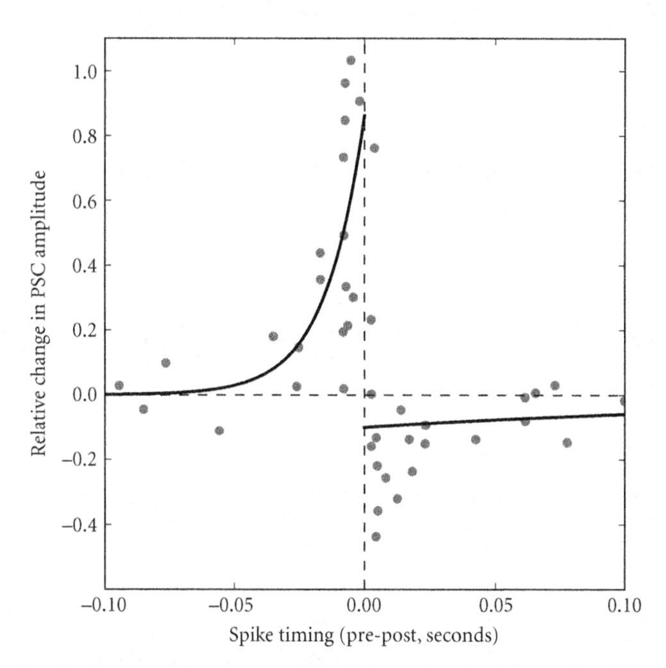

FIGURE 6.8 Spike-timing-dependent plasticity (STDP) experimental evidence. This figure shows the effect of changing the relative timing of pre- and postsynaptic spikes on synaptic efficacy. The percentage change in the postsynaptic current (PSC) amplitude at 20 to 30 min after a repetitive spiking input (60 pulses at 1 Hz). This change is plotted against relative spike timing (presynaptic spike minus postsynaptic spike) during the input. The figure shows that when the spikes generated by a neuron regularly came after presynaptic spikes, the amplitude of subsequent PSCs at that synapse increased, and when the spikes came before the presynaptic spikes, the amplitude decreased. The black lines indicate exponential fits to the data, which are commonly used as an "adaptation function" that summarizes this effect. (Adapted from Bi & Poo [1998] with permission.)

In addition, there has long been evidence that there are homeostatic mechanisms in individual neurons that are important for learning (Turrigiano & Nelson, 2004). Intuitively, these mechanisms ensure that neurons do not increase their weights indefinitely: if two neurons firing together causes them to wire more strongly, then there is no reason for them to stop increasing their connection strength indefinitely. Homeostatic mechanisms have been suggested that ensure that this kind of synaptic saturation does not occur. Some have suggested that the total connection strength a neuron can have over all synapses is a constant. Others have suggested that neurons monitor their average firing rate and change connection strengths to ensure that this rate stays constant. Still others have suggested that a kind of "plasticity threshold," which changes depending on recent neuron activity and determines whether weights are positively or negatively increased for a given activity level, is the mechanism responsible (Bienenstock et al., 1982). It is still unclear what the subcellular details of many of these mechanisms might be, but their effects have been well established experimentally.

Trevor Bekolay, a member of my lab, has recently devised a learning rule that incorporates many of these past insights, and is able to both reproduce the relevant physiological experiments and learn high-dimensional vector transformations. The mathematical details of this rule can be found in Appendix D.4. This rule is based on two past rules, the well-known BCM rule that characterizes a homeostatic mechanism (Bienenstock et al., 1982) and a recently proposed spike-driven rule from my lab that learns using postsynaptic activity in combination with error information (MacNeil & Eliasmith, 2011). Trevor has noted that his rule is well described as a homeostatic, prescribed error sensitivity (hPES) rule.

The hPES rule simultaneously addresses limitations of both standard Hebbian learning rules and STDP. In particular, unlike most standard rules, hPES is able to account for precise spike time data, and unlike most STDP (and standard Hebbian) rules, it is able to relate synaptic plasticity directly to the vector space represented by an ensemble of neurons. That is, the rule can tune connection weights that compute nonlinear functions of whatever high-dimensional vector is represented by the spiking patterns in the neurons. Unlike past proposals, hPES is able to do this because it relies on an understanding of the NEF decomposition of connection weights. This decomposition makes it evident how to take advantage of high-dimensional error signals, which past rules have either been unable to incorporate or attempted to side-step (Gershman et al., 2010). Thus, hPES can be usefully employed in a biologically plausible network that can be characterized as representing and transforming a vector in spiking neurons–precisely the kinds of networks we find in the SPA (Section 6.7 provides a tutorial using this rule).

To demonstrate the biological plausibility of hPES, Figure 6.9 shows that the rule reproduces both the timing and frequency results determined in STDP experiments. It is notable that it is not necessary to separately consider spike triplets to capture the frequency effects. Thus, the proposed rule is much simpler than that in Pfister and Gerstner (2006), while capturing the same experimental results. This makes it clear that the rule is able to function appropriately in a spiking network, and has a sensitivity to spike timing that mimics our best current characterization of learning *in vivo*. More than this, however, we would like to demonstrate that the rule can learn useful functions.

Like many other rules, this rule can incorporate error information coming from other parts of the brain to affect the weight changes in a synapse. There

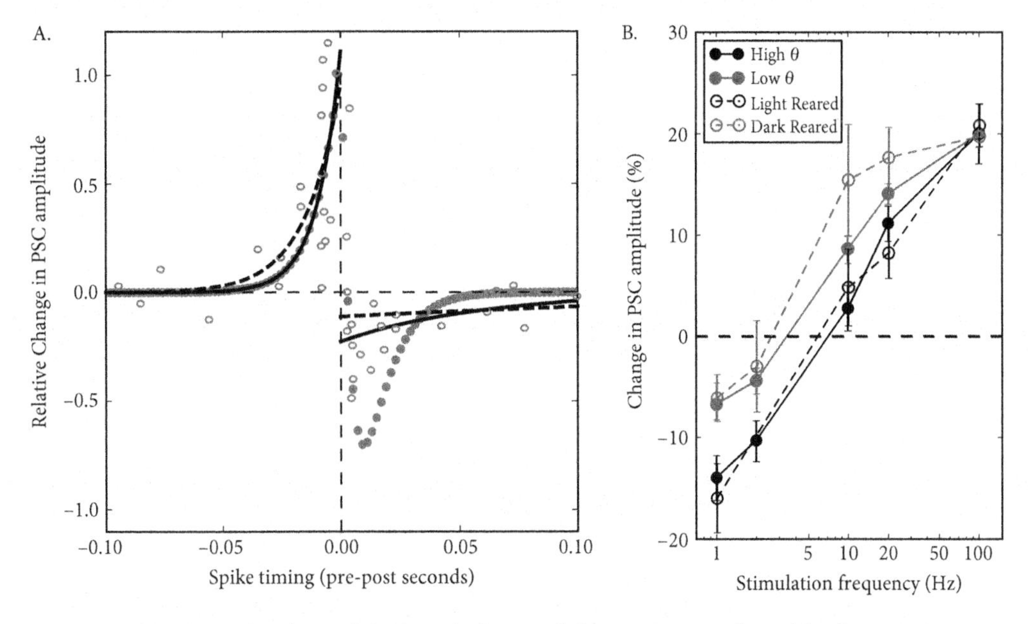

FIGURE 6.9 Reproduction of STDP timing and frequency results with the hPES rule. **A.** Spike-timing effects using hPES. Simulated data points are filled circles. Exponential fits to these data points are solid black lines. Data from Figure 6.8 are reproduced (open circles, dashed lines) for comparison. The simulated experiments capture the experimental data well, although they are less noisy. **B.** Data (open circles, dashed lines) and simulations (filled circles, solid lines) for frequency-dependent effects for STDP (data from Kirkwood et al. [1996] with permission). Each line shows the increasing efficacy as a function of changing the stimulation frequency. Black versus gray colors compare low versus high expected background firing (controlled by rearing animals in the dark and light, respectively). The hPES rule captures these effects well. All points are generated with a post-pre spike time difference of 1.4 ms.

is good evidence for this kind of modulatory learning in several brain areas. For example, the stability of the oculomotor integrator depends on retinal slip information, but only after it has been processed by other parts of brainstem (Askay et al., 2000; Ikezu & Gendelman, 2008, Ch. 5). More famously, modulatory learning is also subserved by the neurotransmitter dopamine in cortex and basal ganglia. Specifically, there is strong evidence that some dopamine neurons modulate their firing rate both when rewards that are not predicted occur and when predicted rewards do not occur (Hollerman & Schultz, 1998). Consequently, it has been suggested that dopamine can act as a modulatory input to cortex and parts of the basal ganglia (e.g., striatum), helping to determine when connections should be changed to account for unexpected information in the environment. It is now well established that

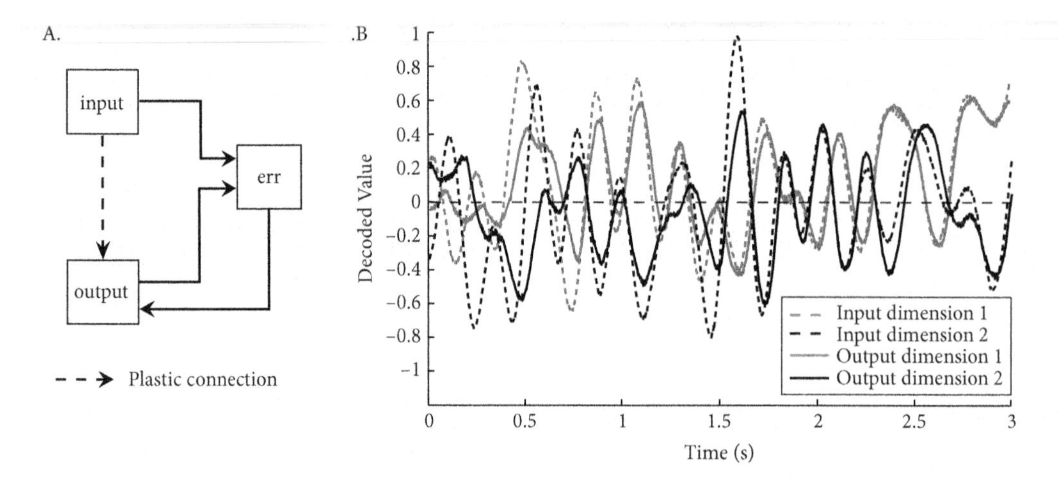

FIGURE 6.10 A network learning a 2D communication channel. **A.** The network consists of an initially random set of connections between the input and output populations that use hPES to learn a desired function (dashed line). The "err" population calculates the difference, or error, between the input and output populations, and projects this signal to the output population. **B.** A sample run of the network as it learns a two-dimensional communication channel between the input and the output. Over time, the decoded value of the output population (solid lines) begins to follow the decoded value of the input population (dashed lines), meaning that the network has learned to represent its input values correctly.

this kind of learning has a central role to play in how biological systems learn to deal with contingencies in their world (Maia, 2009). Unsurprisingly, such learning is important in SPA models that seek to explain such behavior (*see* Section 6.5).

So, although the source of the error signal may be quite varied, it is clear that biological learning processes can take advantage of such information. To demonstrate that hPES is able to do so, Figure 6.10 shows a simple example of applying this rule to a vector representation, with feedback error information. This network learns a two-dimensional communication channel. Figure 6.10A shows the structure of a circuit that generates an error signal and learns from it, and Figure 6.10B shows an example run during which the output population learns to represent the input signal. To keep things simple, this example includes the generation of the error signal in the network. However, the error signal could come from any source, internal or external, to the network.

This example demonstrates that the rule can learn a communication channel, but of course much more sophisticated nonlinear functions are important for biological cognition. A particularly important transformation in the SPA is circular convolution–recall that it is used for binding, unbinding, and content-sensitive control. Given the central role of convolution in the architecture, it is natural to be concerned that it might be a highly specialized, hand-picked, and possibly unlearnable transformation. If this was the case, that would make the entire architecture much less plausible. After all, if binding cannot be learned, we would have to tell a difficult-to-verify evolutionary story about its origins. Fortunately, binding can be learned, and it can be learned with the same learning rule used to learn a communication channel.

Figure 6.11 shows the results of hPES being applied to a network with the same structure as that shown in Figure 6.10A, which is given examples of binding of three-dimensional vectors to generate the error signal. These simulations demonstrate that the binding operation we have chosen is learnable using a biologically realistic, spike-time-sensitive learning rule. Further, these results show that the learned network can be more accurate than an optimal NEF network with the same number of cells.[2]

This simple demonstration does not completely settle the issue of how binding is learned, of course. For example, this network is learning the circular convolution of three-dimensional, not 500-dimensional, semantic pointers. And, notice that it takes about 200 s of simulation time to do so. This is actually quite fast, but we know that as the dimensionality of the vectors being bound increases, the length of time to learn convolution increases quickly, although we do not yet know the precise relation.

As well, there are important questions to answer regarding how the error signal would be generated, how the error signals can be targeted to the appropriate systems that perform binding, and so on. In fact, it should not be surprising that telling a story about how binding is learned might be difficult–many of the more cognitive behaviors that require binding are not evident in people until about the age of 3 or 4 years (e.g., analogy [Holyoak & Thagard, 1995], false belief [Bunge, 2006], multiplace relations in language [Ramscar & Gitcho, 2007], etc.). Consequently, I would argue that this demonstration should allay concerns that there is anything biologically or developmentally implausible about choosing circular convolution (or nearly any kind of vector multiplication) as a binding operation, but it still leaves open the majority of questions related to the cognitive development of binding.

[2] The fact that the learned network is better than "optimal" should not be too surprising. Optimal weights are computed with known approximations–for example, that the tuning curves are smooth, and that the rate model responses are the same as the spiking responses. It strikes me as interesting, and worth exploring in more detail, how NEF models may be improved with the fine-tuning provided by appropriate learning rules. This is explored briefly in MacNeil and Eliasmith (2011).

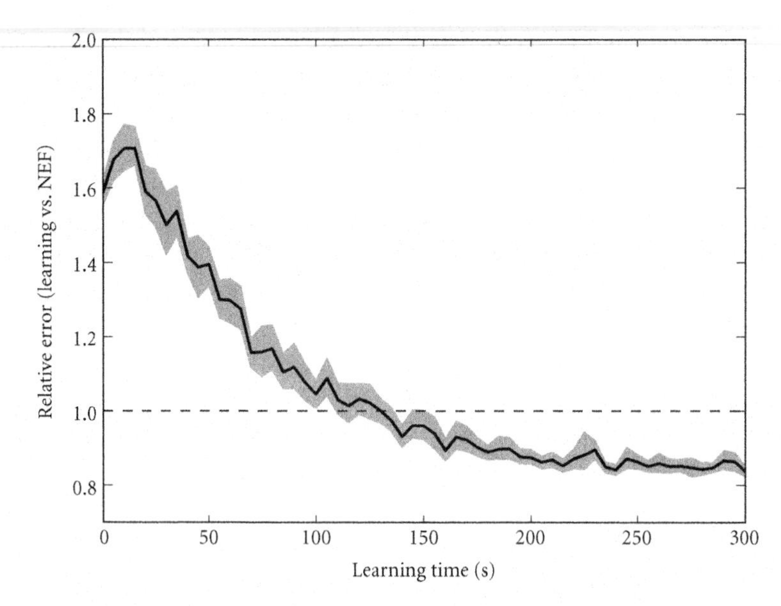

FIGURE 6.11 Learning circular convolution of vectors. This graph shows that hPES can learn a multidimensional, nonlinear vector function. Specifically, the network learns the binding operator (circular convolution) used in the SPA. The input is two 3D vectors, and the output is their convolution. An error signal is generated as in Figure 6.10. As can be seen, the learned network eventually does better than the one whose weights are optimally computed using the NEF methods. Both networks have 1200 neurons; 20 versions of each were run. Relative error is calculated with respect to the time and trial averaged NEF error. The gray band indicates the 95% confidence interval.

6.5 ■ Example: Learning New Actions

The examples of learning that I have presented so far can be best characterized as learning how to manipulate representations. That is, given some representation, a random initial transformation, and information about a desired transformation in the form of an error signal, the learning rule will update the random initial transformation to approximate the desired transformation. This kind of learning may be common and appropriate for much of cortex (e.g., when learning to identify a new object, or extract specific features out of the input). However, it does not directly address a central concern for animals: learning how to act so as to gain benefits from the environment without explicit guidance.

In this section, I describe some recent work, also from Trevor Bekolay, that demonstrates how hPES can be integrated directly into the model of the basal ganglia discussed in Section 5.2. Extending the basal ganglia model with hPES supports the tuning of cortical-striatal connections. This allows the system to adapt to changing reward contingencies in the environment. That is, it allows the system to learn new ways to control its behavior, depending on what is most rewarding–a basic biological behavior.

It is now well established that the basal ganglia play a central role in this kind of "instrumental conditioning" (*see* Maia [2009] for a review). Instrumental conditioning was introduced as a means of understanding animal behavior by psychologists in the nineteenth century. This phenomenon is now generally thought of as a kind of "reinforcement learning," which captures any learning that can be characterized as an agent choosing actions in an environment to maximize rewards. Reinforcement learning is often thought of as a subfield of machine learning, and so has inspired many explicit algorithms for attempting to solve this important problem. Famously, some of the algorithms of reinforcement learning have been mapped to specific neural signals recorded from animals involved in instrumental conditioning (Schultz et al., 1997).

As briefly mentioned in the previous section, the neurotransmitter dopamine is often singled out as carrying error signals relevant for such learning. More specifically, dopaminergic neurons in the substantia niagra pars compacta (SNc) and ventral tegmental area (VTA) have been shown to mediate learning in the cortical-striatal projection (Calabresi et al., 2007). These neurons fire in a manner that seems to encode a reward prediction error (Bayer & Glimcher, 2005). That is, their firing increases if the animal gets an unexpected reward, and is suppressed if the animal gets no reward when it is expecting one. This information is precisely what is needed for the animal to determine how to change its future actions to maximize its reward.

Conveniently, we can augment the previous basal ganglia model (Fig. 5.1) to include these additional dopamine neurons and their relevant connections to striatal neurons (*see* Fig. 6.12). Because dopamine is largely modulatory, the general functioning of the model is not affected in any way, but we can now explore various means of exploiting the dopamine signal (by using the hPES rule, which can be error driven) to learn new actions given environmental contingencies. Because dopamine is a modulatory neurotransmitter, we can think of it as providing a channel of input that does not directly cause neurons to fire more or less but, rather, informs synapses about recent rewards.

For present purposes, I examine the hPES rule in the context of simple but common reinforcement learning tasks used in animal experiments called "bandit tasks." These tasks are so named because the rewards provided are similar to those given by slot machines (i.e., "one-armed bandits"). For example, in a two-armed bandit task, a rat enters a maze like that shown in Figure 6.13A, and must decide whether to take its chances getting rewarded at the left or right reward sites. The reward sites have different probabilities of providing a reward, so the animal is faced with a fairly subtle decision. The rat

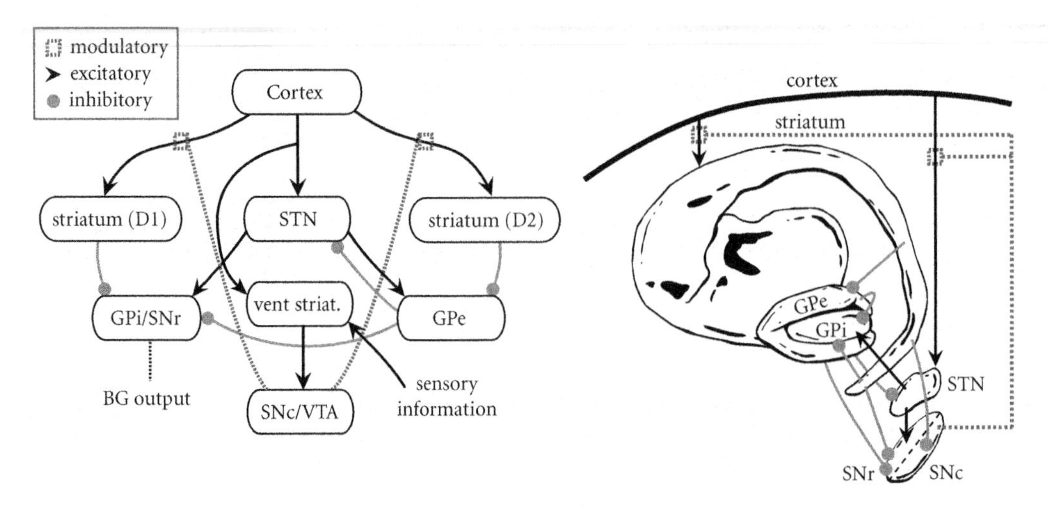

FIGURE 6.12 The basal ganglia with dopaminergic projections and ventral striatum included. The SNc and VTA, which contain efferent dopamine neurons, have been added to the basal ganglia circuit from Figure 5.1. The modulatory dopamine connections are shown with dotted lines. These are typically taken to carry information about reward prediction error, based on current sensory information. Ventral striatum is also distinguished as it encodes values for current and previous states, as well as the reward. This information is then used through SNc/VTA to generate the prediction error that is encoded in the modulatory dopamine signal.

is never guaranteed a reward but, rather, must be sensitive to the probability of reward.

To examine the animal's ability to track these reward probabilities, the likelihood of reward is manipulated during the task. As shown in Figure 6.13B, the animal adjusts its behavior to more often select the site that is rewarded with the greatest frequency. The model does the same thing, adjusting its choices of which way to turn to reflect learned reward contingencies. This occurs in the model because the model adjusts which cortical state matches the precondition for the actions by changing the connection weights between cortex and striatum (i.e., those that implement the $\mathbf{M}_b$ matrix in the model), based on the dopamine signal. In short, it constantly updates the utilities of different cortical states based on feedback. The rule itself works much the same as it did when learning convolution–an error signal is used to update the the weights to compute a transformation between input (cortical state) and output (utility).

Because the ventral striatum receives a dopamine signal that indicates prediction errors, it is a natural structure to record single-cell responses during such tasks. As shown in Figure 6.14, single neurons in the ventral striatum have responses that vary depending on which phase of the task the animal is

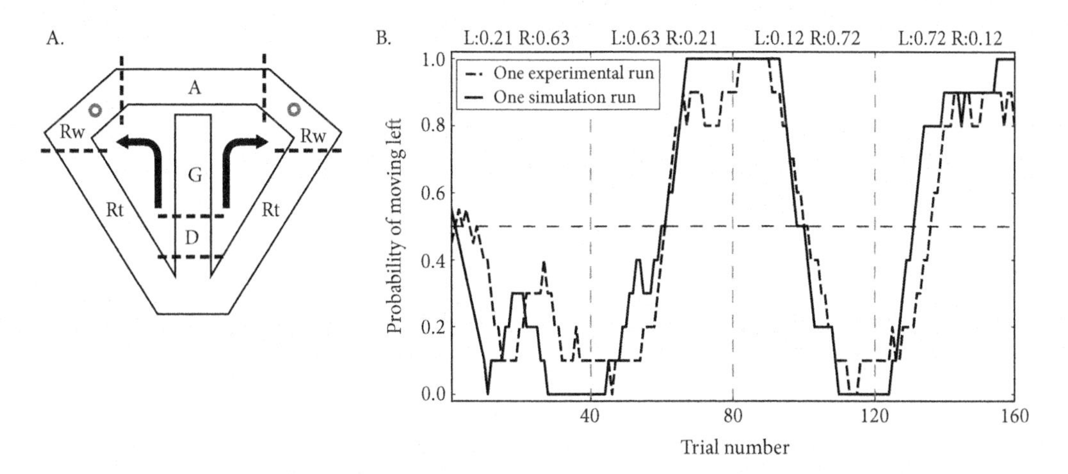

FIGURE 6.13 The two-armed bandit task. **A**. The maze used to run the task. The rat waits at D during a delay phase. A bridge between G and A then lowers, allowing the animal to move. The animal reaches the decision point A during the approach phase and turns either left or right. Reward is randomly delivered at Rw during the reward phase. Finally, at Rt the animal returns to the delay area during the return phase. **B**. Experimental behavioral data from a single trial (dashed line), and model behavioral data from a single trial (solid line). The values above the graph indicate the probability of getting a reward at the left and right reward sites. The animal and model clearly learn to favor the more rewarded site as the reward probabilities change over the course of the trial. The probability of moving left is calculated by a moving average of the last 10 trials. Experimental data and A. adapted from Kim et al. (2009) with permission.

currently in. These same kinds of response are evident in the neurons in the model, as also shown in that figure.

Although the task considered here is simple, the model is clearly consistent with both behavioral and neural data from this task. Given that the basal ganglia model is nearly identical to that used for implementing much more sophisticated action selection in other contexts (e.g., the Tower of Hanoi), these results are an encouraging indication that the same learning rule employed here can support the learning of more sophisticated state/action mappings as well. So, I take this model to show that the SPA can naturally incorporate reinforcement signals to learn new actions in a biologically realistic manner.

However, this is only one of the many kinds of learning accomplished by the brain. As well, this example does not show how low-level learning rules might relate to high-level cognitive tasks. In the next section, I consider a more cognitive application of the same rule.

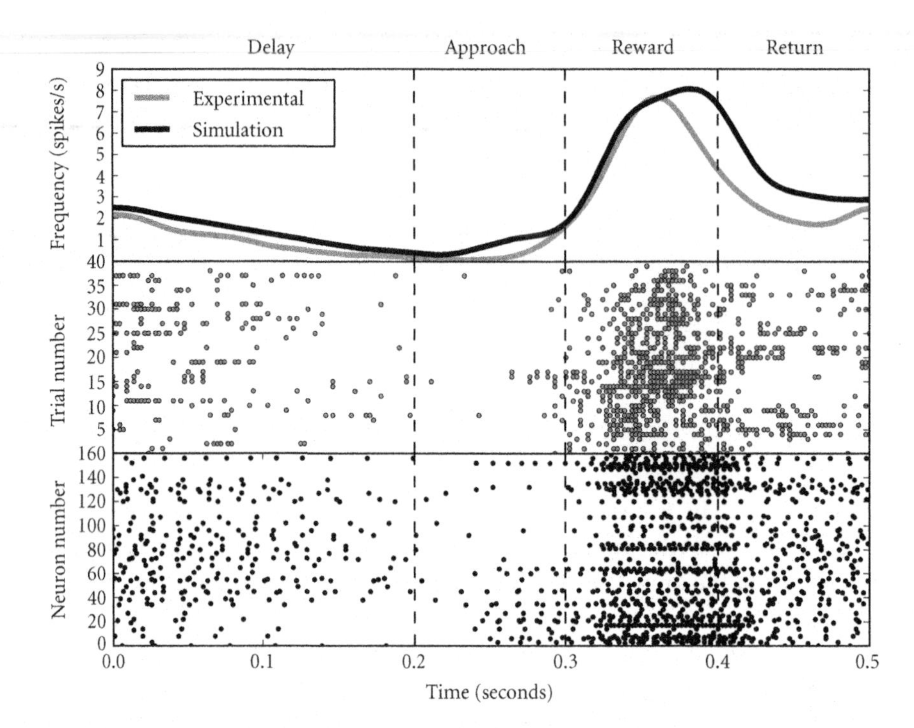

FIGURE 6.14 A comparison of spiking data from the rat striatum and the model. Middle panel: Spike train data over several trials recorded from a single neuron in the ventral striatum. Bottom panel: Spike trains of several neurons in a single trial from the ventral striatum of the extended basal ganglia model (*see* Fig. 6.12). Each dot is a single spike for both panels. Top panel: Both sets of spiking data filtered with a 15-ms Gaussian filter. The model and data share several properties, such as a slight decrease in firing during the end of the approach phase and vigorous firing during the reward phase with a decrease to delay phase levels during the return phase. Figure from Stewart et al. (2012), data from Kim et al. (2009) with permission.

6.6 ■ Example: Learning New Syntactic Manipulations

I have considered how, in an abstract sense, the hPES rule can be used to learn the cognitive binding operation fundamental to the SPA. And I have considered how, in a specific non-cognitive case, hPES can be used in the basal ganglia to learn appropriate actions given changing environmental contingencies. In this section, I consider how this same rule can be used in a specific cognitive task to support learning of context-dependent reasoning.

In my discussion of the Raven's Progressive Matrices (RPM) in Section 4.7, I provide an example of induction that shows how to build a system that learns a new rule based on past examples. I emphasize there that no connection weights are changed in the system, despite this performance. Rather, the system relies on working memory to store and update a transformation over the presentation of several examples. This approach has the advantage of being both very rapid and immediately accessible for further induction. However, it has the disadvantage that once the task is complete, the induced rule is essentially forgotten. In people, however, effective reasoning strategies can be stored in long-term memory and used in the distant future.

In this section, I consider a similar kind of learning to that in the RPM, but with an emphasis on addressing this longer-term behavior (Eliasmith, 2004, 2005b). For variety, I consider a different task: the Wason card selection task (Wason, 1966). This task is challenging because it requires symbolic reasoning and strategy changes across contexts. In the Wason task, participants are given a conditional rule of the form "if P, then Q" (*see* Fig. 6.15). They are then presented with four cards, each of which has either P or not-P on one side,

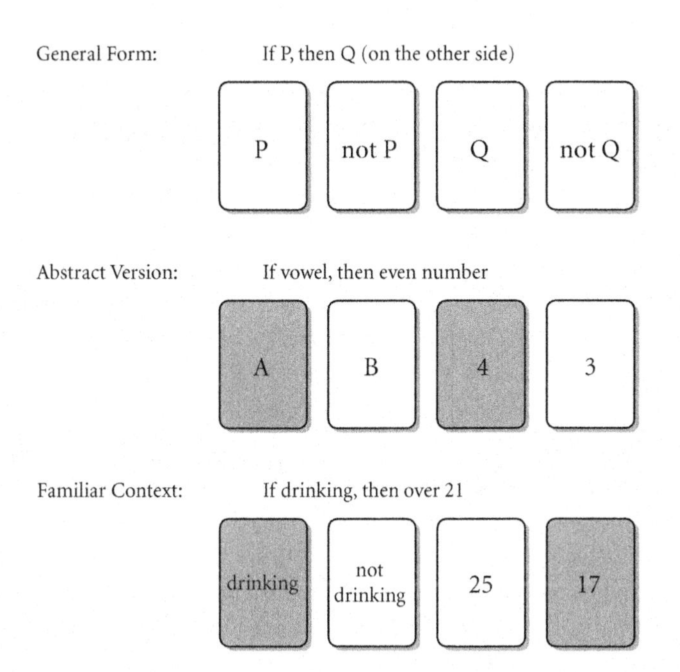

FIGURE 6.15 The Wason selection task. Four cards are presented to the participant. Each card has information on both sides. A rule is stated that relates the information on both sides of the cards (e.g., if there is a vowel on one side, then there is an even number on the other side). The participant must select the cards to flip over that confirm whether the rule is being followed. The top row shows the general pattern of cards used in the task. Because the rules are always material conditionals, the correct response is always to flip over the first and last cards. However, people respond differently given different contents (showing a "content effect"). Darkened cards indicate the most common selections for different rules.

and either Q or not-Q on the other. The visible sides of each card show P, not-P, Q, and not-Q. The task is for the participant to indicate all of the cards that would have to be flipped over to determine whether the conditional rule is true (i.e., is consistent with all four cards). The logically correct answer is to select the cards showing P and not-Q.

There are interesting psychological effects in such a task. If the task is given in an abstract form, such as "if a card has a vowel on one side, then it has an even number on the other," then the majority of participants choose P (the vowel) and Q (the even number) (Cheng & Holyoak, 1985; Oaksford & Chater, 1994), which is incorrect. However, if the task is presented using familiar content (remember that most participants are undergraduates), such as "if a person is drinking beer, then that person must be over 21 years old," then the majority choose P (drinking beer) and not-Q (under 21) (Cox & Griggs, 1982), which is correct. The change in mean performance of subjects is huge, for example, going from only 15% correct on an abstract task to 81% correct on a familiar version (Johnson-Laird et al., 1972). In other words, structurally identical tasks with different contents lead to different performance. Explaining this "content effect" requires understanding how symbolic manipulation strategies may change based on the content of the task.

Initial explanations of Wason task performance were based on just the familiarity of the contexts–drinking being more familiar than arbitrary letter-number relations. However, later explanations suggested that the distinction between deontic situations (i.e., situations related to social duty) and non-deontic situations was of greater importance. In deontic situations the rule expresses a social or contractual obligation. In such situations, human performance often matches the logically correct choices, regardless of familiarity or abstractness (Sperber et al., 1995). To explain this, Cosmides (1989) developed social contract theory (SCT), a view inspired by evolutionary psychology. Social contract theory proposes that natural selection has produced special-purpose, domain-specific mental algorithms for solving important recurring adaptive problems. Alternatively, Cheng and Holyoak (1985) proposed that people reason in all cases using neither context-independent rules of inference nor memory of specific experiences but, rather, using abstract, context-sensitive knowledge structures induced from ordinary life experiences (some of which relate to obligation and some of which do not): they called these knowledge structures "Pragmatic Reasoning Schemas" (PRS).

Both of these theories use context sensitivity to explain why different formulations of the selection task elicit different responses. However, the theories differ in the source of these rules: PRS are induced through experience, whereas SCT algorithms are genetically endowed. Indeed, Cosmides (1989) has challenged any theory based on induction to lay out the mechanistically defined domain-general procedures that can take modern human experience as statistically encountered input, and produce observed domain-specific performance in the selection task as output. Theoretical responses to this challenge have been made, but none present a neural mechanism. In this section, I propose such a mechanism that is consistent with PRS. And I suggest that this mechanism meets the challenge that Cosmides has offered to domain general hypotheses.

In short, this mechanism consists of the combination of induction as in the RPM model and learned associations that relate contexts to appropriate syntactic transformations. Both the induced transformation and the context-to-transformation mapping are encoded into a long-term associative memory using the hPES rule. Figure 6.16 indicates the network architecture

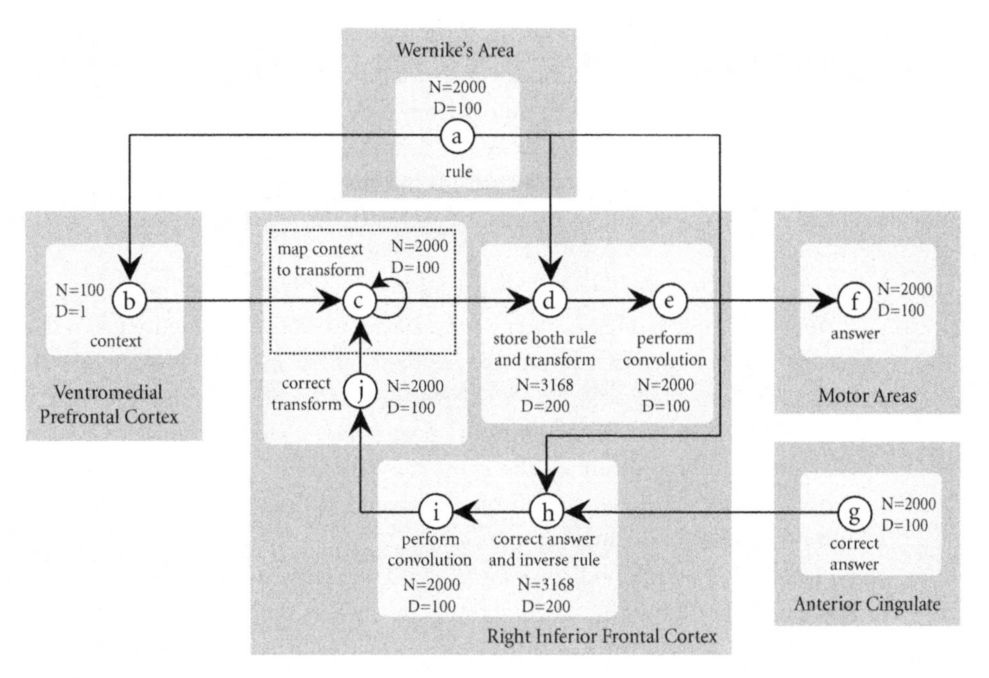

FIGURE 6.16 The Wason task network. Circles indicate neural populations (N is the number of neurons, D is the dimensionality of the value represented by the neurons). In this network *a* represents the input rule, *b* represents the context (based on the semantics of *a*), *c* associates the context with a transformation (i.e., a reasoning strategy), *d* and *e* apply the strategy (i.e., compute the convolution of the rule and the transformation), *f* is the result of the transformation (i.e., the answer), *g* is feedback indicating the correct answer, *h* and *i* determine the relationship between the correct answer and the rule for a given example (i.e., convolve the inverse of the answer with the rule), and *j* represents the determined transformation for a given example. Arrows indicate neural connections. Population *c* is recurrently connected to provide an error signal for the learning rule. The dotted rectangle indicates the population where associative learning occurs.

and function. The neuroanatomical labels on various parts of the model are supported by a variety of imaging studies (e.g., Goel, 2005; Parsons & Osherson, 2001; Parsons et al., 1999; Kalisch et al., 2006; Adolphs et al., 1995; *see* Eliasmith [2005b] for details).

The inductive process underlying this model is similar to that in Section 4.7, inducing the correct transformation in a given context. Essentially, b represents the context information and c represents the model's current guess at an appropriate transformation. The hPES changes the connections that map context onto the appropriate transformation, so as to match the feedback from j (i.e., by minimizing the error, which is the difference between the output of c and j).

The learning part of the circuit should be familiar in that it employs an error feedback signal to inform the modification of connection weights on an input signal. Here, the feedback is provided by the difference between j and c, and the input signal is the context from b. As we saw earlier, the hPES rule can use this circuit structure to learn arbitrary vector transformations. Consequently, we should not be surprised that the model can successfully learn the mapping from context to reasoning strategy using a biologically realistic rule.

To demonstrate that the learning itself functions as expected, results of performing a simple context-strategy association are shown in Figure 6.17. The context signal in this case is one dimensional and the vector it is mapped to is only six dimensional for simplicity. Nevertheless, the ability of the rule to map different contexts to different output vectors (i.e., different transformations) is evident. In the full model, the context is a semantic pointer derived from the rule to be manipulated, and the representations and transformations are 100-dimensional semantic pointers as well.

To specify the full model, we must define the semantic pointers that are in the lexicon and how rules are encoded. In the lexicon, the 25 base semantic pointers are chosen randomly and then bound as needed for combinatorial concepts (e.g., $\mathbf{notOver21} = \mathbf{not} \circledast \mathbf{over} \circledast \mathbf{21}$). The rule representations used in the model are structured semantic pointers of the form:

$$\mathbf{rule} = \mathbf{relation} \circledast \mathbf{implies} + \mathbf{antecedent} \circledast \mathbf{vowel} + \mathbf{consequent} \circledast \mathbf{even}.$$

To perform the basic task, the model must learn the appropriate transformation in each of the two contexts. I label the two contexts "abstract" (for unfamiliar rules like the arbitrary letter-number pairings) and "facilitated" (for familiar rules like those related to drinking). Figure 6.18 shows a simulation that includes learning and recall in both contexts. The first half of the simulation is a training period during which the model is presented with a rule and the appropriate response. The context is determined by examining the content of the rule (i.e., summing the antecedent and consequent lexical items), and the appropriate response is determined by direct feedback from

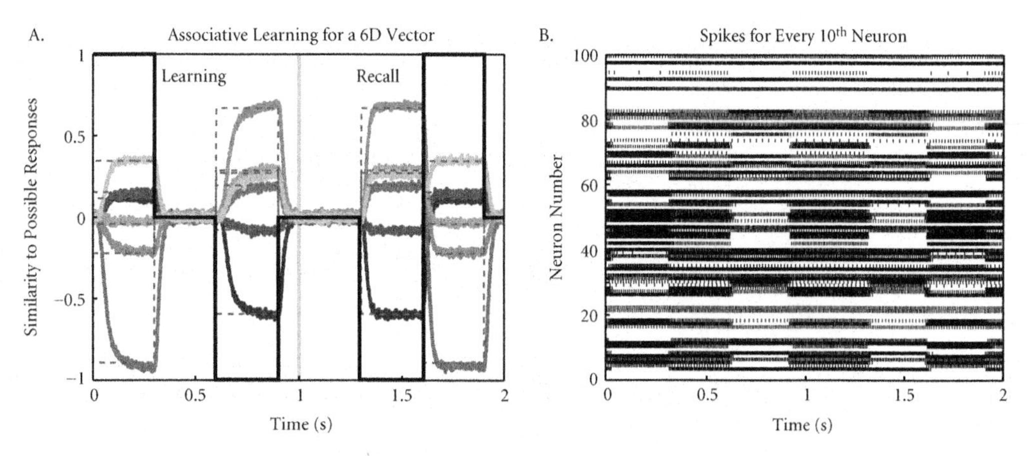

FIGURE 6.17 Associative learning using hPES in a spiking network for 6D vectors. **A.** The thick black line indicates the input context signal. There are three contexts, +1, −1, and 0. The grayscale lines each indicate the represented value of one element of a 6D vector. The dotted lines indicate the ideal answers. During the first half of the simulation, two random vectors (coming from *j* in Fig. 6.16) are being learned under two different contexts. During the second half of the simulation, only the context signal is applied to the population, which then responds with the vector encoded in that context. The network successfully associates the context signal with a close approximation to the original random vector. **B.** The spike trains of the neurons in the associative memory population (for the first half of the run, and every tenth neuron). There are 1000 neurons in this population. The synaptic weights are initialized to random values.

the environment. The model is then tested by presenting one rule at a time, without learning or feedback. As shown, the model learns the appropriate, distinct responses under the different abstract and facilitated contexts.

It may seem odd that the model is taught the "logically incorrect" response during the abstract context (i.e., the appropriate response is deemed to be P and Q). However, I believe that this is plausible because I take this answer to be learned as a result of past experiences in which a biconditional interpretation of the "if-then" locution has been reinforced (Feeney & Handley, 2000). The supposition is that when parents say, for example, "If you eat your peas, then you get some ice cream," they actually mean (and children learn) a biconditional relation (i.e., if they don't eat their peas, they don't get ice cream). In the Wason task, the "correct" answer is always assumed to be a material conditional (i.e., if they don't eat their peas, then they may or may not get ice cream). I am assuming that the bi-conditional interpretation is a default that is used in the abstract, unfamiliar context. In the facilitated context, the expected response is the "logically correct" answer **alcohol** and **not** ⊛ **over21**,

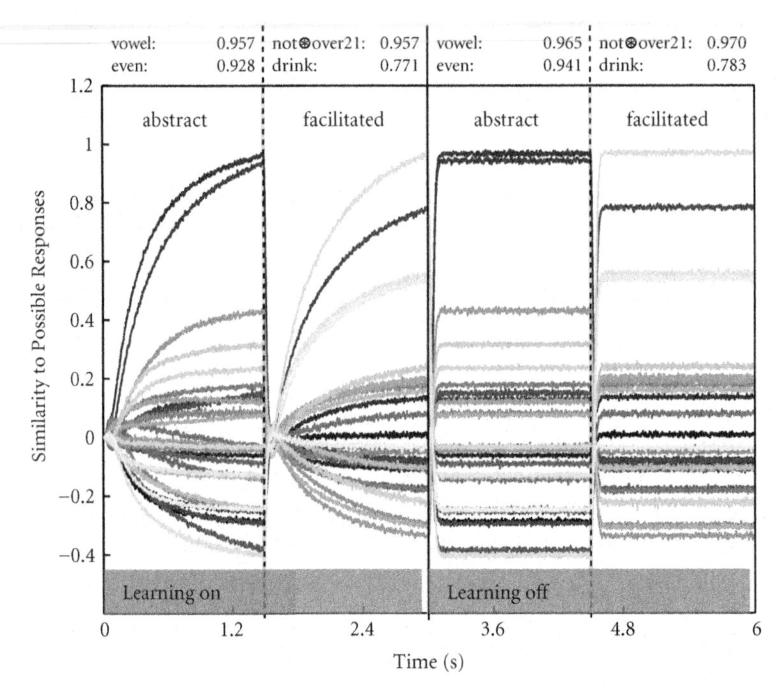

FIGURE 6.18 Model results for learning and applying inferences across two contexts. The time-varying solid lines indicate the similarity of the model's decoded spiking output from population f to all possible responses. The vector names and numerical similarity values of the top two results (over the last 100 ms of that part of the task) are shown above the similarity plot. The current context is labeled as abstract or facilitated. After learning, the system reliably performs the correct, context-dependent transformation, producing the appropriate response. This performance is 95% reliable ($n = 20$) over independent simulations.

because this is learned through experiences where the content-specific cues indicate that this specific rule needs a material conditional interpretation. This characterization is consistent with PRS.

Nevertheless, the result shown in Figure 6.18 is not especially exciting, because it could be taken to merely show that we can teach a neural net-like model to memorize past responses and reproduce them in different contexts. That has been done before. But, much more is going on in this model. Specifically, we can show that here, as in the RPM model (*see* Section 4.7), the system has the ability to syntactically generalize within a context. Thus, it does not simply memorize and reproduce past responses but, rather, learns the structure of the transformation and can use that information to produce novel responses. Syntactic generalization, as I discussed earlier, has often been identified as a hallmark of cognition.

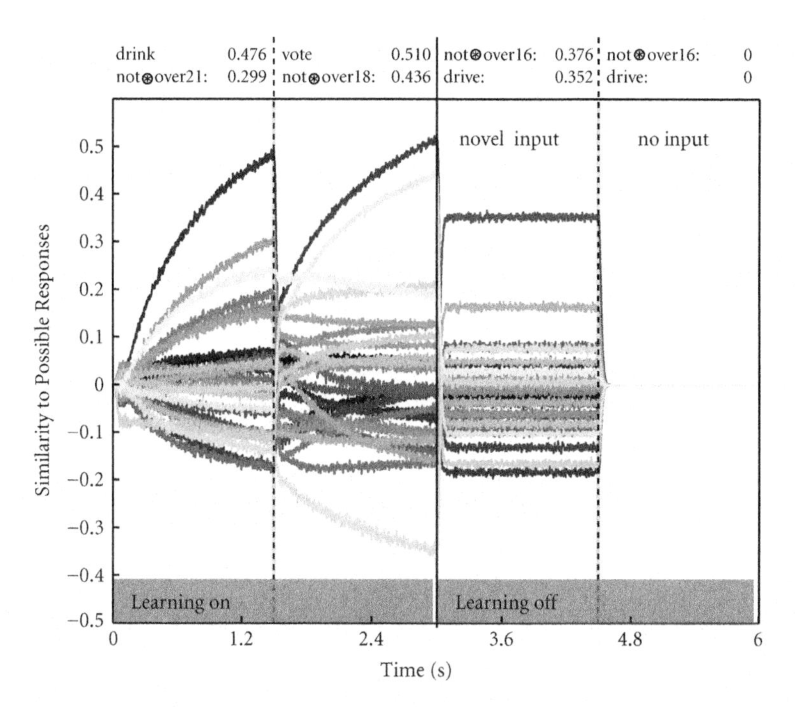

FIGURE 6.19 Simulation results demonstrating syntactic generalization within a context. The format is the same as described in Figure 6.18. In the first half of the simulation, two different facilitated rules are presented, and the relevant transformation is learned. In the third quarter, a novel facilitated rule is shown and the model correctly infers the appropriate responses. The last quarter shows that the model does not respond without input.

For the second simulation, I have defined several rules, at least three of which are of the facilitated form (e.g., "if you can vote, then you are over 18"). During the learning period, two different examples of a facilitated rule are presented to the model. Learning is then turned off, and a novel rule is presented to the model. This rule has similar but unfamiliar content and the same structure as the previous rules. The results of the simulation are shown in Figure 6.19. As evident there, the model is able to correctly reason about this new rule, providing the P and not-Q response. So, the model clearly has not memorized past responses, because it has never produced this response (i.e., **not** ⊛ **over16** + **drive**) in the past. This is a clear case of syntactic generalization, as it is the syntax that is consistent between the past examples and the new input rule, not the semantic content.

The semantic content is still playing an important role in the model, because it is determining that the context of the new rule is similar to the already encountered rules. However, the inference itself is driven by the syntactic structure of the rule. Two example rules are shown because multiple examples of the combination of semantics and syntax are needed to allow them to be appropriately separated. This is consistent with the fact that performance on the Wason task improves with instruction (Rinella et al., 2001).

Because this model is implementing a psychological hypothesis at the neural level, there are several ways that we can further examine the model. For example, I have suggested that one key feature of models developed with the SPA is robustness to both background firing noise and neuron death. Because semantic pointer representations are distributed across a group of neurons, performance should degrade gracefully with damage.

To demonstrate this in the current model, we can randomly remove neurons from the "rule and transformation" population (population d in Fig. 6.16). Figure 6.20 shows the effect of removing neurons after training but before testing on the task shown in Figure 6.18. On average, accurate performance occurs even with an average of 1,221 (variance $= 600$, $n = 500$) neurons removed out of 3,168.

We can also explicitly test the robustness of the model to intrinsic noise rather than neuron death. In this case, the model functions properly with up to 42% Gaussian noise used to randomly vary the connection weights.[3]

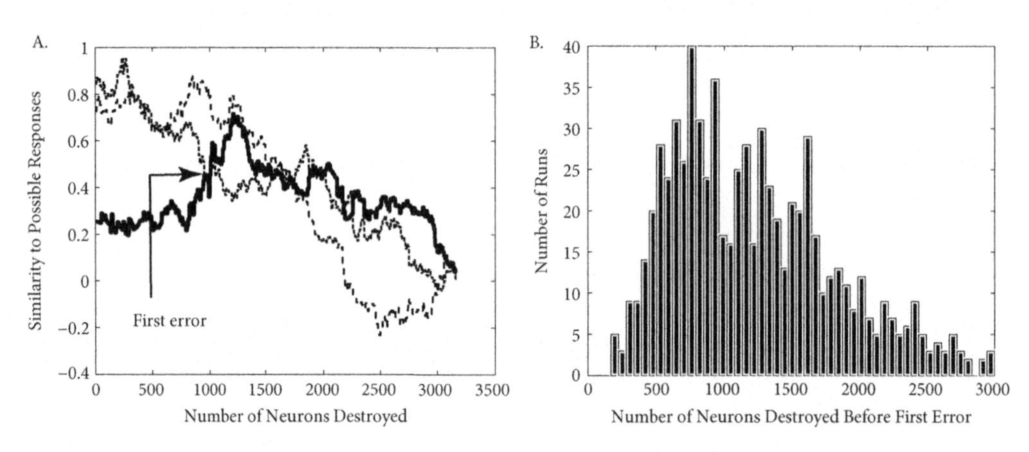

FIGURE 6.20 Performance of the Wason model after random destruction of neurons in population (d) in Figure 6.16. **A.** The effect of continually running the model while random neurons are destroyed until none are left. The lines show the similarity of the three symbols whose semantic pointers most closely match the model's answer. An error is made if either one of the original two top responses falls below the third. **B.** A histogram of the number of neurons destroyed before the first error for 500 experiments (mean $= 1{,}221$).

[3] Each synaptic connection weight was independently scaled by a factor chosen from $N(1,\sigma^2)$. The model continued to function correctly up to $\sigma^2 = 0.42$.

Random variation of the weights can be thought of as reflecting a combination of imprecision in weight maintenance in the synapse, as well as random jitter in incoming spikes. Eliasmith and Anderson (2003) have explored the general effects of noise in the NEF in some detail and shown it to be robust (*see also* Conklin & Eliasmith [2005] for a demonstration of robustness in a model of rat subiculum). In short, these results together demonstrate a general robustness found in SPA models. Conveniently, such biologically plausible robustness allows close comparison of NEF and SPA models to micro-lesion (MacNeil & Eliasmith, 2011) and large-scale lesion data (Lerman et al., 2005) if it is available.

In many ways, the robustness of this Wason task model is a result of its low-level neural implementation. But, as mentioned earlier, the model is also implementing a cognitive hypothesis (i.e., that domain general mechanisms can learn context-sensitive reasoning). As a result, the model can also be examined in a very different way, relating it to more traditional cognitive data. For example, we can use the model to gain some insight into how humans are expected to perform when they are given examples of novel syntactic transformations. Qualitatively, the model is consistent with evidence of practice effects in reasoning tasks (Rinella et al., 2001). However, we can also more specifically examine the performance of the model as a function of the number of examples. In Table 6.1 we can see that generalization improves rapidly with the first three examples and then improves less quickly. We can thus predict that this would translate behaviorally as more rapid and more consistent responses after three examples compared to one example.

Beyond these behavioral predictions, the model also embodies a specific prediction for cognitive neuroscience–namely, that the central difference between the abstract and content facilitated versions of the Wason task is the context signal being provided from the VMPFC, not the mechanism of

TABLE 6.1 *Improvement in Generalization With Learning History*

# of examples shown	Generalization accuracy
1	−0.10
2	0.05
3	0.23
4	0.26

Accuracy is the difference between the semantic pointer representation of the correct answers and the largest incorrect answer, scaled by the largest incorrect answer. A number below 0 indicates that the wrong answer is produced, and larger numbers indicate greater separation between the correct and incorrect answers.

reasoning itself. This is very different from the SCT suggestion that evolutionarily distinct reasoning mechanisms are necessary to account for performance differences on these two versions of the task (Cosmides, 1989). Experiments performed after the presentation of an early version of this model (Eliasmith, 2004) have highlighted the VMPFC as the locus of cognitive processing differences on the two versions of the task (Canessa et al., 2005), supporting this suggestion.

This neuroimaging result highlights why I believe that this model meets Cosmides' original challenge to provide a plausible domain general mechanism able to explain task performance while being inductive. In particular, Cosmides has suggested that the inductive evidence during the normal course of human development cannot serve to explain the superior performance on the content-facilitated task because the relative amount of evidence would not favor the correct response. However, the inductive mechanism provided here does not rely on the relative amount of evidence available in different contexts. Rather, it relies on a minimum threshold of available evidence within a given context (i.e., enough for the learning rule to induce the relevant transformation). As a result, this mechanism is: (1) domain general, because the mechanism is identical across contexts (only the VMPFC context signal changes); (2) inductive, because it learns based on past examples; and (3) able to account for context effects on the Wason task.

Finally, although this model is closely allied to PRS, unlike PRS it does not assume that the transformations used to solve the problem should be limited to a theoretically predetermined group of "schemas" that are the same for all subjects (e.g., the "permission" schema, the "obligation" schema, etc.). Rather, each individual will solve the problem using his or her own estimation of the correct transformation in the given context, as determined by his or her personal experience and learning history in that context. The contexts that can distinguish behavior are thus not restricted to clear-cut theoretical distinctions (e.g., deontic/non-deontic) but, rather, are driven by the similarity of current context signals to past contexts. That is, the similarity space provided by context signals is mapped onto a transformation space. This mapping is discovered by the model based on past examples and may not neatly align with our theoretical classifications of context. This is consistent with findings that successful performance of the task does not align with intuitive theoretical distinctions (Oaksford & Chater, 1996; Almor & Sloman, 1996) and can vary with past experience (Rinella et al., 2001).

This example concludes my discussion of learning and indeed my presentation of the SPA. However, although I have presented all the parts of the SPA, specifying a collection of parts does not satisfy my main goal of suggesting an integrative architecture for biological cognition. As a result, the next chapter brings the pieces together and presents a single model that is able to incorporate a variety of the behaviors and principles we have seen to this point. My motivation is to demonstrate that the SPA examples we have seen are not task-specific, one-off models but, rather, are components of a coherent, unified picture of brain function.

6.7 ■ Nengo: Learning

Accounting for synaptic plasticity is crucial for any characterization of brain function. In this tutorial, I demonstrate how to include one kind of synaptic plasticity in Nengo models. By default, the learning template in the toolbar in Nengo's GUI uses the hPES rule. Nengo supports a much wider variety of learning rules, but many of these require use of the scripting interface. I will not consider scripting until the next tutorial.

The implementation of learning employed by the Nengo template uses an error signal provided by a neural ensemble to modulate the connection between two other ensembles. This template can be used to learn a wide variety of transformations; here, we begin with consideration of how to learn a simple communication channel.

- Open a blank Nengo workspace and create a new network named "Learning."
- Create two new default parameter ensembles named "pre" and "post," with one dimension and 50 nodes each.
- Drag the "Learned Termination" template into the network. In the template constructor, name the error ensemble "error" and give it 100 neurons. Set *Name of pre ensemble* to "pre," *Name of post ensemble* to "post," and *Learning rate* to "5e-7" (i.e., 0.0000005). Press *OK*.

The template adds a new ensemble named "error" and connects the "pre," "post," and "error" ensembles together. The connection between the "pre" and "post" ensembles is initialized with random weights and is the one that will be learned. However, to learn a transformation, we need to establish an error signal using connections from the ensembles in the network. A simple way to generate this error is to take the difference between the output of the post ensemble and the correct transformation being performed elsewhere. Here, we can compute this difference in the input to the "error" ensemble.

- Add a termination to the "error" ensemble. Name the termination "pre," set the input dimension of this termination to 1, set the weight to 1, and *PSTC* to 0.02. Unless otherwise noted, all *PSTC* values in this tutorial should be set to this value.
- Add a second termination to the "error" ensemble. Name the termination "post," set the input dimension of this termination to 1, and set the weight to −1.

- Project from the "X" origin of the "pre" ensemble to the "pre" termination on the "error" ensemble and likewise project from the "X" origin of the "post" ensemble to the "post" termination.

The two terminations added to the error ensemble result in subtraction of the value represented by the post ensemble from the value represented by the pre ensemble. Consequently, a representation of the difference will be available in the error ensemble itself. Minimizing this difference will move the post ensemble towards representing the same value as the pre ensemble, creating a communication channel.

To complete this network, we need to add a few extra elements. First, we need an input function to generate data to send through the communication channel.

- Create a new function input. Name the input "input," set *Output Dimensions* to 1, and click the "Set Functions" button.
- From the drop-down list, select "Fourier Function" and click *Set.*
- Set *Fundamental* to 0.1, *Cutoff* to 10, *RMS* to 0.5, and *Seed* to 1.
- Finalize the function setup by clicking *OK* on all the open dialogs.
- Right-click the created function and select "Plot function." Set the index to 0, start to 0, increment to 0.01, and end to 10. Click *OK.*

The function that is graphed is the randomly varying input that will be provided to the communication channel. The function is defined by randomly choosing frequency components. The details of how the function is chosen are not important, so long as it provides input over the range of the communication channel.[4] The *Seed* parameter can be changed to draw different random values.

- Close the graph of the input function and add a termination named "input" to the "pre" ensemble. The termination should have one dimension, and a weight of 1.
- Project from the function input to the "input" termination on the "pre" ensemble.

To test the learning in the system, it is useful to be able to enable and disable learning as needed. We can add a gate to the error population that allows us to do this.

- Add a gate to the network from the toolbar. Name it "gate," set the *Name of gated ensemble* to "error," set the *Number of neurons* to 100, and set *Gating PSTC* to 0.01. Click *OK.*

[4] Here, the function is chosen as a random, Gaussian white noise with a mean of zero. *Cutoff* is the highest frequency, *Fundamental* determines the frequency spacing, and *RMS* is the average power in the signal.

- Add a function input named "switch," and set the function to be a *Constant Function* with a value of 0. (Be sure to click the "Set Functions" button to switch from using a *Fourier Function* to a *Constant Function.*) Click *OK* until the input function is added.
- Add a termination to the "gate" ensemble named "switch" that accepts one-dimensional input with a connection weight of 1.
- Create a projection between the "switch" function and the "gate" ensemble.

This gate provides us with the ability to turn the error signal on and off. When the error signal is off, it sends a value of zero to the post ensemble, and so no further learning occurs.

Finally, we will add an "actual error" node to the network that calculates error directly (not using neural computations), so we can easily monitor the effects of learning.

- Create a new "default" ensemble named "actual error." Set the number of nodes, dimensions, and radius each to 1.
- Right-click the "actual error" ensemble and click on Mode -> Direct. This population will now no longer use neurons to compute any functions it is involved in.
- Add a "pre" termination to the "actual error" population that has a weight of 1 and a "post" termination that has a weight of -1 (as before, with the "error" population).
- Project from the "pre" and "post" population "X" origins to their respective terminations on the "actual error" population.
- Open *Interactive Plots.*
- Display value plots for each ensemble except the "gate" (four plots). Recall that the scroll wheel re-sizes plots.
- Right-click and hide the "actual error" ensemble because this ensemble is not part of the network, but only for analysis (the plot will remain).
- Right-click the post population and select "connection weights -> pre_00."

The displayed connection weight grid is a visualization of connections between the "pre" and "post" ensembles. Red squares represent excitatory connections and blue squares represent inhibitory ones; the intensity of the color indicates the strength of the connection.

- Right-click "switch" and select "control." This can be used to toggle the gate on the error population. Because we set the value of the "switch" function to zero, the gate is initially closed, no error signal will be created, and thus no learning will occur.

- Right-click "input" and show the value plot. Right-click it again and select "control."
- Run the simulation.

The "actual error" plot should show a large amount of error between the two populations because no learning is occurring. The connection weight values should also not change in the grid visualization. Grab and move the input controller, and notice that the "post" population essentially ignores this input.

- Pause and reset the simulation.
- Set the value of the "switch" input to 1 using the slider.
- Run the simulation.

After an initial period of large error, the error will noticeably reduce after about 10–15 simulation seconds, although the exact length of time depends on the various random elements in the network (e.g., starting weights). Watching the connection weight graph should allow you to notice changes in many of the weights, although the largest changes happen at the beginning of the simulation (don't forget that you can pause the simulation and drag the slider to review as much of the simulation as you have recorded; the default is 4 seconds). Often these weight changes are subtle.

To convince yourself that the system did learn, you can compare the graphs of the "pre" and "post" values. If you now move the controller on the input function, you will notice that the "post" graph tracks the "pre" graph, unlike before learning. When you reset the simulation, the connection weights will be reset, so you can try any of these manipulations multiple times.

Extending the learning network to learn functions only requires changing the error signal used to modulate the plastic connection.

- Save your interactive plot layout for later use by expanding the settings panel at the bottom of the window and clicking the save icon. Close the interactive plot window.
- Return to the Nengo workspace and create a new decoded origin on the "pre" ensemble. Set its name to "square" and the number of output dimensions to 1, then click "Set Functions."
- Set the function to "User-defined Function," click *Set*, and type "x0*x0" as the expression. Click *OK* until the origin is added.
- Remove the projections from the "X" origin of "pre" to the two error ensembles (after dragging the connection away, right-click on the end of the projection line to remove it), and replace them with projections from the "square" origin.
- Restart the *Interactive Plots*, and run the simulation with the "switch" input set to 1.

This network learns connection weights between pre and post that compute the square of the input to pre rather than a communication channel. A helpful demonstration of this can be performed by grabbing the "input" slider and moving it slowly from 1 to -1 and watching the "post" value plot. It will trace out a typical quadratic function.

The "square" origin that is being used to generate the error signal driving this learning acts as described in the Section 3.8 tutorial. Rather than decoding it directly, however, we use it to generate an error signal that directs learning between the "pre" and "post" ensembles. Of course, in a real brain the error signal may be computed by a completely different neural subsystem (*see* MacNeil & Eliasmith [2011] for a detailed example).

Notice that the weights you started with in this second simulation are the same as those you ended with after learning the communication channel. This highlights one interesting effect of learning with synaptic weights–namely, it may take the network a rather long time to correct its earlier training. Negative input values are mapped very differently in these two cases, so you should expect there to be noticeable errors until they are eventually retrained. To learn the square from reinitialized random weights, you have to reset the simulation before running the square origin simulation. This can be done by hitting the reset simulation button in the interactive plots at any time. A pre-built version of this network can be loaded from <Nengo Install Directory>/Building a Brain/chapter6/learn.py.

7. THE SEMANTIC POINTER ARCHITECTURE

C. Eliasmith and X. Choo

7.1 ■ A Summary of the Semantic Pointer Architecture

I have spent the last four chapters presenting the central elements of the Semantic Pointer Architecture (SPA). Because that presentation has been lengthy and somewhat detailed, it may seem fragmented. As a result, it may also not provide a good sense of how the architecture yields a unified characterization of biological cognition. The purpose of this chapter is to explicitly tackle the issue of unification by presenting a single model that includes the features of the SPA presented to this point. Consequently, in this chapter I focus on demonstrating that a single, unified SPA model can perform a wide variety of cognitive (and non-cognitive) tasks.

Let me begin by providing a brief review of the SPA. I have attempted to graphically summarize a standard SPA subsystem in Figure 7.1. I think of this "standard subsystem" as a schema, whose details will depend on the specific area of the brain to which we are mapping it. For example, the number of levels in the hierarchy, the nature of the transformations, the kinds of control, the flow of error information, and so forth will all depend on whether we are considering vision, audition, motor control, working memory, or more abstract functions. Regardless, we can begin to get a "whole brain" view if we consider linking such subsystems into a larger system, as shown in Figure 7.2.

Figure 7.2 is a higher-level schema that captures the organizational structure of SPA models. Not all models will have all of the same components, although ideally no models will have conflicting components. My intent is, after all, to provide a means of constructing unified models of biological

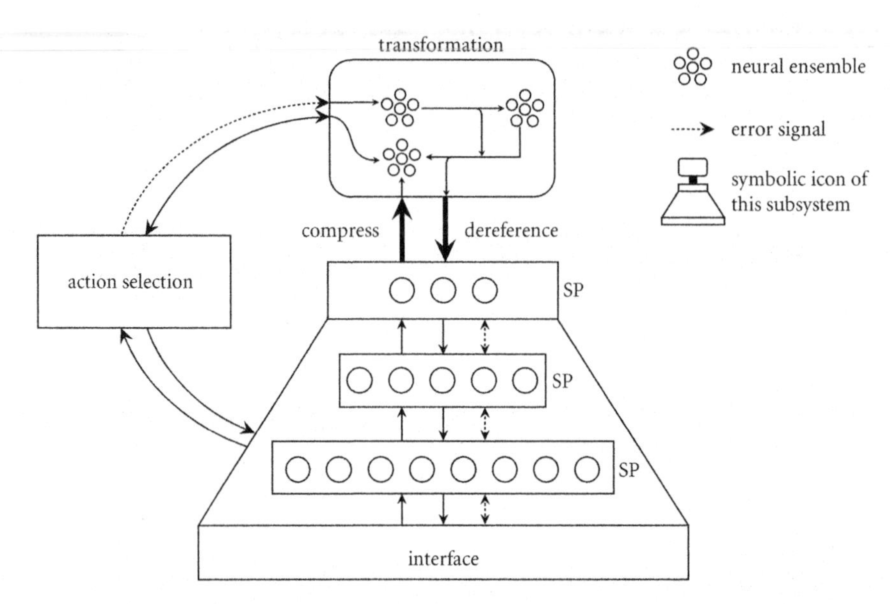

FIGURE 7.1 A schematic for a subsystem of the SPA. At the "interface," a high-dimensional representation enters the subsystem from the environment or another subsystem. It is then compressed, usually through a hierarchical structure, generating semantic pointers. Moving up or down in this hierarchy compresses or dereferences the semantic pointer representations respectively. Throughout the hierarchy, the generated semantic pointers can be extracted and transformed by other elements of the system (rounded box). All transformations are updateable by error signals, some of which come from the action selection component, and some of which may be internally generated. The action selection component influences routing of information throughout the subsystem.

cognition. Proposing and filling in the details of such a higher-level schema is essentially how I think we ought to build a brain.

That being said, it should be evident that my description of the SPA does not make "filling in the details" a simple matter. The SPA itself clearly does not specify a complete set of subsystems and their functions. It may be helpful to think of the SPA as providing something more like a protocol description than an architecture–it describes how neural systems can represent, transform, transfer, and control the flow of information. Even so, the previous chapters do include specific commitments to particular processes occurring in identified parts of the system, which is why I have chosen to call it an architecture.

It is, however, an architecture in development (*see* Section 10.5 for further discussion). It is thus more useful to provide example applications of the architecture than an application-independent specification. In the next

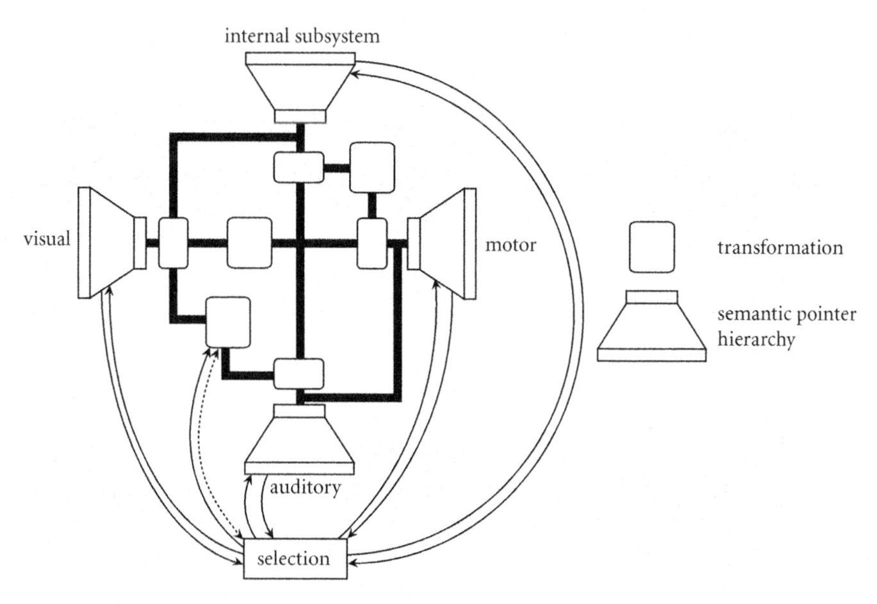

FIGURE 7.2 A schema of SPA models. This figure consists of several interacting subsystems of the type depicted in Figure 7.1. Dark black lines are projections carrying representations between parts of the system, whereas thinner lines indicate control and error signals. An "internal subsystem" is included to highlight the critical role of systems concerned with functions like working memory, encoding a conceptual hierarchy, and so forth. Not all information flow is intended to be captured by this schema.

section, I provide the final example in the book. Here, unlike with past examples, my purpose is to bring together the parts of the architecture I have described. As a result, I focus on integrating the various behaviors we have encountered to this point.

7.2 ■ A Semantic Pointer Architecture Unified Network

In the interest of brevity, I refer to this model as Spaun (SPA Unified Network). We have recently described this model in the journal *Science*, although in less detail than I provide here (Eliasmith et al., 2012). Spaun is a direct application of the SPA to eight different tasks, but in a limited domain. In particular, all of the tasks are defined over input that consists of digits from zero to nine, plus some additional task-control symbols (*see* Figure 7.3A). This input domain has the advantage of including significantly variable, real-world input, while also providing reasonably limited semantics that the model must reason about.

Spaun's output drives an arm that draws its response to the given task. Figure 7.3B shows examples of motor output. In short, we can think of Spaun as having a single, fixed eye and a single, two-joint arm. The eye

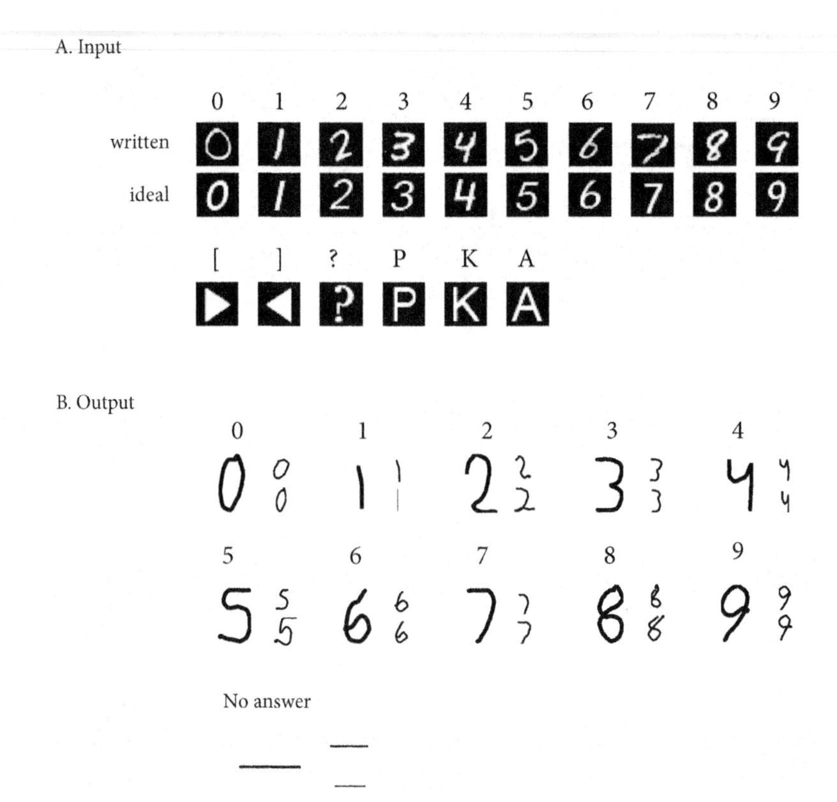

FIGURE 7.3 Input and output with examples for the Spaun model. **A.** The complete set of possible inputs with handwritten (top row) and typographically ideal (second row) examples for each input. Human-generated input is taken from the publicly available MNIST database. Letters and symbols on the bottom row are used to encode task structure. Specifically, leftward and rightward pointing arrows are used as brackets, "?" indicates that an answer is expected, "A" indicates that the next number specifies which task to perform, and "P" and "K" indicate whether Spaun is to identify the "position" or "kind" of number in the question-answering task (*see* Section 7.3.6). **B.** A set of example outputs drawn by Spaun. Each output is shown with three examples to demonstrate the variance of the arm drawings. Outputs include the digits 0–9 and a horizontal dash indicating that none of the other responses is deemed appropriate.

does not move, but rather, the experimenter changes the image falling on it by showing different inputs over time. To begin a specific task, Spaun is shown the letter "A" followed by a number between zero and seven, indicating the current task. The subsequent input will then be interpreted by the model as being relevant to the specified task and processed accordingly, resulting in arm movements that provide Spaun's response to the input in the identified task. Example videos of Spaun performing these tasks can be found at `http://nengo.ca/build-a-brain/spaunvideos`.

I have chosen tasks that cover various kinds of challenges that face a biological cognitive system. Table 7.1 shows the tasks, a brief description, and a brief justification for choosing them. In short, I have attempted to choose tasks that demonstrate that this integrated model preserves the functionality discussed independently in previous chapters. As well, these tasks cover simple demands faced by all biological systems (e.g., reinforcement learning and recognition), as well as tasks known to be routinely solved only by humans (e.g., rapid variable creation and fluid reasoning). The tasks also cover behaviors that deal mainly with perceptual and motor processing (e.g., copy drawing and recognition), and tasks that depend more on the manipulation of structured and conceptual representations (e.g., counting and question answering). Together these tasks place a challenging set of demands on a biologically realistic model system.

As mentioned, the purpose of Spaun is to address all of these tasks in a single model. I take this to mean that the model cannot be externally changed during any of the tasks or between any of the tasks. Additionally, the only input to the model can be through its perceptual system, and the only output can be through its motor system. So, the representational repertoire, background knowledge, cognitive mechanisms, neural mechanisms, and so forth, must remain untouched while the system performs any of the tasks in any order.

To give a sense of a typical run of the model, let me consider the counting task. Starting the task consists of showing Spaun a letter/number pair (e.g., "A 4") to indicate which task it is expected to perform on the upcoming input (e.g., counting). The next input provided is a bracketed number (e.g., "[4]"), which indicates from where the counting should start. The third input is a bracketed number, indicating how many times to count (e.g., "[3]"), and the final input is a "?," indicating that a response is expected. The system then proceeds to solve the task and generates motor commands that produce a number indicating its response (e.g., "7"). At this point Spaun waits for further input. It is then up to the experimenter to provide another task letter/number pair, followed by input, at which point the system again responds, and so on.

The architecture of Spaun is shown in Figure 7.4. Figure 7.4B demonstrates that the model is a specific instance of the SPA architecture schema introduced in Figure 7.2. Figure 7.4A shows the anatomical areas included in Spaun. Table 7.2 justifies the mapping of Spaun's functional architecture onto the neuroanatomy. The Spaun model consists of three hierarchies, an action selection mechanism, and five subsystems. The hierarchies include a visual hierarchy that compresses image input into a semantic pointer, a motor hierarchy that dereferences an output semantic pointer to drive a two degree-of-freedom arm, and a working memory that compresses input to store serial position information (*see* Section 6.2). More precisely, the working memory

TABLE 7.1 *A Description and Justification for the Eight Example Tasks Performed by Spaun.*

Name (task identifier)	Description	Justification
Copy drawing (A0)	The system must reproduce the visual details of the input (i.e., drawn responses should look like the inputs).	Demonstrates that low-level visual information (i.e., input variability) can be preserved and is accessible to drive motor responses.
Recognition (A1)	The system must identify the visual input as a member of a category. Drawn responses need not look like inputs.	Demonstrates that input variability can be "seen through." Allows the identification of stable concepts.
Reinforcement learning (A2)	The system must learn environmental reward contingencies.	Demonstrates reinforcement learning in the context of a large-scale model.
Serial working memory (A3)	The system must memorize and reproduce an ordered set of items.	Demonstrates the integration of serial memory into a cognitive model. Needed for several other more complex tasks.
Counting (A4)	The system must be able to count silently from a presented starting position for a given number of steps.	Demonstrates flexible action selection. Demonstrates knowledge of relations within the conceptual domain.
Question answering (A5)	The system must answer questions about the contents of structured representations in working memory.	Demonstrates the ability to construct (bind), manipulate, and extract information flexibly from internal representations.
Rapid variable creation (A6)	The system must identify syntactic patterns and respond to them rapidly (i.e., without connection weight changes).	Demonstrates a neural architecture able to meet these cognitive demands. Hadley (2009) argues this cannot currently be done.
Fluid reasoning (A7)	The system must solve problems analogous to those on the Raven's Progressive Matrices.	Demonstrates redeployment of cognitive resources on a challenging fluid reasoning task.

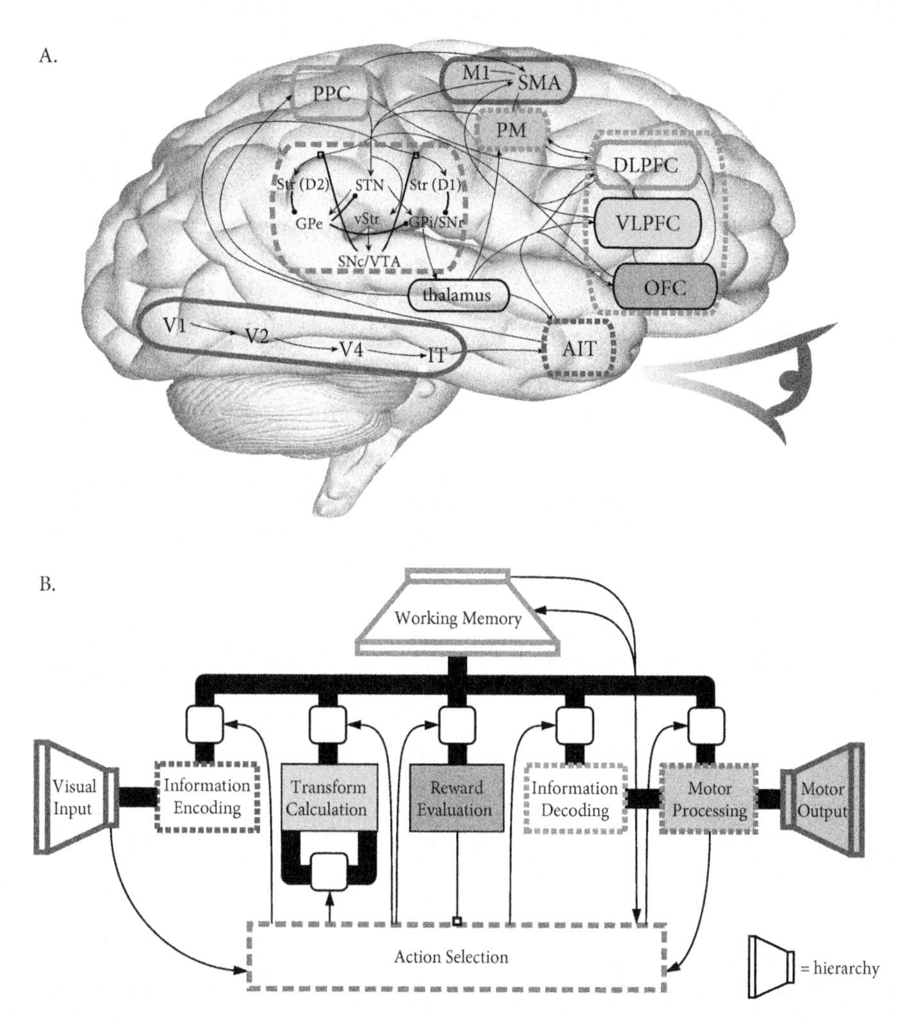

FIGURE 7.4 The architecture of the Spaun model. **A.** The anatomical architecture of the model. Abbreviations and justification of the mapping of these structures to function are in Table 7.2. Lines with circular endings indicate inhibitory GABAergic projections. Lines with square boxes indicate modulatory dopaminergic connections exploited during learning. **B.** The functional architecture of the model. This is a specific use of the schema depicted in Figure 7.2. Thick lines indicate information flow between elements of cortex, thin lines indicate information flow between the action selection mechanism and cortex, and rounded boxes indicate gating elements that can be manipulated to control the flow of information within and between subsystems. The circular end of the line connecting reward evaluation and action selection indicates that this connection modulates connection weights. Line styles and fills indicate the mapping to the anatomical architecture in A. See text for details. (Adapted from Eliasmith et al. [2012] with permission.)

TABLE 7.2 *The Function to Anatomical Mapping Used in Spaun, Including Anatomical Abbreviations Used in the Task Graph Figures*

Functional element	Acronym	Full name and description
visual input	V1	primary visual cortex: the first level of the visual hierarchy, tuned to small oriented patches of different spatial frequencies (Felleman & Van Essen, 1991b)
	V2	secondary visual cortex: pools responses from V1, representing larger spatial patterns (Felleman & Van Essen, 1991b)
	V4	extrastriate visual cortex: combines input from V2 to recognize simple geometric shapes (Felleman & Van Essen, 1991b)
	IT	inferior temporal cortex: the highest level of the visual hierarchy, representing complex objects (Felleman & Van Essen, 1991b). In the graphs, this shows the spike pattern of the neurons encoding the low-dimensional (50D) visual representation (*see* Section 3.5).
information encoding	AIT	anterior inferior temporal cortex: implicated in representing visual features for classification and conceptualization (Liu & Jagadeesh, 2008)
transform calculation	VLPFC	ventrolateral prefrontal cortex: area involved in rule learning for pattern matching in cognitive tasks (Seger & Cincotta, 2006)
reward evaluation	OFC	orbitofrontal cortex: areas involved in the representation of received reward (Kringelbach, 2005)
information decoding	PFC	prefrontal cortex: implicated in a wide variety of functions, including executive functions, task tracking, and manipulation of working memory (D'Esposito et al., 1999)
working memory	PPC	posterior parietal cortex: involved in the temporary storage and manipulation of information, particularly visual data (Owen, 2004; Qi et al., 2010)

TABLE 7.2 *(cont.)*

	DLPFC	dorsolateral prefrontal cortex: temporary storage (working memory) and manipulation of higher-level data related to cognitive control (Hoshi, 2006; Ma et al., 2011)
action selection	Str (D1)	striatum (D1 dopamine neurons): input to the "direct pathway" of the basal ganglia (Albin et al., 1989)
	Str (D2)	striatum (D2 dopamine neurons): input to the "indirect pathway" of the basal ganglia (Albin et al., 1989)
	Str	striatum: on the graphs, Str indicates all the input to the basal ganglia, reflecting the values of various cortical states (*see* Section 5.2)
	STN	subthalamic nucleus: input to the "hyperdirect pathway" of the basal ganglia (Nambu et al., 2002)
	VStr	ventral striatum: involved in the representation of expected reward to generate reward prediction error (Van Der Meer et al., 2010)
	GPe	globus pallidus externus: part of the "indirect pathway" projects to other components of the basal ganglia to modulate their activity (Gurney et al., 2001)
	GPi/SNr	globus pallidus internus and substantia nigra pars reticulata: the output from the basal ganglia (Albin et al., 1989). Labeled GPi on the graphs, the neurons corresponding to the chosen action are inhibited from firing (*see* Section 5.2).
	SNc/VTA	substantia nigra pars compacta and ventral tegmental area: relay signal from ventral striatum as dopamine modulation to control learning in basal ganglia connections (Hollerman & Schultz, 1998)
routing	thalamus	thalamus: receives output from the basal ganglia, sensory input, and coordinates/monitors interactions between cortical areas (Gisiger & Boukadoum, 2011)

TABLE 7.2 *(cont.)*

motor processing	PM	premotor cortex: involved in the planning and guidance of complex movement (Weinrich et al., 1984; Halsband et al., 1993)
motor output	M1	primary motor cortex: generates muscle-based control signals that realize a given internal movement command (Todorov, 2004)
	SMA	supplementary motor area: involved in the generation of complex movements (Halsband et al., 1993; *see* Section 3.6)

Adapted from Eliasmith et al. (2012).

component also provides stable representations of intermediate task states, task subgoals, and context. The action selection mechanism is the basal ganglia model described in Chapter 5.

The five subsystems, from left to right, are used to (1) map the visual hierarchy output to a conceptual representation as needed ("information encoding"); (2) extract relations between input elements ("transformation calculation"); (3) evaluate the reward associated with the input ("reward evaluation"); (4) map output items to a motor semantic pointer ("information decoding"); and (5) control motor timing ("motor processing"). Several of the subsystems and hierarchies consist of multiple components needed to perform the identified functions. For example, the working memory subsystem includes eight distinct memory systems,[1] each of which can store semantic pointers. Some employ compression and some do not, but I have grouped them all here into a single subsystem for simplicity.

Notably, the semantic pointers used in the model vary in size. Specifically, they are 50D at the top of the visual hierarchy, 54D at the top of the motor hierarchy, and 512D everywhere else. Overall, the model uses about 2.5 million spiking neurons. As a result, the model requires significant computational resources (e.g., 24 gigabytes of RAM) and takes approximately

[1] I should note that I'm using "memory system" to denote a general neural architecture and functionality–namely, a recurrent network that stores information over time in its activity. These systems may not always be mapped to what is traditionally identified as a "working memory" area of the brain. This is because it is important to store information such as task set, control state, as well as stimuli features when being able to flexibly perform many different tasks. Storage of task-related representations are often called "executive control" rather than "working memory" *per se*, though how this division maps to the brain is a matter of significant debate.

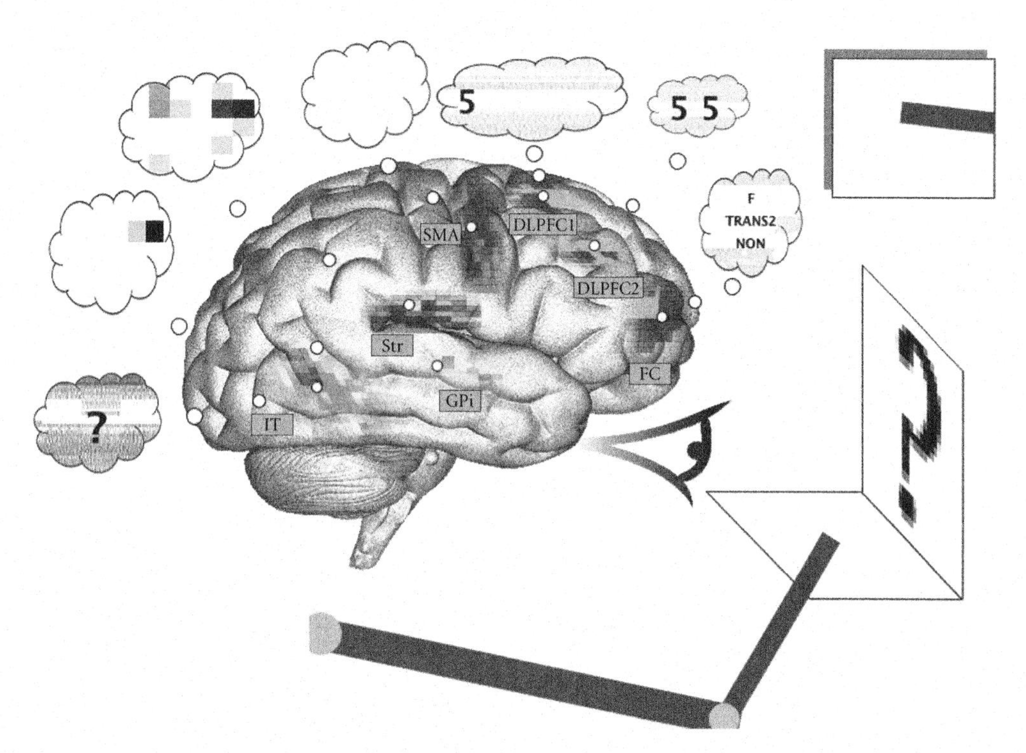

FIGURE 7.5 A screen capture from the simulation movie of the fluid reasoning task. The input image is on the right, a top-down view of the arm is in the top right corner. Spatially organized, low-pass filtered neuron activity is mapped to relevant cortical areas (neurons with similar tuning are near one another). Thought bubbles show spike trains and the results of decoding those spikes in overlayed text. For the vStr the thought bubble shows decoded utilities of possible actions, and in the GPi the selected action is darkest. Anatomical labels are included for reference, but not in the original movie.

2.5 hours to run 1 second of simulation time.[2] Additional details necessary to fully re-implement the model, a downloadable version of the model, and videos demonstrating the tasks described below can be found at `http://nengo.ca/build-a-brain/spaunvideos`. Figure 7.5 shows a screen capture of an example video.

Notice that the model's architecture is not tightly tied to the set of tasks being implemented *per se* but, rather, captures more general information processing considerations. That is, Spaun does not have subsystems specific to tasks, but, rather, specific to generally useful information processing functions. This helps make the model naturally extensible to tasks beyond those

[2] More recent versions of Nengo have full GPU support and better memory management, so runtimes are rapidly decreasing. In addition, we are working with other groups to implement SPA models on hardware designed specifically to run neural models.

considered here. As a result, I believe Spaun demonstrates that the SPA provides a general method for building flexible, adaptive, and biologically realistic models of the brain that can go far beyond the tasks considered here. Before returning to this point, let me consider the specific tasks identified earlier and give a sense of Spaun's performance on them.

7.3 ■ Tasks

Each of the following sections describes one of the tasks in more detail, showing input and output examples and providing summary data on performance. Example runs are provided in a standard graphical format. This format includes a depiction of the specific input and output for a given run, as well as spike rasters and semantic pointer similarity graphs from selected parts of the model, plotted over time. These graphs provide examples of concurrent neural and behavioral data flow through Spaun during the various tasks.

Table 7.2 identifies the labels used on the y-axis of these graphs. The labels are used to identify inputs, outputs, and which anatomical region each row of the plot is associated with. The inputs in the "stimulus" row are images of hand-written or typed letters, numbers, and symbols like those shown in Figure 7.3A. Each input element is shown for 150 ms and separated by 150 ms of blank background. Outputs in the "arm" row are drawn digits, examples of which are shown in Figure 7.3B. The drawn digits are depicted over the 150 to 300 ms time during which the motor cortex drives the rest of the motor hierarchy. The exact time depends on the task being performed. There are two main kinds of neural activity plots in these graphs: spike rasters and similarity plots. The spike raster plots are generated by randomly selecting 2000 neurons from the relevant population and then discarding any neurons with a variance of less than 10% over the run. This is similar to the kind of selection that occurs when an experimentalist only records from neurons with "interesting" responses during a given task. For legibility, individual spikes are not plotted, but instead the density of spikes is mapped to a grayscale image (i.e., low-pass filtered) and plotted.

Similarity plots show the dot product between the representation in the spike raster plot of the same name and relevant semantic pointers in Spaun's vocabulary. This provides a more readable time-varying graph of the contents of parts of the model during the simulations.

7.3.1 Recognition

The recognition task requires Spaun to classify hand-written input, and use that classification to generate a motor response, reproducing Spaun's default written digit for that category. To perform classification, Spaun takes the semantic pointer produced by the visual hierarchy and passes it through

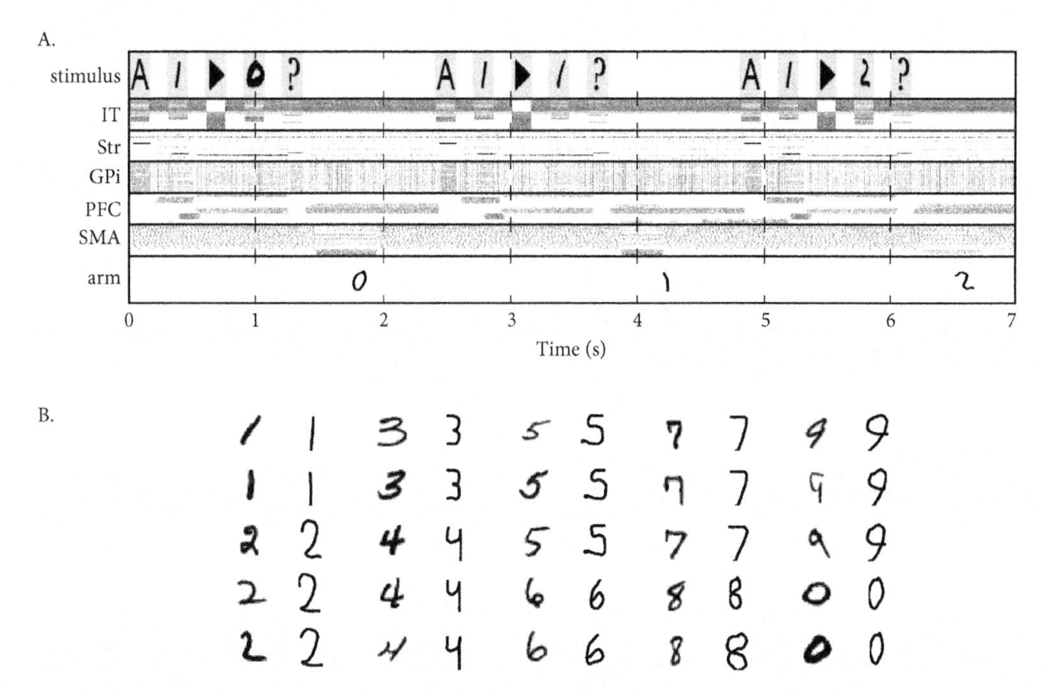

FIGURE 7.6 Digit recognition by Spaun. **A.** Spaun recognizing three digits in a row, with accompanying spike rasters. **B.** Several examples of Spaun responses to each digit type. Notice that the outputs are similar, regardless of the visual properties of the input. This is because the recognized category, not the specific visual characteristics of the input, is driving the response.

an associative, clean-up-like structure that maps it to a canonical semantic pointer for that category. This pointer is then routed to the motor system to draw the relevant number.

Three successive categorizations are shown in Figure 7.6A over time. As shown in Figure 7.6B, the motor responses are not sensitive to variations in the original hand-written input digits. Recognition accuracy of the model is 94% on untrained data, which compares well to human accuracy on this task (about 98%; Chaaban & Scheessele, 2007).

Successful recognition of hand-written input demonstrates Spaun's ability to "see through" the variability often encountered in naturalistic input. In short, it shows that variable input can be categorized, and that those categories can effectively drive action.

7.3.2 Copy Drawing

The copy-drawing task consists of requiring Spaun to attempt to reproduce the visual features of its input using its motor system. The task consists of showing a hand-drawn input digit to Spaun and having the model attempt to reproduce the digit by drawing. Spaun must generate the motor command based on remembered semantic pointer information, as the image itself is removed from view.

Solving this task relies on a mapping between the representation generated at the highest level of the visual hierarchy (i.e., IT, which encodes a 50D semantic pointer space) and the representation driving the highest level of the motor hierarchy (i.e., SMA, which encodes a 54D semantic pointer space). To employ the correct mapping, Spaun first recognizes the digit as above and then uses a map for that category to generate the 54D motor command from the 50D visual representation. Specifically, the motor command is an ordered set of coordinates in 2D space that drive the motor system to draw with the arm. The mapping for each category is linear and is learned based on example digit/motor command pairs (five pairs for each category). Learning this linear mapping is done before running the Spaun model and is the same for all runs.

Figure 7.7 shows two example runs, both for drawing the number two. Additional examples of only the input/output mappings are shown in Figure 7.8. Critically, the visual features of the input are captured by Spaun's drawn responses. The reason the mapping from the visual semantic pointer to the motor semantic pointer is linear is because of the sophistication of the compression and decompression algorithms built into the perceptual and motor systems. In general, it would be incredibly difficult to attempt to map directly from pixel-space to muscle-space–that is, to determine how a given image should determine joint torques so as to approximately reproduce a given image. This function no doubt exists, but a direct mapping would be highly nonlinear. However, the compression hierarchies in Spaun are able to successfully extract and exploit the visually and motorically salient features, turning the final visuomotor mapping into a simple linear one.

Spaun's ability to perform the copy-drawing task demonstrates that it has access to the low-level perceptual (i.e., deep semantic) features of the representations it employs and that it can use them to drive behavior. Together, the copy-drawing and recognition tasks show that semantic pointers provide representations able to capture sensory features as well as behave categorically. In addition, the tasks demonstrate that the SPA can generate models that move smoothly between shallow and deep semantic features, depending on task demands.

7.3.3 Reinforcement learning

The reinforcement task is a three-armed bandit task, analogous to the two-armed task discussed in Section 6.5 but more challenging. In this case, when Spaun begins the task, it selects one of three possible actions, drawing a 0, a 1, or a 2. It is then shown a 1 if it receives a reward for that action or a 0 if it

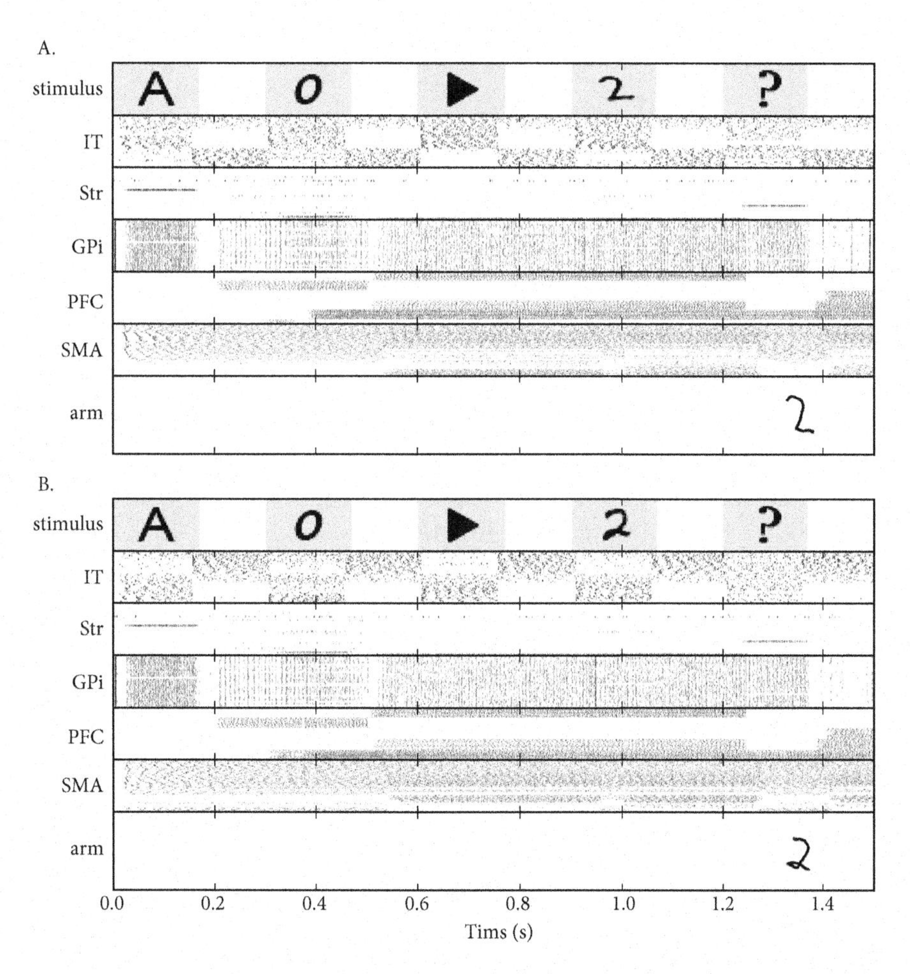

FIGURE 7.7 Copy drawing by Spaun. Two different examples of copy drawing the number two. **A.** Reproduction of a two with a straight bottom. (Reproduced from Eliasmith et al. (2012) with permission.) **B.** Reproduction of a two with a looped bottom.

receives no reward. It then chooses its next action, and so on until there is a new task. The rewards are randomly given with a probability of 0.72 for the "best" action and 0.12 for the other actions. The best action changes every 20 trials.

The STDP rule and its application are the same as discussed in Section 6.5. The basal ganglia in this case selects the number to be chosen as usual and also routes the visual feedback information such that it is interpreted as a reward signal driving OFC and subsequently SNr/VTA at the appropriate times in the task.

Figure 7.9A demonstrates the beginning of an example run for 12 trials. This figure shows the expected reduction in the reward prediction error as the

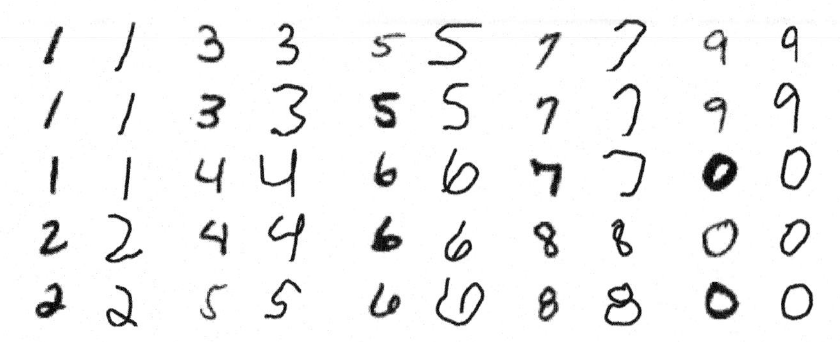

FIGURE 7.8 Example pairs of inputs and outputs for the copy-drawing task. As can be seen, low-level visual features (such as straight versus looped bottoms, slant, stroke length, etc.) of the examples are approximately reproduced by Spaun. This can be contrasted with Figure 7.6B, where visual properties specific to a given input are not reproduced.

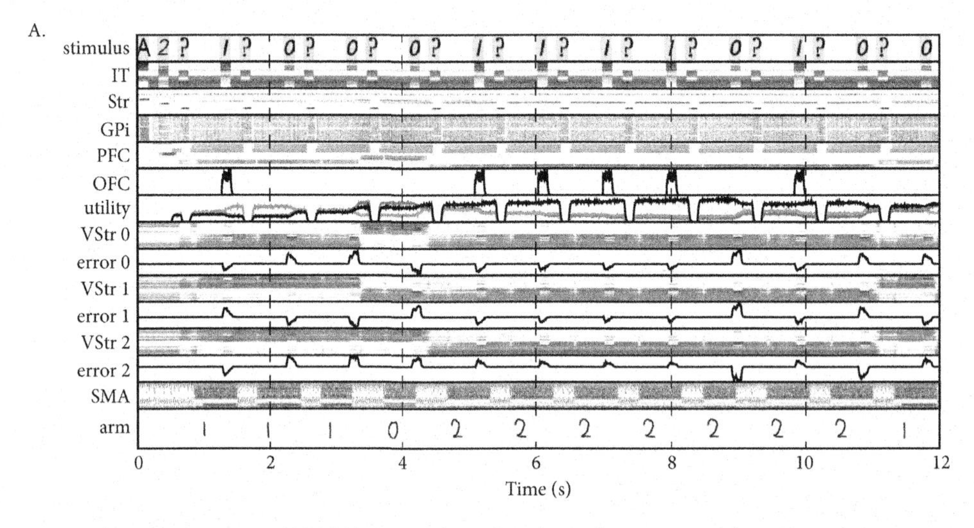

FIGURE 7.9 Reinforcement learning in a three-armed bandit task in Spaun. **A.** The reward and behavioral time course of a run. Here the best action was to draw a 2. After some incorrect guesses, this contingency is learned by the model at the beginning of the trial. However, two unlucky rewards at the end of the trial (at 9 s and 11 s) cause the "utility" trace (a decoding of the Str activity) to decrease, and hence the model picks a 1. The reward prediction error signal is shown separately for each of the three possible actions (this can also be thought of as a reward vector, which is the decoding of vStr activity). As can be seen, "error 2" decreases as the trial proceeds, until the unlucky rewards occur. **B.** A behavioral run of 60 trials. Every 20 trials, the probability of reward for the three choices changes, as indicated at the top of the graph. The probability of choosing each action is indicated by the continuous lines. These probabilities are generated by averaging over a five-trial window. Reward delivery during the run is indicated by the "x" marks along the top of the graph. This is analogous to the two-armed bandit task discussed earlier (*see* Section 6.5). (Reproduced from Eliasmith et al. [2012] with permission.)

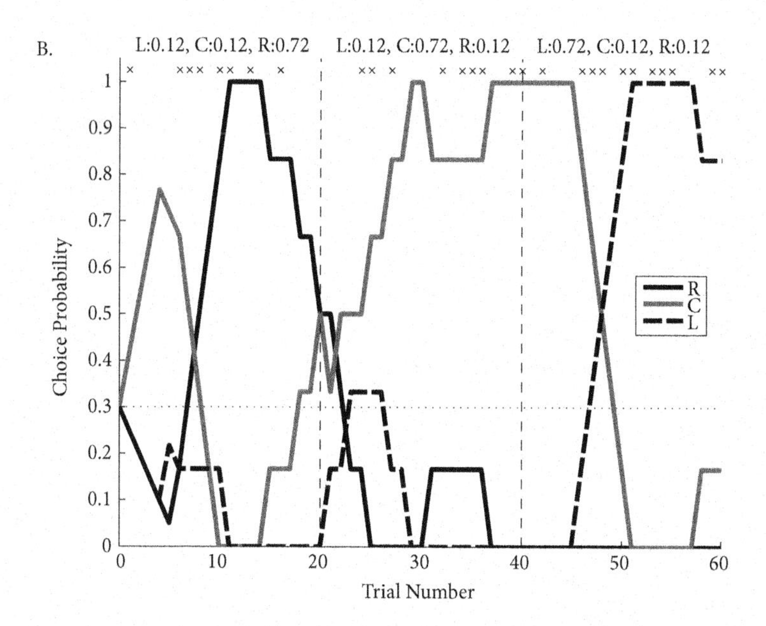

FIGURE 7.9 *(cont.)*

model learns to expect reward after its choice of a 2. Figure 7.9B shows just the behavioral data over a run of 60 trials (20 for each "arm" of the task). As can be seen, the model is able to adjust its choice behavior based on the contingent reward provided by its environment.

Spaun's replication of this reinforcement learning behavior demonstrates that the model is adaptive, yet stable throughout the tasks (i.e., performance of any other task after this one remains the same). More generally, Spaun's performance shows that it can learn about environmental contingencies and adjust its behavior in response and that it does so in a way that matches known neural and behavioral constraints.

7.3.4 Serial Working Memory

In the serial working memory task, Spaun is presented with a list of numbers that it is expected to memorize and then recall. The implementation is similar to that described in Section 6.2. Notably, this is the first task that explicitly requires binding (although the previous tasks use the same working memory subsystem).

The memory trace representations used by Spaun are of the form:

$$\begin{aligned}\mathbf{trace}_i =& \rho\mathbf{Episodic}_{i-1} + (\mathbf{position}_i \circledast \mathbf{item}_i) + \\ &\gamma\mathbf{WorkMem}_{i-1} + (\mathbf{position}_i \circledast \mathbf{item}_i)\end{aligned}$$

where the γ and ρ parameters are set to the same values as before (*see* Section 6.3), and i indexes the elements in the list. This representation differs from that for the full model in Section 6.2 by not including a separate representation of the unbound item. This removes the ability of the system to perform free recall, but reduces the computational resources needed to run the model.

Figure 7.10 shows two example runs of the model. The encoding of the input into working memory is evident in the similarity graphs in both cases. In Figure 7.10A, the recall is successful. However, in Figure 7.10B, we see that recall fails on the fourth item in this longer list. Nevertheless, Spaun is able to recall elements after the failure, demonstrating the robustness of this working memory system.

Figure 7.11 shows that the model retains the primacy and recency effects of the earlier model, which matches well to human behavioral data. However, the model also clearly has a tendency to forget more quickly than humans on this short timescale task. Conservatively, the model accounts for 17 of 22 data points, where all misses are on the longest two lists and 4 of the 5 misses are on the longest list. Recall that the non-Spaun version of this model accurately matched several working memory tasks with slower presentation times (*see* Section 6.2). The present lack of a match to this much more rapid presentation of items suggests that embedding the original model in a more complex system changed its behavior slightly,[3] or that there are interesting differences in short versus long timescale working memory data that has been previously overlooked. In either case, this provides a nice demonstration of how building integrated models might change our understanding of even well-studied cognitive phenomena.

However, in general the generation of human-like memory accuracy curves in Spaun demonstrates that its working memory performance maps well to known behavior. So, although some details may differ, performance on this task shows that the model discussed in detail in Section 6.2 maintains its relevance when integrated into Spaun. Because working memory is central to several subsequent tasks, this task also shows that Spaun is performing these tasks with appropriate cognitive constraints.

7.3.5 Counting

The counting task demonstrates the generation of flexible, structured sequences of actions. One interesting difference between this task and the flexible action sequences discussed in Section 5.6 is that the internal representations being sequenced are structured. Specifically, the representations for numbers are constructed through recursive binding. This proceeds as follows:

[3] Although this is difficult to be certain of without confidence intervals on the original data.

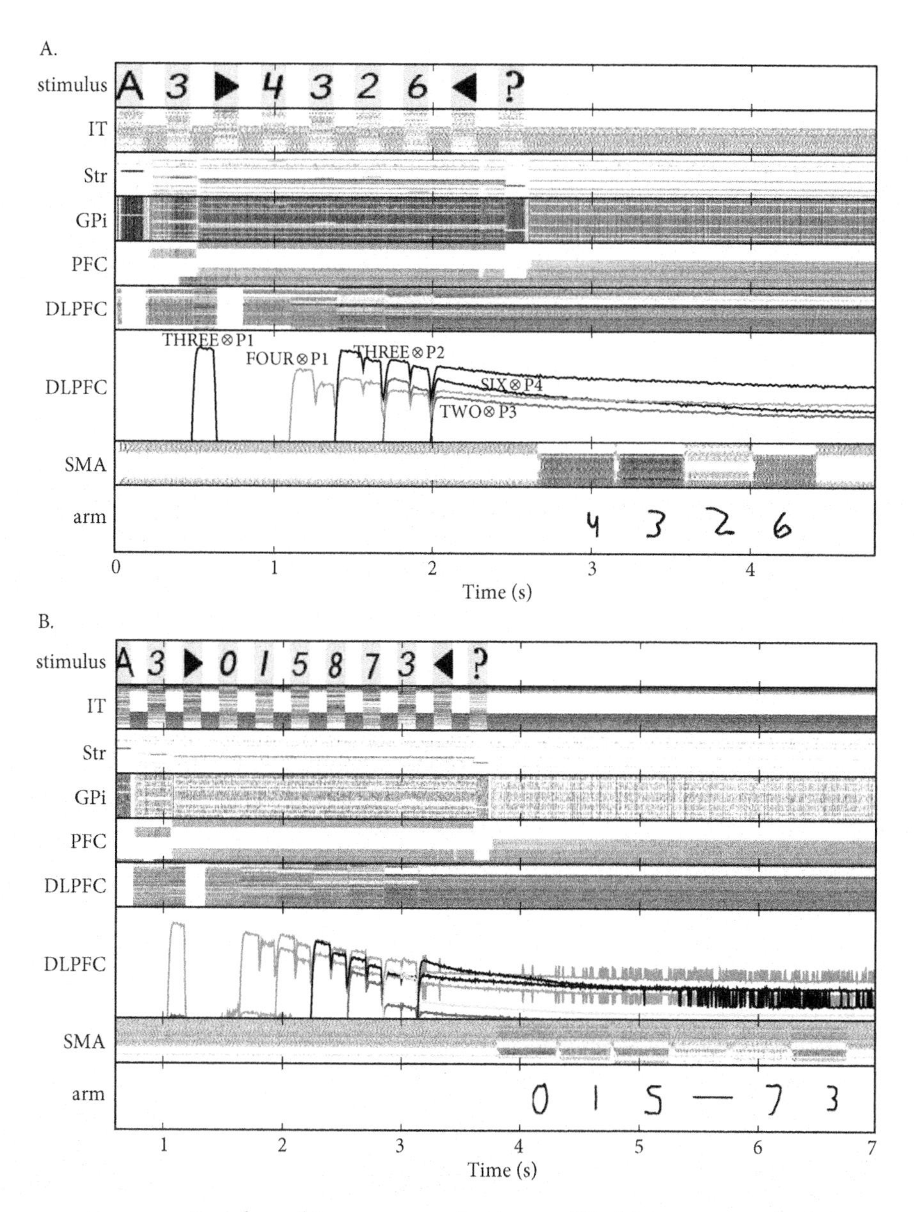

FIGURE 7.10 Serial working memory in Spaun. Working memory is shown here both as a spike raster and a similarity graph. As can be seen in the similarity graph, the first item to enter memory is discarded, as it indicates the task to be performed. The subsequent remembered items are then encoded by their position in the input list. The relative similarities and the overall memory decay is evident in the similarity graph. **A.** An example successful serial recall trial. **B.** An example error trial, with an error in the fourth digit, but successful recovery of subsequent digits.

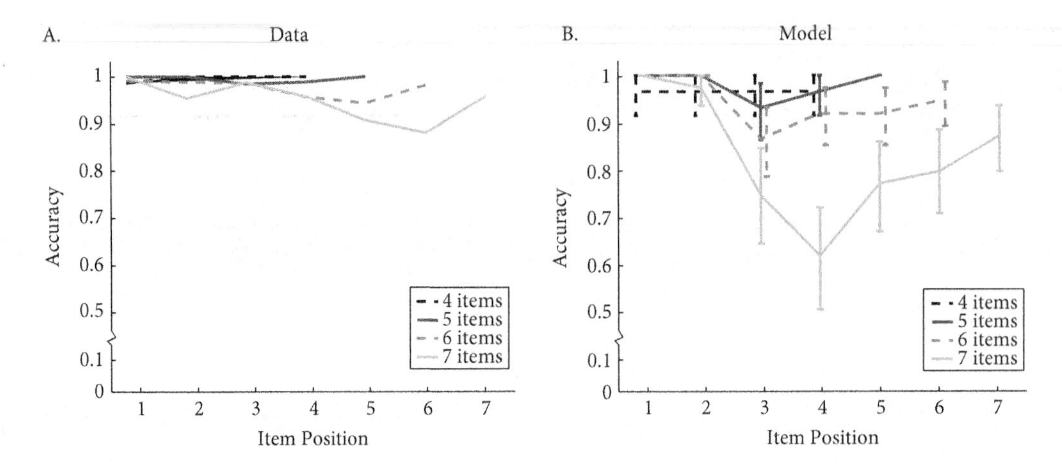

FIGURE 7.11 Recall accuracy curves for serial working memory in Spaun. Primacy and recency effects are observable in Spaun as in the OSE model (Fig. 6.3). Error bars are 95% confidence intervals over 40 runs per list length. Human data for a rapid digit serial recall task is shown in A (data from Dosher [1999] only means were reported). Comparing the two, 17 of 22 human means are within the 95% CI of the model. (Reproduced from Eliasmith et al. [2012] with permission.)

1. Choose a random unitary vector,[4] call that **one**.
2. Choose a second random unitary vector, call that **addOne**.
3. Construct the next number in the sequence through self-binding, i.e., **two** $=$ **one** $\circledast$ **addOne**.
4. Repeat.

This process will construct distinct but related semantic pointers for numbers. Crucially, this method of constructing number representations encodes the sequence information associated with numerical concepts. The relationship of those conceptual properties to visual and motor representations is captured by the clean-up-like associative memory introduced earlier that maps these semantic pointers onto their visual and motor counterparts (and vice versa). As a result, visual semantic pointers are associated with conceptual semantic pointers, which are conceptually inter-related through recursive binding. This representation serves to capture both the conceptual semantic space and the visuomotor semantic space and link them appropriately. I should note that these are the representations of number categories used in all previous tasks as well, although it is only here that their inter-relation becomes relevant.

[4] A unitary vector is one that maintains a length of 1 when convolved with itself.

The input presented to the model for the counting task consists of a starting number and a number of positions to count–for example, "[2] [3] ?." The model then generates its response–for example, "5." This kind of silent counting could also be thought of as an "adding" task. In any case, given the representations described above, counting is a matter of recognizing the input and then having the basal ganglia bind the base unitary vector with that input the given number of times to generate the final number in the sequence. Given a functioning recognition system, working memory, and action selection mechanism, the task is relatively straight forward.

Figure 7.12A shows Spaun performing this task over time. As is evident there, several independent working memories are employed to track the target number of counts (DLPFC2), the current count value (DLPFC1), and the number of counts that have been performed (DLPFC3). Consequently, the internal action sequences are flexible, in that they are determined by task inputs, yet they are structured by the relations between the concepts being employed. Figure 7.12B demonstrates the expected linear relationship between subvocal counting and response times (Cordes et al., 2001). In addition, Spaun's count time per item (419 ± 10 ms) lies within the human measured range of 344 ± 135 ms for subvocal counting (Landauer, 1962). However, the variance in the model's performance is lower. Nevertheless, the variance increases with count length, reproducing the expected Weber's law relation between variance and mean in such tasks (Krueger, 1989).

Successful counting demonstrates that flexible action selection is effectively incorporated into Spaun. It also shows that the model has an understanding of order relations over numbers, and can exploit that knowledge to produce appropriate responses.

7.3.6 Question Answering

The question-answering task demonstrates that Spaun has flexible access to new information that it encodes in its internal representations. To demonstrate this flexibility, I introduce two special letters, "K" (kind) and "P" (position), for indicating which of two specific questions we are asking of the model. For example, we might present "[1 1 2 3] P [2]," which would be asking: What is in position 2? The answer in this case should be "1." In contrast, "[1 1 2 3] K [2]" would be asking: Where is the kind "2?" In which case, the answer should be "3."

This task again combines many of Spaun's abilities, with a particular emphasis on unbinding and clean-up. After recognition, the items are sent to working memory for encoding into an internal representation. The basal ganglia then monitor subsequent input, using it to determine what kind of routing is appropriate to answer the question. The final input digit is then compared

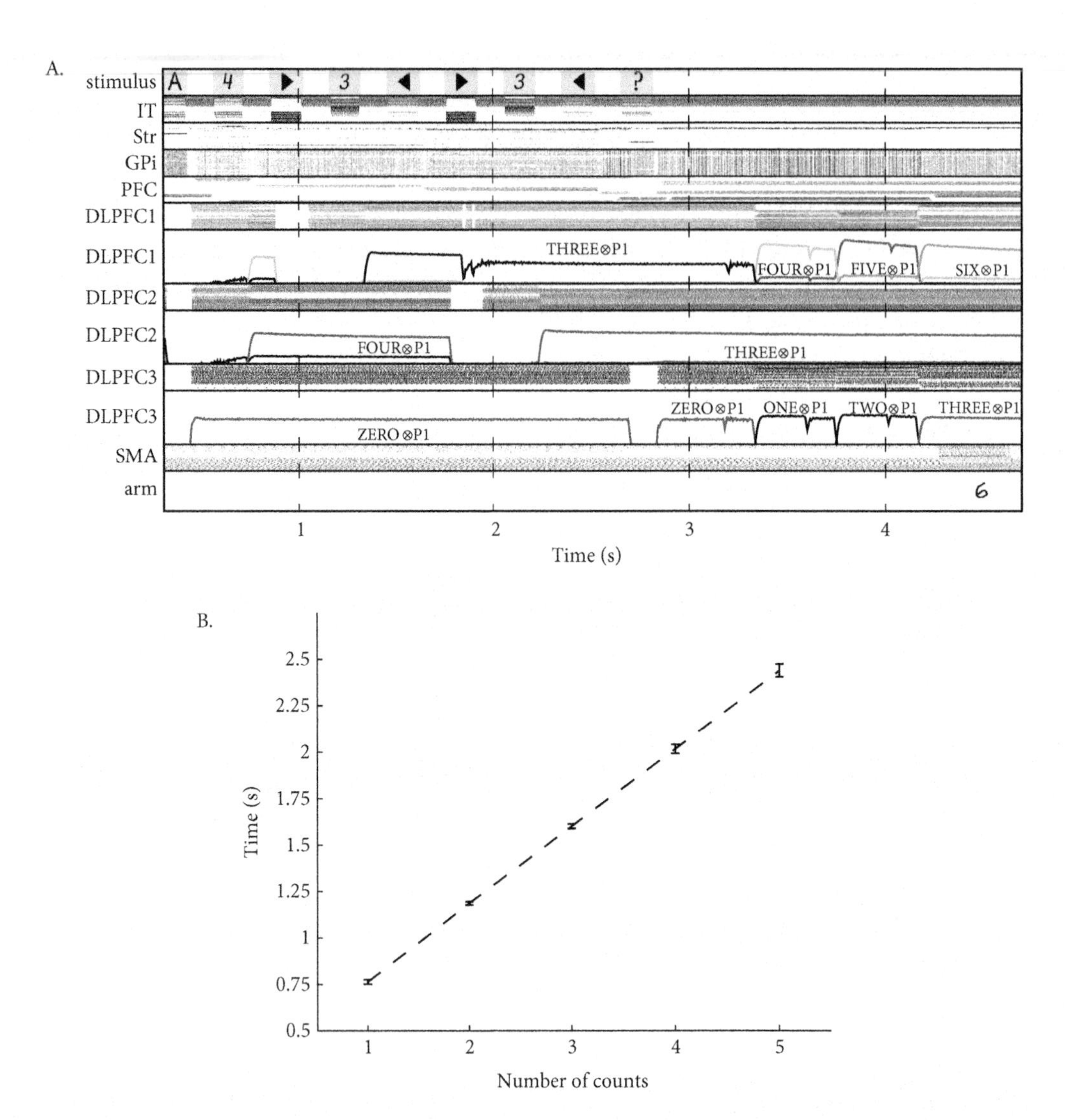

FIGURE 7.12 Counting in Spaun. **A.** Performance of the model over time. The model is first shown a starting digit, in this case 3. It is then shown a number to count by, in this case also 3. It then counts silently, as seen in the DLPFC1 similarity plot. DLPFC2 stores the target number of counts, and DLPFC3 tracks the number of counts that have been performed. When DLPFC2 and DLPFC3 are equal, the task is terminated and the motor response is generated. **B.** The time it takes to count as a function of the number of times to count in a run. The slope and mean (419 ± 10 ms) are consistent with human data (Landauer, 1962), although the variance is much lower. An increase of variance with increasing mean (Weber's law) is also evident. (Reproduced from Eliasmith et al. [2012] with permission.)

to the representation in working memory in the appropriate manner, resulting in the answer to the provided question.

Figure 7.13A demonstrates a "position" question, and Figure 7.13B demonstrates a "kind" question. Unsurprisingly, the accuracy of the response is affected by the properties of working memory. Consequently, we can predict that if the same experiment is performed on people, then longer input sequences will see an increased error rate when asked questions about the middle of the sequence (*see* Fig. 7.14). However, the kind of question being asked does not affect the accuracy. It would be interesting to examine human performance on the same task.

The ability of Spaun to answer questions about its input shows that it has flexible, task-dependent access to the information it encodes. This ability arises directly from the combination of the serial working memory shown in task 3 and flexible, structured action sequences shown in task 4.

7.3.7 Rapid variable creation

In a recent paper, Hadley (2009) argued forcefully that variable binding in connectionist approaches remains an unsolved problem. His arguments have crystallized a traditional concern expressed previously by Jackendoff (2002) and Marcus (2001). Although Hadley has acknowledged that many solutions have been proposed, he has highlighted that *rapid* variable creation and deployment, in particular, are inconsistent with all current proposals that are not simply neural implementations of a classical solution. However, rapid variable deployment is deftly performed by humans. Consider a central example from Hadley:

Training Set

Input: *Biffle biffle rose zarple.* Output: *rose zarple.*
Input: *Biffle biffle frog zarple.* Output: *frog zarple.*
Input: *Biffle biffle dog zarple.* Output: *dog zarple.*

Test Case

Input: *Biffle biffle quoggie zarple.* Output: ?

Hadley has suggested that this task requires rapid variable creation because the second to last item in the list (i.e., rose/frog/dog/quoggie) can take on any form, but human cognizers can nevertheless identify the overall syntactic structure and identify "*quoggie zarple*" as the appropriate response. So it seems that a variable has been created that can receive any particular content and that will not disrupt generalization performance.

Hadley has argued that learning rules cannot be used to solve this problem, given the known constraints on the speed of biological learning. Weight changes do not happen quickly enough to explain how these brief inputs can

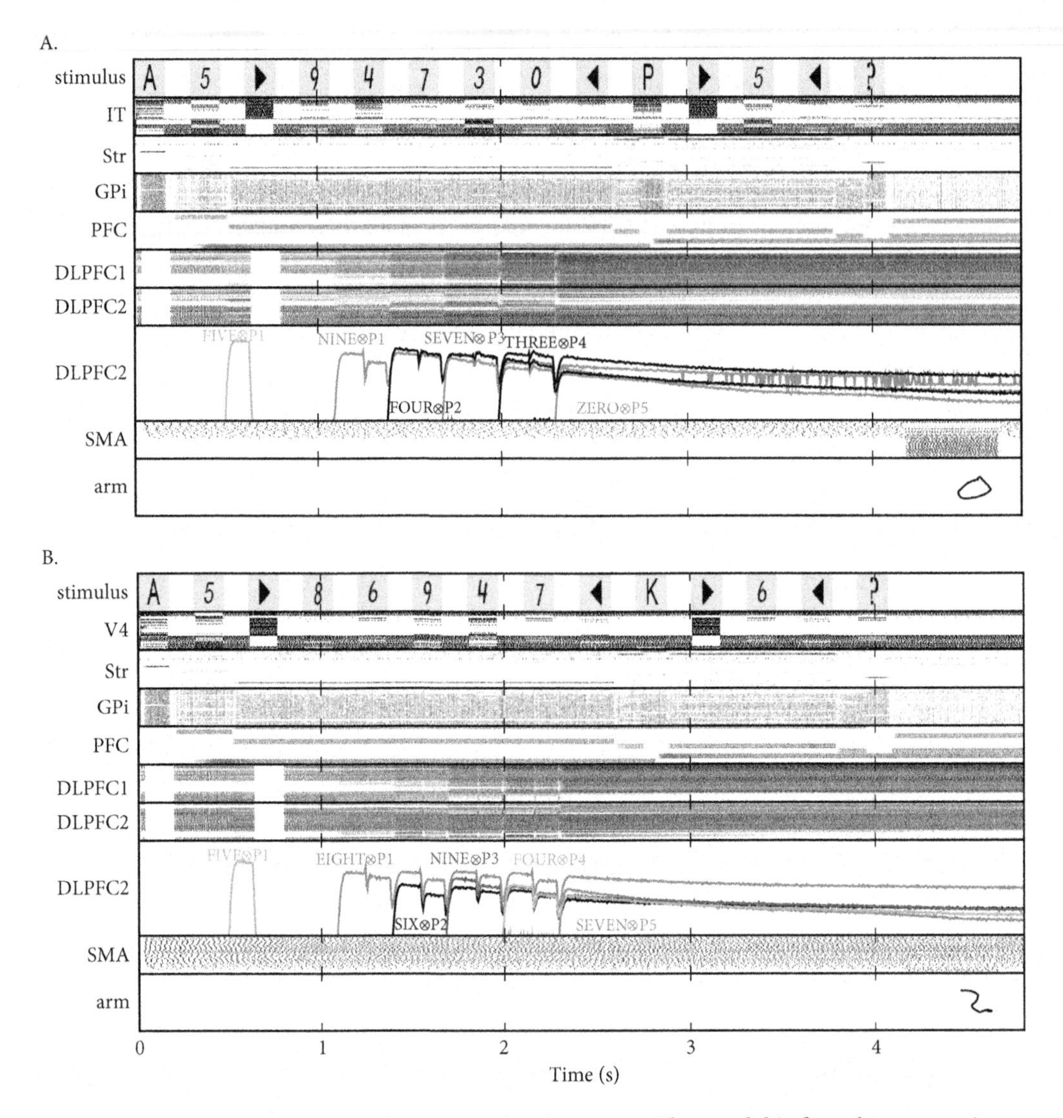

FIGURE 7.13 Question answering in Spaun. The model is first shown a string of digits. It is then shown either a "P" for position or "K" for kind. It is then shown a digit, which it interprets as asking about the position or kind as previously indicated. The motor system then draws the response. The encoding and decay of the string of digits about which the question is answered can be seen in the DLPFC2 similarity graphs. **A.** A "P" example run, where it is asked "What is in position 5?" The model's response is zero. **B.** A "K" example run, where it is asked "Where is the kind 6?" The model's response is (position) "2."

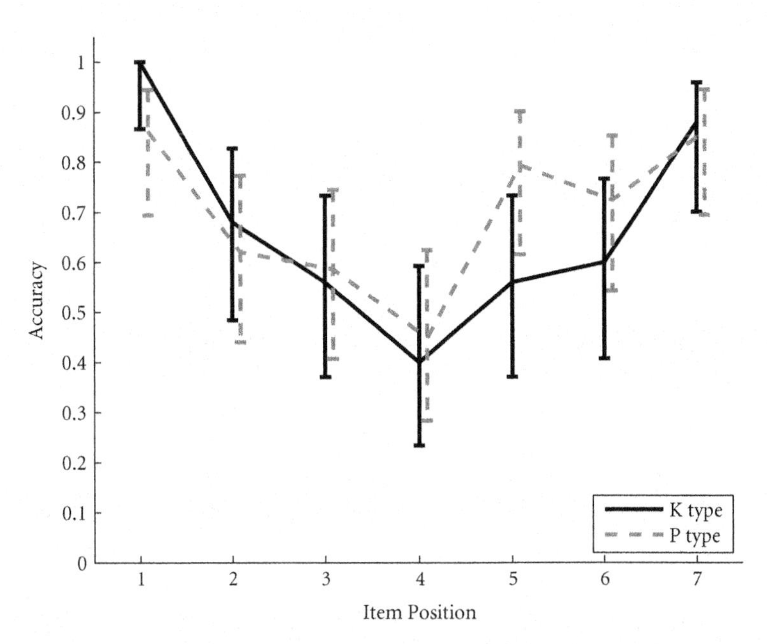

FIGURE 7.14 Question-answering error rates. Primacy and recency effects are seen in the accuracy of the question answering in Spaun because working memory is the main determinant of accuracy. The type of question does not affect the accuracy. Error bars indicate 95% confidence intervals for 25 queries of each position and each type of question. (Reproduced from Eliasmith et al. [2012] with permission.)

be successfully generalized within seconds. He has also argued that approaches such as neural blackboard architectures (NBAs; *see* Section 9.1.3) would need to pre-wire the appropriate syntactic structures. Finally, he has suggested that VSAs cannot solve this problem. This is because, as he has correctly noted, the binding operation in VSAs will not provide a syntactic generalization. However, the SPA has additional resources. Most importantly, it has inductive mechanisms employed both in the Raven's example (*see* Section 4.7) and the Wason example (Section 6.6). The Raven's example, in particular, did not rely on changing neural connection weights to perform inductive inference. As a result, we can use this same mechanism to perform the rapid variable creation task.

To demonstrate rapid variable creation, Spaun needs to reason over complex structured input. For example, an input sequence might be "[1 1 2 3] [2 3] [1 1 5 3] [5 3] [1 1 6 3] [6 3] [1 1 1 3] ?." The expected response in this case is "1 3." The parallel to Hadley's example with words should be clear. Each set of two bracketed items is an input/output pair. The input item consists of a constant repeated twice, a variable item, and another constant. The output item consists of the appropriate transformation of the given input. More generally, the input consists of a combination of variables and constants, and the task is to determine which to preserve in a response.

Spaun is able to solve this general kind of task by performing induction on the presented input/output pairs to infer the transformation rule in VLPFC that is being applied to the input. Figure 7.15A shows Spaun's behavior on a stimulus sequence with the same structure as that proposed by Hadley. Figure 7.15B shows the model's response for a simpler, but analogous case, which has the same general form of input (i.e., some combination of variables and constants). Notably, in either case, Spaun's response is extremely fast, occurring less than 1 second after the end of the input. This easily meets Hadley's requirements for counting as "rapid" (i.e., less than 2 seconds from the last example presentation; Hadley [2009] p.517).

Because this task has not been run on human subjects, it is difficult to provide direct quantitative comparisons between Spaun and behavioral data. Nevertheless, the task is included here to make an important theoretical point: Spaun's ability to successfully perform the rapid variable creation task demonstrates that the nonclassical assumptions behind the SPA do not prevent it from exhibiting central cognitive behaviors. In other words, Spaun shows that biologically realistic, connectionist-like architectures with nonsymbolic representations can account for prototypical symbolic behavior, contrary to the expectations of many past researchers (e.g., Hadley, 2009; Marcus, 2001; Jackendoff, 2002; Fodor & Pylyshyn, 1988).

7.3.8 *Fluid Reasoning*

The final task I consider is fluid reasoning (*see* Section 4.7). The purpose of including this task is to demonstrate the ability of the model to perform a new cognitive task by re-deploying already existing components. That is, for this example, there are no new aspects of the architecture to introduce at all. The only difference between this task and past ones is that the control structure in the basal ganglia is augmented to include three additional mappings (about 900 neurons).

The input to the system is in the form of a typical Raven's matrix, although it is shown serially. This is appropriate, as eye movement studies of human subjects show left-to-right reading of the matrices is the most common strategy (Carpenter et al., 1990). For example, a simple case is "[1] [1 1] [1 1 1] [2] [2 2] [2 2 2] [3] [3 3] ?," where the expected output is "3 3 3." The main difference between this task and the rapid variable creation task is the structure of the input. Specifically, here induction is only driven by pairwise comparisons within rows, so the middle input sequence is employed twice: once as post-transformation and once as pre-transformation.

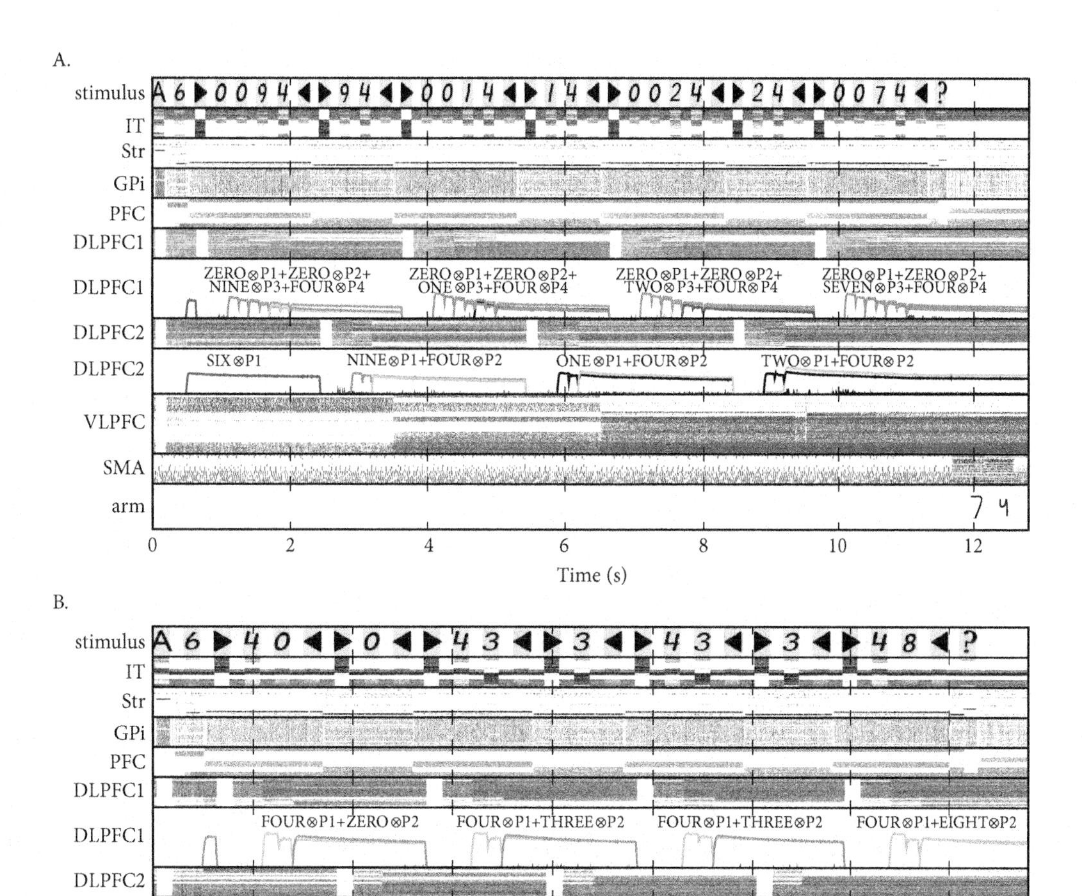

FIGURE 7.15 Rapid variable creation in Spaun. The stimulus is a set of input/output pairs in which some elements are variables and some are constant, or ignored. There can be any number of input/output pairs of any length. DLPFC1 encodes the example input; DLPFC2 encodes the corresponding example output. VLPCF is computing a running estimate of the induced transformation between input and output given the examples. **A.** An example run with the same structure as that provided by Hadley. **B.** An example run of a different but analogous task requiring a similar kind of inference to solve. (Reproduced from Eliasmith et al. [2012] with permission.)

The performance of Spaun on two example matrices is shown in Figures 7.16A and B. Spaun solves such matrices regardless of the particular items used. Notice that in Figure 7.16B, it is relying partly on its knowledge of the relationship between numbers to infer the pattern (which is not necessary to solve patterns like those in Figure 7.16A). Spaun can also successfully identify other patterns, such as decreasing the number of items or the count (results not shown).

It is somewhat difficult to compare the model to human data because the specific matrices being used are different. Perhaps more critically, the task itself is different. Here, the model must *generate* the correct answer, but in the Raven's matrices, only a *match* to 1 of 8 possible answers must be indicated. Nevertheless, here is a rough comparison: humans average 89% correct (chance is 13%) on the Raven's matrices that include only an induction rule (5 of 36 matrices; Forbes, 1964). Spaun is similarly good at these tasks, achieving a match-adjusted success rate of 88%.[5] This match is at least in the appropriate range. Perhaps more relevant for comparison to human data is the performance of the closely related model on the actual Raven's Matrices, which is discussed in Section 4.7.

Regardless of how convincing (or not) we take this comparison to be, the important conceptual point is that this example clearly demonstrates how identical elements of the architecture can be redeployed to perform distinct tasks (*see* Anderson, 2010). That is, only changing the control structure is necessary to extend the overall functional abilities of the architecture.

7.3.9 Discussion

Although the above set of tasks are defined over limited input and a simple semantic space, I believe they demonstrate the unique ability of the SPA to provide a general method for understanding biological cognition. Notably, it is straightforward to add other tasks into the Spaun architecture that can be defined over structured representations in the same semantic space, as demonstrated by the final task. For example, including additional forms of recall, backward counting, *n*-back tasks, pattern induction, and so on would result from minor additions to the state-action mappings in the basal ganglia. In addition, Spaun has the rudimentary beginnings of a mechanism for constructing its own state-action mappings based on environmental feedback. Consequently, a wide variety of tasks should be accessible in this simple semantic space with conceptually minor additions to the model.

[5] The raw accuracy rate of Spaun is 75% (95% CI of 60%–88%). If we assume that Spaun's answer on the error trials (i.e., trials in which it does not *produce* the correct answer) would *pick* the correct match to one of the eight available alternatives 50% of the time, then the match-adjusted rate is 75%+25%*50% = 88%.

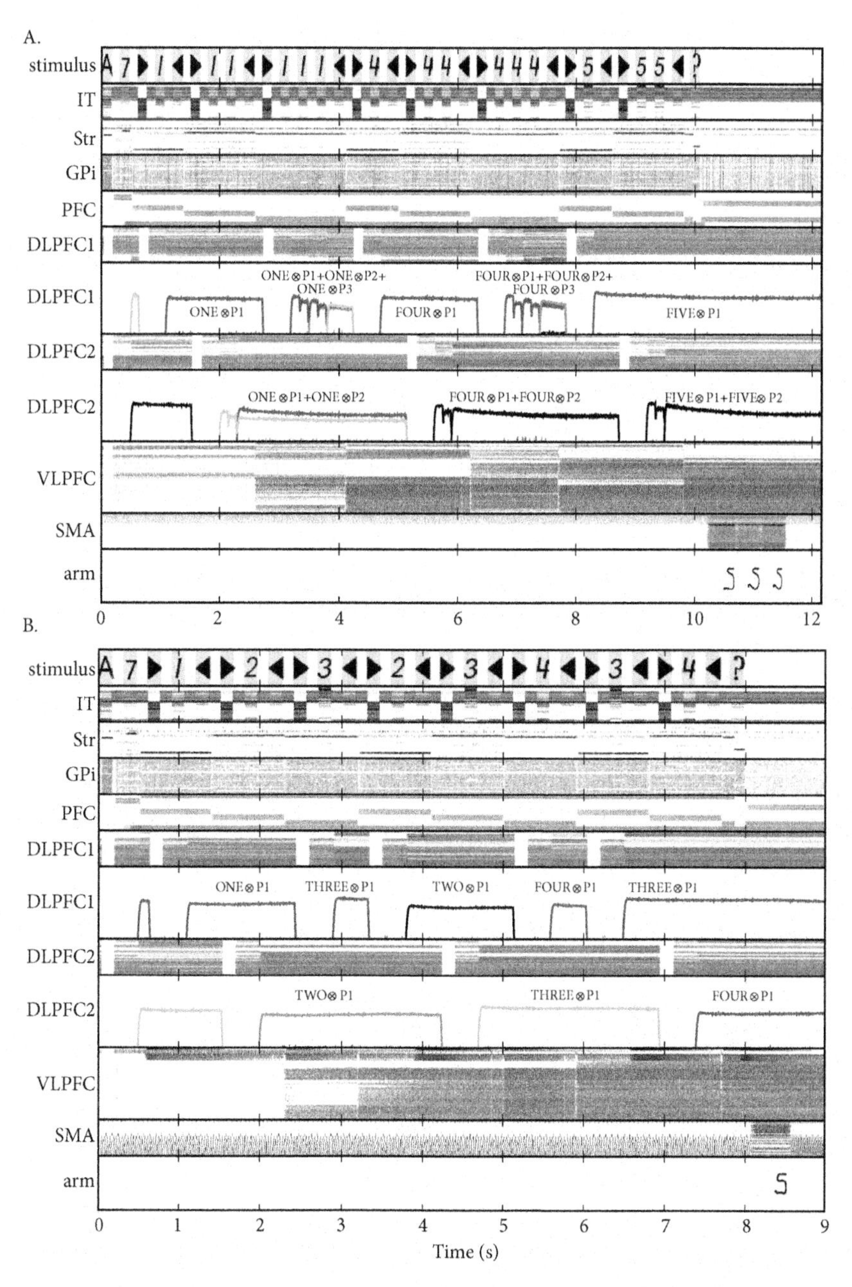

FIGURE 7.16 Fluid reasoning in Spaun. The spiking activity encoding the currently inferred rule is shown in the VMPFC row. This is a running average of the inverse convolution (i.e., the inferred transformation) between representations in DLPFC1 and DLPFC2, as appropriate. **A.** The time course of Spaun's activity while inferring that the pattern in the input is "increase the number of elements by one." **B.** The time course of Spaun's activity while inferring that the pattern is "add one to the current value." In this case, Spaun is producing a result that is never explicitly in its input. (Reproduced from Eliasmith et al. [2012] with permission.)

However, focusing on the performance of the model on distinct tasks, as I have done to this point, does not highlight what I take to be the main lessons we can learn from this model. Rather, I take the interesting aspects of Spaun to be more evident when considering all of the tasks together. Perhaps the most interesting feature of Spaun is that it is *unified*. Thus the performance on each task is less important than the fact that all of the tasks were performed by the exact same model. No parameters were changed nor parts of the model externally "rewired" as the tasks varied. The same model can perform all of the tasks in any order.

Such unification is further evident because the representations used throughout the model are the same, as are the underlying computational principles and methods of mapping to neural spikes. As a result, this single model has neuron responses in visual areas that match known visual responses, *as well as* neuron responses and circuitry in basal ganglia that match known responses and anatomical properties of basal ganglia, *as well as* behaviorally accurate working memory limitations, *as well as* the ability to perform human-like induction, and so on. Spaun is thus both physically and conceptually unified.

In addition, it is interesting that Spaun's architecture does not have one subsystem for each task. Rather, it exploits different combinations of available resources in different ways. Consequently, adding the fluid reasoning task after having all the other tasks requires only an additional 900 neurons in the basal ganglia and thalamus (those neurons control the new organization of available resources). This kind of scalability would not be possible without a unified architecture.

Spaun is also reasonably flexible. It performs context-sensitive interpretation of its input–that is, it uses the same input "channel" to identify tasks, receive task-relevant input, and identify queries. It limitedly adapts to environmental contingencies, such as rewards. It can rapidly learn new patterns in its input and employ them in producing a novel response. It can redeploy the same architectural elements in different ways to achieve different ends. Spaun is thus a behaviorally flexible system.

Spaun is also robust. There is a lot of intrinsic noise in the model because of the use of spiking neurons. Additionally we can destroy many of the cells and observe a graceful degradation in performance (analogous to the Wason model in Section 6.6). It is robust to the many variations in the handwriting used to drive the system. It continues to function when it makes errors in recall, induction, recognition, and so on (*see*, e.g., Fig. 7.10). Similarly, we can intentionally provide "invalid" input, and the system continues to function (*see* Fig. 7.17). Spaun is thus robust to both random and systematic variability.

Finally, it is important that Spaun is only an *instance* of the SPA. Critically, the methods used to construct this simple model are not restricted to the semantic or conceptual space of numbers. Consequently, increasing the

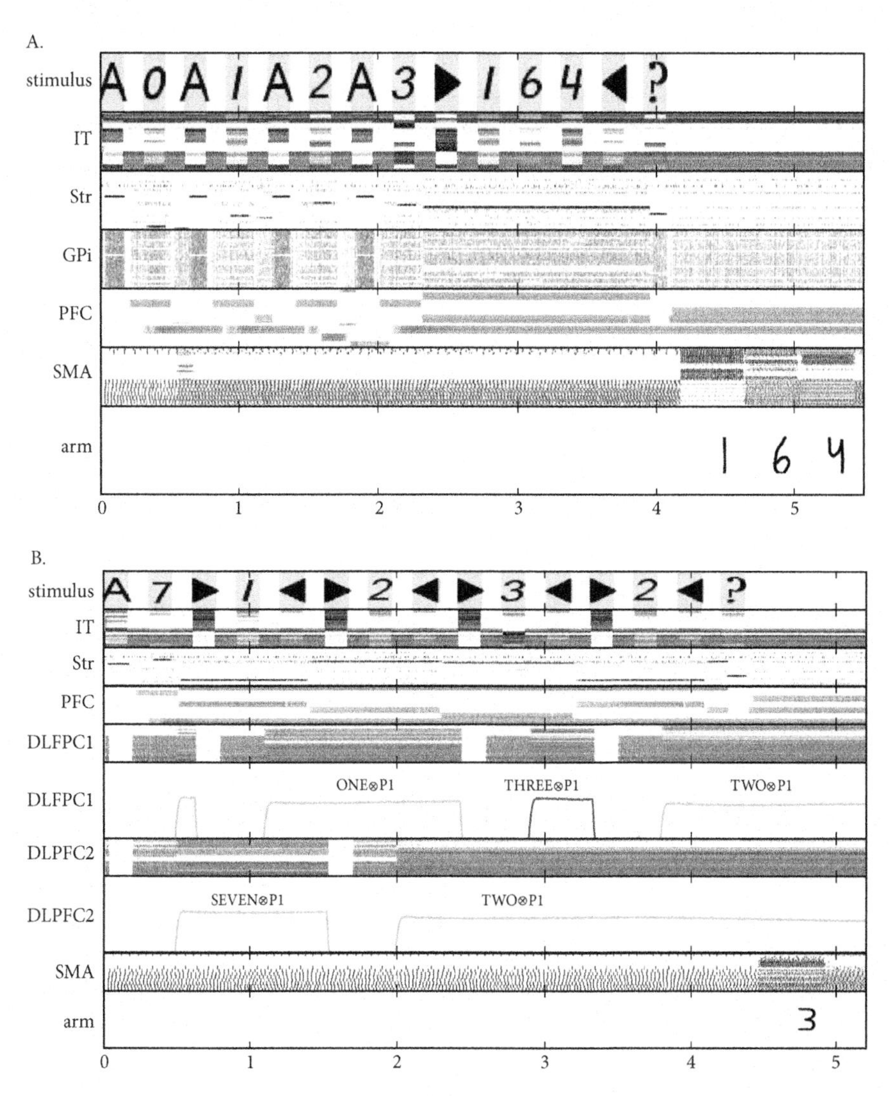

FIGURE 7.17 Invalid input in Spaun. **A.** A series of four tasks is identified before any valid input is provided. Nevertheless, Spaun performs the last identified task (serial working memory) correctly. **B.** Spaun is asked to complete a Raven's matrix halfway through the second row. It uses the information provided to give a best guess, which in this case is correct. **C.** A completely invalid question is asked in the question-answering task (there is no "zero" position). Spaun does not respond but proceeds normally to the subsequent task. Such examples help demonstrate the robustness of the model. (Reproduced from Eliasmith et al. [2012] with permission.)

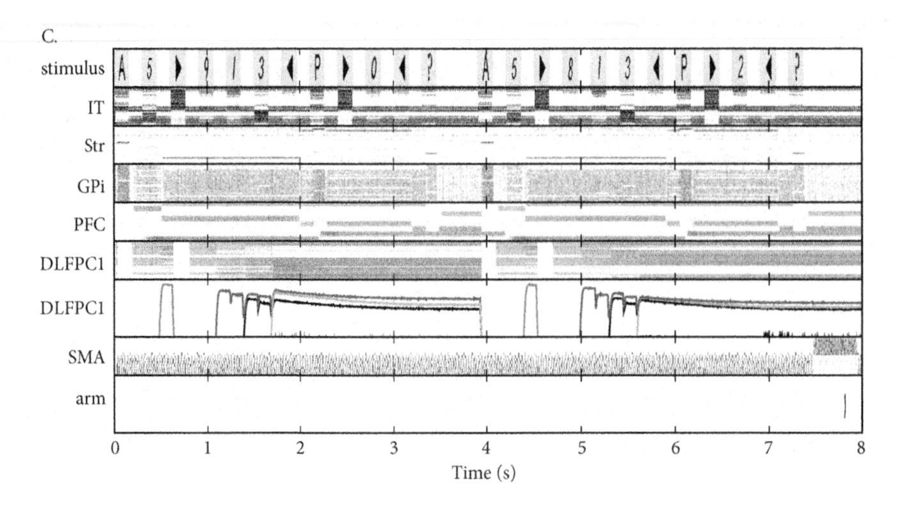

FIGURE 7.17 (*cont.*)

complexity of the input space, conceptual representations, or output systems can be accomplished through direct application of the same principles used in this simple case. Overall, the purpose of Spaun is to show how SPA models can be generated, *in general*, using the principles described throughout the book.

Of course, Spaun is also just a meager beginning. There are many more ways it could be unified, flexible, and robust. I discuss several of these in Chapter 10, which identifies some central challenges for the SPA. Nevertheless, I believe Spaun represents an exciting beginning to developing models of biological cognition. Perhaps as time goes on, a community of experts in biological cognition will develop that identifies empirically grounded principles to extend such models in myriad ways. As I explore in subsequent chapters, I am hopeful that the SPA can play a role in this process because I believe it has distinct advantages over other approaches.

However, one of its advantages does not fit naturally into discussions of current approaches to cognitive modeling. So, I consider that next.

7.4 ■ A Unified View: Symbols and Probabilities

As outlined in Chapter 1, one traditional means of distinguishing symbolic and connectionist approaches is that the former is taken to be most appropriate for describing high-level planning and rule following, and the latter is taken to be most appropriate for capturing lower-level pattern recognition and statistical kinds of behavior (e.g., classification, pattern completion, etc.). These strengths are typically attributed to the reliance on symbols and symbolic structures by classical approaches, and the reliance on statistically learned distributed representations by connectionist approaches. In this section, I want to consider how semantic pointers can be simultaneously interpreted as symbols and probabilities and how processing in the SPA can be similarly interpreted as both rule following and statistical inference.

To help make this point, it is useful to contrast the SPA with a more symbolic approach. As I will discuss in some detail in Section 9.1.1, the ACT-R cognitive architecture is perhaps the most successful and influential cognitive modeling architecture currently on offer. Many aspects of the SPA have been inspired by the successes of ACT-R. For example, both ACT-R and the SPA characterize the basal ganglia as the locus of certain kinds of action selection. However, there are some critical differences between the approaches. The most obvious difference is that representations in the SPA are distributed vector representations, whereas most representations in ACT-R are symbolic. I have often glossed over this difference by characterizing the SPA as manipulating symbol-like representations for ease of exposition. However, ultimately these different representational commitments result in differences in computational power. So, rather than thinking of semantic pointers as symbol-like, I would now like to consider how one can think of them as being probability distribution-like.

To give a sense of why probability distributions can be more computationally powerful than symbols,[6] notice that a probability distribution more completely describes the possible states of the world than a symbol. Typically, a symbol is taken to have no special internal structure. Its semantics are determined by relations to other symbols (e.g., Harman, 1982) or its relations to the external world (e.g., Fodor, 1998). In contrast, probability distributions, like any vector, can be treated in ways that draw directly on their internal structure. Let us consider an example.

Suppose we have the symbol "dog," and we also have a probability distribution that represents the class of "dog," $\rho(g)$. We can define a rule like, "If something is a dog, then it is furry." In a symbolic system, like ACT-R, the presence of the "dog" symbol will result in the presence of the "furry" symbol given this rule. There is only one "dog" symbol and it leads to only one "furry" symbol (unless we add more rules). In contrast, in a probabilistically driven system, an inference rule will relate *distributions*, which can take on many values. For example:

$$\rho(f) = \int_g \rho(f|g)\rho(g)\,dg$$

[6] Claims of pure computational power differences are difficult to establish and typically uninteresting. Rather, I have in mind a resource-limited notion. That is, if we think of the activity of a given population of 100 neurons as indicating that a particular symbol is being used or not, or if we think of 100 neurons as representing the value of a vector in a vector space, then can we compute a different set of functions? If one set of functions is larger than the other, then I would call that representation more computationally powerful.

provides a means of determining the probability that something is furry, $\rho(f)$, knowing the probability it is a dog and how those distributions relate, $\rho(f|g)$. The internal structure of the resulting "furry" distribution will depend on the internal structure of the "dog" distribution. In fact, only in the special case where $\rho(f|g)$ is equal to a kind of identity function will this rule be equivalent to the symbolic one. Consequently, the variety of rules that can be defined over a probability distribution is much greater than those that can be defined over a symbol; for probability distributions, rules may or may not depend on the details of the structure of that distribution.[7]

In fact, this increase in computational power has been recognized, and explored by the ACT-R community. They have worried about "partial matching" of productions (if-then rules) to input symbols in the past (Lebiere & Anderson, 1993). However, the vast majority of ACT-R architectures and models do not employ this technique. And, those that do have implemented a fairly limited form of it (i.e., only some aspects of a rule can partially match). The SPA, in contrast, includes partial matching as a central kind of computation, one that acts to perform statistical inference.

Recall that in the SPA there is a mapping from cortex to striatum, which I have discussed as determining the "if" part of an "if-then" rule (*see* Section 5.4). This mapping is, in fact, a partial matching operation. As mentioned in my earlier description, if we think of the map from cortex to striatum as a matrix $\mathbf{M}_b$ and the cortical state as a vector $\mathbf{x}$, then the result of $\mathbf{M}_b\mathbf{x}$ is a vector in which each element is a value that represents how well $\mathbf{x}$ matches the "if" part of each rule represented in $\mathbf{M}_b$. That is, the "degree of match" is exactly the result of the projection of cortex into striatum in the SPA. Continuing through the basal ganglia, that result is then used, via a winner-take-all-like operation, to determine the appropriate action, which is then mapped to a new cortical state $\mathbf{x}'$. This process is depicted in Figure 7.18A.

So, under a symbolic view of the SPA, we might say that the semantic pointer (as a symbol) represented in cortex is used to match a rule in the basal ganglia, resulting in the selection of the action associated with that rule. Under a more vector-based view of the SPA, we would say that the semantic pointer (as a vector) is measured against familiar states in the basal ganglia, resulting in the selection of a new cortical state most appropriate under the given measure.

[7] Again, this is not a mathematical fact but, rather, a conceptual one. That is, this observation depends on what we take to be basic cognitive states. If we decided to suggest that each dimension in a semantic pointer is represented by many distinct symbols (one for each distinguishable value along the dimension), then we could say that the SPA is a symbolic architecture. In such a case, there is no interesting difference in the set of computations we can describe using either formalism. However, such a claim does violence to the typical use of "symbolic" in the behavioral sciences.

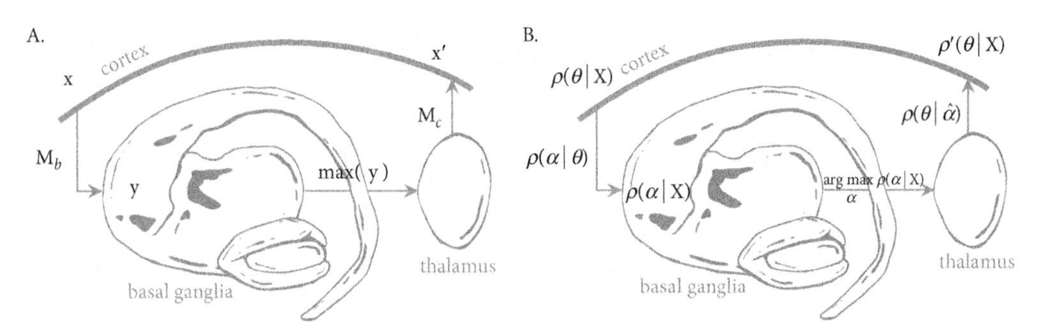

FIGURE 7.18 Two SPA interpretations of the cortical-basal ganglia-thalamus loop. **A.** The matrix-vector characterization of basal ganglia that I have previously discussed in terms of implementing if-then rules (*see* Section 5.6). **B.** An equivalent probabilistic characterization of basal ganglia function. This characterization helps identify how the SPA goes beyond traditional production systems, and hence unifies symbolic and connectionist approaches to cognitive modeling. See text for details.

Interestingly, because the SPA employs high-dimensional vectors as its central representations, this exact same process has a very natural statistical interpretation, as shown in Figure 7.18B. Mathematically, all objects (scalars, functions, fields, etc.) can be treated as vectors. So, we can simply stipulate that (normalized) semantic pointer vectors are probability distributions. But, this does not provide us with an actual interpretation, mapping specific distributions onto brain states. Consequently, we must be careful about our probabilistic interpretation, ensuring that the already-identified neural computations are still employed, and that the distributions themselves make biological sense.

Perhaps surprisingly, the interpretation provided here lets us understand the cortical-basal ganglia-thalamus loop as implementing a standard kind of Bayesian inference algorithm known as Empirical Bayes. Specifically, this algorithm relates perceptual observations (**X**), parameters of a statistical model of that input (θ), and so-called hyperparameters (α) that help pick out an appropriate set of model parameters. Essentially, identification of the hyperparameters allows the brain to determine a posterior distribution on the model parameters based on the observation. We can think of this distribution as acting like a prior when relating observations and model parameters. More practically, we can think of the hyperparameters as expressing what parameters are expected based on past experience.

The basic idea behind this interpretation is that cortex takes sensory measurements and attempts to determine a model (or "explanation") for those measurements. In short, it infers θ given **X**, using past experience that helps determine the form of various priors. However, this inference is approximate and difficult, so it employs the basal ganglia to iteratively improve the estimate of that distribution using learned relations between environmental regularities, especially in instances where the cortical model is inaccurate. The resulting distribution can then be used by the motor system to generate samples that are consistent with the model, or that distribution can be used to predict future states, or help interpret new sensory measurements.

To be specific, let me go through the steps depicted in Figure 7.18B. Before determining the initial distribution in the diagram, $\rho(\theta|\mathbf{X})$, we begin with a likelihood over model parameters, $\rho(\mathbf{X}|\theta)$, where $\mathbf{X}$ are the observations, and cortex has learned how to generate an initial guess of the probability distribution for those observations given various values of the parameters θ.[8] Cortex initially estimates a distribution of how likely various parameter settings are given the observations. Typically, this would be done using Bayes' rule:[9]

$$\rho(\theta|\mathbf{X}) \propto \rho(\mathbf{X}|\theta)\rho(\theta).$$

However, the difficulty of computing this distribution exactly is precisely why we need to employ the basal ganglia. Consequently, the first pass through cortex will only provide a rough estimate of this distribution. This provides a starting point analogous to default routing being used to generate the initial cortical state $\mathbf{x}$ in the matrix-vector characterization.

The mapping into the striatum from cortex is defined by $\rho(\alpha|\theta)$, which encodes the learned relationship between the hyperparameters and the parameter values. In virtue of mapping through this distribution, the result in striatum relates the observations directly to these hyperparameters:[10]

$$\rho(\alpha|\mathbf{X}) = \int_\theta \rho(\alpha|\theta)\rho(\theta|\mathbf{X})\, d\theta,$$

which is analogous to $\mathbf{y}$ in the matrix-vector interpretation. Very loosely, we can think of α as standing for "action."

Now, as before, a maximization operation is assumed to be performed by the rest of the basal ganglia circuit. In inference, this is known as maximum *a posteriori* (MAP) inference and is used to determine the specific value of α (i.e., $\hat{\alpha}$) that maximizes the original distribution $\rho(\alpha|\mathbf{X})$.

This result is then sent through thalamus and back to cortex, where the thalamic-cortical mapping is used to determine a new distribution of the

[8] To interpret a likelihood, you can imagine that the observations $\mathbf{X}$ are fixed and that the probability varies as the parameters, θ, are changed.

[9] As is common, I am ignoring the denominator on the right-hand side, $\rho(\mathbf{X})$, both because it is a constant and because it is very difficult to compute. In short, it does not change the answer we compute except by a scaling factor.

[10] I am assuming α and $\mathbf{X}$ are conditionally independent given θ.

parameters θ given the inferred value of $\hat{\alpha}$–that is, $\rho(\theta|\hat{\alpha})$. Once back in cortex, this distribution can be used to update the estimate of the distribution of the parameters given the data and the original likelihood:

$$\rho'(\theta|\mathbf{X}) \propto \rho(\mathbf{X}|\theta)\rho(\theta|\hat{\alpha}),$$

which completes the loop. Typically in such inference algorithms, the loop is performed several times to improve the quality of the estimate of $\rho(\theta|\mathbf{X})$, as it is known to get more accurate on each iteration. Such iteration need not be employed, however, if time is of the essence.

Further, the $\rho(\theta|\mathbf{X})$ distribution is very useful because it can be employed to support many further inferences. For example, it can be used to determine how likely a new observation x_{new} is given past observations:

$$\rho(x_{new}|\mathbf{X}) = \int_{\theta} \rho(x_{new}|\theta)\rho(\theta|\mathbf{X})\, d\theta.$$

Or, it can be used to drive motor action:

$$\rho(c|\mathbf{X}) = \int_{\theta} \rho(c|\theta)\rho(\theta|\mathbf{X})\, d\theta$$

where c is the motor command to execute, and so $\rho(c|\theta)$ would encode the previously learned mapping between a model of the world and motor commands. Critically, this new characterization of the SPA processing loop preserves the role of the basal ganglia in action selection and provides a natural interpretation of reinforcement learning in the striatum (i.e., to determine α). More generally, this probabilistic interpretation of the loop as an iterative inference process is consistent with all aspects of my previous discussions of the basal ganglia.[11]

Let me step away from the formalism for a moment to reiterate what I take to be the point of this exercise of re-interpreting the cortex-basal ganglia-thalamus loop. Recall that a central distinction between the symbolic approach and connectionism is that the former has strengths in language-like rule processing and the latter has strengths in probabilistic inference. I have spent most of the book showing how semantic pointers can be understood as implementing language-like rule processing and am here reiterating that they can also be understood as implementing statistical inference (I had suggested this only for perception and motor control earlier, in Chapter 3). As a

[11] Three brief technical points for those who might be concerned that a probabilistic characterization assumes very different cortical transformations than before. First, the convolution of two probability distributions is the distribution of the sum of the original variables, so the previous binding operation is well characterized probabilistically. Second, the superposition and binding operations do not significantly alter the length of the processed variables. Specifically, the expected length of two unit length bound variables is one, and the superposition is well-normalized by the saturation of neurons. Third, the multiplication and integration operation I have used several times is identical to matrix-vector multiplication.

consequence, I believe that these two consistent and simultaneously applicable interpretations of the SPA highlight a fundamental unification of symbolic and connectionist approaches in this architecture. And, it begins to suggest that the SPA is in a position to combine the strengths of past approaches. Further, this kind of unification is not available to approaches, like ACT-R, which do not preserve the internal structure of the representations employed.

This, however, is only one preliminary comparison of the SPA to past work. In the second part of the book, I provide more thorough comparisons.

7.5 ■ Nengo: Advanced Modeling Methods

In the tutorial from Section 5.9, we built a network with thousands of spiking neurons that encodes structured representations, transforms these semantic pointers, and includes dynamics that route information and model working memory. Despite these many features, the network is still relatively simple compared to most models presented in the book, all of which pale in comparison to biological brains. In this tutorial I do not introduce a more sophisticated network but focus instead on some useful methods for efficiently constructing and testing larger models.

Most important for large projects is the ability to generate networks in Nengo using scripts. The scripting interface can be used to automate every interaction with Nengo that could be executed with the graphical user interface. It can also be used to extend Nengo's capabilities to work with third-party software or with your own libraries. This tutorial briefly covers some aspects of the scripting library used to build large semantic pointer architecture models. For scripting tutorials that cover more basic utilities, such as those for creating individual ensembles with customized tuning curves or integrating Nengo with Matlab®, please see the online documentation at `http://nengo.ca/documentation`. I also encourage you to read the many scripts included with Nengo in the "/demo" and "/Building a Brain" directories, which can serve as useful examples of using more Nengo scripting functions.

- Open a blank Nengo workspace and select View→Toggle Script Console to open the scripting console.
- Type "`run Building a Brain/chapter7/spa_sequence.py`." This is the same as opening the file through menus or the toolbar.
- Double-click the "Sequence" network to open it. Double-click the "state" subnetwork to open it. Click the "buffer" population.
- Type "`print that.getDimension()`" in the script console and hit enter.

The console will display the dimensionality of the population (i.e., 16) that is currently selected (identified with the keyword "that").

- Type "`print that.getEncoders()`" and hit enter.

The console will now display an array that lists of all the encoders in the population: one encoder for each neuron and 16 values for each encoder. The full list of functions you can access through scripting in this kind of manner is listed on the Nengo website at `http://nengo.ca/javadoc/` in standard JavaDoc format. The full Java API can take significant time to become familiar with, so we have written several Python classes to help make scripting easier. Before describing those, let us take a look at the model.

- Open the network in the *Interactive Plots* viewer and run the simulation.

This model proceeds through a set of five rules that are cyclic (i.e., $A \rightarrow B \rightarrow \ldots \rightarrow E \rightarrow A \rightarrow \ldots$). This can be thought of as a sequence of memorized transitions, like the alphabet. This sequential rehearsal network should be familiar from Section 5.4, where it was first introduced. Recall that basal ganglia activation indicates which rule's input condition most closely matches the inputs at a given time. The thalamus then implements the actions specified by the active rule (reflected in the "Rules" display).

When the model runs, we can watch the basal ganglia progress through the rules by watching the best matched item in the "buffer" change over time. The "Utility" box in the visualizer shows the similarity of the current basal ganglia input with the possible vocabulary vectors. The "Rules" box shows which transition is currently being followed (rule "A" is A->B).

- In your operating system's file system, find and open Nengo's install directory. This is the directory where you unzipped Nengo's files in the first tutorial.
- Open the "Build a Brain" folder, located in Nengo's install directory.

The files located in this folder (and the demo folder) are all Python scripts. They can be opened with any text editor, although good specialized Python editors are freely available on the internet.

- Open the file "chapter7/spa_sequence.py" in a text editor.

The file you've just opened contains the instructions required to build the "Sequence" network. These instructions should match the listing given below.

```
1  from spa import *     #Import SPA related
                         packages
2
3  class Rules:          #Define mappings for BG and
                         Thal
4      def A(state='A'):
5          set(state='B')
6      def B(state='B'):
```

```
 7              set(state='C')
 8          def C(state='C'):
 9              set(state='D')
10          def D(state='D'):
11              set(state='E')
12          def E(state='E'):
13              set(state='A')
14
15      class Sequence(SPA):    #Extend the imported
                                 SPA class
16          dimensions=16       #Dimensions in SPs
17          state=Buffer()      #Create a working
                                 memory/cortical element
18          BG=BasalGanglia(Rules())    #Set rules
                                         defined above
19          thal=Thalamus(BG)           #Set thalamus
                                         with rules
20          input=Input(0.1,state='D') #Define the
                                         starting input
21
22      seq=Sequence()                  #Run the class
```

The first line of the file imports packages that contain other scripts that provide methods that simplify network construction in Nengo, which are called in this script. The next block of code defines the "Rules" class, which sets the rules that the basal ganglia and thalamus will instantiate. The rule definitions all follow the following format:

```
def RuleName(inputState='InputSP'):
    set(outputState='OutputSP')
```

RuleName sets the name of the rule that will appear in the interactive graphs. *InputSP* is the name of a semantic pointer in the vocabulary that an input from the ensemble named *inputState* should match for the rule to be applied. *OutputSP* is the name of a semantic pointer in the vocabulary that is sent to the ensemble *outputState* when the rule is applied. Note that in the spa_sequence.py demo, both *inputState* and *outputState* are set to the same ensemble, which is simply named "state." The rules given in the demo specify a chain of rules that loop through the states: when "A" is the best matched state, "B" is sent as an input; when "B" is the best matched state, "C" is sent; and so on.

The "Sequence" class that is defined below the "Rules" class inherits several functions from the "SPA" class, which is defined in the code that was imported (you can examine this code in the "python" directory of the Nengo installation). The first line in the class defines how many dimensions each semantic pointer will have in the model (16 in this example). The "Sequence" class then defines four objects: a buffer, a basal ganglia component, a thalamus component, and an input.

The buffer is an integrator that serves as a working memory component and it is assigned to a variable named "state." The basal ganglia and thalamus act as described in Chapter 5. Note that the class "Rules" is passed to the basal ganglia constructor, and the basal ganglia variable "BG" is passed to the thalamus constructor on the following line. These steps are required to create basal ganglia and thalamus components that conform to the specified rules.

The last object created in this class is an input function. The syntax for creating an input function is as follows: `Input(duration, target='SP')`. The first parameter, *duration*, specifies how many seconds the input will be presented for, the parameter *target* gives the population that will receive input from the function, and "SP" is the name of the semantic pointer that will be given as input.

The four objects created in the "Sequence" class are assigned to variables named "state," "BG," "thal," and "input," and these variable names are used to name the items created in the Nengo network when the script is run. The final line of the script runs the "Sequence" class, which generates all the network objects described above and connects them together to form the "Sequence" network.

Because we have a script that will generate the network automatically, it is easy to make adjustments to the network and test them quickly. For example, we can replicate the result shown in Figure 5.7 of Section 5.4, where the sequence is interrupted by constant input.

- Save the "spa_sequence.py" file as "spa_sequence_tutorial.py," to avoid overwriting the demo script accidentally.
- Edit the line declaring the input object to read `input=Input(10, state='A')`.
- Save the "spa_sequence_tutorial.py" file.
- Return to a blank Nengo workspace and open the scripting console.
- Type "run Building a Brain/chapter7/spa_sequence_tutorial.py" and press enter.
- Open the network with *Interactive Plots* and run the simulation.

The input to the state buffer is now presented for 10 seconds rather than for the previous 0.1 seconds. As reported in Chapter 5, this prevents the activation of the second rule in the sequence despite its high utility because the input value constantly drives the network back to its initial state. The solution described in the earlier chapter was to introduce routing to control the flow of information.

- Open the file "spa_sequencerouted.py" located in the Building a Brain/chapter7 folder in your text editor.

The "spa_sequencerouted.py" file is identical to the "spa_sequence.py" file with the exception of changing the network name, and the following lines:

```
    def start(vision='START'):
        set(state=vision)
...
    vision=Buffer(feedback=0)
    input=Input(0.4, vision='0.8*START+D')
```

The first change is the introduction of a new rule named "start." Unlike the other rules, the "start" rule does not assign a fixed semantic pointer to the "state" ensemble when the rule is activated. Rather, it moves the value stored in the "vision" ensemble into the state ensemble. The "vision" ensemble itself is created in the "Routing" class. Notably, the `feedback=0` setting in this "Buffer" call makes the "vision" ensemble not have any recurrent connections, and hence the ensemble just represents its state while being driven by input.

The last change sets the initial input to contain the "START" semantic pointer (added to a starting semantic pointer) and connects this input to the new "vision" ensemble, rather than projecting directly to the "state" ensemble. The presence of the "START" semantic pointer in the "vision" ensemble will cause the start rule to copy the input from the "vision" ensemble to the "state" ensemble. The utility of the "start" rule will almost always be lower than the utility of the sequence rules because the input is a version of "START" that has been "blurred" by whichever letter has been chosen as the starting point (the letter "D" in the demo script) and an explicit scaling of 0.8. Consequently, the start rule will only transfer information from the visual area when no other action applies So, even though the input is applied for 400 ms, it has no effect after about 100 ms, as discussed in Section 5.6.

- Return to a blank Nengo workspace.
- Open the "spa_sequencerouted.py" script in the "Building a Brain/chapter7/" directory into the workspace.
- Run the "Routing" network in the *Interactive Plots.*

You will notice that it works fine, but of course the input is only presented for 0.1 seconds.

- Edit the line declaring the input object to read `input=Input(10, vision='0.8*START+D')`, to show the input for a long time. Save this edited version of the script as "spa_sequencerouted_tutorial.py."
- Return to a blank Nengo workspace.

- Open the "spa_sequencerouted_tutorial.py" script in the "Building a Brain/chapter7/" directory into the workspace.
- Run the "Routing" network in the *Interactive Plots*.

With the routing, the network now ignores inputs from the visual area after the initial "start" rule is completed. The network will thus proceed to cycle through the rules despite there being a constant visual input, as desired. Recall that the purpose of this tutorial is not to introduce the specific function of this network. Rather, it is to demonstrate how such a network can be described with a few short lines in a script file. By adding more rules, buffers, gates, and so on, these scripting techniques can be used to implement reasonably sophisticated models such as the Tower of Hanoi and Spaun. Such models allow us to ask interesting research questions.

Model Extensions

We built the Routing network using a scripting library that specified the network in terms of abstract rules, but it is often useful to have scripts that can create ensembles of neurons and adjust their parameters directly. To demonstrate this, the next part of this tutorial involves the creation of a simple clean-up memory to process the state of the Routing network.

- Add the following lines to the bottom of your "spa_sequencerouted_tutorial.py" file, or use the completed file "spa_sequencerouted_cleanup.py."

```
import hrr
vocab = hrr.Vocabulary.defaults[model.
   dimensions]
pd = [vocab['A'].v.tolist()]
cleanup = model.net.make('cleanup A',
   neurons=100, dimensions=1)
model.net.connect(model.state.net.network.
   getOrigin('state'), cleanup, transform=pd)
```

The first two lines of this code listing are required to obtain a reference to all the vocabulary vectors that have already been defined for the network. The code then retrieves the vector associated with the name "A" in line 3 and creates a scalar ensemble named "cleanup A" in line 4. The last line creates a projection from the "state" origin of the state ensemble to the new clean-up ensemble using a transformation matrix, "pd," specified by the vocabulary vector "A." This means that each component of the state vector is multiplied by the corresponding component of the "A" vector and summed to produce a one-dimensional quantity represented by the clean-up ensemble.

This operation is equivalent to a dot product between the state vector and the defined vocabulary vector "A."

- Return to a blank Nengo workspace.
- Open the edited "spa_sequencerouted_tutorial.py" script in the "Building a Brain/chapter7/" directory into the workspace.
- Open the Routing network in the *Interactive Plots* viewer.
- Ensure the value of the "cleanup A" ensemble is being graphed (pull up the plot by right-clicking on the background if needed).
- Run the simulation.

The value of the "cleanup A" population should rise only when the value of the state ensemble matches the "A" vector. This is a very simple implementation of a clean-up memory that only computes the similarity of a single vector to the represented state.

- Return to an empty Nengo workspace.
- Open the "spa_sequencerouted_cleanupAll.py" script in the "Building a Brain/chapter7" directory.
- Run the network with the Interactive Plots viewer and let it continue for about 2 seconds and pause it.

This simulation demonstrates a clean-up memory that includes all of the vectors in the network's vocabulary. Each dimension of the clean-up population represents the similarity of a vocabulary vector to the value of the state ensemble. This clean-up is still rudimentary and could be improved by adding a thresholding function or by adding mutual inhibition between the possible responses. If you look at the code for this example, you will find that it is very similar to the previous file. The main difference is that where the previous script referred to a specific vector, "A," this script contains a loop that processes all of the vocabulary items contained in the list "vocab.keys," as shown in the code excerpt below.

```
pd = []
for item in vocab.keys:
    pd.append(vocab[item].v.tolist())
```

You may also have noticed that the new script file contains code for recording data. This code uses the nef.Log object to create a special node within the network that writes all of the data it receives to a comma-separated values (CSV) file.

- Find the file *NengoDemoOutput.csv* located in the Nengo install directory.
- Open this file using a spreadsheet program such as Microsoft Excel, Calc, or a text editor.

The CSV file contains a spreadsheet with five columns. The first column lists the simulation time and indicates when the data were recorded (here, every millisecond of simulation time). The second column is labeled "cleanup." This column records the decoded vector values from the "cleanup" node. The third column is the same information for the "state" node. The fourth column, labeled "cleanup_spikes," is a vector with a count of the number of spikes emitted by each neuron since the last time data were recorded. Finally, the fifth column "state_vocab" records the similarity of the node's output to the vectors in the current vocabulary. This should be reminiscent of the headings of the semantic pointer graphs in *Interactive Plots* mode. This logging utility can be used to quickly specify which data to record from any ensemble.

Nengo also includes an analogous utility for input. This utility can be used by constructing a network and calling the `net.read_inputs` function with a CSV file name. Any column in that CSV file can then be connected to any node in the network to provide prespecified input to that node (e.g., `net.connect('inputState', state)`). Sometimes it is also useful to be able to have the input to the network depend on its output in a reasonably complicated way (perhaps for simulating an environment). For this purpose, Nengo supports a class called a SimpleNode. This class allows arbitrary Python code to be implemented in the node and connected directly to any point in a Nengo network. All of these utilities are described in much more detail in the online documentation at `http://nengo.ca/docs/html/`.

My main purpose in mentioning these elements of Nengo is to note that constructing, sending input to, and recording data from SPA models can all be done in a programmatic manner. Consequently, the methods I have introduced in these tutorials can be used to efficiently construct research-quality models in Nengo. In the final tutorial at the end of the next chapter, I provide an overview of the steps involved in constructing such models.

PART II

IS THAT HOW YOU BUILD A BRAIN?

8. EVALUATING COGNITIVE THEORIES

8.1 ■ Introduction

I have completed my description of the Semantic Pointer Architecture. Along the way, I have mentioned several unique or particularly crucial features of the SPA and provided a variety of examples to highlight these attributes. However, I have not had much opportunity to contextualize the SPA in the terrain of cognitive theories more generally. This is the task I turn to in the next three chapters.

As you may recall from Chapter 1, cognitive science has been dominated by three main views of cognitive function: the symbolic approach, connectionism, and dynamicism. In describing and comparing these views, I argued for the idea there is a general consensus on criteria for identifying cognitive systems within the behavioral sciences. I dubbed this set of criteria the Core Cognitive Criteria (CCC; *see* Table 8.1), and briefly enumerated them at the end of Section 1.3. At the time, I made no attempt to justify this particular choice of criteria, allowing me to turn directly to the presentation of the SPA.

In this chapter, I return to the CCC to make it clear why each is crucial to evaluating cognitive theories. In the next chapter, I describe several previously proposed architectures for constructing cognitive models. This allows me to both introduce the state-of-the-art in cognitive modeling and to provide a context for evaluating the SPA, which I do at the end of that chapter. In the final chapter, I suggest several ways in which the SPA may cause us to reconsider central cognitive concepts, including "representation," "dynamics," "inference," and "concept," and I discuss future challenges and directions for the SPA.

TABLE 8.1 *A Reproduction of Table 1.1 of the Core Cognitive Criteria (CCC) for Theories of Cognition.*

1. Representational structure
   a. Systematicity
   b. Compositionality
   c. Productivity (the problem of variables)
   d. The massive binding problem (the problem of two)
2. Performance concerns
   a. Syntactic generalization
   b. Robustness
   c. Adaptability
   d. Memory
3. Scientific merit
   a. Triangulation (contact with more sources of data)
   b. Compactness

8.2 ■ Core Cognitive Criteria

The CCC are "core" in the sense that they are the result of distilling the most typical sorts of criteria that researchers employ in deciding whether a system is cognitive. I believe it is also appropriate to use these criteria to evaluate cognitive architectures. That is, if a proposed architecture is likely to produce and explain systems that can satisfy these criteria, then I consider it a good architecture. Although I have said little about what it takes to be a cognitive architecture *per se*, I believe I have cast the net broadly enough to satisfy reasonable definitions from proponents of any of the three standard views (for discussions and history of the notion of a "cognitive architecture," *see* Thagard, 2011, or Anderson, 2007).

In the remainder of this section, I rely heavily on the terminology introduced in Chapter 1. Consequently, it may prove helpful to review Sections 1.2 and 1.3, to recall the distinguishing features, metaphors, and theoretical commitments of the symbolic approach, connectionism, and dynamicism. For ease of reference, the CCC themselves are reproduced in Table 8.1.

8.2.1 Representational Structure

Historically speaking, theories of cognition have been most thoroughly evaluated using criteria that focus on representational structure. Perhaps this is because representational commitments can be used to easily distinguish among dynamicism, connectionism, and the symbolic approach. As a consequence, the following discussion is closely related to ongoing debates among these views.

8.2.1.1 Systematicity

The fact that cognition is systematic–that there is a necessary connection between some thoughts and others–has been recognized in several different ways. Evans (1982), for example, identified what he called the Generality Constraint. This is the idea that if we can ascribe a property to an object, then we can ascribe that same property to other objects, and other properties to the original object (e.g., anything "left of" something can also be "right of" something). For Fodor and Pylyshyn (1988), the same observation is captured by their systematicity constraint. This constraint is, in brief, that any representational capacities we ascribe to a cognitive agent must be able to explain why our thoughts are systematic. They must explain, in other words, why if we can think the thought "dogs chase cats," then we can necessarily also think the thought that "cats chase dogs."

They argue that a syntactically and semantically combinatorial language is able to satisfy this constraint. A combinatorial language is one that, constructs sentences like "dogs chase cats" by combining atomic symbols like "dogs," "chase," and "cats." As a result, if one of those symbols is removed from the language then a whole range of sentence-level representations are systematically affected. For example, if we remove the symbol "cats," then we can think neither the thought "dogs chase cats" nor the thought that "cats chase dogs."

Although many non-classicist researchers disagree with the assumption that only a combinatorial language with atomic symbols can explain human systematicity, the observation that our thoughts are systematic is widely accepted. This is likely because systematicity is so clearly evident in human natural language. As a result, whatever representational commitments a cognitive architecture has, it must be able to describe systems that are appropriately systematic.

8.2.1.2 Compositionality

If a representation is compositional, then the meaning of that representation is determined by its structure and the meaning of its constituents (Fodor & Pylyshyn, 1988). Most formal languages respect this constraint. So, if I have a word like "fish," which takes a set of objects in the world, and I have a word like "pet," which determines a relation between an owner and an animal, then a combination like "pet fish" would indicate the "fish" objects that lie in the "pet" relation to an owner. So, the meaning of the combination is a simple function of the two constituents. At first glance, this may seem to reflect semantics of natural languages.

However, there is good evidence that concepts in natural language often do not combine compositionally. To return to the previous example, our notion of a "pet fish" does not, in fact, seem to be a simple function of our notions of a "pet" and a "fish." For example, although a prototypical pet is a dog, and a prototypical fish is a bass, a prototypical pet fish turns out to be a goldfish, which has no obvious semantic connection to either a dog or a bass (Osherson & Smith, 1981). One might want to argue that "logically speaking" the locution "pet fish" simply identifies the class of fish that are pets. However, the goal of cognitive systems research is to understand the best examples of cognitive systems that we have, not to pre-specify the way we think such systems *should* work.

Specifying semantic constraints on cognitive systems in advance ignores the often contingent nature of natural language semantics–a semantics that forms the basis of the flexibility of our behavior. One simple example of this contingency can be found in the dictionary definitions of "ravel" and "unravel." They turn out to be the same. This, of course, is surprising because the prefix "un-" usually reverses the meaning of a root. In short, the mapping between our combinatorial syntactic languages and semantics seems to be much more complex and contingent than is allowed for by simple compositionality. The way in which semantic information is combined, or even if it is combined, in the face of different syntactic structures remains largely mysterious. As a consequence, most cognitive systems researchers reject natural language as being compositional, in anything like the way a formal language is.

Nevertheless, understanding how words are composed–semantically as well as syntactically–to create complex representations is crucial to understanding cognitive systems. It is just that simple compositionality, as originally proposed by Fodor and Pylyshyn, will not always do. As a result, explaining observed compositionality effects remains an important criteria for a cognitive theory to address. Real cognitive systems sometimes draw on a wealth of experience about how the world works when interpreting compositional structures. Although we have much to learn about human semantic processing, what we do know suggests that such processing can involve most of cortex (Aziz-Zadeh & Damasio, 2008), and can rely on information about an object's typical real-world spatial, temporal, and relational properties (Barsalou, 2009).

I want to be clear about what I am arguing here: my contention is simply that ideal compositionality is not always satisfied by cognitive systems. There are, of course, many cases where the semantics of a locution (e.g., "brown cow") is best explained by the simple, idealized notion of compositionality suggested by Fodor and Pylyshyn. The problem is that the idealization misses a lot of data (such as the examples provided). Just how many examples are missed can be debated, but in the end an understanding of compositionality that misses fewer should clearly be preferred.

Thus, determining how well an architecture can define systems that meet the (non-idealized) compositionality criteria will be closely related to determining how the architecture describes the processing of novel, complex representations. We should be impressed by an architecture that defines systems that can provide appropriate interpretations that draw on sophisticated models of how the world works–but only when necessary. That same theory needs to capture the simple cases as well. No doubt, this is one of the more challenging, yet important, criteria for any architecture to address.

In sum, compositionality is clearly important, but it is not only the simple compositionality attributed to formal languages that we must capture with our cognitive models. Rather, the subtle, complex compositionality displayed by real cognitive systems must be accounted for. The complexity of compositionality thus comes in degrees, none of which should be idealized away. An architecture that provides a unified description spanning the observed degrees of compositional complexity will satisfy this criteria well.

8.2.1.3 Productivity

Ideally, productivity is the ability of a language to generate an infinite variety of valid sentences with a finite set of words and a finite grammar. In many ways, it was precisely this property of language that Chomsky (1959) relied on in his highly effective critique of behaviorism–a critique largely seen as a major turning point in the cognitive revolution. Fodor and Pylyshyn (1988) have identified productivity as a third central feature of cognition. Similarly, Jackendoff (2002) has identified the third challenge for cognitive theories as one of explaining the "problem of variables." This is the problem of having grammatical templates in which any of a variety of words can play a valid role. Typically, these variables are constrained to certain classes of words (e.g., noun phrases, verbs, etc.). Despite such constraints, there remains a huge, possibly infinite, combination of possibly valid fillers in some such templates. As a result, Jackendoff has described the existence of these variables as giving rise to the observed productivity of natural language.

The claim that productivity is central for characterizing cognition has received widespread support. Productivity is, however, clearly an idealization of the performance of any actual cognitive system. No real cognitive system can truly realize an infinite variety of sentences. Typically, the productivity of the formalism used to describe a grammar outstrips the actual productivity of a real human cognizer. For example, there are sentences that are "grammatically" valid that are neither produced nor understood by speakers of natural language. One way of generating examples of such sentences is to use a grammatical construction called "recursive center embedding." An easily understood sentence with one such embedding is "The dog the girl fed chased the boy." A difficult to understand example is "The dog the girl the boy bit fed chased the boy," which has two embeddings. As we continue to increase the number of embeddings, all meaning becomes lost to natural language speakers. The reason, of course, is that real systems have finite processing resources.

The two most obvious resource limitations are time and memory. These are, in fact, intimately linked. As we try to cram more things into memory, we run out of space, and what was previously in memory is forgotten more quickly. When we are trying to process a complex sentence in "real-time," adding items (such as embedded clauses) to memory eventually causes a failure in recall because we run out of resources. As a result, we are not able to make sense of the input. In some sense, this is merely the observation that cognitive systems are limited in their capacity to comprehend valid input.

In short, real-world cognitive systems have a *limited* form of productivity: albeit one that allows for a high degree of representational flexibility and supports the generation and comprehension of fairly complex syntactic structures. Although cognitive systems are representationally powerful, perfect productivity is not a reasonable expectation and hence is not a true constraint on cognitive architectures. It is, as with compositionality, a matter of degree. Natural measures of the degree of productivity of a system include the depth, length, and number of variables of a structure that can be successfully manipulated. Such measures both capture the sense that there is a limit on productivity and acknowledge that a great deal of representational flexibility is provided by cognitive representations. A good cognitive architecture should provide resources that match both the power and limits of real productivity.

8.2.1.4 The Massive Binding Problem

Jackendoff (2002) has identified his first challenge for cognitive modeling as "the massiveness of the binding problem." He has argued that the binding problem, well-known from the literature on perceptual binding, is much more severe in the case of cognitive representations. He has suggested that this is so because, when constructing a complex syntactic structure, many "parts" must be combined to produce the "whole." Jackendoff provides simple examples that demand the real-time encoding of phonological, syntactic, and semantic structure. Not only must all of these separate structures be encoded, but Jackendoff has argued that each of these structural elements must be inter-related, compounding the amount of binding that must be done. It is clear to Jackendoff that traditional accounts of perceptual binding (e.g., synchrony) have not been designed for this degree of structural complexity.

Given that there is a massive amount of binding in cognitive systems, it is not surprising that in some circumstances the same item may be bound more than once. Jackendoff has identified this as a separate challenge for connectionist implementations of cognition: "the problem of two." Jackendoff's example of a representation that highlights the problem of two is "The little star's beside a big star" (p. 5). Identifying this as a problem is inspired by the nature of past connectionist suggestions for how to represent language. Such suggestions typically ascribe the representation of a particular word, or concept, to a group of neurons. Thus, reasoned Jackendoff, if the same concept appears twice in a single sentence, then it will not be possible to distinguish between those two occurrences, because both will be trying to activate the same group of neurons.

Although the problem of two poses certain challenges to understanding cognition as implemented in neurons, it is not obviously separate from the more general considerations of systematicity and binding. After all, if an approach can bind a representation for "star" to two different sentential roles, then it should solve the problem of two as well. Consequently, I have placed the problem of two under this criterion, because I take it to relate to "binding" more generally.

To summarize, Jackendoff sees the scalability of proposed solutions to binding as being essential to their plausibility. The degree to which a proposed architecture is cognitive is directly related to how well it can generate appropriately large structures, in the same length of time and with the same kind of complexity as observed in real cognitive systems. So, we can see the "massiveness of binding" challenge as emphasizing the practical difficulties involved in constructing a suitably productive system. This problem thus straddles the distinction I have made between representational concerns and performance concerns. Regardless of where it is placed on the list of cognitive criteria, it highlights that a good cognitive architecture needs to identify a binding mechanism that scales well.

8.2.2 Performance Concerns

Criteria related to representational structure focus on the theoretical commitments that underwrite cognitive architectures. In contrast, criteria related to performance concerns are directed more towards the implementation of cognitive systems.

8.2.2.1 Syntactic Generalization

Like the massive binding problem, syntactic generalization straddles representational and performance concerns. In short, this criterion identifies one of the most crucial kinds of performance we should expect from a system that satisfies the representational criteria. Namely, the ability to generalize based on the syntactic structure of representations. We have already encountered this notion in earlier example models from Sections 4.7, 6.6, and 7.2.

Let me consider another example. Suppose you are told that you must figure out what "triggle" means given the following information:

- "The square being red triggles the square being big" implies that the square is big when the square is not red; also
- "The dog being fuzzy triggles the dog being loud" implies that the dog is loud when the dog is not fuzzy.

Now I ask the question: What does "the chair being soft triggles the chair being flat" imply? If you figured out that this implies that the chair is flat when it is not soft, you have just performed a difficult case of syntactic generalization. That is, based on the syntactic structure presented in the two examples, you determined what the appropriate transformation to that structure is in order to answer the question. Evidently, the problem of rapid variable creation (*see* Section 7.3.7) is just a simpler form of syntactic generalization. Although the emphasis in the case of rapid variable creation is on speed, the problem being solved is essentially one of identifying syntactic structure and using it in a novel circumstance.

There are, in fact, much simpler examples of syntactic generalization, such as determining when contractions are appropriate (Maratsos & Kuczaj, 1976). However, the example I provided above shows the true power of syntactic generalization. That is, it demonstrates how reasoning can be sensitive to the structure, not the meaning, of the terms in a sentence. Taken to an extreme, this is the basis of modern logic, and it is also why computers can be used to perform reasonably sophisticated reasoning.

It should not be surprising, then, that if we adopt the representational commitments of computers, as the symbolic approach does, syntactic generalization becomes straightforward. However, if we opt for different representational commitments, as do both dynamicism and connectionism, then we must still do work to explain the kinds of syntactic generalization observed in human behavior. In short, syntactic generalization acts as a test that your representational commitments do not lead you astray in your attempt to explain cognitive performance.

I should note here, as I did for several of the criteria in the previous section, that human behavior does not always conform to ideal syntactic generalization. As described in Section 6.6, there are well-known effects of content on the reasoning strategies employed by people (Wason, 1966). In the preceding example, this would mean that people might reason differently in the "dog" case than in the "chair" case, despite the fact that they are syntactically identical. So again, we must be careful to evaluate theories with respect to this criterion insofar as it captures human performance, not with respect to the idealized characterization of the criterion. In short, if people syntactically generalize in a certain circumstance, then a good cognitive architecture must be able to capture that behavior. If people do not syntactically generalize, then the architecture must be able to capture that as well.

8.2.2.2 Robustness

Syntactic generalization deals with performance on a specific, although ubiquitous sort of cognitive task. Robustness, in contrast, deals with changes of performance across many cognitive tasks. "Robustness" is a notion that most naturally finds a home in engineering. This is because when we build something, we want to design it such that it will continue to perform the desired function regardless of unforeseen circumstances or problems. Such problems might be changes in the nature of the material we used to construct the object (e.g., the fatiguing of metal). Or, these problems might arise from the fact that the environment in which the object is built is somewhat unpredictable. In either case, a robust system is one that can continue to function properly despite these kinds of problems and without additional interference from the engineer.

In some respects, the hardware of modern digital computers is extremely robust. It has been specifically built such that each of the millions of transistors on a chip continues to perform properly after many millions of uses. This robustness stems partially from the fact that the voltage states in a transistor are interpreted to be only on or off, despite the fact that the voltage varies ± 5 V. As a result, if an aging transistor no longer reaches 5 V, but is above zero, it will usually be considered to be acting properly. The cost of this robustness is that these machines use a lot of power: The human brain consumes about 20 W of power; much less impressive digital computers, like the one sitting on your desk top, use tens or hundreds of times more power.

Nevertheless, computers are well designed to resist problems of the first kind–degradation of the system itself. In fact, these same design features help resist some of the second kind of problem–environmental variability. Specifically, electromagnetic fluctuation, or heat fluctuations, can be partially accommodated in virtue of the interpretation of the transistor states. However, "interesting" environmental variability is not accounted for by the hardware design at all. That is, the hardware cannot run a poorly written program. It cannot use past experience to place a "reasonable" interpretation on noisy or ambiguous input. This limitation is not solely the fault of the hardware designers. It is as much a consequence of how software languages have been designed and mapped onto the hardware states.

All of this is relevant to cognitive science because the symbolic approach has adopted some of the central design assumptions underlying the construction and programming of digital computers. But, it seems that biology has made different kinds of trade-offs than human engineers in building functional, flexible devices. Brains use components that, unlike transistors, are highly unreliable (often completely breaking down) but that use very little power. Consequently, the kinds of robustness we see in biological systems is not like that of computers. So, when connectionists adopted a more brain-like structure in their theoretical characterization of cognition, much was made of the improved robustness of their models over their symbolic competitors. Indeed, robustness was not of obvious concern to a symbolic proponent, because the hardware and software on which their simulations ran (and whose theoretical assumptions they had taken on board) essentially hid such concerns from view.

However, robustness concerns could not be ignored for long when connectionists began explaining certain kinds of phenomena–such as the "graceful degradation" of function after damage to the brain–that seemed beyond the reach of the symbolic approach (Plaut & Shallice, 1994). Additionally, adopting more brain-like architectures made it possible for connectionists to explain the performance of cognitive systems on tasks such as pattern completion, recognition of noisy and ambiguous input, and other kinds of statistical (as opposed to logical) inference problems.

But, connectionist models continued to make "un-biological" assumptions about the nature of the implementation hardware. Most obvious, perhaps, was that connectionist nodes were still largely noise free. Nevertheless, early connectionist models made the important point that matching the observed robustness of cognitive systems is at least partly tied to their specific implementation. More generally, these models established that robustness was a relevant criterion for characterizing cognitive architectures.

In sum, as a criterion for a good cognitive architecture, robustness demands that an architecture supports models that continue to function appropriately given a variety of sources of variability, such as noisy or damaged component parts, imprecision in input, and unpredictability in the environment.

8.2.2.3 Adaptability

Adaptability has long been one of the most admired features of cognitive systems. It is, in many ways, what sets apart systems we consider "cognitive" from those that we do not. Simply put, adaptability is exhibited by a system when it can update its future performance on the basis of past experience. This, of course, sounds a lot like learning, and indeed the two terms can often be used interchangeably. However, as discussed earlier, in cognitive science (especially connectionism), learning often refers specifically to the changing of parameters in a model in response to input. Adaptability, however, goes beyond this definition. There are, for example, nonlinear dynamical systems that can exhibit adaptable behavior without changing any of the parameters in the system (the fluid intelligence model in Section 4.7 is one example). Indeed, many dynamicist models rely on precisely this property.

Notice also that adaptability can be provided through sophisticated representational structures. A prototypical symbolic model can respond to input it has never seen before based solely on the syntactic structure of that input, often producing reasonable results. As long as the rules it has been programmed with, or has learned, can use structure to generalize, it will be able to exhibit adaptability: one example, of course, is syntactic generalization.

So, adaptability has been identified and explored by all past approaches to cognition. Although the kinds of adaptability considered are often quite different (chosen to demonstrate the strengths of the approach), all of these types of adaptability are exhibited by real cognitive systems to varying degrees. That is, exploitation of nonlinear dynamics, tuning of the system parameters to reflect statistical regularities, and generalization over syntactic structure are all evident in cognitive systems. Consequently, a theory that accounts for adaptability in as many of its various guises as possible will do well on this criterion.

8.2.2.4 Memory

For a system to learn from its experience, it must have some kind of memory. However, it is by no means obvious what the nature of memory is, either behaviorally or mechanistically. Consequently, it is unclear what specific components an architecture needs, or how such components might internally function, to explain the varieties of memory.

Past approaches have treated the issue of memory in very different ways–as symbolic databases, as connection weights between nodes, as slowly varying model parameters, and so forth. Ultimately, the adequacy of a cognitive architecture on this criterion is going to be determined by its ability to address the wide variety of data that relate to memory. There are a vast number of possible phenomena to account for when considering memory, but for the sake of brevity I focus on just two aspects of long-term and working memory that seem especially crucial for cognitive behaviors: manipulation of contents and capacity. In particular, it is useful to contrast the high manipulability but relatively limited capacity of working memory with the relatively static nature but enormous capacity of long-term memory.

Cognition often necessitates the manipulation of complex, compositional structures. Such manipulation can demand significant resources from both long-term and working memory. This is because the structures themselves must be stored in long-term memory, so as to be available for application to a cognitive task. In addition, when such structures are employed, intermediate steps, final results, and the original structure itself may need to be manipulated in working memory.

The capacity limitations on human working memory are quite severe: it seems to be limited to only about four items (Cowan, 2001). However, those items are immediately available for guiding goal selection, for being transformed, and for other kinds of occurrent processing (Miyake & Shah, 1999). Thus, it has often been suggested that working memory is crucial for allowing cognitive systems to bring their general expertise to bear on the current problem at hand (Baddeley, 2003). This means that any cognitive architecture must directly address the relation between long-term and working memory.

The nature of long-term memory itself remains largely mysterious. Most researchers assume it has a nearly limitless capacity, although explicit estimates of capacity range from 10^9 to 10^{20} bits (Landauer, 1986). So, although not limitless, the capacity is probably extremely large. Significant amounts of work on long-term memory have focused on the circumstances and mechanisms underlying consolidation (i.e., the process of preserving a memory after it has been initially acquired). Although various brain areas have been implicated in this process (including hippocampus, entorhinal cortex, and temporal cortices), there is little solid evidence of how specific cellular processes contribute to long-term memory storage.[1] Nevertheless, the functional role of long-term memory has remained central to our understanding of cognitive systems. Consequently, any cognitive theory must specify its role and, more importantly, its interaction with other parts of the cognitive system.

In this woefully incomplete consideration of memory, I have suggested that a good cognitive architecture must account for the functions of, and relationships between, working and long-term memory. These considerations are among the many reasons why researchers typically include "memory" in the list of important capacities to consider when building cognitive models. As I suggested at the beginning of this section, the real test of architectures against this criteria will be the breadth and depth of their ability to explain memory-related empirical results.

8.2.2.5 Scalability

As I noted previously in Section 1.3, all behavioral scientists ultimately have a large and complex system as their target of explanation. As a result, the simple cognitive models typically proposed in contemporary research must be scaled up to truly become the kinds of cognitive explanations we would like. The reason that scalability is a criterion for cognitive architectures is that there are many difficulties hidden in the innocuous sounding "scaled up."

It is notoriously difficult to predict the consequences of scaling. This is tragically illustrated by the case of Tusko the elephant. In 1962, Louis West and his colleagues decided to determine the effects of LSD on a male elephant to determine whether it would explain the phenomenon of "musth," a condition in

[1] Long-term potentiation (LTP) is often taken to be a candidate, but experiments explicitly demonstrating its role in long-term memory consolidation outside of hippocampus are lacking.

which elephants become uncontrollably violent (West et al., 1962). Because no one had injected an elephant with LSD before, they were faced with the problem of determining an appropriately sized dosage. The experimenters decided to scale the dosage to Tusko by body weight, based on the known effects of LSD on cats and monkeys.

Unfortunately, 5 minutes after the injection of LSD the elephant trumpeted, collapsed, and went into a state resembling a seizure. He died about an hour and a half later. Evidently, the chosen means of scaling the dosage had the effects of administering a massive overdose that killed the 7000 pound animal. If the dose had been scaled based on a human baseline, it would have been 30 times lower. If it had been scaled based on metabolic rate, it would have been about 60 times lower. If it had been scaled based on brain size, it would have been about 300 times lower. Clearly, the dimension along which you characterize an expected scaling is crucial to determining expected effects. The lesson here for cognitive theories is that scaling can seldom be fully characterized as "more of the same," because we may not know which "same" is most relevant until we actually scale.

A second reason to take scaling seriously, which is more specific to functional architectures, comes from considering the complexity of potential interactions in large systems. As Bechtel and Richardson have forcefully argued, decomposition and simplification is an essential strategy for explaining complex systems (Bechtel & Richardson, 1993). It is not surprising, then, that most cognitive theorists take exactly this kind of approach. They attempt to identify a few basic functional components or principles of operation that are hypothesized to characterize "full-blown" cognition. Because we are not in a position to build models of "full-blown" cognition, the principles are typically applied in a limited manner. Limitations are often imposed by abstracting away parts of the system, highly simplifying the task of interest, or both. When subsequently faced with a different task to explain, a new model employing the same principles is constructed and new comparisons are made. Sometimes, different models employing the same architecture will use few or no overlapping components

Such a strategy is problematic because it skirts the main challenge of building complex systems. As any engineer of a complex real-world system will tell you, many of the challenges involved in building such systems come from the *interactions* of the component parts. As has been well established by disciplines such as chaos theory and dynamical systems theory, the interactions of even simple components can give rise to complex overall behavior. To ensure that the hypothesized cognitive principles and components can truly underwrite a general purpose cognitive architecture, simple, task-specific models must be integrated with one another. This integration will result in scaling up the overall system, which is essential for simultaneously accounting for a wide variety of tasks.

Because scalability is difficult to predict from simpler examples, the weakest form of scalability is *scalability in principle*. Scalability in principle amounts to demonstrating that there is nothing in your assumed characterization of cognition that makes it unlikely that you could scale the theory to account for full-blown cognition. This form of scalability is weak because it could well be the case that a crucial dimension for predicting scalability has been missed in such an analysis.

A far more significant form of scalability is *scalability in practice*. That is, actually building large models of large portions of the brain that are able to account for behaviors on many tasks, without intervention on the part of the modeler. Such scaling can be extremely demanding, both from a design standpoint and computationally. However, this simply means that being able to construct such models makes it that much more convincing that the underlying architecture is appropriate for characterizing real cognitive systems.

8.2.3 Scientific Merit

The final set of criteria I consider are those that relate to good scientific theories, regardless of their domain. Although most introductions to the philosophy of science discuss anywhere from five to eight different properties that make for a good theory (Quine & Ullian, 1970; McKay, 1999), here I consider only two: triangulation and compactness. I have picked these because they have proved to be the most challenging for cognitive theories (Taatgen & Anderson, 2010; Anderson, 2007). And, given the very large number of tunable parameters typical of current cognitive models, these two criteria are especially critical (Roberts & Pashler, 2000).

8.2.3.1 Triangulation (Contact With More Sources of Data)

The behavioral sciences are replete with a variety of methods. Some methods characterize the opening and closing of a single channel on a neural membrane, whereas others characterize activity of an entire functioning brain. In some sense, all of these methods are telling us about the functioning of the same underlying physical structure–the brain. So, if we take a cognitive theory to be a theory of brain function, then constraints from all such methods should be relatable to our theory.

Unfortunately, this ideal seems to be largely unrealized by contemporary cognitive theories. Perhaps this is because cognitive science researchers have generally been placed into traditional academic disciplines like psychology, neuroscience, computer science, and so on. As a result, the conventional methods of one discipline become dominant for a given researcher, and so his or her work becomes framed with respect to that specific discipline. In some sense, the identification of "cognitive science" as a multidisciplinary but unified enterprise is an attempt to overcome such a tendency. Nevertheless, cognitive theories seem to often have a "home" in only one or perhaps two of the subdisciplines of cognitive science.

A truly unified theory of cognitive function, in contrast, should have clear relationships to the many disciplines of relevance for understanding brain function. Most obviously this would be made evident by such a theory being able to predict the results of experiments in any relevant discipline. In short, the more kinds of data that can be used to constrain and test a theory, the better. We should, after all, be impressed by a model that not only predicts a behavioral result but also tells us which neurons are active during that task, what kind of blood flow we should expect in the relevant brain areas, what sort of electrical activity we will record during the task, how we might disrupt or improve performance on the task by manipulating neurotransmitters within the system, and so on. Although the interdisciplinary nature of the behavioral sciences can prove to be one of the most daunting aspects of performing research in the area, it also provides one of the most stringent tests of the quality of a purported theory of cognition.

We must be careful when applying this criterion to not simply count the number of successes of a theory, as this would bias us toward existing theories. Rather, we need to consider the spread of explanatory successes provided by a theory. Triangulation is about the breadth of a theory and its ability to account for many of the different kinds of data relevant to our understanding of cognitive systems. Thus a theory does better on this criterion if it addresses more kinds (not strictly a greater volume) of data.

8.2.3.2 Compactness

Although I have called this criteria "compactness," it often goes by the name of "simplicity." In fact, I think the former better indicates what is important about theories that are deemed good. It is not the fact that they are "easy" or "simple" theories that makes them good–they may indeed be quite difficult to understand. Rather, it is that they can be stated comprehensively and in a highly succinct manner that makes them good. Often, this kind of succinctness is possible because the theory employs mathematics, a language whose terms and relationships are well defined, and that can succinctly express complex structure. Although mathematics itself does not supply theories of the world (because the terms in mathematical expressions must be mapped onto the world), it *is* good for clearly expressing the relationships between theoretical terms.

One reason compact expressions of a theory are highly valued is because they make it difficult to introduce unnoticed or arbitrary changes in a theory when employing it in different contexts. If a cognitive theory changes when moving from simple cognitive tasks to more complex ones, there is little sense to be made of it being a single, compact theory. For example, if we must introduce a parameter in our theory for no reason other than to fit new data, then the additional complexity introduced by that parameter is poorly motivated from a theoretical standpoint. Similarly, if we must change our mapping between our theory and the world depending on the circumstances (e.g., a "time step" in a model is not always the same amount of real time), then the statement of our theory should include a description of how to determine what the appropriate mapping is making the theory less compact. In short, any unprincipled "fitting" of a model to data pays a price in compactness. Consequently, a theory will do well on this criterion if it can be stated succinctly, and if that statement remains consistent across all of its applications.

Notably, the two criteria related to scientific merit combine to provide opposing constraints. The result is a demand for theories that explain a wealth of data, but that do so compactly. This, not surprisingly, is an ideal that is typical for scientific theories.

8.3 ■ Conclusion

As I have said before, this list of criteria is no doubt incomplete. However, I suspect that any theory that had nothing to say about one of these criteria would be deemed worse than one that did, all else being equal. I have also attempted, no doubt unsuccessfully, to identify these criteria in a non-theory-laden way. For example, even if we don't think there are representations, we still ought to think that the system has compositional behavior, because we have operationalized such behavior, and hence can measure it. In general, most of these criteria relate directly to measurable properties of cognitive systems, so hopefully satisfying them comes with reasonably few pre-theoretic constraints.

Recall that these CCC are largely inspired by the questions behavioral scientists have been asking about cognition over the last 50 years (*see* Section 1.3). Undoubtedly, such questions have been driving decisions about the kinds of data to collect, and to do so they must be far more specific than the CCC I have identified. One result of this specificity seems to be that much of our knowledge about cognitive systems is highly empirical–we know lots about how people perform on various kinds of memory tasks, but we do not know why. That is, we can *describe* regularities in behavior, but we have not characterized the underlying principles or mechanisms well enough to really understand the genesis of that behavior. As a result we do not really understand how the system would function under a much wider variety of circumstances than those we have explicitly tested. Using the CCC to evaluate cognitive approaches can help emphasize the importance of also striving for systematic, theoretical characterizations of cognition, to complement our empirical understanding.

Unfortunately, as I argue in the next chapter, current approaches do not generally do very well on more than a few of these criteria (*see* Table 9.1 for a summary). Pessimistically, we might think that the last 50 years of research in the behavioral sciences has taught us that a satisfactory cognitive theory is missing. More optimistically, we might think that we are in a better position to evaluate cognitive theories than we were 50 years ago. And, there are clearly many theories to evaluate.

8.4 ■ Nengo Bonus: How to Build a Brain–a Practical Guide

In previous tutorials, I have provided examples of specific models that can be built with Nengo. However, these do not directly address a much more difficult question: How do you construct a brand new model? As many modelers will tell you, it takes a lot of practice to become good at creating new models. In this tutorial, I attempt to provide some tips and suggestions on this process by working through the development of a novel model of perceptual decision-making.

The collective experience of modelers in my lab has led us to identify three main steps to consider when building a model:

1. Identify the inputs and outputs–This helps determine the boundaries of the model. Often, they will be neural populations that represent the driving signals and response values for the model.
2. Identify the required internal variables–Collectively, these determine the state space of the model. They determine the variables that functions can be defined over. Often, one or more internal neural populations will represent them in the model.
3. Identify the functions to be realized–These functions determine the connections between neural populations, transforming the representations in those populations as needed. These give the model its behavior.

I recommend initially going through these steps on paper before starting to actually build the model in Nengo. This can take the form of a box and arrow diagram, where boxes are variables and arrows are functions, even roughly specified. Of course, as with all design processes, multiple passes may be needed before settling on a final model. Further, at each stage, the neural anatomy and physiology should be considered and will offer particular constraints on reasonable variables and the topology of the network.

A Model of Perceptual Decision-Making

The best way to understand these steps is to see them in action. So, for this tutorial, I have chosen to construct a simple model of perceptual decision-making. Importantly, I am only concerned with creating a model of this task; I am not concerned with establishing how good (or bad) this model is. As a result, the following discussion should only be taken as an example of the process involved in the construction of new models.

A common experiment used to probe perceptual decision-making is one in which subjects are shown a set of semi-coherent, randomly moving dots (Shadlen & Newsome, 1996). Subjects, usually monkeys or humans, must determine the direction that has the most coherent motion and indicate that direction through a key press or eye movement. Shadlen and Newsome (2001) have suggested that the gathering of evidence regarding direction of motion of such stimuli occurs in the lateral intraparietal (LIP) area in the monkey. Consequently, the model I develop will be of that area. Neurons in LIP have physiological properties typical of much of cortex (it is usually assumed that responses from pyramidal cells provide most of the data), show sensitivity to movement direction, and receive projections from motion area MT, which will be an input to the model (Colby et al., 1993; Roitman & Shadlen, 2002).

This preliminary description allows us to rough characterize the model: we need it to take input indicating the current observed direction of motion, track that information over time, and then at some point (when sufficient information has been gathered) produce an output indicating the estimated direction of motion. In fact, this approach is common in existing mathematical models of decision-making (*see* Heekeren et al., 2008, for an overview). These tend to rely on an "accumulator" to gather information regarding a perceptual stimulus such as motion (e.g., Usher & McClelland, 2001). An accumulator is what I have been referring to as an integrator (*see* Section 2.2.3). In such models, an accumulator is used to add up input information over time, until the resulting value reaches some threshold level, at which point the system produces an output to trigger the relevant response.

Step 1: Inputs and Outputs

Given this high-level functional and anatomical outline of the model, we can begin to move to specifics. Specifying inputs and outputs helps define the boundaries of the model, determining the brain areas and processing to be included and excluded. In our case, specifying inputs and outputs will answer questions such as: How much of the vision system will be part of the model? Should the entire motor system used to generate a response also be included? I generally recommend starting with as small and manageable a scope as possible, as other components (such as many of the models described in this book) can often be added later.

For this model, using the motion area MT as input will keep the model simple. So, the model will not consider how the visual system combines temporally changing retinal information to generate an estimate of the overall motion direction at any instant in time. Rather, a population of neurons in MT will project into the model (in LIP), providing instantaneous motion information.

To fully characterize the model's input, two constraints must be satisfied. First, the input must be specified in terms of a vector. Second, we must specify how that vector is represented by neurons. That is, we must determine the distribution of encoders and other neural properties. In this case, the input is the direction and speed of motion, which is natural to represent as a two-dimensional vector, $\mathbf{u}(t)$. Further, I will assume that the MT neurons represent this input as is common in cortex: each neuron has a preferred direction vector (its encoder), chosen from an even distribution over the input space $\mathbf{u}$, and there is a wide variety of background firing rates, sensitivities, and so on (*see* Chapter 2).

To create this population of input neurons, we can use either the standard drag-and-drop Nengo interface or the scripting interface. In either case, we can use default neuron properties for the most part. The number of neurons in this population is somewhat arbitrary: more neurons leads to more accuracy (*see* Section 2.2.1). A general rule of thumb is that 50 neurons per dimension gives fairly good accuracy. In this model, we will include a fairly large degree of random noise both to emphasize the robustness of the model and to produce spike trains that reflect a realistic degree of neural variability.

Below is a script implementing the model as described to this point. It includes an input population that will have a slider that can be used to set the MT value (i.e., direction and magnitude of stimulus motion) while the model is running:

```
import nef
net=nef.Network('2D Decision Integrator')
input=net.make_input('input', [0,0])
MT=net.make('MT', neurons=100,
   dimensions=2, noise=10)
net.connect(input, MT, pstc=0.01)
net.add_to_nengo()
```

We must now specify the output. In particular, we must decide what form this output will take. In past work, it has typically been proposed that there is a set number of integrators: the same number as there are possible directions of motion that the subject must distinguish (*see*, e.g., Bogacz et al., 2007; Krajbich & Rangel, 2011). Each of these integrators has its own separate output to the rest of the brain. This is somewhat problematic, as it suggests that the brain adds and subtracts accumulators depending on the task (a process not explained by these models) and that or the brain has some maximum number of separate accumulators, most of which are not employed for simpler tasks (hence wasting potentially valuable resources).

However, the use of vector-based representations suggests an alternative approach. Rather than having integrators for each possible direction of motion, a single two-dimensional integrator can be used. This integrator would be used no matter what the task demands are, as it effectively integrates in every direction simultaneously. This is not only simpler from a theoretical standpoint, but also more neurally efficient, as all the neurons in the integrator contribute to any such accumulation task to the degree they are relevant.

Adopting this approach, the output is a single ensemble representing a two-dimensional vector. More specifically, if the decision threshold has not yet been reached (i.e., if the length of the integrated vector is less than some threshold value), then the output should be zero. When the decision threshold is reached, the output should be the integrated value itself.

As with the input, after deciding what is being represented by the output (a two-dimensional vector), we must also specify how it will be represented by neurons. We could adopt the same approach as for the input: an evenly distributed representation. However, in this situation our specification of the output suggests a particular variation on the typical neural representation. Notice that the output neurons never need to represent a small vector (other than zero), because if the integrated value is small (has a length less than some threshold), then the output is supposed to be zero. We can take this into account when specifying the neural representation by adjusting the background firing rates of the neurons. Noticing this, and knowing about the potential solution, is an example of how experience plays an important role in designing such models.

Recall from Figure 2.8 that we normally assume that neural tuning curves are evenly distributed, giving good representational accuracy across their full range. However, if a model requires neurons that are better at representing a specific range of values (e.g., above some threshold), then the intercepts of those tuning curves can be adjusted. In this case, if all of the tuning curve intercepts are between 0.3 and 1 (rather than the default range of -1 to 1), then values smaller than 0.3 will result in no neural firing. If no output neurons are firing, then the output value being represented is zero. Notably, this is a situation where the implementational details strongly affect the behavior of the model. Indeed, our model makes a strong neurological prediction as to the firing properties of these output neurons: they should have close to zero background firing, and only a small subset of those neurons will start firing as the decision is made.

Below is a script line that adds output neurons to the model using the same default parameters as the MT input, except for the intercepts. This line explicitly sets the output threshold to 0.3.

```
output=net.make('LIP output', neurons=100,
   dimensions=2, noise=10, intercept=(0.3,1.0))
```

Step 2: Internal Variables

The next step is to define the neural populations needed to represent states that are internal to the model. These are the states, $x(t)$, that we will use to define behaviorally relevant functions in the next step. Given the simplicity of this model, the only internal component that is needed is a two-dimensional integrator, presumed to reside in LIP. Our integrator is defined exactly as in Section 2.2.3, and so the default neuron parameters are again used, providing an even representation of the state space (i.e., a unit circle). Because these neurons must store information over time, we will need many of them to have a stable representation. For this reason, we can increase the number of neurons from 100 to 200 (doubling the accuracy). The following script line defines an appropriate neural population:

```
LIP=net.make('LIP', neurons=200,
   dimensions=2, noise=10)
```

The "noise" parameter determines how much current noise is injected while the simulation runs. This helps capture the observed spike time variability of neurons in this area.

Step 3: Functions

We are now in a position to define connections between our specified neural populations to realize the relevant functions for our model. In this case, most connections are determined by the fact that the core of the model is an integrator. As discussed in Section 2.2.3, an integrator requires two connections: here, the input from MT to LIP and the feedback connection from LIP to LIP. As mentioned in that section, there is a formula (equations B.10 and B.11) for determining the functions to be computed by those connections, given the particular dynamical system to be implemented (*see* Appendix B.3).

To exploit that formula, the mathematical function must be expressed in the standard form:[2]

$$\dot{\mathbf{x}}(t) = \mathbf{A}\mathbf{x}(t) + \mathbf{B}\mathbf{u}(t)$$

where, in this case, $\mathbf{x}(t)$ is the state variable stored in the integrator and $\mathbf{u}(t)$ is the velocity input from MT. To implement an integrator, we want $\mathbf{x}(t)$ to stay constant when there is no input at all (so $\mathbf{A} = 0$), and otherwise add the input $\mathbf{u}(t)$ to the current state value (so $\mathbf{B} = \mathbf{I}$; *see* Section 2.2.3).

[2] This is for the simple linear case; *see* Appendix B.3 for the nonlinear case.

According to our formula, this means that this input connection should compute the function

$$f(\mathbf{u}(t)) = \tau\mathbf{B}\mathbf{u}(t)$$
$$= \tau\mathbf{u}(t),$$

and the feedback connection should compute the function

$$f(\mathbf{x}(t)) = (\tau\mathbf{A} + \mathbf{I})\mathbf{x}(t)$$
$$= \mathbf{x}(t).$$

For both functions, τ is the post-synpatic time constant for the neurotransmitter used in the feedback connection. For stability reasons, these feedback connections are generally modeled as NMDA connections with a time constant of 100 ms (Seung, 1996). This is much longer than AMPA connections (10 ms) that I assume elsewhere in the model. A longer time constant is well suited for stabilizing information over time, whereas shorter time constants are well suited for values that can change rapidly.

Because the functions we have identified are simple (they are just linear scalings of their input), they can be easily implemented in the scripting system as follows:

```
net.connect(MT, LIP, weight=0.1, pstc=0.1)
net.connect(LIP, LIP, pstc=0.1)
```

Finally, a connection is needed between LIP and its output. As discussed previously, this function should be $\mathbf{x}(t)$ if the length of the $\mathbf{x}(t)$ vector is greater than a threshold (set to 0.3), and otherwise it should be zero. Mathematically, this can be written as

$$\mathbf{r}(t) = \begin{cases} \mathbf{x}(t) & \text{if } \|\mathbf{x}(t)\| > \theta \\ 0 & \text{otherwise} \end{cases} \tag{8.1}$$

However, we do not need to compute this function directly in the connection weights alone because of our previous specification of the output. Recall that when we specified the output, we adjusted the intercepts of the tuning curves such that any value with a magnitude less that 0.3 would cause no firing in the population, leading to a represented value of zero. In other words, the neural properties themselves are performing the desired calculation. This means that the connection implementing equation 8.1 can simply be written as $f(\mathbf{x}) = \mathbf{x}$. If we assume this connection uses AMPA receptors, as above, then it can be created with the following line in the script:

```
net.connect(LIP, output, pstc=0.01)
```

Model Behavior

Here is the complete model script as we have defined it:

```
1  import nef
2
3  net=nef.Network('2D Decision Integrator')
4  input=net.make_input('input', [0,0])
5
6  MT=net.make('MT', 100, 2, noise=10)
7  LIP=net.make('LIP', 200, 2, noise=10)
8  output=net.make('output', 100, 2,
9      intercept=(0.3,1), noise=10)
10
11 net.connect(input, MT, pstc=0.01)
12 net.connect(MT, LIP, weight=0.1, pstc=0.1)
13 net.connect(LIP, LIP, pstc=0.1)
14 net.connect(LIP, output, pstc=0.01)
15
16 net.add_to_nengo()
```

The first line imports the Python "nef" package, which provides the programming interface to Nengo. The next line creates the network object that contains all of the model's elements. Line 4 creates a 2D input to drive MT, with initial values of zero on both dimensions. Lines 6 through 8 create the neural populations with the specified number of neurons, dimensions, and noise. Notice that line 8 specifies the intercepts, allowing for the efficient implementation of equation 8.1. Lines 10 through 13 form the synaptic connections within the network, computing the desired functions and using the specified time constants. These connections use the default transformation matrix (i.e., the identity matrix) scaled by the weight parameter, if it is provided. The final line adds all of the defined elements and connections to the Nengo GUI.

- To load this script, click the folder icon in the top left corner. From the Nengo install directory, navigate to /Building a Brain/chapter8 and select 2D_decision_integrator.py. Click *Open*.
- To run the network, click the *Interactive Plots* icon.
- Once the network is displayed, press the play button, let it run for about 0.5 seconds, and press pause.

With no input, you will notice that MT and LIP do not move much, and that there is little activity from the output population.

- Right-click on the sliders and set the value to [−0.5, 0.5]. Click the play button.

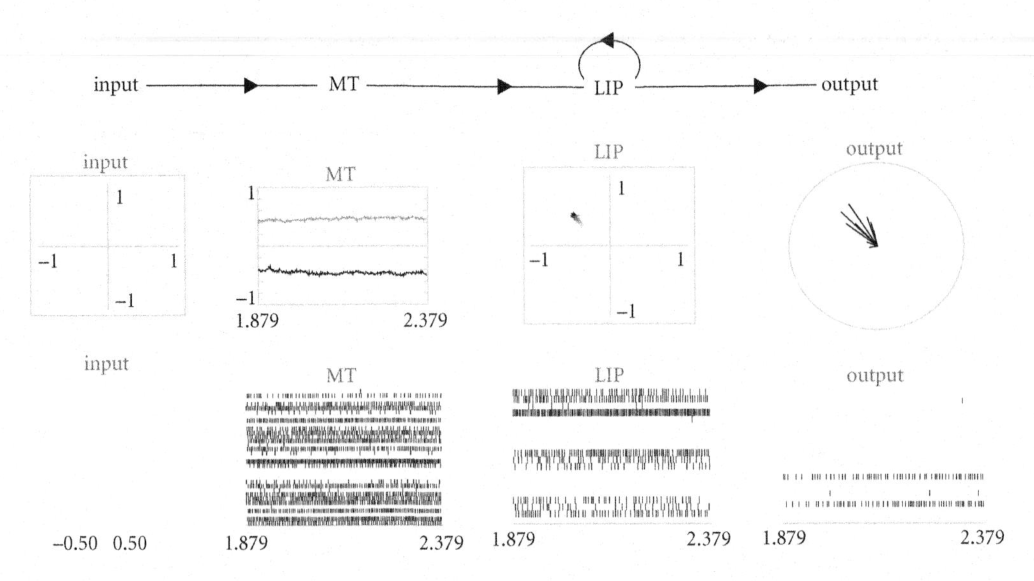

FIGURE 8.1 Activity of a 2D decision-making circuit. These are the results at the end of the simulation described in the main text. The top row shows the network topology. The middle row shows the input in the **x** space, the MT representation of **x** over time, the LIP representation of accumulated evidence, and the response of output neurons along their preferred directions. The corresponding input sliders and spike rasters are shown along the bottom row.

You will now notice that the input moves rapidly to the top left corner, and MT encodes that position over time (*see* Fig. 8.1). LIP slowly moves toward that same corner as it accumulates evidence that there is sustained motion in that direction. When LIP eventually gets close enough to the corner, the neurons in the output population will begin to fire, as shown in the preferred direction vector plot. The neurons with responses clustered in the appropriate direction will be most active. These neurons can thus drive a motor response in that same direction.

This circuit will reproduce many of the properties of standard noisy accumulator models, including appropriate reaction time distributions, more rapid responses to stimuli with more coherence (i.e., a larger magnitude input), and straightforward generalizations to other kinds of perceptual information. In addition, this model will account for neuron dynamics and spike patterns, and does not need to be changed regardless of the number of targets. However, good matches to data will depend on more careful selection of some parameters, such as input signal weights and response thresholds. More interestingly, the breadth of the model can also be improved in many ways.

Iteration

Perhaps the most important goal to keep in mind when constructing a cognitive model is to keep it simple. What I have described so far is such a beginning. Starting simple puts you in a good position to progressively improve the model by both revisiting past assumptions in light of new information and by adding additional elements to the model. In this section I discuss several ways in which our decision-making model could be extended. Many of these extensions are included in models posted on the Nengo website.

One obvious way to improve the model is to have the model begin with the same stimulus as is actually presented to subjects in this experiment (i.e., random dot stimuli). This necessitates constructing a model of the visual motion pathway that proceeds from movies of motion stimuli to vectors indicating global motion input, as assumed in the simple model above. Aziz Hurzook in my lab has proposed such a model (Hurzook, 2012). This kind of extension is important for ensuring that the proposed mechanism functions appropriately with realistic stimuli.

In addition to improving the perceptual aspects of the model, it is also possible to improve the cognitive control aspects. Including the basal ganglia in the model and defining mappings from cortical states to control states would make the model more complete. For example, we could define these two mappings to allow the model to control its own performance of the task:

1. If [there is no activity in MT], then [reset (inhibit) LIP neurons]
2. If [there is activity in the output neurons], then [route response to motor output].

The first rule implicitly notes that the default state of the network is for the neurons to be integrating their input (i.e., when there is MT activity). The second rule allows the model to include the possibility of not immediately routing the accumulated response, which is important for response flexibility. Further, if such a set of rules were to be embedded in a larger model, such as Spaun, then each rule could include an additional element that would determine whether the current task context was appropriate for applying these rules.

It is also possible to improve the motor aspects of the model by transforming the output directions into actual eye movement commands. This would require the addition of a frontal eye field area that would use the LIP output to help generate motor signals to move the eyes.

A generalization that could be used to make new predictions would be to increase the number of dimensions being integrated. Monkeys and humans, after all, live in three spatial dimensions and are likely capable of making motion judgements in all three directions, although they are seldom tested in this way in experimental settings. The same model structure would be a natural extension to this case, where each of the populations would represent 3D vectors.

Finally, it is possible to extend the model to account for more complex tasks that introduce extra elements, such as rewards, delays, and priming. Pursuing any of these suggested extensions–or simply gathering additional information regarding neuron response properties, anatomical connections, or ways in which the information captured by the model is used–will require iterating over these model construction steps. In addition, difficult decisions need to be made regarding analysis of the data produced by the model. Often, keeping the analysis as similar as possible to that reported in a paper to which the model is being compared is most desirable.

Although this concludes the Nengo tutorials in the book, there are many additional models, examples, and videos, along with extensive documentation on the Nengo website (`http://nengo.ca`). I invite you to explore these and contribute to the online discussion, construction and evaluation of models of biological cognition.

9. THEORIES OF COGNITION

9.1 ■ The State of the Art

In Chapter 1, I recounted a history of cognitive science in which there are three main contenders: the symbolic approach, connectionism, and dynamicism. From that discussion, it may seem natural to conclude that each paradigm is likely to be equally well represented by state-of-the-art cognitive architectures. However, once we begin to examine implemented architectures, it becomes clear that this is not the case. By far the dominant paradigm underwriting contemporary, functioning cognitive architectures is the symbolic approach.

In fact, it is somewhat of an embarrassment of riches when it comes to symbolic cognitive architectures. In a number of recent surveys listing more than 20 different architectures, all but a handful are symbolic (Pew & Mavor, 1998; Ritter et al., 2001; Taatgen & Anderson, 2010).[1] To be fair, many of these architectures include methods for integrating some connectionist-type components into their models (e.g., Cogent, Cooper, 2002) but typically as something of an afterthought. However, there are some architectures that consider themselves to be explicitly hybrid architectures (e.g., Clarion, Sun, 2006; and ACT-R, Anderson, 2007). Inclusion of dynamicist components in cognitive architectures is even more rare.

Arguably, the most successful and widely applied cognitive architecture to date is the ACT-R architecture (Anderson, 2007), which relies on symbolic representations and incorporates connectionist-like mechanisms in the

[1] *See* `http://en.wikipedia.org/wiki/Cognitive_architecture` for an accessible but highly incomplete list.

memory system. The current 6.0 version is one in a long line of architectures that can trace their methodological assumptions back to what many consider the very first cognitive architecture, the General Problem Solver (GPS; Newell & Simon, 1976). Consequently, ACT-R shares central representational and processing commitments with most symbolic architectures, and thus is a natural choice as a representative of the paradigm. Although ACT-R contains some neurally inspired components, making it something of a hybrid approach, the core representational assumptions are undeniably symbolic.

Another well-known symbolic architecture is Soar (Rosenbloom et al., 1993). It is currently in version 9 and focuses on generating functional cognitive models that address large-scale and long-term knowledge acquisition and exploitation. Its successes are many.[2] However, unlike ACT-R, Soar has not been used to explore the biological plausibility of its central assumptions. Thus ACT-R is unique among symbolic approaches for having been mapped to both behavioral and neural data (Anderson, 2007). Because I am centrally interested in *biological* cognition, these features of ACT-R make it an important point of comparison for the SPA. In many ways, the ACT-R architecture embodies the best of the state-of-the-art in the field.

There are several other contemporary architectures that are significantly more biologically inspired, connectionist approaches to cognitive modeling. Consequently, they are also natural to compare to the SPA. These include architectures that use various mechanisms for constructing structured representations in a connectionist network, such as Neural Blackboard Architectures (van der Velde & de Kamps, 2006), which use local integrators, and SHRUTI (Shastri & Ajjanagadde, 1993), LISA (Hummel & Holyoak, 2003), and DORA (Doumas et al., 2008), which use synchrony.

Another influential connectionist-based approach to structure representations focuses less on specific implementations and more on proposing a broad theoretical characterization of how to capture symbolic (specifically linguistic) processing with distributed representations and operations (Smolensky & Legendre, 2006 a,b). The architecture associated with this work is known as ICS and shares central commitments with the SPA regarding structured representations.

Other connectionist approaches focus less on representational problems and deal more with issues of control and learning in cognitive tasks. Leabra is an excellent example of such an approach, which has had much success mapping to reasonably detailed neural data (O'Reilly & Munakata, 2000). In addition, there has been a recent effort to combine Leabra with ACT-R (Jilk et al., 2008) as a means of simultaneously exploiting the strengths of each. This, again, provides an excellent comparison to the SPA.

[2] *See* `http://sitemaker.umich.edu/soar` for access to publications, code, and examples.

Finally, there are some dynamicist approaches to cognitive modeling that have been gaining prominence in recent years (Schöner, 2008). Perhaps the best known among these is Dynamic Field Theory (DFT), which employs a combination of dynamicist and neurally inspired methods to model cognitive behaviors (e.g., Schöner & Thelen, 2006). The focus of DFT on time, continuity, and neural modeling provides a useful and unique comparison for the SPA as well.

In the remainder of this chapter, I consider each of these approaches in more detail, describing their strengths and some challenges each faces. This discussion will provide background for a subsequent evaluation of these theories with respect to the CCC. And, it sets the stage for an explicit comparison between this past work and the SPA.

Before proceeding, it is worth explicitly noting two things. First, there is important work in cognitive science that has been very influential on the architectures described here, including the SPA, which I do not discuss in detail for lack of space or lack of a full-fledged architecture specification (e.g., Newell, 1990; Rogers & McClelland, 2004; Barsalou, 2003). Second, because I am considering six architectures, each receives a relatively short examination. This means that my discussion is somewhat superficial, although hopefully not inaccurate. As a consequence I encourage you to follow up with the provided citations for a more detailed and nuanced understanding of each of the architectures under consideration.

9.1.1 Adaptive Control of Thought-Rational

Adaptive control of thought-rational, or ACT-R, is perhaps the best developed cognitive architecture. It boasts a venerable history, with the first expression of the architecture being as early as 1976 (Anderson, 1976). It has been very broadly applied, characterizing phenomena from child language development (Taatgen & Anderson, 2002) to driving (Salvucci, 2006), and from the learning of algebra (Anderson et al., 1995) to categorization (Anderson & Betz, 2001).

The many successes and freely available tools to build ACT-R models have resulted in a large user community developing around this architecture (*see* `http://act-r.psy.cmu.edu/`). This, in turn, has had the effect of there being many "sub-versions" of ACT-R being developed, some of which include central components from what were originally competing architectures. For example, a system called ACT-R/PM includes "perceptual-motor" modules taken from EPIC (Kieras & Meyer, 1997). This has led to proponents claiming that the ACT-R framework allows the creation of "embodied" cognitive models (Anderson, 2007, p. 42). Because a central goal of ACT-R is to provide descriptions of "end-to-end behavior" (Anderson, 2007, p. 22), this is an important extension of the basic ACT-R framework.

The basic ACT-R architecture consists of a central procedural module that implements a production system and is bidirectionally connected to seven other modules (i.e., the goal, declarative, aural, vocal, manual, visual, and imaginal buffers). Each of these modules has been mapped to an area of the brain, with the procedural module being the basal ganglia, and hence responsible for controlling communication between buffers and selecting appropriate rules to apply based on the contents of the buffers. A central constraint in ACT-R is that only a single production rule can be executed at a time. It takes about 50 ms for one such production rule to "fire," or be processed (Anderson, 2007, p. 54).

Because of its commitment to a symbolic specification of representations and rules, most researchers take ACT-R to be a largely symbolic cognitive architecture. However, its main proponent, John Anderson, feels no particular affinity for this label. Indeed, he has noted, that it "stuck in [his] craw" when he was awarded the prestigious Rumelhart prize in 2005 as the "leading proponent of the symbolic modeling framework" (Anderson, 2007, p. 30). In his view, ACT-R is equally committed to the importance of subsymbolic computation in virtue of at least two central computational commitments of the architecture. Specifically, "utilities" and "activations" are continuously varying quantities in the architecture that account for some central properties of its processing. For example, continuous-valued utilities are associated with production rules to determine their likelihood of being applied. As well, utilities slowly change over time with learning, determining the speed of acquisition of new productions. Anderson takes such subsymbolic commitments of the architecture to be at least as important as its symbolic representational commitments.

As well, these subsymbolic properties are important for the recent push to map ACT-R models to fMRI data, because they play a crucial role in determining the timing of processing in the architecture. In his most recent description of ACT-R, Anderson has provided several examples of the mapping of model responses to the bold signal in fMRI (Anderson, 2007, pp. 74–89, 119–121, 151). This mapping is based on the amount of time each module is active during one fMRI scan (usually about 2 s). The idea is that an increase in the length of time a module is active results in increased metabolic demand, and hence an increased BOLD signal, picked up by the scanner. Although the quality of the match between the model and data varies widely, there are some very good matches for certain tasks. More importantly, the inclusion of neural constraints, even if at a very high level, significantly improves the plausibility of the architecture as a characterization of the functions being computed by the brain.

Several critiques have been leveled against the ACT-R architecture as a theory of cognition. One is that the architecture is so general that it does not really provide practical constraints on specific models of specific tasks

(Schultheis, 2009). Before the explicit mappings to brain areas of the modules, this criticism was more severe. Now that there is a more explicit implementational story, however, there are at least some constraints provided regarding which areas, and hence which modules, must be active during a given task. However, these constraints are not especially stringent. As Anderson has noted, the mapping to brain areas is both coarse and largely a reiteration of standard views regarding what the various areas do (Anderson, 2007, p. 29).

For example, the ACT-R mapping of the declarative module to prefrontal areas is neither surprising nor likely to be generally accurate. The prefrontal cortex is extremely large and is differentially involved in a huge variety of tasks, sometimes in ways not suggestive of providing declarative memories (Quirk & Beer, 2006). So, it is not especially informative to map it to only one (i.e., declarative memory) of the many functions generally ascribed to the area. As well, it would be surprising if all retrieval and storage of declarative memory could be associated with prefrontal areas, given the prominent role of other structures (such as hippocampus) in such functions (Squire, 1992).

Similarly, the visual module is mapped to the fusiform gyrus, despite the acknowledgment that there are many more areas of the brain involved in vision. Although fusiform no doubt plays a role in some visual tasks, it is somewhat misleading to associate its activation with a "visual module" given the amount of visual processing that does not activate fusiform.

In addition, the mapping between an ACT-R model and fMRI data often has many degrees of freedom. In at least some instances, there are several mapping parameters, each of which is tuned differently for different areas and differently for different versions of the task in order to realize the reported fits (Anderson et al., 2005). This degree of tuning makes it less likely that there is a straightforward correspondence between the model's processing time and the activity of the brain regions as measured by BOLD. Thus, claims that ACT-R is well constrained by neural data should be taken with some skepticism. In fairness, Anderson (2007) has noted that these mappings are preliminary, has acknowledged some of the concerns voiced here, and has clearly recognized the fMRI work as only the beginning of a more informative mapping (pp. 77, 248).

Another criticism of the ACT-R approach is that it is a largely non-embodied approach to cognition. For example, the claim that ACT-R/PM provides for truly "embodied" models seems to overreach the actual application of this version of ACT-R, even with its specialized extensions (Ritter et al., 2007). The perceptual and motor areas of this architecture do not attempt to capture the full processing of natural images or the control of a high degree of freedom limbs. Rather, they typically provide lengths of time that such processing might take, capturing high-level timing data related to processing, such as attentional shifts. Applications of such models are typically to tasks such as the effect of icon placement and other elements of interface design on reaction times (although *see* Ritter et al., 2007, for a more sophisticated application). Visual representations in such models are typically specified as symbolic lists of features (e.g., "gray circle"; Fleetwood & Byrne, 2006). Although the notion of "embodiment" can clearly come in degrees, and these attempts surpass those of GPS and Soar, it seems disingenuous to suggest that such models are "embodied" in the sense typically intended by theorists of embodied cognition (Haugeland, 1993; Clark, 1997). Consequently, ACT-R has not yet achieved its laudable goal of accounting for end-to-end behavior.

Nevertheless, ACT-R continues to develop. And, there are continuing attempts to ground the architecture in biological structure. Of particular note is the on going attempt to integrate ACT-R with an architecture I consider shortly, called Leabra. I discuss this combination in more detail in my discussion of Leabra. However, it is helpful to see what kinds of challenges for ACT-R this integration is expected to help alleviate. For example, the integration is intended to answer questions such as "How can 'partial matching' of production rules that 'softens' traditional production conditions operate?" "How can the learning of utility parameters that control production selection be grounded in plausible assumptions about the nature of feedback available?" and "How can context influence production matching and selection beyond the explicit specification of precise conditions?" (Jilk et al., 2008, p. 211). What seems to tie these considerations together are ways of making symbols less rigid (*see also* my discussion in Section 7.4). In the end, then, the commitment of ACT-R to symbolic representations seems to be hampering its explanation of the subtleties of cognitive behavior, as well as its mapping to neural structure.

In conclusion, the many successes of ACT-R are undeniable, and the desire to constrain the architecture further with detailed neural considerations is understandable and to be commended. However, it is not yet clear if the commitment of ACT-R to symbolic representations, production rules, and a production system will, in the end, allow it to develop in the way its proponents foresee.

9.1.2 Synchrony-Based Approaches

When faced with the problem of implementing structured representations in a neural substrate, one of the basic decisions that must be made is how to implement the "binding" of representations to one another. In the next three sections, I describe three different suggestions for how this might be done in a cognitive architecture. I begin, in this section, by discussing synchrony-based approaches.

One of the earliest suggestions for how syntactic binding might occur was borrowed from a suggestion in visual neuroscience that feature binding is the result of the synchronization of spiking activity across neurons representing different aspects of the visual image (von der Malsburg, 1981). This suggestion has been imported into cognitive modeling by Shastri and Ajjanagadde (1993) in their SHRUTI architecture. More recently it has seen perhaps its most sophisticated neural expression in the LISA architecture from Hummel and Holyoak (2003) and the more recent DORA architecture from Doumas et al. (2008).

In LISA, synchrony is used as a direct method for neurally representing structured relations (Hummel et al., 1994; Hummel & Holyoak, 1997, 2003). In DORA, close asynchrony is employed instead (Doumas et al., 2008). In both architectures, a structured representation is constructed out of four levels of distributed and localist representations. The first level consists of localist "subsymbols" (e.g., mammal, furry, male, etc.). The second level consists of localist units connected to a distributed network of subsymbols relevant to defining the semantics of the second level term (e.g., dog is connected to furry, mammal, etc.). The third-level consists of localist "subproposition" nodes that bind roles to objects (e.g., dog+*verb-agent*, or dog+*verb-theme*, etc.). The fourth and final level consists of localist proposition nodes that bind subpropositions to form whole propositions (e.g., dog+chase-agent and cat+chase-theme). LISA and DORA are interesting cases because they are clearly connectionist networks, but they implement a classical representational scheme because all elements of the structures are explicitly tokened whenever a structure is tokened. In this sense, they are precisely the kind of implementation that could underwrite the symbolic representations of ACT-R. For clarity, I restrict the remainder of my discussion to the LISA architecture (*see* Fig. 9.1), though the main points hold for both LISA and DORA.

Hummel and Holyoak (2003) have suggested that LISA can perform relational inference and generalization in a cognitive architecture that they argue is "psychologically and neurally realistic" (p. 220). They demonstrate LISA on several examples that show both inference and learning of new schemas from past examples. Their work focuses on analogy, and hence a central feature of the architecture is its ability to "map" one structured representation to another. For example, given the prior knowledge that Bill has a Jeep, Bill wanted to go to the beach, and thus Bill drove his Jeep to the beach, LISA can infer in a new instance in which John has a Civic and John wants to go to the airport, that John will drive his Civic to the airport (p. 236). Hummel and Holyoak proceed to show how this basic ability can explain a myriad of psychology experiments, and argue that LISA "provides an existence proof that [modeling human relational inference] can be met using a symbolic-connectionist architecture in which neural synchrony dynamically codes variable bindings in [working memory]" (p. 245).

It is clear that their architecture can capture some features of human reasoning. It is less clear that the model uses "neurally realistic" mechanisms and assumptions. Hummel and Holyoak have been careful to note that each of their localist units is intended to be a population of neurons: "we assume

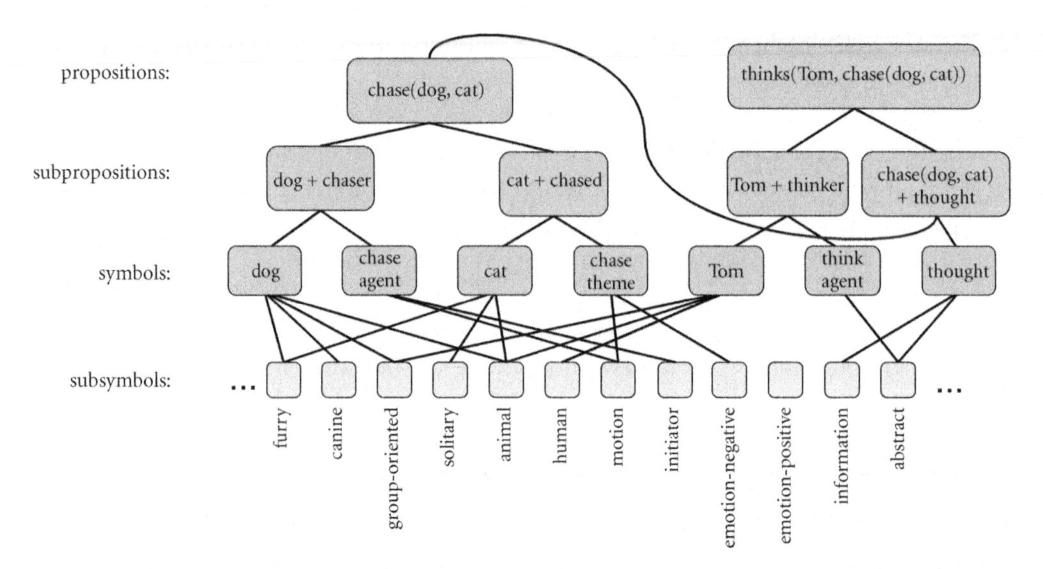

FIGURE 9.1 The LISA architecture. Boxes represent neural groups and lines represent synaptic connections. Shown are just those neural groups needed to represent "dogs chase cats" and "Tom thinks that dogs chase cats." (Based on Hummel & Holyoak [2003], figure 1, reproduced from Stewart & Eliasmith, 2011a.)

that the localist units are realized neurally as small populations of neurons (as opposed to single neurons), in which the members of each population code very selectively for a single entity" (p. 223). Because of this selectivity, each such population only represents one object, subproposition, or proposition. They have noted that this kind of representation is essential to the proper functioning of their model.

But, this limits the ability of the model to scale to realistic representational requirements. Because a separate neural group is needed to represent each possible proposition, the neural requirements of this system grow exponentially. To represent any simple proposition of the form *relation*(*agent*, *theme*), assuming 2000 different relations and 4000 different nouns, 2000*4000*4000 = 32,000,000,000 (32 billion) neural *populations* are needed. This is unrealistic, given that the brain consists of about 100 billion *neurons*, can handle a much larger vocabulary (of approximately 60,000 words), and needs to do much more than represent linguistic structures (see Appendix D.5 for details). So, although their representational assumptions may be appropriate for the simple problems considered in their example simulations, such assumptions do not allow for proper scaling to truly cognitive tasks. These same assumptions hold for DORA.

A separate aspect of LISA that is neurobiologically problematic is its use of neural synchrony for performing binding. There have been many criticisms of the suggestion that synchrony can be effectively used for binding. For example, after considering binding, O'Reilly and Munakata (2000) have concluded that "the available evidence does not establish that the observed synchrony of firing is actually used for binding, instead of being an epiphenomenon" (p. 221). Similarly, from a more biological perspective, Shadlen and Movshon (1999) have presented several reasons why synchrony has not been established as a binding mechanism and why it may not serve the functional role assigned by proponents.

Further, synchrony binding in LISA is accomplished by increasing real-valued spike rates together at the same time. Synchrony in LISA is thus very clean and easy to read off of the co-activation of nodes. However, in a biological system, communication is performed using noisy, individual spikes, leading to *messy* synchronization, with detailed spectral analyses necessary to extract evidence of synchronization in neural data (Quyen et al., 2001). These same idealizations are employed for asynchronous binding in DORA. Thus it is far from clear that using more neurally realistic, spiking, noisy nodes would allow either LISA or DORA to exploit the mechanism it assumes.

However, there is an even deeper problem for these models. The kind of synchronization they exploit is nothing like what has been hypothesized to exist in biological brains. In LISA, for example, synchronization occurs because there are inhibitory nodes connected to each subproposition that cause oscillatory behavior when the subproposition is given a constant input. That oscillation is then reflected in all units that are excitatorily connected to these subpropositions (i.e., propositions and themes/verbs/agents). Therefore, binding is established by constructing appropriately excitatorily connected nodes, and the oscillations serve to highlight one such set of nodes at a time.

So, the representation of the binding *results in* synchronization patterns: this essentially puts the bound cart before the synchronous horse. Synchronization in the neurobiological literature is supposed to *result in* binding (*see*, e.g., Engel et al., 2001). For example, if "red" and "circle" are co-occurring features in the world, then they are expected to be synchronously represented in the brain, allowing subsequent areas to treat these features as bound. Thus the synchronization patterns result in binding, not the other way around. An analogous concern holds for DORA, where the precise asynchronization is determined by desired bindings, which are reflected in the network wiring. Consequently, the neural plausibility of both LISA and DORA is not supported by current work on synchronization (and asynchronization) in the neurosciences.

9.1.3 Neural Blackboard Architecture

LISA and DORA suffer from severe scaling problems because they rely on dedicated resources for each possible combination of atomic symbols. As a result, attempts have been made to avoid this problem by positing more flexible, temporary structures that can underwrite symbolic binding. One recent approach is that by van der Velde and de Kamps (2006), who have proposed a binding mechanism that they refer to as the Neural Blackboard Architecture (NBA).

Blackboard architectures have long been suggested as a possible cognitive architecture (Hayes-Roth & Hayes-Roth, 1979; Baars, 1988), although they tend to find their greatest application in the AI community. Blackboard architectures consist of a central representational resource (the blackboard) that can be accessed by many, often independent, processes that specialize in different kinds of functions. The blackboard, along with a controller, provides a means of coordinating these many processes, allowing simpler functional components to work together to solve difficult problems.

Given their AI roots, implementations of these architectures tend to be symbolic. However, the growing interest in trying to relate cognitive architectures to brain function, coupled with concerns about scalability of some neural architectures, has led van der Velde and de Kamps to propose their methods for implementing blackboard-like representations in a connectionist network. Like proponents of ACT-R and LISA, they are interested in demonstrating that some aspects of the symbolic representational assumptions underlying the architecture can be given a neurally plausible characterization.

In their presentation of the NBA, van der Velde and de Kamps explicitly describe it as a means of addressing Jackendoff's challenges (also captured by the representational CCC; Section 9.2.1). To avoid the possibility of an exponential growth in the number of neurons needed for structure representation, the NBA uses a smaller set of "neural assemblies" that can be temporarily associated with particular basic symbols. Larger structures are then built by binding these assemblies together using a highly intricate system of neural gates. This not only allows for representation of basic propositional logic, but also of full sentential structures.

For word assemblies to be bound to structure assemblies, there is a connection topology that links the assemblies and is able to have sustained, slowly decaying activation. It is the activation of that connection structure (a "memory circuit") that determines the binding. A separate "gating circuit" determines which role (e.g., *agent* or *theme*) a particular word assembly is connected to in the structure (*see* Fig. 9.2).

The use of neural assemblies that can be temporarily bound to a particular symbol greatly reduces the number of neurons required for this method. Nevertheless, as with LISA, each of the localist units is actually intended to be a group of neurons, meaning that the architecture ignores the details of neural spiking activity, and instead deals with an abstract, real-valued neural "activity." If we assume that these neural groups consist of an average of 100 neurons, and that the vocabulary is limited as in the LISA example

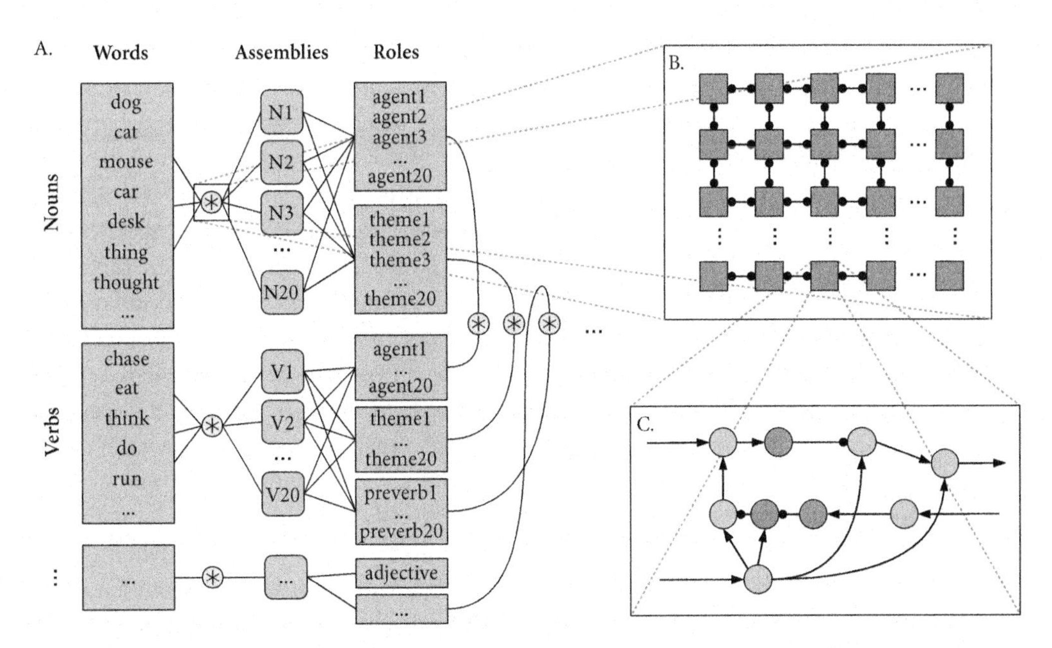

FIGURE 9.2 The Neural Blackboard Architecture. Neural groups representing words in **A** are connected by mesh grids **B** built using the neural circuit **C**. Excitatory connections end with arrows, and inhibitory connections with circles. (*See* van der Velde & de Kamps [2006] for more details; reproduced from Stewart & Eliasmith, 2011a.)

(6000 terms), the NBA can represent any simple structured proposition of the form *relation*(*agent, theme*) by employing approximately 960,000,000 neurons. For a more realistic, adult-sized vocabulary of 60,000 symbols, around 9,600,000,000 neurons, or about 480 cm^2 of cortex, is needed for simple two-element relations (*see* Appendix D.5 for details). In sum, approximately one-fifth of the total area of cortex is needed, which is much larger than the size of known language areas (*see* Appendix D.5).

Although this is a slightly more reasonable number of neurons than in LISA, it does not suggest good scaling for the NBA. This is because these resources are required just for the *representation* of simple structures, not the processing, transformation, updating, and so forth, of these structures. And, there remains the issue of representing (processing, transforming, etc.) more sophisticated structures with more grammatical roles. The focus of the NBA on representation is a reasonable starting point, but it means that there is little provision for characterizing a general cognitive architecture. To fully specify a blackboard architecture, you need to also specify the controller, and the other processes interacting through the blackboard–each of these will require more neural resources.

There are other reasons to be concerned with the proposed representational structure, beyond the problems with scaling and function. For example, the gating structures (*see* Fig. 9.2B, C) are highly complex and carefully organized, such that the loss or miswiring of one connection could lead to catastrophic failure of the system. There is no evidence that microlesions should be expected to have these sorts of effects. On the contrary, lesions underlying syntactic deficits are quite large, typically larger than for other deficits (Hier et al., 1994). Further, connecting all nouns to all noun assemblies and all noun assemblies to all verb assemblies requires long distance and highly complete connectivity (within the entire 480-cm^2 area), whereas most real cortical connections are much more local and sparse (Song et al., 2005). As well, the evidence cited in the original paper to support the necessary specific structures merely demonstrates that some individual inhibitory cells in visual cortex synapse on other inhibitory cells in the same layer (Gonchar & Burkhalter, 1999). This does not render the NBA plausible in its details.

Notably, like LISA and DORA, the NBA is also best seen as an implementation of classical symbolic representation. Each word, structural element, and binding is explicitly represented in a specific location in the brain. As such, it is another means of addressing ACT-R's problem of relating symbolic representations to neural implementation. However, the proposed implementation makes unrealistic assumptions about the connectivity and robustness of the neural substrate. Finally, even with those assumptions, it seems unlikely that the architecture will scale to the level of human cognition with the resources available in the brain.

9.1.4 The Integrated Connectionist/Symbolic Architecture

A third approach to characterizing binding in a neurally-inspired architecture is that employed by Vector Symbolic Architectures (VSAs; sometimes called "conjunctive coding"). As mentioned in my earlier discussion, the SPA employs a VSA-based approach that has a conceptual forerunner in the tensor product representations suggested by Smolensky (1990). In an excellent and ambitious book, Smolensky and Legendre (2006a) have described how tensor products can be used to underwrite language processing and have suggested what they call the Integrated Connectionist/Symbolic Cognitive Architecture (ICS). Coupled with the representational considerations of ICS are Optimality Theory (OT) and Harmonic Grammar (HG), which provide functional characterizations of linguistic processing as soft-constraint satisfaction. As a result, the ICS is largely characterized in the context of linguistics, especially phonology.

Specifically, HG is used to demonstrate that tensor products can have sufficiently complex embedded structure to do linguistic processing. As such, HG provides a link between the purely symbolic OT and the implementation of such a theory in a connectionist network. Notably, both HG and OT are used in the context of localist networks. Although Smolensky and Legendre have described how the localist HG networks could be systematically mapped to networks employing distributed representations, the actual models are neither run nor optimized with distributed representations (*see*, e.g., Smolensky & Legendre [2006a] chapter 11, where the relationship is best described). Instead, the soft constraint problems are solved using the localist networks.

Regardless, the work has been widely appreciated for its comprehensive and detailed account of the relations among language processing, parallel constraint satisfaction, and connectionism. The ICS provides a clear, quantitative description of how specific constraints on language processing can be used to explain a wide variety of linguistic phenomena including phonology, grammaticality judgments, and semantic–syntactic interactions. The ICS is clearly a significant achievement, and it provides by far the most comprehensive connectionist account of language currently available. Further, the methods are systematic and rigorous, meaning their application to new phenomena is clear and the predictions are testable. In short, the ICS goes far beyond the SPA or any other architecture in its description of the functions needed to explain linguistic behavior.

That being said, however, there are some concerns about considering the ICS as a general account of biological cognition. For one, the biological plausibility is untested and unconvincing as ICS currently stands. As mentioned, simulations are typically localist, and it is not evident how populations of neurons could perform the same functions without being "wired up" differently for each optimization (e.g., to parse each sentence). Relatedly, ICS includes no description of control structures that could "load" and "manipulate" language as quickly as it happens in the human system. So despite the isomorphic mapping between localist and distributed networks, it remains a major challenge for the ICS to show how a distributed network can actually perform all the steps needed to process language in a biologically relevant way.

As with past approaches I have considered, biological plausibility is related closely to scaling issues, a known challenge for the ICS's preferred method of binding: the tensor product. As mentioned earlier, tensor products scale badly because the tensor product of two N dimensional vectors results in an N^2 length vector. So, embedded structures become exponentially large and unwieldy. For example, to represent structures like "Bill believes John loves Mary" using 500 dimensional vectors (as in the SPA) would require 12 billion neurons, somewhere between a quarter and a half of the neurons in cortex (*see* Appendix D.5)–as with the NBA, this is only for representation. In their book, Smolensky and Legendre have explicitly noted the utility of a compression (or "contraction") function like that employed by all other VSAs to address this scaling issue: "the experimental evidence claimed as support for these models suggests that contracted tensor product representations of some kind may well be on the right track as accounts of how people actually represent structured information" (p. 263). However, they do not subsequently use such representations.

Using uncompressed tensor products seems to be motivated by Smolensky and Legendre's deep theoretical commitment to connecting a description of the brain as a numerical computer with the description of the mind as a symbolic computer. As a result, their approach has been praised by Anderson (2007) for its focus on how to do complex processing with the proposed representation: "The above criticism is not a criticism of connectionist modeling per se, but . . . of modeling efforts that ignore the overall architecture . . . in the Smolensky and Legendre case, [their approach] reflects a conscious decision not to ignore function" (pp. 14–15). The approach appeals to Anderson because Smolensky and Legendre hold a view of cognitive levels that is compatible with Anderson's suggestion that the symbolic approach often provides the "best level of abstraction" for understanding central aspects of cognition (2007, p. 38). In their discussion of the ICS, Smolensky and Legendre have explicitly noted their commitment to an isomorphism between their representational characterization in terms of vectors and a characterization in terms of symbols (Smolensky & Legendre, 2006b, p. 515). If you use *compressed* representations, however, the isomorphism no longer holds. Compressed representations lose information about the symbolic structures they are taken to encode, so they can quickly diverge from an "equivalent" symbolic representation.

This commitment to isomorphism is not a throw-away aspect of the ICS: "The formal heart of these [ICS] principles is a mathematical equivalence between the fundamental objects and operations of symbolic computation and those of specially structured connectionist networks" (p. 147). Ironically, perhaps, this is the same equivalence that ACT-R researchers have given up in their concern to match behavioral data, and why Anderson does not consider ACT-R to be a (purely) symbolic approach. In any case, if this equivalence could be realized, it would provide ACT-R with a direct means of relating its symbolic representations to neuron-like units. This, of course, is the goal that has been shared by all three connectionist approaches we have seen to this point.

But, all three have run into the same two problems: scaling and control. Control relates to how the system itself can construct, manipulate, decode, and so forth, the representations employed. The scaling issue relates to difficulties in mapping standard symbolic representations into neural ones. Perhaps the control problem could be solved with time, but I suspect that the scaling issue is symptomatic of a deeper problem.

Specifically, I believe that the failure to scale stems from a fundamental problem with attempting to neurally implement a classical representational scheme: If we require each atom within a structure to appear in the representation of the overall structure (as per Fodor & Pylyshyn, 1988), then we will either have an exponential explosion in the number of neurons required (as in LISA and the ICS) or a complex and brittle system for temporarily binding and rebinding terms that is incompatible with known neural constraints (as in the NBA). In other words, implementing a classical (i.e., explicitly compositional) representational scheme directly makes for unscalable systems when considering actual neural constraints. Perhaps, then, the problem lies in the "top-down" representational constraints assumed by symbolic architectures to begin with. Perhaps it is a mistake to think that a good characterization of biological cognition will come from an implementation of a classical symbolic system.

9.1.5 Leabra

The last connectionist cognitive architecture I consider has taken a very different approach to characterizing cognition. Rather than focusing on representation, the Local, Error-driven and Associative, Biologically Realistic Algorithm (Leabra) focuses on control and learning (O'Reilly & Munakata, 2000). The Leabra method is intended to be used to learn central elements of a cognitive architecture. The algorithm is able to account for results traditionally explained by either back-propagation or Hebbian learning, as it combines these two kinds of learning within its algorithm.

Proponents of Leabra have employed it widely and have convincingly argued that it has helped develop our understanding of gating of prefrontal representations, as well as reward, motivation, and goal-related processing in cognitive control (O'Reilly et al., 2010). One recent application of Leabra is to the Primary Value and Learned Value (PVLV) model of learning (O'Reilly et al., 2007), which simulates behavioral and neural data on Pavlovian conditioning and the midbrain dopaminergic neurons that fire in proportion to unexpected rewards (it is proposed as an alternative to standard temporal-difference methods).

This learning model has been integrated into a model of the brain that includes several areas central to cognitive function. The result is called the Prefrontal-cortex Basal-ganglia Working Memory (PBWM) model. PBWM employs PVLV to allow the prefrontal cortex to control both itself and other brain areas in a task-sensitive manner. The learning is centered around subcortical structures in the midbrain, basal ganglia, and amygdala, which together form an actor/critic architecture. The authors demonstrate the model on several working memory tasks, perhaps the most challenging being the 1-2-AX task (O'Reilly & Frank, 2006; Hazy et al., 2007).

This task consists of subjects being shown a series of numbers and letters, and their having to respond positively to two possible target patterns. The first is a 1 followed by AX, and the second is a 2 followed by BY. There may be intervening letters between the number and its associated letter string. There may also be non-target stimuli, to which no response is given. This is a difficult task because the subject must keep in mind what the last number was, as well as what the associated next target and current stimulus pattern is, all while ignoring irrelevant stimuli. Nevertheless the model successfully performs the task.

The general structure (e.g., basal ganglia gating of prefrontal cortex) of the PBWM model is supported by various sources of empirical evidence. For example, Frank and O'Reilly (2006) administered low doses of D2 (dopamine type-2 receptor) agents cabergoline (an agonist) and haloperidol (an antagonist) to normal subjects. They found that the "go" and "no-go" learning required by the PBWM model on the 1-2-AX tasks was affected as expected. Specifically, cabergoline impaired, whereas haloperidol enhanced, go learning from positive reinforcement, consistent with presynaptic drug effects. And, cabergoline also caused an overall bias toward go responding, consistent with postsynaptic action.

It is clear from such examples of past success that Leabra has taken a very different approach to modeling cognitive behavior than the other approaches I have considered. Although the representations in the tasks employed are not as complex as those used by other connectionist architectures, the mapping onto brain structures, the ability to address control, and the inclusion of learning mechanisms are significant strengths of this work. It is perhaps not surprising that Leabra-based approaches are often lauded for their level of biological realism.

Nevertheless, there remain central features of Leabra that cast some doubt on how well the proposed mechanisms capture what we know about low-level brain function. For example, these methods again use rate neurons, which makes it difficult to compare the results to single cell spike train data recorded from the brain structures of interest. This is especially important when it comes to detailed timing effects, such as those considered in Section 5.7. Thus, capturing precise neural dynamics seems highly challenging.

Perhaps more importantly, there are basic computational methods underlying Leabra that are of dubious plausibility. The most evident is that Leabra directly applies a *k*-Winner-Takes-All (kWTA) algorithm, which is acknowledged as biologically implausible: "although the kWTA function is somewhat biologically implausible in its implementation (e.g., requiring global information about activation states and using sorting mechanisms), it provides a computationally effective approximation to biologically plausible inhibitory dynamics" (`http://grey.colorado.edu/emergent/index.php/Leabra`; *see also* O'Reilly & Munakata [2000], pp. 94–105 for further discussion). In other words, the actual dynamics of the system are replaced by an approximation that is computationally cheaper on digital computers.

The problem is that capturing precise neural dynamics now seems out of reach. The approximation Leabra employs is much cleaner and faster than the actual dynamics, distancing the method from comparison to neural data. For example, although the basal ganglia is often characterized as a kWTA computation, explicitly simulating this process is critical to capturing many subtleties. As described in earlier SPA simulations (*see* Section 5.7), the timing of behavior depends crucially on the amount of value difference between items, the complexity of the actions, neurotransmitter dynamics, and so on. Although approximations can be practically useful, it is essential to clearly demonstrate that the approximation is accurate in situations in which it is employed. This is generally not the case for Leabra models, making its connection to neural dynamics, in particular, unclear.

Similar concerns arise when we consider the Leabra algorithm itself (pseudocode is available at `http://grey.colorado.edu/emergent/index.php/Leabra`). Much is currently known about plausible learning mechanisms in the brain (*see* Section 6.4). However, in the Leabra algorithm, there are several steps that are implausible to implement in a real neural network–that is, in a network that does not have a "supervisor" able to control the flow of information through the network. For example, the algorithm includes combining two different kinds of learning at each step, one that takes some amount of time to settle and another that does not. If there can be different numbers of learning steps within one overall simulation step, then it seems likely that there would be unexpected dynamic interactions without Leabra's externally controlled implementation. As well, units are clamped and unclamped by the supervisor. Again a neurally plausible implementation would have to demonstrate how such changes of state could be intrinsically controlled.

I should note that these concerns with biological plausibility are much different than those for other approaches I have considered. Here, the concerns are related to specific neural mechanisms more than to representational assumptions. This is largely because Leabra models typically do not have very sophisticated representations. Proponents do give some consideration to binding, although it is in the context of simple vision tasks and has not been extended to complex structured representations. Thus Leabra does not offer a full cognitive architecture on its own.

This limitation of Leabra has been duly noted by its proponents, and has resulted in the recent push to combine ACT-R and Leabra that I mentioned earlier (Section 9.1.1). This combination is natural because both approaches share some overall architectural commitments (basal ganglia as a controller, cortex as working memory, etc.). In addition, ACT-R has more sophisticated representations and Leabra has more realistic biological mechanisms (Jilk et al., 2008). In Section 9.1.1, I outlined some of the concerns about ACT-R that Leabra is expected to help address. Similarly, ACT-R is expected to help

Leabra by answering questions such as "How can new procedural learning piggyback on prior learning, rather than using trial and error for each new behavior?" and "How can verbal instructions be integrated and processed?" (p. 211). In short, Leabra needs language-like representations and ACT-R needs biological plausibility.

Unfortunately, there has not been a lot of progress in combining the approaches. I believe this is largely a representational problem–that is, proponents have not effectively integrated symbolic and neuron-like representations. Instead, different kinds of representations are associated with different parts of the brain: "we associate the active buffers, procedural production system, and symbolic representations from ACT-R with the prefrontal cortex and basal ganglia, while the graded distributed representations and powerful learning mechanisms from Leabra are associated with the posterior cortex" (Jilk et al., 2008, p. 213). Although Jilk et al. explicitly worry about the "interface" between these kinds of representations, they do not have a solution for easily allowing these distinct brain areas to communicate within a unified representational substrate. As a result, the methods have currently not been *integrated*, but more accurately "attached." Consequently, it is difficult to argue that the resulting models solve the kinds of problems they were supposed to address by combining complementary strengths. I believe it is fair to say that the desired unification of biological realism and symbolic processing has so far remained elusive for Leabra and ACT-R.

9.1.6 Dynamic Field Theory

My examination of the state-of-the-art to this point has addressed only symbolic and connectionist approaches. However, since the mid-1990s there has been a continuous and concerted effort to construct neurally inspired cognitive models using a more dynamicist approach (Schöner, 2008). The methods behind the approach are known as Dynamic Field Theory (DFT) and are based on earlier work by Shun-ichi Amari on what he called "neural field equations" (Amari, 1975). These equations treat the cortical sheet as a continuous sheet of rate neurons that are coupled by inhibition and excitation functions. They thus model a "metric dimension" (e.g., spatial position, luminance, etc.) over which neural activity is defined. Much early work focused on the kinds of stable patterns (attractors) that could be formed on these sheets, and under what conditions different patterns formed. In essence, this was one of the earliest integrator network models of neural stability.

In the hands of modern DFT researchers, these networks of stable activity patterns are connected in various configurations to map onto more sophisticated behavior, such as control of eye movements, reach perseveration, infant habituation, and mapping spatial perception to spatial language categories.

Interestingly, talk of "representations" is not uncommon in this work, despite a general resistance in dynamicist approaches to discussing representation (van Gelder & Gelder, 1995; Port & van Gelder, 1995). For example, Schöner (2008) has commented that "localized peaks of activation are units of representation" (p. 109), and Mark Blumberg has spoken of "representation-in-the-moment" in his introduction to DFT (Blumberg, 2011). Nevertheless, there is also a strong emphasis on embodiment of the models, with many being implemented on robotic platforms and an uncompromising description of the dynamics of the models.

Some of the best known applications of DFT have been to infant development. For example, Schöner and Thelen (2006) have presented a DFT model of infant visual habituation. The model consists of two interacting and coupled neural fields. The first represents the activation that drives "looking," and the second represents the inhibition that leads to "looking away," or habituation. The model is presented with simulated visual input of varying strengths, distances, and durations and is able to simulate the known features of habituation, including familiarity and novelty effects, stimulus intensity effects, and age and individual differences. Other DFT models capture the development of perseverative reaching in infants (Thelen et al., 2001) or the improvement in working memory through development (Schutte et al., 2003).

More recent work has addressed more cognitive tasks, such as object recognition (Faubel & Sch, 2010), spatial memory mapping to spatial language (Lipinski et al., 2006), and speech motor planning (Brady, 2009). The speech motor task model, for example, is constructed out of two neural fields by mapping the input/output space to a first sheet of neural activity, and then having a second-order field learn how to map the current first-order state into a desired next first-order state. This is a nice example of a simple hierarchical dynamical system that is able to capture how static control signals can aid the mapping between appropriate dynamic states.

Such examples demonstrate that, like Leabra, the DFT approach uses working memory models as an important component of building up more cognitive models. Unlike Leabra, however, DFT is focused on ensuring that the low-level dynamics are not compromised by introducing algorithmic methods into their simulations. However, DFT, unlike Leabra, is often not mapped to specific anatomical structures and types of neurons.

There are several reasons to be concerned with the neural plausibility of DFT. For one, DFT relies on non-spiking single cell models. Spikes can introduce important fluctuations into the dynamics of recurrent networks, and so ignoring them may provide misleading results regarding network dynamics. As well, the assumption in DFT that all neurons are physiologically the same is not borne out by neural data. As described earlier, neurons possess a wide variety of responses, and are generally heterogeneous along a variety of dimensions. Because the DFT models are homogeneous, they are difficult to compare in their details to the kinds of data available from single neuron or neural population recordings.

But again, as with Leabra, there are more important representational concerns. Perhaps the most crucial drawback is that there are no methods for introducing structured representations into the approach. Hence, most "cognitive" applications of DFT address fairly minimally cognitive tasks–clearly not the kind of tasks addressed by more traditional cognitive models, such as ACT-R. It is somewhat difficult to argue that these last two approaches are general cognitive architectures given their lack of ability to address high-level cognitive function. Although these approaches have not made unscalable assumptions in trying to incorporate symbolic representations, they have also not provided an alternative that will allow their methods to be applied to the kinds of tasks for which symbolic representations are needed. In short, it is unclear that they provide routes to biological *cognition*.

9.2 ■ An Evaluation

My description of the state-of-the-art suggests that there are two classes of neurally relevant cognitive models. Those that are driven by classical considerations in cognitive science regarding representation (i.e., ACT-R, LISA, NBA, and ICS) and those that are driven by more connectionist, dynamical, and biological considerations (Leabra and DFT). Unfortunately, the former have sacrificed biological constraints, being unmapped (or unmappable) to neural details. The latter have sacrificed cognitive constraints, being unable to explain behavior sensitive to structured representations. Overall, then, it seems that past approaches are biological *or* cognitive, but not both.

This unfortunate divide shows up in many ways. For example, the more biological models are better connected to neuroscientific data, thus doing better on the triangulation criterion. The more cognitive models are better able to address all four criteria related to representational structure. As well, the more biological models tend to be more robust, whereas the more cognitive models can account for central performance features, like syntactic generalization. To support these observations, and to provide a more detailed evaluation of the current state-of-the-art, in this section I return to the CCC and consider how each criterion applies to these architectures. I want to be clear that my evaluation is in no way intended to detract from the important contributions of this past work. Rather, my purpose is to highlight points of difference, thus accentuating the relative strengths and weaknesses of the various approaches that have been taken to understanding biological cognition to date.

9.2.1 *Representational Structure*

a. Systematicity Because symbolic approaches to cognitive representations are systematic, so are the approaches that attempt to implement a symbolic representational system directly. Of the approaches I considered, this includes ACT-R, LISA, the NBA, and the ICS. The other two approaches, Leabra and DFT, do not traffic in symbolic representations and hence do not have an obvious kind of representational systematicity. If, however, we take systematicity to also identify behavioral regularities, then it seems clear that both of these approaches have systematic processes and hence some systematic behavior. For example, if Leabra can react appropriately to "1 appears before AX" then it can also react appropriately to "AX appears before 1." Similarly for DFT: If it can identify that "the square is left of the circle," then it can also identify that "the circle is left of the square."

I take it, however, that most cognitive theorists understand systematicity more in the classical sense. Hence, the former approaches, which are able to employ representations that account for the processing of language-like structure, do better with respect to this criteria.

b. Compositionality Like systematicity, compositionality seems equally well captured by ACT-R, LISA, the NBA, and the ICS. However, in this case, it is not clear that they capture the "right kind" of compositionality. Recall from my discussion of this criteria earlier that compositionality seems to be an idealization of the way in which meaning actually maps to linguistic structure. None of these approaches provides an especially detailed account of the semantics of the symbols that they employ. Hence, they do not explain compositional semantics of the kind found in locutions like "pet fish." However, they all equally account for purely compositional semantics, given their ability to implement a classical representational architecture.

In general, these approaches have little say about semantics, let alone semantic compositionality. It may seem that the ICS, because it employs highly distributed representations, should be better able to capture the semantics of the employed representations. Smolensky and Legendre (2006a) have considered some cases of more grounded spatial representations, but in the end they noted that the "semantic problem ... will not be directly addressed in this book" (p. 163). The NBA and ACT-R do not discuss semantics in much detail. Although the proponents of LISA do, the semantics employed consist only in sharing a few subsymbolic features that are chosen explicitly by the modeler (e.g., the "Bob" and "John" symbol nodes are both linked to a "male" subsymbol node and hence are semantically similar). Thus, LISA does not provide any kind of general account of semantics, grounding, or semantic composition significantly beyond standard classical accounts.

Again, Leabra and DFT do not capture classical compositional semantics because they do not implement structured representations. Nevertheless, proponents of Leabra consider some aspects of linguistic processing (O'Reilly & Munakata, 2000). However, none of the discussed linguistic models employ Leabra-based architectures. Rather, they are largely demonstrations that the underlying software can implement past connectionist models of language.

Consequently, there is no suggestion that Leabra itself can capture semantics or compositionality effects.

DFT models that directly address semantics are not common, although some characterize the mapping between simple spatial words and perceptual representations (Lipinski et al., 2006). These provide rudimentary grounded semantics. More generally, however, the DFT commitment to embodiment brings with it the suggestion that representational states in these models have more grounded semantics than are found in other approaches to cognition. The claim that language is highly influenced by the embodied nature of the language user is a common one in some approaches to linguistics (Gibbs Jr., 2006; however, *see* Weiskopf [2010] for reservations). These views are allied with the DFT approach. However, despite such theoretical suggestions that semantics may be addressed effectively by DFT, there are no models that capture general semantic or compositionality effects.

Overall, current approaches seem to either account for classical compositional semantics or have the potential to ground semantics without relating this grounding to structured, language-like representations. Of the past approaches, the ICS seems to have the representational resources to do both, but the grounding of its distributed representations is not considered. Clearly, much remains to be done to integrate simple classical compositional semantics with more sophisticated semantics grounded in detailed perceptual and motor representations.

c. Productivity (the problem of variables) Again, systems that implement classical representational schemes are able to be productive, at least in principle. In my previous discussion of productivity, I emphasized that doing well on this constraint meant matching the limits of real productivity. In this respect, ACT-R and LISA do better than the NBA and the ICS. LISA, for example, places an explicit constraint on the number of items that can be in working memory while an analogy is being considered. This constraint plays no small role in its ability to match a variety of psychological results. As well, some ACT-R models manipulate a decay parameter that determines how quickly the activation of a production in declarative memory goes to zero. This can do an effective job of explaining working memory limitations in many tasks.

Because Leabra and DFT do not use language-like representations, they do not have much to say about productivity. If, instead, we take Jackendoff's more general "problem of variables" seriously, then it becomes clear that much work remains to be done by both of these approaches. Most of the applications of DST and Leabra avoid describing how anything like a structured representational template can be employed by their approaches. Although Leabra's behavior has been characterized in terms of such templates (e.g., in the 1-2-AX task), such templates are learned anew each time the model encounters a new scenario. Thus it is not clear how the learned relations can be manipulated as representational structures in themselves–that is, as more than just behavioral regularities. Jackendoff, of course, has argued that to explain the flexibility of human cognition, such structures are essential.

In many ways, Jackendoff's characterization of the productivity problem explains some of the scaling problems we find in LISA, the NBA, and the ICS. In all three cases, it is the wide variety of items that can be mapped into the syntactic variables that result in unwieldy scaling when considering productivity at the level of human cognition. LISA, for example, requires nodes dedicated to all the possible bindings between variables and values. As described earlier, this results in an exponential explosion of nodes that soon outstrips the resources available in the brain. The NBA needs to hypothesize complex and fragile binding circuits. The ICS has very poor scaling when such structures get even marginally large.

So, although implementations of classical representations can explain productivity in principle, the representational assumptions they make to do so result in implausible resource demands. This is true despite the fact that some provide useful accounts of the limits of productivity through other mechanisms (typically working memory limitations). The approaches that do not employ language-like representations have little to say about productivity.

d. The massive binding problem (the problem of two) The "massive binding problem" is obviously a scaled-up version of the "binding problem." All of the implementations of classical representation that I have discussed have provided solutions to the binding problem. In each case, I considered their ability to scale in the context of linguistic representations. And, in each case, I demonstrated that the proposed representational and binding assumptions were not likely to be implemented in a biological system (*see* Appendix D.5 for details).

ACT-R, in contrast, is silent on the question of what the neural mechanism for binding is. Although this allows it to address this particular criterion well (as binding is essentially "free"), it will do poorly on subsequent criteria because of this agnosticism.

Again, DFT and Leabra have little to say about this particular issue because they do not describe methods for performing language-like binding in any detail.

So, although all of the neural implementations of classical representation are able to solve the problem of two (because they all posit a viable mechanism for small-scale binding), none of them scale up to satisfactorily address the massive binding problem. The approaches to biological cognition that do not account for binding also do poorly with respect to this criterion. Only ACT-R successfully addresses it, with the qualification that it has little to say about what mechanisms underwrite such binding.

9.2.2 Performance Concerns

a. Syntactic generalization As I mentioned in my earlier discussion of syntactic generalization (Section 8.2.2.1), a commitment to a classical representational structure automatically provides syntactic generalization in principle. However, there are two important caveats to this observation. First, being able to syntactically generalize in principle does not mean being able to do so in practice. As I discussed earlier (Section 7.2), current approaches are not able to meet known constraints on neural mechanisms and the observed speed of generalization (Hadley, 2009). The second caveat is that there are clear content effects in human performance of syntactic generalization (*see* Section 6.6).

With this more pragmatic view toward syntactic generalization in mind, the four proposed approaches that adopt a classical representational method do not meet this criterion well. ACT-R, for example, suffers from being able to only provide an *ad hoc* solution to the content effects that are observed. Because there are no implicit semantic relationships between the symbols employed in ACT-R (i.e., we cannot compare "dog" to "cat" by performing a simple operation, such as a dot product, on those items), any content effects have to be explained in virtue of there being separate, specific rules for different symbols. In view of such concerns, there has been work employing LSA-based semantic spaces in ACT-R (Budiu & Anderson, 2004), but the semantic information is generated outside of the ACT-R framework. Thus, if a new word is encountered by the model, it cannot be incorporated into its semantic representational space without re-running a costly LSA analysis.

LISA, the NBA, and the ICS are unable to meet the temporal constraints. As mentioned in Section 7.3.7, Hadley (2009) has considered these constraints in detail, and has concluded that such approaches do not have the ability to solve the problem of rapid variable creation. Essentially, LISA does not have a mechanism for adding new nodes to its localist network. Similarly, the NBA does not have a method for introducing and wiring up a new word or structure into its fairly complex architecture. The ICS also does not provide a characterization of how the system can itself generate representations that can be used to perform syntactic generalization on-the-fly. I would add that it is also not obvious how ACT-R would handle a completely novel symbol and be able to generate a new production to handle its interpretation.

Perhaps unsurprising given their representational commitments, neither Leabra nor DFT have been shown to perform syntactic generalization.

In sum, although ideal syntactic generalization can be performed by any of the approaches that employ classical representational schemes, the practical examples of such generalization in human behavior are not well accounted for by any of the current approaches.

b. Robustness I concluded my earlier discussion of robustness with the observation that this criterion demands that the models resulting from a cognitive theory should continue to function regardless of changes to their component parts, or variability in their inputs and environment (Section 8.2.2.2). I also described how, historically, the more connectionist approaches had a better claim to robustness, often being able to "gracefully degrade" under unexpected perturbations. One question that arises, then, is whether the connectionist implementations of a classical representational scheme preserve this kind of robustness.

From my earlier descriptions of LISA and the NBA, it is evident that neither of these approaches is likely to remain robust in the face of neural damage. For example, even minor changes to the complex wiring diagrams posited by the NBA will rapidly reduce its ability to bind large structures. Similarly, binding in LISA is highly sensitive to the removal of nodes, as many individual nodes are associated with specific bindings or concepts. As well, for LISA, a reliance on synchrony is subject to serious concerns about the effects of noise on the ability to detect or maintain synchronous states (Section 9.1.2). Noting that each of the nodes of these models is intended to map to many actual neurons does not solve this problem. There is little evidence that cortex is highly sensitive to the destruction of hundreds or even several thousand neurons. Noticeable deficits usually occur only when there are very large lesions, taking out several square centimeters of cortex (1 cm^2 is about 10 million neurons). If such considerations are not persuasive to proponents of these models, then the best way to demonstrate that the models are robust is to run them while randomly destroying parts of the network, and introducing reasonable levels of noise. In short, robustness needs to be empirically shown.

ACT-R's commitment to symbolic representations suggests that it, too, will not be especially robust. Of course, it is difficult to evaluate this claim by considering how the destruction of neurons will affect representation in ACT-R. But there are other kinds of robustness to consider as well. As I have mentioned, it is a concern of proponents of the approach that they do not support partial matching (Section 9.1.1). This means, essentially, that there cannot be subtle differences in representations–the kinds of differences introduced by noise, for example, that affect the production matching process. Or, put another way, there cannot be "close" production matches that could make up for noise or uncertainty in the system. So, ACT-R is neither robust in its processes nor its representations.

The robustness of the ICS is difficult to judge. If the system were consistently implementable in a fully distributed manner, then it should be quite robust. But, because control structures are not provided, and most of the actual processing in current models is done in localist networks, it is not clear how robust an ICS architecture would be. So, although the distributed representations used in tensor products should be highly robust to noise, the effects of noise on the ability of the system to solve its optimization problems in distributed networks has not been adequately considered. As a consequence, we might expect that an ICS system could be quite robust, but little evidence is available that supports that expectation, especially in the context of a reconfigurable, distributed system.

We can be more certain about the robustness of DFT. Attractor networks are known to be highly robust to noise and reasonably insensitive to the disruption of individual nodes participating in the network. It is likely, then, that the dynamics intrinsic to most DFT models will help improve the robustness of the system overall. One concern with robustness in DFT models is that it is well known that controlling dynamical systems can be difficult. Because there are no "control principles" available for DFT, it would be premature to claim that any DFT model will be robust and stable. So, although the representations seem likely to be robust, it is difficult to determine, in general, if the processes will be. Essentially the architecture is not well-specified enough to draw general conclusions.

Similarly, for Leabra, the reliance on attractor-like working memories will help address robustness issues. In many Leabra models, however, the representations of particular states are highly localized. Because the models are seldom tested with noise and node removal, it is again difficult to be certain either way about the robustness of a scaled-up version of these models. The current theoretical commitments, at least, do not preclude the possibility that such models could be robust.

Overall, then, state-of-the-art approaches do not perform especially well with respect to this criterion. It seems that systems are either highly unlikely to be robust or that there is little positive evidence that generated models would continue to function under reasonable disturbances caused by neuronal death, input variability, or other sources of noise.

c. Adaptability Adaptability comes in many different guises (*see* Section 8.2.2.3). Syntactic generalization, for example, is a kind of adaptability in which syntactic representations are used to reason in unfamiliar circumstances. As discussed earlier, "in principle" syntactic generalization is captured by all of the classical representational approaches. However, rapid generalization, exemplified in the problem of rapid variable creation, is not well handled by any of these approaches except perhaps ACT-R.

Perhaps the most familiar form of adaptability is learning. Many state-of-the-art approaches consider learning of one kind or another. ACT-R models, for example, are able to generate new productions based on the use of past productions to speed processing. This is done through a kind of reinforcement learning (RL) rule. Leabra, too, has much to say about reinforcement learning, and hence is quite adaptable in this sense. Because DFT has focused on models of child development, it too has dealt explicitly with the adaptability of the system it is modeling. And, finally, LISA includes a kind of learning to extract schemas from past experience. Only the ICS and the NBA have little to say about adaptability in this sense.

As indicated in my discussion in Section 6.4, there are many different kinds of learning to be considered when characterizing biological cognition. Although many approaches have *something* to say about learning, they often address only one specific kind of learning (e.g., RL in ACT-R, Hebbian learning in DFT, and schema extraction in LISA). Ideally, many of these different kinds of learning should be integrated within one approach. On this count, Leabra is notable for its inclusion of Hebbian, RL, and backprop-like forms of learning within its methods.

One other aspect of adaptability that is especially salient for LISA, the NBA, the ICS, and DFT is that of task switching. Because these approaches are silent regarding the operation of a control system that can guide the flow of information through the system, they have little to say about the evident ability of cognitive systems to move seamlessly between very different tasks. Each of these approaches tends to construct different models for different tasks. Hence, it is unclear to what degree such approaches will support this kind of observed adaptability.

Task-sensitive adaptability may also be a concern for Leabra and ACT-R. Although control is given serious consideration in Leabra, it is often limited to control within a specific task. It is unclear whether the architecture can support a more general account of information routing across a variety of tasks. Similarly, ACT-R models are usually tightly focused. Of course, this concern comes in a matter of degrees. The highly complex tasks of some ACT-R models (e.g., driving) may involve completing many subtasks, necessitating significant consideration of control. However, integrating many diverse ACT-R models within one functioning system is not common practice.

In conclusion, I take it that all state-of-the-art architectures have considered some forms of adaptability. However, none of the architectures capture the variety of timescales and mechanisms that are responsible for adaptation in biological cognition. One adaptation problem that provides a significant challenge for all considered approaches except ACT-R is the problem of rapid variable creation.

d. Memory Out of the many kinds of memory, I highlighted long-term and working memory for special consideration. I noted that long-term memory has a high capacity and that working memory is rapid. I suggested that characterizing each, and their relationship, is critical to capturing occurrent cognitive processing.

The ACT-R approach explicitly distinguishes long-term and working memory using the terms "declarative memory" and "buffers," respectively. Indeed, ACT-R has been used to describe many different kinds of memory phenomena. Consequently, it does well on this criterion. However, one central drawback that remains in ACT-R's account of long-term memory lies in its inability to relate complex perceptual long-term memories to the system's understanding of the world. The declarative memory that ACT-R has includes many facts about the world that can be stated symbolically. These are typically determined by the programmer of the model. However, this misses how the symbols used in those descriptions relate to the perceived and manipulated external world. That is, there is no representational substrata that captures the variability and subtlety of perceptual experience. So the account of long-term memory provided by ACT-R is partial and semantically superficial.

LISA and the NBA both have working memory-like mechanisms, although neither captures serial order effects. In the NBA these mechanisms are used to support temporary binding of symbols. In LISA, they account for the capacity limitations of working memory observed in people. However, both have poor accounts of long-term memory. LISA's minimal semantic networks could be considered a kind of long-term memory. However, they do not capture either retrievable facts about the world or more subtle perceptual representations like those missing from ACT-R. The NBA does not discuss long-term memory at all, although perhaps the word nodes could be considered long-term memory. However, like LISA, these do not encode factual knowledge or relate to sensory experience. The ICS discusses neither.

Leabra and DST are both centrally concerned with working memory phenomena. As a result, they have good characterizations of working memory in a biologically inspired substrate, although neither effectively accounts for serial effects. Both have little to say about long-term memory, however. This, perhaps, should be qualified by the observation that Leabra is able to learn procedures for manipulating its representations, and these may be considered long-term memories. As well, DFT considers the development of various kinds of behavior over time that would presumably persist into the future. In both cases, however, these are clearly very limited forms of long-term memory when compared to the declarative system of ACT-R, or to the kinds of perceptual and motor semantic representations that are crucial for central aspects of biological cognition.

e. Scalability Geoff Hinton recently wrote: "In the Hitchhiker's Guide to the Galaxy, a fearsome intergalactic battle fleet is accidentally eaten by a small dog due to a terrible miscalculation of scale. I think that a similar fate awaits most of the models proposed by Cognitive Scientists" (Hinton, 2010, p. 7). In short, I agree.

I highlighted several challenges relating to scalability in Section 8.2.2.5. These included knowing the dimension along which to evaluate scaling, knowing how to manage interactions in large systems, and being able to establish scalability in practice, rather than only in principle. I have obliquely addressed some of these issues in my discussion of previous criteria. For example, the massive binding problem clearly relates to the ability to scale binding assumptions up to complex representations. Similarly, my concerns about the ability of many approaches to be able to address multiple tasks in a single model is closely related to issues of accounting for interactions in a complex system. In both cases, I was somewhat pessimistic about the probable success of current approaches.

In the context of the previous analysis of scalability, this pessimism is reinforced by the inability of several approaches to even meet "in principle" scalability constraints. For example, each of LISA, the NBA, and the ICS seem unlikely to be able to support representations of average complexity over a human-sized vocabulary of about 60,000 words without overtaxing available resources in the human brain (*see* Appendix D.5). DFT and Leabra do not address binding issues in any detail, and hence it is difficult to be optimistic about their abilities to scale as well. Although ACT-R can claim to be able to encode many complex representational structures, attempts to relate its underlying representational assumptions to a neural architecture have so far failed. This does not bode well for its scalability in the context of the physical resources of the brain either.

As I discussed in the section on adaptability, it is also difficult to be optimistic about the scalability of the functional architectures associated with most of the approaches. The ICS, LISA, the NBA, DFT, and Leabra do not specify the majority of the control processes necessary for implementing a functional system that is able to perform many cognitive tasks in a single model. Again, on this count ACT-R is in a much better position. This is not only because ACT-R has been applied to more complicated tasks but also because, unlike the other approaches, ACT-R provides principles that help specify how it can be applied to novel complex tasks.

Nevertheless, even in the ACT-R architecture there are few actual examples of one model performing many distinct tasks. Thus, when we turn to consideration of "in practice" scalability, it is clear that we have a long way to go in our understanding of biological cognition. As mentioned, most of these approaches construct different models for each different task to which they are applied. Constructing integrated systems that can perform many different tasks using the exact same model is largely unheard of. And, as the story of

Tusko the elephant reminds us, thinking that "more of the same" will solve this problem is a dangerous assumption.

Echoing Hinton, this is perhaps the performance criterion that holds the greatest challenges for current cognitive architectures.

9.2.3 Scientific Merit

Criteria related to scientific merit are less often discussed in cognitive science than might be expected from a scientific discipline.[3] Perhaps this is because it is all too obvious that cognitive theories should respect such criteria. Or perhaps it is deemed too difficult to effectively address such criteria early in the discipline's development. Or perhaps both. In any case, I strongly believe that, as Taatgen and Anderson (2010) have put it: "Eventually, cognitive models have to live up to the expectations of strong scientific theories" (p. 703).

a. Triangulation (contact with more sources of data) Proponents of ACT-R have been explicitly concerned with trying to improve the ability of their methods to contact more sources of data, in particular biological data. Their most successful efforts have been aimed at connecting various models to fMRI data, as I described earlier. This same goal lies behind recent attempts to integrate ACT-R and Leabra.

It is clear that ACT-R is well constrained by behavioral data such as reaction times, choice curves, eye movements, and so on. But, as I discussed earlier (Section 9.1.1), despite the efforts undertaken to relate ACT-R to neural data, the quality of that relationship is highly questionable. For example, by far the majority of brain areas have not been mapped to elements of the ACT-R architecture–eight have been. As well, the number of free parameters used to fit fMRI data is often very high. And, the mapping is not one based on a physiological property of the model (e.g., the activation level of a module) but, rather, a temporal one (amount of time spent using a module). Essentially, brain areas are treated as being on or off.

Perhaps we can dismiss these as minor concerns that will be addressed by future work on ACT-R. None of these are, after all, fundamental flaws. But, a deeper problem lies in the fact that it is unclear how *other* neural data can be related to ACT-R models. With a commitment to symbolic representation, and no theory on how such representations relate to single neuron activity, most medium or fine spatial and temporal scale methods cannot be related to ACT-R models (including electroencephalogram [EEG], magnetoencephalography [MEG], electrocorticography, local field potential [LFP] recordings, multi-unit recordings, single-unit recordings, calcium imaging, etc.). Given the increasing power and sophistication of such methods, this does not

[3] Although those who agree with Searle's (1984) quip–"A good rule of thumb to keep in mind is that anything that calls itself 'science' probably isn't" (p. 11)–may be less surprised.

bode well for the future of ACT-R. Perhaps this explains the urgency with which proponents are pursuing methods of overcoming this major conceptual hurdle.

Although ACT-R proponents have not embraced a particular implementational story, many such stories have been offered, as I have recounted. However, the neural implausibility of each of LISA, NBAs, and the ICS makes them unlikely to be able to ultimately provide a connection between detailed neural methods and a classical representational system like ACT-R. All suffer from over-taxing available neural resources. As well, to the best of my knowledge, none have made explanatory or predictive connections to detailed neural data. The NBA, for example, has not been related to high-resolution MEG, fMRI, LFP, or other experiments that might provide insight into its proposed connectivity and function. LISA has not predicted or explained expected patterns of synchrony that have been measured during various tasks. The ICS has not been correlated with MEG, fMRI, or EEG experiments run during language processing. And none have been tested against the plethora of data available from experiments on related mechanisms in non-human animals.

Many of these same criticisms can be leveled against DFT. Although the models are run on abstract "neural sheets," and plots of the dynamics of neural activity can be provided, these are generally not compared directly to available neural data (although *see* Jancke et al. [1999]). For example, there have been many single-cell experiments performed during working memory tasks in various mammals. Such experiments do not provide evidence for the idealized homogeneous responses and highly stable dynamics found in DFT models. Indeed, a recent theme in some of this work is to explain why the dynamics in working memory areas are so *unstable* and why neural responses are so *heterogeneous* (Romo et al., 1999; Singh & Eliasmith, 2006; Machens et al., 2010). Similarly, many of the developmental processes addressed by DFT models are observed in non-human animals as well, so relevant single-cell data are often available for such models. Nevertheless, it is uncommon to find an explicit connection between these models and the neural data.

By far the most biologically driven of the approaches I have considered is Leabra. This is nowhere more evident than in the consideration of its contact with neural data. As I described (Section 9.1.5), Leabra has been used to predict the effects of different kinds of neurotransmitter agonists and antagonists. It has also been used to predict the effects of aging and disease on cognitive function caused by variations in underlying neural mechanisms. It has also been employed to explain higher-level neural data like fMRI. And, unlike ACT-R, in Leabra neural activity is used to match to fMRI activation (Herd et al., 2006).

That being said, Leabra is not able to match certain kinds of detailed single-cell data for several reasons. First, because it models "average" neurons, it is missing the typically observed neural heterogeneity. This is crucial because it is becoming increasingly evident that the precise distribution of neural responses can tell us much about the function of a population of cells (Eliasmith & Anderson, 2003, chapter 7; Machens et al., 2010). As well, Leabra uses rate neurons and so is unable to provide spike timing data, as I discussed in more detail earlier. Relatedly, concerns I voiced about the use of kWTA in place of actual neural dynamics make it less able to address a variety of dynamical neural phenomena in any detail. Nevertheless, Leabra has been more directly related to neural data than other approaches.

However, for both DFT and Leabra, their more convincing connection to low-level neural data has been achieved at the price of being less able to explain higher-level behavior. This can be seen in the desire of proponents of Leabra to connect their work to ACT-R. The enviable breadth of application of ACT-R is not achievable within Leabra because it does not provide theoretical principles that determine how the low-level activity patterns in Leabra relate to the symbol manipulation exploited in ACT-R. This is true despite the explicit attempts to marry these two approaches I mentioned previously. Similarly, in DFT the "most cognitive" behaviors addressed are the identification of simple spatial relations from visual stimuli and the tracking of occurrent objects. These are not comparable to the "most cognitive" behaviors addressed by the likes of ACT-R.

In sum, there are no state-of-the-art approaches that convincingly connect to a broad range of empirical results regarding biological cognition. Rather, most approaches have a preferred kind of data and, despite occasional appearances to the contrary, are not well constrained by other kinds of data. For example, Leabra and DFT stand out as being better connected to neural details than the other approaches. Unfortunately, this has resulted in their being less well connected to high-level behavioral results.

b. Compactness Perhaps all of the considered approaches should be praised for their compactness because they are each stated rigorously enough to be simulated. This is not a common characteristic of much theorizing about brain function (although it is, thankfully, becoming more so). Nevertheless, there are clearly differing degrees to which the considered approaches are rigorous, systematically applied, and non-arbitrarily mapped to the brain.

I believe ACT-R is the most compact of currently available theories. It is quantitatively stated; it has some central, unvarying parameters (e.g., 50 ms per production); and the mapping of the functional description to the brain is clearly stated. In addition, it includes a consistent set of control structures and representational commitments, which means that different applications of the architecture retain its central features. Perhaps the greatest source of arbitrariness stems from the fact that the declarative memory, which contains the relevant rules and facts, is specified by the modeler. It has been suggested that this essentially provides the architecture with unlimited flexibility (Schultheis,

2009). However, that is only a problem if other commitments of the architecture do not limit this flexibility, but they do. The 50-ms assumption provides some limitations on what can be done within a given time frame, assuming the complexity of productions is kept within reason. As well, if we insist on the specification of models that are able to perform many tasks without changing the model, interactions between components will force some degree of consistency among specified rules. Unfortunately, different ACT-R models are seldom integrated in this manner. Nevertheless, the architecture provides the resources necessary to do so.

Approaches such as LISA, the NBA, the ICS, and DFT, which do not specify a general control structure, are in a difficult position with respect to compactness. These methods currently need the modeler to design a wide variety of elements prior to modeling each particular task. LISA requires a specification of the wiring to support the relevant bindings and processing, as well as a specification of the representations and semantics needed for a task. The NBA and the ICS require similar specifications. Analogously, DFT requires identification of the number and interaction of the attractor networks and their relation to input and output stimuli.

Of course, specifying things is not the problem. The problem is that design decisions are often different from model to model because the architecture itself does not provide general principles for constructing models. To be fair, some such specifications are more principled than others. For example, mapping from a harmony network to a distributed equivalent is mathematically defined in the ICS. In contrast, the particular choice of subsymbolic nodes to specify semantics and the particular wiring to perform a task in LISA and the NBA is left to the modeler. Of course, these criticisms could be addressed if there was a single, canonical model that captured performance on many tasks at once. Then, a given model would suffice for a broad range of tasks, and the charge of "arbitrariness" would be weak. Unfortunately, such canonical models are not on offer. These concerns are clearly related to the scalability issues I discussed earlier.

Leabra has something of a more principled control structure, although at a very high level of abstraction. Nevertheless, the coupling of Leabra-based models, such as PVLV and PBWM, is to be commended for being systematically used across a variety of models. Attempts are clearly made to keep the many parameters of these models constant, and basic structural assumptions generally remain consistent. However, I believe that ACT-R is more compact than Leabra largely because judgments of compactness depend also on the breadth of application of a theory. Because Leabra does not address a wide variety of cognitive behaviors accessible to ACT-R, but ACT-R does heavily overlap with Leabra in its account of reinforcement learning (although Leabra's account is more thorough), ACT-R seems to be more compact overall. In addition, ACT-R is more systematic in its mapping to temporal phenomena.

When turning to consideration of temporal phenomena in general, most of the considered approaches are disconcertingly arbitrary. That is, perhaps the greatest degree of arbitrariness in all of the current accounts is found in their mappings to time. There are no obvious temporal parameters at all in some accounts, such as the ICS and the NBA. Other approaches, like LISA and Leabra, are prone to use "epochs" in describing timing. Unfortunately, different models often map an "epoch" onto different lengths of "real" time, depending on the context. As well, Leabra relies on basic computational mechanisms that avoid time all together (e.g., kWTA).

In contrast, DFT is fundamentally committed to the importance of time. However, a difficulty arises when we consider how the time parameters in DFT models are related to real time. As with LISA and Leabra, DFT does not specify how to map model parameters to real time, and so different models are free to map model time to real time in different ways.

Finally, ACT-R is less temporally arbitrary because of its insistence on always using 50 ms to be how long it takes to fire a production. However, as I hinted above, the complexity of productions may vary. This means that very simple or very complex rules may be executed in the same length of time. Without a further specification of some constraints on the complexity of productions, the ACT-R mapping to time is not as systematic as is desirable.

To conclude, although all of these approaches are to be commended for rigorously specifying their models, ACT-R and Leabra are the most compact. This stems from greater evidence of their providing a systematic mapping between components of the theory and the system they are modeling. However, all approaches are somewhat arbitrary in their relation to real time.

9.2.4 *Summary*

Evaluating past approaches with respect to the CCC helps make clear the kinds of trade-offs different approaches have made. It also helps highlight general strengths and weaknesses that appear in the field as a whole. I have summarized my discussion in Table 9.1.

It is possible to see some general trends from this table. It is clear, for example, that symbol-like representation and biological detail remain somewhat orthogonal strengths, as they have been historically. It is also evident that each approach would greatly benefit from pursuing larger-scale models, to more effectively address several criteria including not only scalability but also adaptability, massive binding, and compactness. These same large-scale models could serve to demonstrate the robustness of the approaches, another criterion on which most approaches do poorly.

Finally, this analysis also suggests that the general criteria for good scientific theories could be much better addressed by current approaches. This is an observation that, as I noted earlier, has been made by other theorists in the

TABLE 9.1 *Comparison of Past Approaches With Respect to the CCC, Rated on a Scale from 0 to 5*

	ACT-R	LISA	NBA	ICS	DFT	Leabra
Systematicity	++++	++++	++++	++++	++	++
Compositionality	++	++	++	++	+	-
Productivity	++++	+++	+++	+++	-	-
Massive binding	+++	+	+	+	-	-
Syntactic generalization	++	++	++	++	-	-
Robustness	-	+	+	++	++	++
Adaptability	++	+	-	-	+	++
Memory	+++	++	+	-	++	++
Scalability	+	-	-	-	-	-
Triangulation	+	+	+	+	++	++
Compactness	++	+	+	+	+	++

Columns are in order of least to most biologically plausible.

field. This observation is also consistent with a sentiment sometimes encountered among non-modelers, who suggest that there are "too many free parameters" in most large-scale models. The more compact and data-driven theories of cognition become, the more unreasonable such off-the-cuff dismissals will be.

One of my main motivations for providing this analysis, and the CCC in general, is to help situate the SPA in the context of past work. While presenting the SPA, I have attempted to highlight some similarities and differences between the SPA and other approaches. In the next three sections, I have gathered and related these observations to the CCC in an attempt to give a fuller picture of how the SPA fits into the landscape of current cognitive approaches.

9.3 ■ The Same...

The approaches I have been considering in this chapter span the range of paradigms on offer over the last 50 years of cognitive science. Although I have picked the more biologically relevant approaches from each of the symbolic approach, connectionism, and dynamicism, the main theoretical commitments of all three approaches are well represented in what I have surveyed. I now turn to a consideration of the approach on offer here. Perhaps, it is already obvious that the SPA is all and none of these past approaches. Indeed, my earlier plea to "move beyond metaphors" (Section 1.2) intentionally opened the door to the possibility that no one paradigm would turn out to be right. In this section, I discuss the similarities between each of these past approaches and the methods and commitments of the SPA. This demonstrates that the SPA has borrowed heavily from past work. In the next section, I emphasize the differences.

The SPA shares with LISA, the NBA, and the ICS a commitment to providing a neurally plausible account of cognition that captures symbol-like representation and manipulation. In all four cases, there is an emphasis on the importance of binding to construct structured representations that can then support a variety of syntactic and semantic manipulations. In fact, the preferred mechanism for binding in the SPA comes from the same family as that employed in the ICS. As a result, many of the mathematical results available regarding the ICS are quite informative about representational and transformational resources available to the SPA. Relatedly, the ICS offers a sophisticated story regarding how such representations can be related to a wide variety of linguistic processing. The representational commitments of the SPA are largely consistent with that characterization.

The SPA shares a central commitment to dynamics with DFT. It is no surprise, then, that attractor networks form a basic computational component that is re-used throughout both architectures. This concern for dynamics connects directly to interest in perception-action loops, and accounting for the integration between these aspects of behavior, and more traditionally cognitive aspects. There is little in the SPA that could not be reformulated using terms more familiar to dynamic systems theorists. Control theory and dynamic systems theory use many of the same mathematical tools, concepts, and methods, after all. The common use of temporally laden terminology belies this deeply shared commitment to understanding biological cognition through time.

Turning to Leabra, the focus moves more to neuroscientific details. Leabra's central focus on the biological basis of cognition is shared with the SPA. The similarities between Leabra and the SPA go further; however, there is a general agreement about some basic architectural features. Both consider the cortex-basal ganglia-thalamus loop to play a central role in controlling cognitive function. Both consider the biological details of this architecture when discussing learning of control, as well as control of past learned behaviors. Not surprisingly, then, both have similar relationships to neural data regarding the dopamine system. It is clear that Leabra has much to recommend it. The SPA has been heavily influenced by this and related approaches to basal ganglia function (e.g., Gurney et al., 2001). Overall, the SPA and Leabra share a concern with the biological implementation of adaptive control in cognition.

ACT-R and the SPA share a basic interest in high-level cognitive function. Both propose ways that rule-like behavior, resulting from symbol-like representation, can occur. They share, with Leabra, a commitment to the cortex-basal ganglia-thalamus loop as being the central control structure in explaining such behavior. All three see a similar place for reinforcement learning and hence relate to similar learning data in some respects. As well, both the SPA and ACT-R have something like rule matching at the heart of that adaptive control structure. A related interest of ACT-R and the SPA is found in their focus on the timing of behaviors. Specifically, behavioral-level reaction times are relevant to the functioning of both architectures. Finally, both approaches are very concerned with understanding the mapping of the cognitive architecture to the underlying neural structures. Hence, both are interested in relating their approaches to the biological basis of cognition.

Given these similarities, there are many ways in which the SPA can be considered a combination of commitments of the ICS, DFT, ACT-R, and Leabra. Although ACT-R and Leabra alone do not seem to have been broadly integrated, adding the ICS to such an attempt provides an important extra ingredient for integration. The challenge, after all, was to relate the representations of ACT-R to those of Leabra. The ICS provides a rigorous and general mapping between the two. In addition, the basic control structure is similar between ACT-R and Leabra, and, because the ICS and DFT are silent on the issue of control, there are no conceptual conflicts preventing this broader integration. The SPA can thus be seen as an attempt to build the necessary bridges between these approaches in a constructive and neurally responsible way, while incorporating some conceptual insights from DFT regarding dynamics and embodiment.

In the first chapter, I suggested that successfully breaking from the dominant metaphors in cognitive science may lead to a new kind of theory, as happened in the wave-particle debate in physics. The intent of such an integration is to not only adopt the strengths of each approach, but to overcome the weaknesses. In the next section, I focus on differences between the SPA and past approaches in order to make this case for the SPA. In short, I want to suggest that the SPA is more than the sum of its parts.

9.4 ■ ...But Different

While the SPA shares an interest in understanding symbol-like behavior in a neural substrate, it takes a notably different approach from each of LISA, the NBA, and the ICS. It is, as I mentioned, most similar to the ICS. However, the difference lies in the fact that the result of binding is always compressed in the SPA. As I discussed in Section 9.1.4, this is a crucial difference between the SPA and the ICS. In fact, this is a difference that helps distinguish the SPA from past attempts at implementing symbolic architectures in general. This is because compressive binding guarantees that the SPA does not implement a classical symbolic representational system (*see* Section 4.2).

Because of the compression, there is a loss of information in the encoded representation. This means that there is no isomorphism between the original symbolic structure and the elements of the compressed semantic pointer. Without that isomorphism, classic compositionality does not hold, and there is no clean implementational relationship between the SPA representations and a classical symbolic representation. In short, I have been careful to note that the SPA traffics in "symbol-like" representations, not symbols. They are like symbols in their ability to encode structure, but unlike symbols in their compositional characteristics. It is exactly this "approximation" to symbols that allows the SPA to scale much more plausibly than any of the other approaches (*see* Sections 4.3 and 4.6 and Appendix D.5). The price, of course, is a decrease in accuracy of reconstructing what was encoded. This puts important constraints on the depth of structure that can be encoded. But, as mentioned in my description of the CCC, those same kinds of constraints are relevant for all of the representational criteria: productivity is limited, compositionality is partial, and systematicity is restricted.

This lack of an isomorphism between the neural implementation and the symbolic-level description means that the ICS and the SPA embody different conceptualizations of the relationship between levels of description in cognitive science. I have described both conceptions in some detail (Sections 9.1.4 and 2.3, respectively). It is evident from those discussions that the difference lies in the "messy" mapping between levels allowed by the SPA. This messiness means that clean isomorphisms are unlikely and that neural details will matter for capturing certain kinds of behavior (e.g., recall the example of serial working memory).

Nevertheless, higher-level descriptions can be helpful, if approximate, for answering certain kinds of questions about cognitive function (e.g., what rule was induced in the Raven's task?). Of course, the neural level description is an approximation to an even more detailed model in a similar manner. The pragmatic aspects of "descriptive pragmatism" about levels (*see* Section 2.3) means that the questions we are asking of a model help determine what the appropriate level of analysis and simulation of the model is. Like the ICS, the SPA supports descriptions across a range of levels. But in the SPA this is *not* in virtue of an isomorphism. And, I should add, the SPA covers a wider range of levels than does the ICS.

Although the non-classicism of the SPA approach helps distinguish it from the NBA and LISA, there is perhaps a deeper difference. In the NBA and LISA, the representation of specific items is fixed to a specific part of the network. In a localist network, this describes the standard method of having the "dog" representation be a single node. In the distributed case, this would be some specific subset of nodes being the "dog" representation. In the SPA, the representation of "dog" can be in many different networks at the same time or at different times. This is because that representation can (often, although not necessarily exclusively) be found in the activity patterns of a population of cells. Populations of cells represent vector spaces, and many different items can be represented in a given vector space. At any point in time, the specific representation in a population depends on *transient* activity states. In short, SPA representations are more *mobile* than those in the NBA and LISA.

Perhaps it is helpful to think of the SPA as characterizing neural representations as being somewhere between standard computer representations and standard neural network representations. Standard computers are powerful partly because they do not commit a particular part of their hardware to specific representations. Thus, the same hardware can process representations about many different things. It is the *function* of the hardware, not its representations, that is often taken to define its contribution to the overall system. If we adopt the NBA- or LISA-style methods of representation, then specific populations or nodes are committed to particular representations. The SPA commits hardware to a specific vector space but not to a specific point or a specific labeling of a vector space. This helps it realize some of the key flexibility of traditional computers in being able to represent many things with the same hardware (and represent the same thing with many different bits of hardware). Consequently, the same neural population can compute functions of a wide variety of representations. This is a subtle yet important difference between the SPA and many past connectionist approaches. It is because of this flexible use of neural populations that the SPA can support control and routing of representations. This is an aspect of cognitive function conspicuously absent from LISA and the NBA.

In the previous section I highlighted the shared emphasis on dynamics and embodiment between DFT and the SPA. This is a conceptual similarity, but the practical applications are quite different. The SPA is regularly compared to, and constrained by, detailed single-cell spiking dynamics. As well, the dynamic components of the SPA are intended to be actual neurons, mapping to the observed heterogeneity, tuning, and response properties of these elements of the real system. As such, the parameters of the nodes are measurable properties of the brain. DFT, in contrast, does not provide such a specific mapping to aspects of the physics and functions of neurons.

The SPA also has a broader characterization of the dynamics and interaction of the perceptual, motor, and cognitive systems than is found in DFT. A central reason for this is that the SPA characterizes the dynamics of much higher-dimensional spaces than DFT typically does. Because the SPA distinguishes between the state space and the neuron space in which the state space is embedded, it is possible to understand the evolution of a very complex neuron space with a few degrees of freedom in the state space. This mapping then makes it possible to extend the dimensionality of the state space to even higher dimensions, while keeping the principles that map dynamics to neural representation the same. In many ways, the SPA working memory that stores a 100D semantic pointer is a straightforward extension of the 2D working memory of a DFT sheet of neurons (Eliasmith, 2005a). However, the computations and dynamics available in that higher-dimensional space can be much more sophisticated. Specifically, the principles of the NEF allow explicit nonlinear control to be introduced into these dynamical systems, which opens a host of important possibilities for constructing networks with useful dynamics (e.g., for routing, gain control, binding, controlled oscillation, etc.).

Additionally, the NEF provides the SPA with methods that allow such systems to be constructed without the Hebbian learning typically relied on in DFT models. This allows much more complicated systems to be constructed and compared to data from developed cognitive systems. Nevertheless, to address developmental questions, learning must be used. As I have emphasized in Chapter 6, many varieties of biologically plausible learning are also available in the SPA.

Learning is also a central focus of Leabra and, hence, a point of comparison between the SPA and Leabra. As I have already mentioned, both approaches allow for the characterization of a wide variety of learning results. As well, much of the control architecture is shared between the SPA and Leabra. The main differences between the approaches lie at the extremes of levels of analysis. At the level of individual neurons, the SPA offers a characterization of neural dynamics and individual spike responses not available in Leabra. And crucially, the SPA does not rely on computational approximations like kWTA to produce its predictions. Rather, the detailed dynamics–for example, of basal ganglia function and adaptation–arise out of the low-level dynamics of SPA networks. These low-level dynamics are governed by various neurophysiological properties, such as single-cell currents, neurotransmitters, and receptors.

At the opposite end of the spectrum, the SPA provides an explicit account of symbol-like representation not available in Leabra. This account is integrated with the neural details, the shared control structures, and the shared aspects of learning. It also solves the "interface" problem that has confronted the attempted integration of Leabra with ACT-R (Jilk et al., 2008, p. 213). After all, there is no interface between "symbolic" semantic pointers and "perceptual" or "motor" semantic pointers–semantic pointers are semantic pointers. In many ways, the representational aspects of the SPA provide to Leabra what ACT-R was intended to bring to their integration.

I have described many of the similarities between the SPA and ACT-R, but perhaps the differences are more obvious. The main difference stems from distinct representational commitments. ACT-R begins with symbols, the SPA begins with populations of neurons. Consequently, for ACT-R the problem of mapping symbols to neurons looms large–for the SPA, it was a starting point. I have provided several examples that show how relating the models produced by the SPA to specific neural details comes quite easily to the SPA. I believe that this conceptual benefit of the SPA outweighs the 30-year headstart ACT-R has with respect to building actual cognitive models. Interestingly, many other differences between these approaches result from this switch in representational assumptions.

For example, the SPA allows for a more flexible and powerful characterization of rule application, as I have discussed in detail previously (Section 7.4). In short, the switch in representations has changed *how* rules can be expressed (i.e., as vector transformations), what *kinds* of rules can be expressed (i.e., more statistical mappings), and how available rules can be *matched* to current context (i.e., statistical or partial matching). When applying the SPA to a cognitive task, standard variable-binding rules used in ACT-R are often not necessary. For example, in Section 5.6, there are many rules that simply copy information between elements of the architecture, rather than explicitly including variables in the rules themselves. If this approach proves generally useful, then it may be an important change to how cognitive rules are characterized across many tasks.

It is also notable that the SPA's representational story highlights a conceptual difference between how ACT-R and the SPA treat levels. ACT-R proponents, unlike those of the ICS, suggest that there is a "best" level of description for cognitive systems: "In science, choosing the best level of abstraction for developing a theory is a strategic decision . . . in both [symbolic and connectionist] cases, the units are a significant abstraction . . . I believe ACT-R has found the best level of abstraction [for understanding the mind]" (Anderson, 2007, pp. 38–39). Although Anderson has left open the possibility that the best level is not yet available, it seems to me to simply be a mistake to expect that there is *a* best level for understanding the mind.

I believe that the suggestion that there is a single best level of description will always cry out for an answer to the question: "The best for what purpose?" If this purpose changes, even slightly, then so might the most appropriate level of description. The SPA's focus on *relating* levels, rather than *picking* levels, allows descriptions to be systematically adjusted to meet the relevant purposes for a given question about cognition. This, of course, is central to my notion of descriptive pragmatism. Sometimes the best description will demand reference to single-cell spike patterns or a characterization of the effects of specific neurotransmitters. Other times, the best description will demand the specification of which rule was applied or what sentence was generated by the system. A unified approach, able to relate many such descriptions across many different levels, strikes me as the goal of most scientific enterprises. I am not alone. For example, Craver (2007) has argued in detail that good explanations in neuroscience should contact many specific mechanisms at multiple levels of description and across disciplinary fields (*see also* Bechtel & Richardson, 1993; Bechtel, 2005; Thagard & Litt, 2008).

In general, I take it that the SPA spans the relevant levels of description in a more convincing manner than past approaches, including ACT-R. A major reason that the SPA can employ highly detailed neural representations, as well as abstract symbol-like representations, is that the methods for characterizing representation are flexibly specified. For example, the NEF mapping of vector spaces to neurons does not impose any specific form on the neural nonlinearity. Although most models I have presented use a simple spiking leaky integrate-and-fire model, the complexity of useful single neuron models in the SPA is limited only by the researcher's knowledge and available computational resources. In previous work, we have used adapting LIF neurons (Singh & Eliasmith, 2006), reduced bursting neurons (Tripp & Eliasmith, 2007), and detailed conductance-based neurons (Eliasmith & Anderson, 2003). The fundamental approach remains the same.

Similarly, the amount of detail in the synaptic model is not constrained by the NEF approach. Consequently, if deemed relevant, a researcher can introduce uncertain vesicle release, saturating synaptic conductances, axonal delays, etc. In most of our models we introduce a generic noise term to account for these kinds of uncertainties, but the precise mechanisms leading to such variability can be explicitly included, if necessary, for a desired explanation.

Much of this breadth stems from the SPA's adoption of the NEF, which has been, and continues to be, centrally concerned with neural plausibility. For example, in previous work it has been shown that any network constructed with the NEF can be made consistent with Dale's Principle,[4] hence capturing a major architectural constraint ignored by much past modeling (Parisien et al., 2008). In addition, a wide variety of single-cell dynamics, including adaptation, bursting, Poisson-like firing, and so forth, has been directly incorporated into the methods of the NEF, helping to make it generally applicable to cortical and subcortical systems (Tripp & Eliasmith, 2007). The continuing development of a neurally responsible theoretical substrate for the SPA means that the SPA characterization of biological cognition can continue to incorporate ever more subtle aspects of biology.

Each such added detail improves the realism with which the model's mechanisms approximate those that have been characterized by neuroscientists. And, each such added detail opens a field of neuroscience from which data can be drawn to be directly compared to the model. Further, many such details result in changes in neural spike patterns, one of the most commonly measured properties of neural systems. Thus, model results from the SPA approach are directly comparable to one of the most common sources of neural data. Given our understanding of the relationship between single-cell activity and other methods for recording neural signals, we can use these single-cell models to predict other kinds of data, such as that gathered from local field potentials (LFPs), electroencephalograms (EEGs), fMRI, and so on. Essentially, different kinds of filters can be used to process the model data and

[4] Dale's Principle states that the vast majority of neurons in the brain are either inhibitory or excitatory–not both.

provide comparisons to a wide variety of methods for measuring brains (*see*, e.g., Section 5.8).

In short, the SPA does not identify a settled "level of abstraction" for characterizing cognition. As a result, the SPA is capable of generating the explanations we need: explanations sensitive to the questions they are answering; explanations appealing to the *relevant* mechanisms. The preceding chapters have provided many examples of explanations at different levels of description. But, hopefully, the most salient feature of these examples is that, *taken together*, they provide a *unified* characterization of biological cognition not available from any other approach.

9.5 ■ The SPA Versus the SOA

To conclude this chapter, let me return to the CCC with which I began. In Table 9.2 I have reproduced Table 9.1 that compares state-of-the-art (SOA) architectures and included a column for the SPA. I have already suggested that the SPA might be considered an amalgamation of bits and pieces of past approaches. Nevertheless, returning to the CCC shows how the result may be greater than the sum of its parts.

Of course, the present evaluation is not likely to be considered an objective one. Perhaps only time will tell. But it is worth emphasizing that even without the addition of the SPA, this table highlights several critical challenges for current models of biological cognition. As discussed in Section 9.2.4, biological and symbolic plausibility generally remain orthogonal strengths. As well, models able to autonomously address a variety of diverse tasks–thereby addressing the flexibility of biological systems–are rare. Indeed, most aspects of scalability remain a major challenge. Further, cognitive models are seldom tested for their robustness, despite the robustness of biological cognitive systems being one of their hallmark features. Finally, there are masses of valuable experimental data from all areas of neuroscience that are critical to understanding biological cognition that remain unaddressed by the current SOA. At the very least, current approaches should identify how their preferred approach can, in principle, connect to such data. More usefully, models should be built that explicitly predict such data. These general lessons aside, in the remainder of this section I provide brief justifications for the SPA ratings in this table.

When it comes to systematicity, compositionality, and productivity, the SPA scores at least as well as ACT-R. This is because the SPA's representational commitments allow it to capture the same idealizations of representational capacity as any other approach, while not violating scalability constraints. I would argue that it does better than past approaches because, in not violating those constraints, it imposes reasonable limits on the idealizations of such

TABLE 9.2 *Comparison of All Approaches With Respect to the CCC, Rated on a Scale from 0 to 5*

	ACT-R	LISA	NBA	ICS	DFT	Leabra	SPA
Systematicity	++++	++++	++++	++++	++	++	++++
Compositionality	++	++	++	++	+	–	++
Productivity	++++	+++	+++	+++	–	–	++++
Massive binding	+++	+	+	+	–	–	+++
Syntactic generalization	++	++	++	++	–	–	+++
Robustness	–	+	+	++	++	++	+++
Adaptability	++	+	–	–	+	++	+++
Memory	+++	++	+	–	++	++	+++
Scalability	+	–	–	–	–	–	++
Triangulation	+	+	+	+	++	++	+++
Compactness	++	+	+	+	+	++	+++

Columns are in order of least to most biologically plausible.

criteria. In addition, the SPA provides grounded semantics for its representations, unlike past approaches. However, I have not conclusively demonstrated some of the benefits of these features of the SPA here, so I leave the ratings with respect to these three criteria matched to ACT-R.

Similar considerations apply to the massive binding criterion: The SPA alone makes sense of how the brain can meet such constraints while not violating scalability (*see*, e.g., Section 4.8). Consequently, the SPA is rated equivalently to ACT-R.

For syntactic generalization, the SPA, like ACT-R, provides a mechanism that can reproduce the speed with which such problems are solved by people (*see* Section 7.3.7). Unlike ACT-R, the SPA provides a natural explanation of content effects that follows directly from its account of semantics (*see* Section 6.6). Consequently, the SPA scores higher on this criterion.

The robustness of past approaches is highly variable and often poorly tested. In contrast, the SPA allows standard demonstrations of robustness by the random destruction of neurons (*see* Fig. 6.20), changes in dimensionality and strategy (Rasmussen & Eliasmith, 2011a), and input variability (*see* Section 3.5). Further, the SPA provides a general guarantee of robustness of the components to noise (Section 2.2). Because the SPA relies on the NEF, which was developed to account for a high degree of heterogeneity and expected noise levels found in cortex, it is not surprising that implementations of SPA models are generally robust. In addition, SPA models often take advantage of the robust dynamics found in neural integrators, attractor networks, and other stable dynamical systems. In short, the SPA incorporates robustness at all of its levels of description, from single cells up. Consequently, I have rated the SPA higher on this criterion than other approaches.

I earlier noted that current architectures together cover a broad range of adaptation phenomena but that none covered all of them. Again, I think the SPA is unique in its breadth of coverage of adaptive behavior. It incorporates biologically realistic learning of several kinds (e.g., syntactic generalization, STDP), structures for flexible routing of information (e.g., basal ganglia, cortical routing), and a general reliance on methods of adaptive control (e.g., for motor control and reinforcement learning). Unlike past neural approaches, the problem of rapid variable creation thus poses little challenge for the SPA (*see* Section 7.3.7). Unlike past non-neural approaches, the SPA is able to account for neural adaptability. Thus, I have rated the SPA higher on this criterion.

With respect to the memory criterion, the methods of the SPA have been directly applied to working memory phenomena, resulting in a uniquely good model of serial working memory. I have also provided examples that include various kinds of long-term memory (e.g., the Tower of Hanoi, clean-up, the Wason card task, and the bandit task, among others). Although declarative memory has not been explicitly discussed, long-term memory of semantic pointers and encoding of structures into semantic pointers have both been described. So, overall I do not think that the SPA can claim any special status when it comes to accounting for memory phenomena. However, it also does not have any obvious "in principle" impediment to accounting for particular memory phenomena. In addition, as with most other cognitive phenomena, the SPA account uniquely spans a variety of levels of analysis. Consequently, I have rated the SPA equal to the highest rating on this criterion.

I have made much of the potential scalability of the SPA. However, because it is at an early stage of development, it has not been applied to tasks as complex as those considered by ACT-R. Nevertheless, unlike past approaches, it provides principles for describing such tasks in a scalable manner. As well, in Chapter 7 I have provided a characterization of a single system that addresses a wide variety of simple behaviors within the context of a single model. Consequently, the SPA offers more than just "in principle" scalability. Nevertheless, the SPA, like all other approaches, has much more to do to convincingly address this criterion. Consequently, I have rated the SPA low but slightly higher than past approaches.

Finally, with respect to the two scientific merit criteria, I think the SPA fares especially well. As I have already detailed above, the ability of the SPA to address a large variety of relevant data is perhaps its most distinguishing feature. As well, it is a very compact theory. The three principles of the NEF and the four aspects of the SPA are simple to state (although they can take two books to explain). As well, the SPA does not allow arbitrariness in its mappings to time or to neural structures, in contrast to past approaches.

Perhaps the least constrained aspect of the SPA stems from a lack of knowledge about the interactions between brain structures. The SPA does not yet explicitly fix functional specifications to many parts of the brain. I would argue that this state of affairs reflects our limited knowledge of such functions but that the SPA should serve to help provide such specifications. Nevertheless, this clearly leaves room for a certain amount of arbitrariness in SPA-based models. As with other approaches, construction of more models that perform a wide variety of tasks without any intervention will help to limit this arbitrariness. However, this limitation is shared with other approaches, whereas the non-arbitrariness of temporal constraints and the compact specification of the architecture are unique strengths of the SPA. Consequently, I have scored the SPA slightly higher than past approaches on both criteria.

In sum, the SPA not only provides unique advantages, such as flexible biological plausibility, it does not suffer any of the most salient drawbacks of past architectures. Of course, much remains to be done to truly demonstrate that the SPA can underwrite large-scale biological cognition. And, as I mentioned in the introduction to this section, even if the present comparison of the SPA to the SOA strikes some as overly optimistic, the analysis of the CCC and limitations of current approaches highlights several general areas for improvement in the field. In the next chapter I turn to consideration of some challenges for the SPA, and briefly consider several conceptual consequences of thinking about cognition from the perspective of the SPA.

10. CONSEQUENCES AND CHALLENGES

At several points throughout the book I have hinted at some of the conceptual consequences of the SPA (*see*, e.g., Sections 2.3 and 7.4 on "levels" and a unification of symbols and probabilities, respectively). However, I believe that there are several other consequences of adopting the SPA on which I have not yet touched. For example, if the SPA is truly both an integration and extension of the three standard approaches to cognitive science, then the distinction between these "paradigms" is not a useful one for the progress of the field. After all, it makes little sense to suggest that the symbolic approach is correct if adoption of that view makes it exceedingly difficult to connect our theory to crucial neuroscientific data that constrains cognitive theories. Similarly, claiming that connectionism is the right approach seems unwarranted if we must reject tolerated assumptions of that view (e.g., that there are concept-level localist representations) and significantly alter other parts (e.g., by introducing dynamic spiking networks). Finally, we cannot claim that dynamicism has won the day if it does not provide the resources needed to capture high-level cognitive function or identify systematic methods for relating its models to the neural substrate. Rather, integrating past approaches appropriately may render the classic distinctions between these approaches irrelevant. What remains relevant, however, is determining the best means of theoretically characterizing such an integration.

If the SPA succeeds at affecting such an integration, we would expect it to cause us to re-think central notions in the behavioral sciences. Consequently, in what follows, I presume that the SPA is successful at providing this integration and consider how four core notions may be revised. These notions are: representation, concepts, reasoning, and dynamics (recall that each of these concepts, and others, are listed in Table 1.2, which identifies debates that Thagard [2012] has suggested the SPA can help resolve). I certainly do not intend to do full justice to the wide variety of views regarding these ideas. Rather, I present considerations that I take to be food-for-thought–considerations that help discern the potentially far-reaching consequences of the SPA. The philosopher in me feels guilty for giving such short shrift to such important issues. Nevertheless, I think it is helpful to reflect on a variety of ramifications of adopting the SPA, even if only briefly.

10.1 ■ Representation

In some quarters, the concept of "representation" has been closely identified with classical digital computers and Turing's (1950) notion of computation (e.g., van Gelder & Gelder, 1995; Fodor, 1981). However, behavioral scientists typically have a much more liberal notion, freely talking about representations in neural networks, analog computers, and even in the structure of dynamical system state spaces (Schöner, 2008). In the SPA I have adopted this more liberal view.

This is at least partly because I believe that the SPA can provide an account of cognition that *inter-relates* the many kinds of representations identified in behavioral science research. Like the NEF on which it is based, the SPA is quite flexible in what it calls a representation. Semantic Pointer Architecture representations can be scalar values, functions, vector fields, symbols, images, and so on. Similarly, a wide variety of computations–linear, nonlinear, continuous, discrete, statistical, logical, and so forth–are perfectly consonant with the SPA. In every case, however, mental representations are characterized as some kind of vector, and computations as some kind of vector transformation, implemented in neurons. The representational and computational commitments of the SPA are thus both flexible and unified.

One unique aspect of the SPA characterization of representation is that the vectors typically identified as representations, of which semantic pointers are one common variety, are *not* vectors of neural activity. Unlike the standard connectionist understanding of representation that deals with "activity vectors," the SPA provides *two* mutually consistent perspectives on the activities of neurons. One relates to directly measurable properties of the system, such as spike patterns. The other relates to the underlying vector space that is taken to be represented by those measurable properties. Connectionists typically talk about neither of these.

This distinction between neuron space and state space representation is crucial for bolstering our scientific understanding of neural function. This is because the mapping between the biological wetware and the underlying vector space provides a means of relating explicit, measurable brain function to a more general scientific understanding of the world. For example, it allows us to understand certain neurons in area MT (as measured by brain function) as being sensitive to visually detected velocities (a general physical property). Essentially, such mappings give us a way to connect concise, scientific, and low-dimensional descriptions to the often very high-dimensional neural activity that we record from brains. These mappings can massively reduce the complexity of descriptions of brain function, which is why appealing to the notion of "representation" is useful in the first place.

This way of understanding mental representations has consequences for our neuroscientific and psychological understanding of representation. In neuroscience, taking each cell to have a preferred direction in a high-dimensional state space can go a long way to helping us understand what is often considered the perplexing variety of observed neural responses. For example, in models of working memory, incorporating high-dimensional representations provides for simple explanations of the variety of responses observed during the delay period (Singh & Eliasmith, 2006; Machens et al., 2010). As well, the seeming complexity of remapping in prefrontal cortex (Hoshi, 2006) can be easily understood in the SPA as routing different sources of information into the same high-dimensional vector space (e.g., visual information in one task, and auditory information in another task).

Further, semantic pointers themselves can be seen as a natural generalization of neural representations that rely on preferred directions in a vector space–a kind of representation familiar in neuroscience. Perhaps the most probed representations in the brain are those in primary visual cortex (V1). In V1, individual cells are often characterized as sensitive to specifically oriented bars in particular retinal locations, with their activity falling off as the stimulus either rotates from the preferred orientation, or moves out of the specific retinal location (Kandel et al., 1991, pp. 532–540). It is not uncommon to consider these individual cells as carrying specific content about input stimuli (e.g., oriented bars at a location). In this very particular sense, we can identify each cell with a specific label indicating what it means to the animal (i.e., we can assign it a preferred direction). These "localist" single cells in V1 are often taken to represent coefficients of Fourier-like decompositions of images (Olshausen & Field, 1996).[1] Notably, exactly this kind of decomposition forms the basis of SPA visual representation (*see* Section 3.5). Although this could be interpreted as an unusual type of localist representation, a more natural interpretation arises if we consider a large population of such neurons, in which

[1] Representations similar to those found in V1 have been identified in auditory cortex, the hippocampal complex (including place cells and head direction cells), motor areas, and many other parts of the brain.

case these representations are parts of a distributed representation of an entire visual image.

Semantic pointers are localist and distributed in precisely this same way. In particular, the SPA also extends a V1 type of representation to concepts. That is, at higher levels "localist" single cells are taken to represent coefficients of Fourier-like decompositions of concepts. In both vision and the conceptual space, the neural tuning curve is taken to be one basis for the decomposition, and hence neural activity encodes the currently relevant coefficient. As with V1, looking at a single neuron can provide a very specific characterization of what that neuron means to the animal, but in a broader context we can take population activity to be representing an entire concept (i.e., a semantic pointer). Further, just as there is significant redundancy in V1 single-cell tuning (suggesting that many neurons participate to encode a given coefficient), there is also significant redundancy in the SPA representation of semantic pointers. In short, the SPA applies a representational scheme to symbol-like concepts that is a natural generalization of our current understanding of neural representation.

Turning to the psychological perspective, there are clear consequences of this characterization of mental representation for our understanding of the role symbols play in cognition. We may, in a sense, think of symbol-like representations in the SPA as being "doubly distributed." That is, representations of symbols are not only distributed over a vector space, but that vector space is also distributed over neural activities. I take this as an important strength of the SPA because it allows us to relate high-level characterizations of symbol-based cognition to detailed spiking data. Simultaneously, it allows us to capture the internal structure of such symbols, providing an increase in the variety and subtlety of manipulations that we can define over such representations (*see* Section 7.4).

There are at least three major consequences of this characterization of symbolic representation: (1) the internal structure of symbols is important for characterizing a wide variety of cognitive behavior; (2) concept representations are better characterized as "symbol-like" than as symbols; and (3) mental representations are best thought of as often temporary processing states of activity, rather than as objects that reside in a specific location in the system. The general idea behind the first consequence has been championed by connectionists for quite some time. However, the SPA more systematically characterizes that structure, allowing us to specifically identify sophisticated kinds of transformations that support statistical inference, symbolic processing, and so on (*see* Section 7.4). In addition, the SPA provides a clear relationship between perceptual and conceptual representations that is typically not found in connectionist accounts.

The importance of the second consequence is accentuated by the poor scalability of attempts to implement symbol-based architectures with connectionist methods. Characterizing mental representations as symbol-*like* carries with it a reminder that many cognitive properties, such as systematicity, compositionality, and productivity, are *idealizations* of human performance. Hence, they do not provide straightforward constraints on cognitive architectures. In contrast, the available physical resources of the brain do provide hard constraints. Currently, it seems that these constraints will only be respected if the system traffics in symbol-like representations.

The third consequence goes beyond symbol-like representations, encompassing a general perspective on the brain as a functional system. The suggestion that the typical case of occurrent mental representation is a temporary state follows from the ubiquity of vector *processing* in the SPA. This view lies in direct contrast to the "activation" of anatomically specific representational states. In essence, the SPA suggests that we think about the brain as more of a processor of neural signals rather than as a storehouse of neural representations. Representations are, of course, stored. However, the SPA embodies a shift in emphasis: neural populations are more like a way station for a wide variety of constantly changing representational content, rather than a home base for some small subset of re-activated representations. Embracing this change in perspective provides an understanding of how the brain can be a flexible, adaptive, scalable, and computationally powerful system. This shift also highlights the temporal aspects of neural representation in the SPA. Because the representational contents of a given neural population can change quite dramatically from moment to moment, the dynamics of neural representations must be considered when constructing SPA models. I return to the importance of dynamics shortly.

All three of these consequences help remind us of the fact that most representations in the SPA are a kind of semantic promissory note. Semantic pointers carry with them a means of accessing much more sophisticated semantics than they themselves directly encode. Thinking of central representational structures as not being unified semantic wholes, like symbols, provides a subtler and more biologically plausible view of how mental representation might be organized. The idea that semantics are encoded through compression and dereferencing provides a natural explanation for the various amounts of time and effort it takes to extract semantics of words from experimental subjects.

One final note on the semantics of SPA representations: I have not described a general semantic theory in this book. However, I have described a theory of "neurosemantics" in detail in past work (Eliasmith, 2000, 2006), which I take to be fully compatible with the SPA. Encouragingly, this theory has been used by others to account for semantic phenomena as well (Parisien & Thagard, 2008). Nevertheless, I do not consider traditional issues with the semantics of mental representations here.

10.2 ■ Concepts

Recent theoretical work on concepts has come to the consensus that "in short, concepts are a mess" (Murphy, 2002, p. 492). For some, this merely suggests that more work needs to be done. For others, such as Machery (2009), this leads to the conclusion that "the notion of concepts ought to be eliminated from the theoretical vocabulary of psychology" (p. 4). Machery's argument begins with the observation that there are many different things concepts are used to explain (e.g., motor planning, analogy, categorization, etc.). He then proceeds to suggest that it makes no sense to identify concepts with just one of these functions, as that would be a clearly incomplete account of concepts. He also argues that it has proven untenable to specify a theory that accounts for all of these different uses of the term. He argues that this state of affairs drives us to accepting his "Heterogeneity Hypothesis"–that is, that there are many kinds of conceptual phenomena and that there is no unification of the kinds. The consequence of accepting this hypothesis is that the notion of "concept" should be eliminated from the behavioral sciences.

A central element of this argument is Machery's concern that there is no clear unification behind the notion of a "concept." However, I believe that the SPA demonstrates how a particular kind of mental representation can be used to explain the variety of conceptual phenomena. After all, we have seen explicit examples of how semantic pointers can underwrite motor planning, structure-mapping, categorization, and many other "conceptual" phenomena. As I described in Section 4.8, semantic pointers can contain content from perceptual areas and motor areas, can encode realistic relations as well as allowable dynamic transformations, and so on. And, each subelement of a semantic pointer can itself "contain" many subelements. Critically, these elements are not merely "stuck together" to satisfy our theoretical need for diverse conceptual representations. Rather, these structures can define semantic spaces at a more abstract level than their constituent elements. These more abstract spaces help the system perform various kinds of rapid or approximate comparisons between elements within that space, saving valuable processing time when deep semantic analysis is not needed. As well, these abstract spaces can be transformed to drive motor behavior, to guide cognitive planning, or to interpret perceptual input. So, not only do semantic pointers provide a means of combining a wide variety of information into a single mental representation, but they also provide clear links to behavioral consequences.

Note that this kind of "unification" of concepts is a *functional* (not a representational) one. That is, the presence of a semantic pointer encoding a complex structure does not provide a unified account of concepts on its own. It is only in the context of the entire architecture–decoding, manipulating, generating, and so forth, such representations–that we can provide a unification of conceptual phenomena. Semantic pointers, by definition, do not carry enough information to fully capture the semantics of the thing they are about. Nevertheless, I submit that a functional unification is one that is sufficient to allay Machery's concerns, and hence avoid the conclusion that "concept" is not a useful concept.

A separate consequence of this SPA account of concepts is that although some concepts are highly tied to a specific sensory modality, others are not. As I mentioned in my earlier discussion of semantics (Section 3.3), the SPA can be thought of as a computational account consistent with Barsalou's work on perceptual symbol systems (Barsalou, 1999). However, the SPA account differs from this and other neo-empiricist views of concepts (e.g., Prinz, 2002). Those accounts typically argue that concepts do not provide amodal representations, contrary to what is presumed by most theorists. As Machery (2009) has noted, the empirical evidence offered in support of this conclusion does not, in fact, contradict some amodal accounts. I would like to suggest that semantic pointers can be seen as one such account.

Semantic pointers are sometimes tightly tied to perceptual information, perhaps even for many different modalities, but they need not be. For example, semantic pointers that encode information about statistical relationships between co-occurrences of words (*see* Section 3.4) would have no obvious modal content. Similarly, semantic pointers employed in Spaun to encode sequence information are independent of modality (*see* Section 7.3.5). In short, semantic pointers offer a way of understanding how concepts can preserve access to deeper perceptual information, while still being able to be manipulated independently of *actually* having access to that information, and independently of that information's effect on at least some aspects of the conceptual content.

To conclude this section, I would like to briefly suggest that, taken together, the SPA's combination of (1) access to amodal and modal conceptual information, and (2) a functional unification, allows it to account, in principle, for the four major theories of concepts. I will consider each in only a few sentences–my intent is to be suggestive, not conclusive.

The first is the "classical" view, which proposes that having a concept is having a set of necessary and sufficient conditions. Although generally out of favor, there do seem to be some concepts, such as mathematical ones, that might be best described in this manner. Semantic pointers can capture such concepts by their being defined by purely syntactic relations to other lexical items–for exmple,

$$\mathbf{triangle} = \mathbf{closedShape} \circledast \mathbf{property1} + \mathbf{threeSides} \circledast \mathbf{property2} + \mathbf{straightEdges} \circledast \mathbf{property3}.$$

A second main theory of concepts is prototype theory. The SPA account of prototypes has been discussed in the perceptual case in Section 3.5, where the prototype is the mean of the clustering of the set of examples. In Spaun, this mean was used as the perceptual basis for higher-level concepts. In the more conceptual case, prototypes are naturally accounted for as slot-filler structures, as discussed in Section 4.8.

Third, exemplar theories can be accounted for by having specific perceptual semantic pointers bound into the concept of interest at a higher level (e.g., a specific slot). Or perhaps specific examples of a category can be generated by using the statistical model in early perceptual areas in a top-down manner (as shown in Fig. 3.7). I suspect both mechanisms, and others, are employed by the brain. In any case, the ability of exemplar theory to explain the rapid generation of novel categories that display typicality effects is naturally captured by the representational flexibility of semantic pointers.

Finally, the theory theory of concepts can be accounted for by the fact that semantic pointers can consist of a wide variety of representations, including allowable transformations, relations to other concepts, and so forth, as necessary to encode a "theory." Such representations might include information about which features of the concept are essential, which are perceptual, how an object can change under various transformations, and so on. Nothing in the conceptual phenomena best captured by the theory theory seems to challenge the applicability of semantic pointers to those phenomena.

Again, these considerations are so brief that they can at best be considered suggestive. Nevertheless, they are at least suggestive–that is, even this brief discussion underlines that there is no obvious impediment to the SPA unifying our understanding of concepts. I believe that semantic pointers have the representational flexibility, appropriate contact with perception and action, ability to capture modal and amodal representations, and characterization of symbol-like representations and structure necessary to make sense of the current "mess" in conceptual theorizing. In short, I am suggesting that we can come to a more unified understanding of concepts by systematically employing semantic pointers in a functional architecture.

10.3 ■ Inference

Like "concept," "inference" experiences a very broad application in the behavioral sciences. For example, the term can be used to refer to "subpersonal" processes that automatically generate representations of the world that go beyond the information available to our senses. For example, when I assume that an object has a back side even though I cannot see it, I am often said to be relying on mechanisms of perceptual inference. In other instances, the term is used to characterize effortful reasoning. In some such cases, we do not explicitly know the rules we are using to reason in this way. In other cases, we may know the rules explicitly, such as when we are performing logical inference. Given the variety of inference phenomena, it is tempting to make an argument analogous to that which Machery made about concepts: that the diffuseness of the phenomena suggests we can neither identify a specific phenomenon as being diagnostic of inference nor can we hope to find a unified theory of inference.

Perhaps not surprisingly, my suggestion is that the SPA can play a role in helping us to understand inference by characterizing it in the context of a functional architecture. In other words, I believe that the SPA can help us define unified models, like Spaun, that integrate a variety of observable brain processes in a consistent and coherent way. This, then, allows us to taxonomize the variety of inference phenomena we observe by relating them directly to mechanisms explicated by our model. Thus, the unification of what we label "inference" comes from its many varieties being explained by a unified, underlying, functional architecture.

One reason we might think that such a unification can be provided by the SPA is that it is in the unique position of being able to integrate the more symbolic kinds of inference with the more statistical kinds of inference, and the more linguistic kinds of inference with the more perceptual kinds of inference. In Section 7.4, I argued that the consistent matrix-vector and statistical interpretations that we can apply to the cortical-basal ganglia-thalamus circuit is unique to the SPA. This was partly because the representations in the SPA are symbol-like while retaining significant internal structure.

Perhaps surprisingly, this duality can go a long way toward capturing many of the kinds of inference that we would like to explain. For example, in describing how a semantic pointer comes about through compression (Section 3.5), I could also be taken to be describing how subpersonal perceptual inference can be captured by the SPA. Similarly, when I was discussing the bandit task (Section 6.5), I could also be taken to be discussing how subpersonal states help to infer appropriate actions given changing environmental regularities. When I was describing how we can explain induction in the Raven's matrix task (Section 4.7), I could also be taken to be describing how effortful inference can be used to identify abstract regularities. And finally, language-based logic-like inference and planning are captured naturally in the architecture, as was demonstrated by the Tower of Hanoi (Section 5.8) and Wason card task (Section 6.6) examples. Consequently, the SPA does a reasonable job of naturally accounting for a wide variety of inferential phenomena.

In conclusion I will note that, as in the case of concepts, Spaun's unification of "inference" is woefully incomplete. Although a model like Spaun goes some way to integrating a variety of inferencing processes within the context of a single model, it clearly does not fully capture the complexity of even these example processes. Nevertheless, models like Spaun do suggest that we may be able to understand the wide variety of inference phenomena by employing unified functional architectures like the SPA.

10.4 ■ Dynamics

In past work, I have been quite critical of dynamicism as a cognitive approach (Eliasmith, 1996, 1997, 1998, 2009b). However, I believe that the SPA provides a way of addressing those criticisms, while still embracing the compelling insights of the view. For example, dynamicists generally do not map the parameters and variables in their proposed models to the underlying physical system that is being modeled (i.e., neurons, networks, brain areas, etc.) in any detail (*see*, e.g., Busemeyer & Townsend, 1993). Clearly, the SPA provides a mapping down to the level of individual neurons, allowing us to directly compare the mechanisms that give rise to the dynamics that we observe and our models of those dynamics. Additionally, I and others (e.g., Bechtel, 1998) have criticized the anti-representationalism that is often espoused by dynamicists, something clearly not embraced by the SPA.

More generally, I have expressed serious concerns regarding the approach to dynamics taken by all three standard approaches (*see* Section 9.2.3b). As I discussed there, none of the three approaches provides principles for mapping cognitive models onto independently measurable, low-level system dynamics, such as membrane or synaptic time constants, refractory periods, and so on. In short, unlike past approaches, the SPA is concerned with real-time dynamics "all the way down." It is concerned with the dynamics inherent in the physical body that must be controlled by the nervous system (Section 3.6). It is concerned with the dynamics of recurrently connected spiking attractor networks (Section 6.2). And, it is concerned with the dynamics of complex cognitive processes (Section 5.8). Above all, the SPA is concerned with how all of the various temporal measurements we make of the system, at many levels of analysis, interact to give rise to the variety of dynamics we observe.

One consequence of this focus on real time and real physics is that the SPA is in a position to embrace the constructive insights of researchers who emphasize the embodiment and embeddedness of cognitive systems. I have argued elsewhere that a problematic tendency of these researchers is to blur the lines between agents and environments (Eliasmith, 2009a). Indeed it has been claimed that "nothing [other than the presence of skin] seems different" (Clark & Chalmers, 2002, p. 644) between brain-brain and brain-world interactions. The suggested conclusion is that our characterization of cognition cannot stop at the boundaries of an agent. Ironically, I believe that the very dynamical nature of these systems, which is often appealed to in drawing such conclusions, suggests exactly why there *is* a difference between the inside and outside of the nervous system. In short, the degree and speed of coupling inside the nervous system is generally much greater than that between the nervous system and the body, or the body and the world. One straightforward reason for this is that the body has mass. Hence, body dynamics tend to be considerably different than the dynamics of electrical signal transmission, because such signals have nearly no mass.

This suggestion–that we can rely on differences in dynamics to identify useful system boundaries for scientific exploration–has a further consequence for a long-held position in philosophy called "functionalism." Functionalism is the view that what makes a system the kind of system it is–for example, a mind–is determined by the functions that it computes. On a standard understanding of what counts as computation (e.g., as characterized by Turing machines), functionalism suggests that two systems operating at very different speeds–that is, with very different dynamics–could both be computing the same function (Turing machines ignore the amount of time taken to compute a function). As a consequence, philosophers have sometimes argued that an interstellar gas cloud or a disembodied spirit might have a mind because it might be computing the same function as humans (Putnam, 1975). However, such conclusions clearly do not hold if we think that the length of time it takes to compute a function matters for whether or not that function is being realized. Mathematically, all that I am suggesting is that functions describing cognition be written as not only functions of state but also as functions of time. As I have suggested elsewhere, such a "temporal functionalism" provides a much more fine-grained, and I believe plausible, basis for a taxonomy of cognitive systems (Eliasmith, 2003).

The importance of time for understanding what cognitive functions a biological system can actually compute is illustrated by the discussion of control in Chapter 5. Indeed, the obvious practical importance of control and routing make it hard to dissociate the functions a system can perform from its dynamics. Consider the disastrous consequences of replacing digital telephone switches with human operators. There is little sense to be made of the claim that the "same functions" can be computed in both circumstances (although a standard Turing characterization may not distinguish them). Similarly, the specific dynamics of biological brains can provide important insights into which functions they compute, and how they accomplish those functions.

More subtly, the importance of time is also demonstrated by the SPA characterization of adaptation. As mentioned in Section 6.4, learning is most often associated with changing weights, or constructing new, permanent representations in long-term memory. However, as was made evident by the Raven's matrix task, the ability of past input to influence future response can often be a consequence of controlled recursive dynamics in the occurrent activity of the system. In some cases, as emphasized by Hadley's rapid variable creation task (Section 7.3.7), such dynamics seem the only possible explanation for the observed cognitive phenomena. This places important constraints on what kind of architecture *could* compute the relevant functions, as known constraints on the timing of weight changes precludes the expected story (i.e., that rapid variable creation is explained by weight changes).

In sum, the manner in which dynamics are incorporated into the SPA has conceptual consequences for embodiment, system boundaries, and functionalism. However, the SPA also has practical consequences for contemporary discussions of dynamics in experimental neuroscience. In recent years, there has been an increasing emphasis on the kinds of frequencies and temporal correlations observed during tasks in a wide variety of brain structures (Fries, 2009). Consideration of such phenomena has not played a role in the development of the SPA. Nevertheless, it is an important empirical constraint on SPA models.

For example, Pesaran et al. (2002) carried out working memory experiments while recording from monkey cortex, and performed a spectrographic analysis on the resulting spike patterns. This analysis has been taken to show that during working memory, there is a shift in frequencies, increasing power in the gamma band (i.e., 25–90 Hz) during the delay period. We have performed an identical analysis on the spike trains generated in a standard SPA working memory component. As shown in Figure 10.1, the single-cell and population frequency spectrograms for the model and the data are similar. I'm not aware of any other models reproducing this effect. Some have suggested that such results indicate that gamma band synchronization is a "fundamental mechanism" of neural computation (Fries, 2009). However, I would be hesitant to call such an increase in spectral power "fundamental"[2] as there was no need to consider it to construct models that give rise to the observed patterns. And, nearly identical information processing can be achieved without such patterns (e.g., using more rapidly spiking neurons).

Others have suggested that such patterns are merely epiphenomenal (Tovée & Rolls, 1992). However, such phenomena are unlikely to be strictly "epiphenomenal" (i.e., not causally efficacious), because the presence or absence of such synchronization will drive patterns of currents in the dendrites of neurons receiving this activity, and we know that such patterns can affect neural responses. I am inclined to think that such phenomena are neither fundamental nor epiphenomenal but, rather, they are the consequence of implementing particular information processing operations in a neural substrate. If we are interested in identifying the information processing itself, ignoring details of spectral frequencies may be acceptable. If we want to worry about the details of the processing in the biological system, then we may want to pay attention to such data. The notion of "pragmatic descriptivism" I introduced earlier

[2] Nor is it really a "mechanism." Mechanisms typically have identifiable parts that interact in a manner to produce a regularity (Machamer et al., 2000). Synchronization is the product of a mechanism, not the mechanism itself.

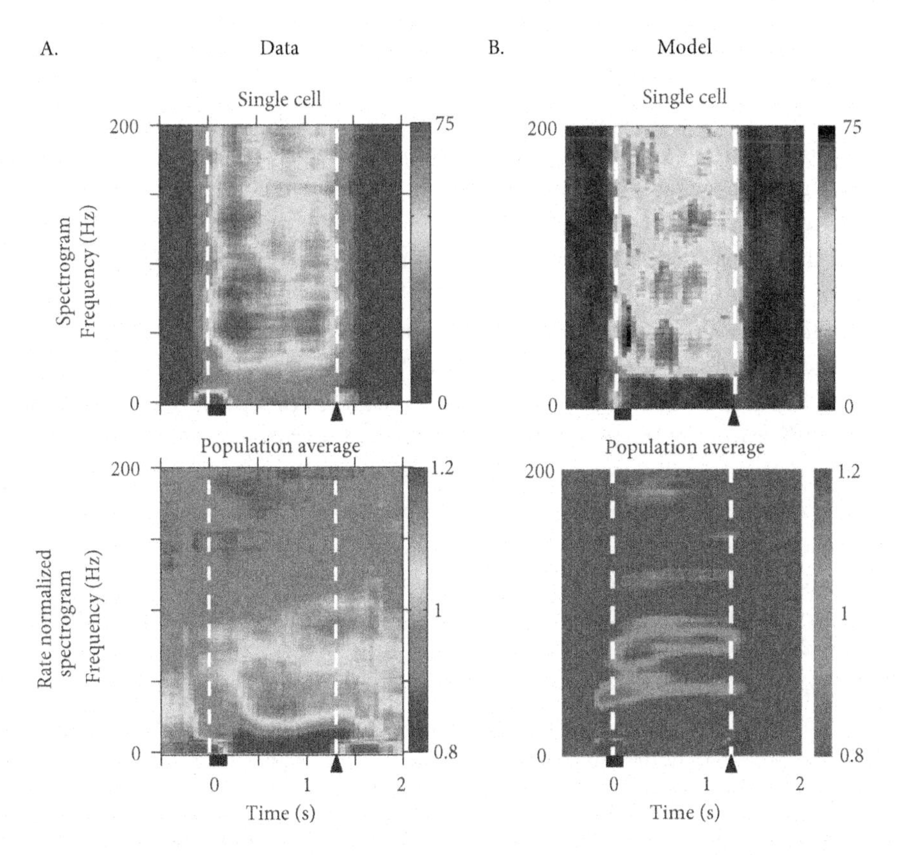

FIGURE 10.1 Frequency profiles during working memory. **A**. Spike spectrograms for a single neuron (top) and the population average (bottom) during a simple working memory task. **B**. The same analysis performed on spikes from an SPA working memory component. Both the model and the data show a shift from minimal frequency response before the stimulus to stimulus-driven low-frequency information (at the first white line) to significantly higher-frequency content during the delay period, with a final reduction after response (at the second white line). Both population averages show frequency content mainly in the gamma range. (Data adapted from Pesaran et al. [2002] with permission. Reproduced from Eliasmith et al. [2012] with permission.)

(Section 2.3) again captures how we should be careful to identify an appropriate level of description, depending on which questions we want answered. My point here is that the flexibility of the SPA lets us adopt a variety of perspectives on neural dynamics.

In conclusion, the SPA incorporation of dynamics provides a uniquely consistent and grounded inclusion of time into models of biological cognition. This has consequences both for a variety of theses in philosophy of mind and more applied debates in contemporary neuroscience.

10.5 ■ Challenges

Although I am optimistic about the contributions that the SPA may make to our understanding of biological cognition, it would be rash to suppose that no challenges remain for the development of the architecture. It is important to keep in mind that the ideas captured in this book represent the beginning, not the end, of a research program. For this reason, it is perhaps worth stating what the SPA is *not*.

First, it is not testable by a single, or small set, of experiments. Being the foundation of a research program, the SPA gains some amount of credibility when it gives rise to successful models. Failures of such models lead to a reconsideration of those specific models and, when systematic, a reconsideration of their foundations. Consequently, the success or failure of the approach will only become evident over many applications to a variety of cognitive phenomena.

Second, the SPA is not a completed theory of mental function: we have not yet actually built a fully cognitive brain (you will not be surprised to learn). Rather, I have described models of perception, action, and cognition. And, I have described these in a way that combining them is reasonably straightforward (*see* the Spaun model, Chapter 7). Nevertheless, there are obviously many *more* behaviors involving these aspects of cognition that I have not discussed.

Third, even theoretically speaking, the coverage of the SPA is uneven. Some aspects of cognition are more directly addressed than others–the SPA is undeniably a work in progress. There is a long list of brain functions that have not even been mentioned in my discussion of the SPA but obviously belong in a general architecture. For example, emotion and stimulus valuation are largely absent. These play a crucial role in guiding behavior and need to be integrated into the SPA, both introducing new areas (e.g., orbital frontal cortex, cingulate, etc.) and interacting with current SPA elements, such as the basal ganglia. As well, perceptual processing in visual, olfactory, tactile, and other modalities has remained largely unaddressed. Also, the role of the cerebellum and many parts of brainstem have not been considered in any kind of detail. The same can be said for the microstructure of the thalamus. In short, the SPA as it stands provides a somewhat minimal coverage of anatomical structures. My defense should not be a surprising one: the SPA is a new approach. I remain hopeful that the methods I have introduced here will generalize to these many other brain areas, but I expect hope to convince no one.

When pressed, I would be willing to concede that the SPA is less of an architecture, and more of an architecture sketch combined with a protocol. That is, the SPA, as it stands, is not yet highly committed to a specific arrangement and attribution of functions to parts of the brain–although some basics are in place. Largely, the SPA specifies how some critical functions can be computed and how effective communication can occur. As a result, a major challenge for the SPA is to fill in the details of this schema.

I believe a natural approach for doing so is to adopt Michael Anderson's "massive redeployment hypothesis" (Anderson, 2010). This is the idea, consistent with the SPA, that neural areas compute generically specifiable functions that are used in a variety of seemingly disparate tasks. The SPA provides a means of understanding how the same neural resources can be driven by very different inputs–that is, by redeploying the resources of cortex using routing strategies. Nevertheless, which areas of the brain perform which specific functions is underspecified by the SPA as it stands.

One of the functions about which the SPA is most explicit is that of the basal ganglia. As initially described, the SPA employs the basal ganglia to perform action selection. However, there is good evidence that if the output of the basal ganglia is removed, actions can still be selected, although less smoothly and quickly. This is consistent with recent hypotheses, supported by anatomical and physiological experiments, that suggest that the basal ganglia is important for novel action sequences and for teaching action selection to the cortex but less important for well-learned actions (Turner & Desmurget, 2010).

The statistical reinterpretation that I provided in Section 7.4 fits very naturally with such a view. After all, in that interpretation it is clear that cortical processing must precede basal ganglia processing and can result in action selection even if basal ganglia does not play its typical role. Specifically, this is because the role of the basal ganglia under that interpretation is more clearly one of refinement than one on which all subsequent cortical processing depends. Indeed, this is also true of the matrix-vector interpretation, but it is less explicit. In both cases, we can think of the architecture as having a kind of "default routing" that allows somewhat inflexible (but rapid) processing to result in reasonably appropriate actions for given inputs. As a result, a major challenge for the SPA is to provide detailed hypotheses regarding such default cortical interactions–another instance of calling for a more comprehensive function specification.

Thinking of the basal ganglia as more of a "teacher" than a "selector" raises aspects of learning that are not yet well integrated into the SPA as well. For example, connections between thalamus and cortex are likely as sensitive to environmental contingencies as those from cortex to striatum, but they have not been made flexible at all in the SPA models presented here. As well, if we think of basal ganglia as performing a corrective role in tuning cortico-cortical connections, then it is important to account for the modulatory effects of basal ganglia projections to cortex, allowing such teaching to occur.

Further, even many of the elements of cortical processing that are included in the SPA need to be significantly extended. For example, visual routing and object recognition should be implemented in the same network, although I have considered them separately in the models presented in Sections 3.5 and 5.5. In addition, there are several challenges regarding clean-up memories (*see* Section 4.6). Not only do we need to better understand which anatomical areas act as clean-up memories in cortex, we also need methods for rapidly constructing such memories, appropriately extending such memories as new information becomes available, and so on. Conspicuously, a model of hippocampus is absent from the SPA. Such a model may be able to account for some of these aspects of clean-up memory. This, of course, remains to be seen.

Even some of the purported strengths of the SPA, like biological plausibility, contain many outstanding challenges. For example, I take the precise nature of time-dependent representation in a biological system to depend on the physical properties of the neurons that underwrite that representation. Consequently, low-level biological properties have representational consequences in the SPA. So, any simplifications that are made in simulations introduce the possibility of mischaracterizing the details of neural representation.

Although the NEF goes a long way to incorporating a variety of low-level biophysical properties of cells, there are some assumptions in the software implementation in Nengo that may need to be revisited as SPA models grow more sophisticated. For example, Nengo typically assumes that all synapses in a particular projection between populations have the same dynamics. However, we know that there is a distribution of time constants and that the processes that determine the precise synaptic dynamics of each cell are variable. It is unclear exactly when having a distribution of dynamics rather than uniform dynamics may matter for the performance of a network.

Additionally, in the models I have presented here, all neurons are point neurons. That is, they have no spatial extent. Consequently, Nengo does not systematically account for the spatial location of synapses on a receiving cell's dendritic tree or soma. There are other spatially sensitive low-level details for which Nengo typically does not account, including dendritic spiking, back propagation of somatic potentials up dendritic trees, and so on. To summarize, SPA models could stand to have more synaptic heterogeneity and could better account for spatial effects. Both of these improvements, however, come with high computational costs.

One reason that such details have not yet been incorporated is because the SPA is more focused on representation in populations of neurons than in single neurons. Nevertheless, even at the population level of analysis, there are many challenges for building SPA models. For example, there is always the challenge of determining the appropriate dimensionality of the space to be represented by a population of cells. The question of whether a high-dimensional space should be represented roughly, or a low-dimensional space represented accurately, does not have a generic answer. Relatedly, the kind of compression or decompression (i.e., change in dimensionality) that should occur as neural representations are processed can vary from area to area and hence needs to be considered carefully for each new component in a model. In general, these observations simply boil down to the fact that many specific representational properties are likely to depend on the particular functions we are characterizing in a given SPA model.

Of course, justifying such function specifications often depends on marshaling empirical evidence in favor of a particular hypothesis. This raises a more practical challenge for Nengo and the SPA. Specifically, it is important to improve the mappings between model data and a variety of measured neural signals. For example, LFPs are often recorded from populations of neurons and compared to individual spikes recorded from the same populations or analyzed on their own. Similarly, ERPs and EEGs, although more removed from individual spiking neurons, should be able to be inferred based on a spiking neuron model. Presumably, such challenges can be met in the same manner as we have done with fMRI (Section 5.8), but this remains to be seen. Developing various "filters" to allow underlying SPA models to produce these kinds of signals is an important challenge.

Returning to the issue of specifying SPA functionality, perhaps the greatest challenges lie in the domain of language. As I mentioned when introducing the notion of binding, understanding linguistic processing requires determining an appropriate way to encode linguistic structure. I chose a particular, simple method that binds the value of variables to the role that the variable plays in the structure. However, there are many questions that arise when attempting to determine how language-like structures should be encoded. For example, should adjectives be bound to nouns, or should they be bound to their own role and added to nouns? Do agent and theme roles need to be bound to the verbs they are used to modify to appropriately preserve semantics? And so on. It remains an enormous challenge to determine which alternative is best for capturing observed cognitive behavior.

Further, the SPA choice of an appropriate binding operation can be challenged. The particular choice of circular convolution for binding is difficult to directly motivate from data. It could well be that one of the other VSAs is ultimately more appropriate for modeling cognitive behavior. Perhaps attempting to implement a sophisticated linguistic approach like Smolensky's (1986) Harmony Theory in the SPA would lead to more stringent constraints on the kind of representation most appropriate for capturing linguistic behavior.

Ultimately, addressing such challenges will force us to confront the general problems of scaling. Undoubtedly, as the number and complexity of model components increases scaling challenges related to integration become critical. Consider, for example, the Spaun model. Despite the fact that it has several anatomical areas integrated in the model, a large number of them are quite simple: reinforcement learning is applied to only a few simple actions; motor control is applied to a single, two-joint arm; perceptual input is at a fixed location; and semantics is limited to the domain of numbers. Either increasing the complexity of any one of these components or adding any further components will bring integration challenges.

Such integration challenges have both a theoretical and a practical side. On the theoretical side, increased complexity is likely to generate unexpected interactions within and among components. Developing an understanding of how the brain ensures stability or appropriate functioning without being able to predict all possible interactions seems crucial. The SPA provides a means of generating sophisticated components, suggests many such components, and describes a protocol for allowing them to interact. However, it does not provide a design methodology that guarantees large-scale stability and lack of pathological behavior–this remains the art in SPA engineering.

More practically, the computational demands of ever larger models are severe. Such demands are also a general consequence of confronting scaling challenges. To improve performance to some extent, we have recently extended Nengo to run on GPUs, taking advantage of their parallel computational resources and significantly speeding up Nengo models. However, the low-power high-efficiency computing necessary to run very large simulations in real time will most likely be realized through more direct hardware implementation. For this reason it is an important challenge to implement SPA models on hardware architectures, such as Neurogrid (Silver et al., 2007) or SpiNNaker (Khan et al., 2008). Our lab is currently working in collaboration with these groups to simulate millions of neurons in real time. However, we are only beginning to address the many challenges that remain for implementing arbitrary SPA models in neuromorphic hardware.

10.6 ■ Conclusion

Despite these and other challenges, I believe that there are compelling reasons to pursue the SPA. First, although it is *not* a completed, unified theory of mental function, it *is* an attempt to move toward such a theory. Such attempts can be useful in both their successes and failures; either way, we learn about constructing a theory of this kind. As I discussed in Chapter 1, some have suggested that such a theory does not exist. But the behavioral sciences are far too young to think we have done anything other than scratch the surface of possible cognitive theories. And, as I have argued in Chapters 8 and 9, I believe the SPA takes us beyond currently available alternatives.

A second, related reason to pursue the SPA is its close connection to biological considerations. A main goal of this work is to show how we can begin to take biological detail seriously, even when considering sophisticated, cognitive behavior. Even if it is obvious that we should use as much available empirical data as possible to constrain our theories in general, actually doing so requires well-specified methods that contact available data as directly as possible. I have, I hope, described specific applications of the SPA that draw on, relate to, and predict empirical data in new and interesting ways (*see* e.g., Sections 3.5, 4.7, 5.7, 6.2, etc.).

A third reason to pursue the SPA is its generality. I have attempted to exhibit the generality of the approach by choosing a broad set of relevant examples that demonstrate, but do not exhaust, the principles at work. And, in much of this chapter, I have speculated about how the SPA can be extended beyond the examples I have provided here. Of course, one book is not enough to adequately address even a small portion of cognitive behavior. In short, my intent is to provide a method and an architecture that opens the way for a much wider variety of work than can possibly be captured in a single book or, for that matter, done in a single lab.

Perhaps the greatest challenge for the SPA is for it to become any kind of serious contender in the behavioral sciences. The field is a vibrant and crowded one. Consequently, I will be happy if reading this book convinces even a handful of researchers that we ought to relate our cognitive theories in more detail to the biological substrate. I will be *very* happy if this book plays a role in increasing the number of cognitive models that are specified in a biological manner. I will be ecstatic if the SPA itself is deemed useful in constructing such models. And, I will be shocked if major elements of the SPA turn out to be right. Although I have attempted to show how the SPA can address a wide variety of the CCC, often better than its competitors, the many challenges that remain for the SPA make it unclear if anything recognizable as the SPA itself will remain as such challenges are addressed.

One thing that has become clear to me is that addressing such challenges for the SPA, or any other architecture, is going to be the work of a community of researchers. This is not the observation that the behavioral sciences are interdisciplinary. Rather, it is the observation that to make real progress on advancing our understanding of biological cognition, groups with diverse expertise will have to work *together*, not just alongside one another. I suspect that the only practical way to do this is to have large-scale integrated theoretical constructs, likely in the form of a computational model or a family of such models, that all of these different experts can test and extend. As a result, there needs to be a degree of agreement on modeling practices, software, databases, and so on. Consequently, my lab continues to spend significant effort developing not only an architecture but tools that go beyond the architecture for helping to distribute, test, and implement biologically realistic large-scale models.

Despite the enormous challenges that remain, I am optimistic. If the field more aggressively pursues the *integration* of results across disciplines by developing coherent methods and tools, then I believe it will be progressively better able to satisfy the CCC. In short, I think that unraveling the mysteries of biological cognition is only a matter of time.

APPENDIX A. MATHEMATICAL NOTATION AND OVERVIEW

This section is intended to provide a brief overview of some mathematical concepts and notation employed in the book.

A.1 ■ Vectors

Vectors are sets of one or more numbers, indicated by a bold lowercase variable–that is, $\mathbf{x} = [x_1, x_2, \ldots, x_D]$. The number of elements, equal to D, is the dimensionality of the vector. For example, a three-dimensional vector might be $\mathbf{x} = [1, 4, 6]$. One-dimensional vectors are also called "scalars."

It is often convenient to draw vectors. They are typically depicted as points, or as arrows from the origin (i.e., the all zero vector) to a point. Figure A.1A shows examples of plotting two-dimensional vectors in this manner.

Vectors can be "stretched" by multiplying them by a scalar, as shown in Figure A.1B. For any vector $\mathbf{x} = [x_1, x_2, \ldots, x_D]$, multiplying by a scalar "a" gives $a\mathbf{x} = [ax_1, ax_2, \ldots, ax_D]$.

Vectors of the same dimension can be added together by adding the elements $\mathbf{x} + \mathbf{y} = [x_1 + y_1, x_2 + y_2, \ldots, x_D + y_D]$, as shown in Figure A.1C. Adding vectors in this way is called "superposition."

Vectors have a length (or "norm"), which can be computed using the Pythagorean theorem. Recall that this theorem tells us that the square of the hypotenuse equals the sum of the squares of the sides in a right-angled triangle. Vectors are the hypotenuse of a right-angled triangle, whose sides are formed by the axes. So, the length of a vector is

$$\|\mathbf{x}\| = \sqrt{x_1^2 + \ldots + x_D^2}$$

as shown in Figure A.1D.

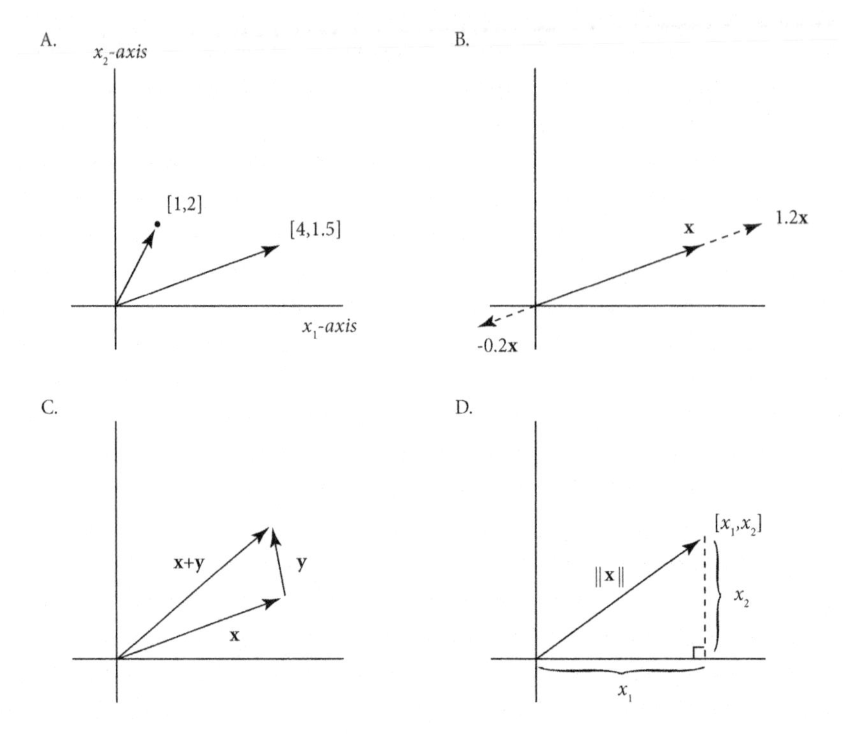

FIGURE A.1 Examples of two-dimensional (2D) vectors, and some operations on them. **A**. 2D vectors plotted as a point and an arrow from the origin. **B**. Multiplying a 2D vector by scaling factors (1.2 and -0.2). **C**. Adding 2D vectors (i.e., "superposition"), $\mathbf{x}+\mathbf{y}=[x_1+y_1, x_2+y_2]$. **D**. Calculating the length of a vector, $\|\mathbf{x}\|=\sqrt{x_1^2+x_2^2}$.

A.2 ■ Vector Spaces

A vector space V is a set of vectors that have addition and scalar multiplication defined, such that standard mathematical properties hold, for example, commutativity (i.e., the order of addition does not matter). Vector spaces also have: a "zero" element that, when added to any vector, results in that vector; an additive inverse for each element that, when added to that element, gives zero; and a "one" element that, when multiplied by any vector, results in that vector.

Intuitively, the points on a number line form a one-dimensional vector space, the points on a standard Cartesian plane form a two-dimensional vector space, and so on (*see* Fig. A.2). A subspace of a vector space V is any set of vectors in V that is also a vector space (*see* Figs. A.2).

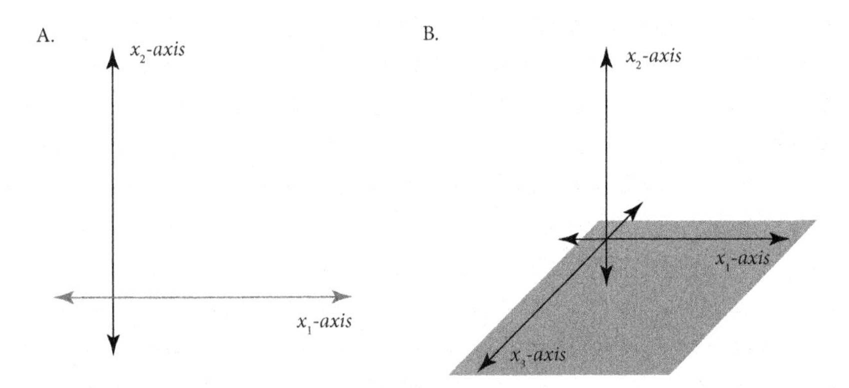

FIGURE A.2 Examples of vector spaces and their subspaces. **A**. A two-dimensional vector space with a one-dimensional subspace, x_1, shown in gray. **B**. A three-dimensional vector space with a two-dimensional subspace, $[x_1, x_3]$, shown in gray.

A.3 ■ The Dot Product

The dot product is an operation between two vectors that is a "projection" of one vector onto the other (*see* Fig. A.3A). As shown, this is depicted by drawing a line at 90° from one vector to the other, mapping along that line, and scaling by the length of the vector being projected on to. The result is a single number (i.e., a scalar) representing the length of the resulting vector.

Algebraically, this operation is written as

$$\mathbf{x} \cdot \mathbf{y} = [x_1 y_1 + \ldots + x_D y_D],$$

which can also be written in summation notation as

$$\mathbf{x} \cdot \mathbf{y} = \sum_{i=1}^{D} x_i y_i.$$

In this notation, which I use occasionally in the book, i acts as a variable that takes on integer values from 1 to D. The expression after the summation sign, $\sum$, is computed for each value of i, and the results are summed.

Crucially, the dot product can also be expressed as a cosine relation, namely,

$$\mathbf{x} \cdot \mathbf{y} = \|\mathbf{x}\| \, \|\mathbf{y}\| \cos\theta$$

(recall that $\|\mathbf{x}\|$ is the length of the vector $\mathbf{x}$). As a result, the dot product is proportional to the angle between the two vectors. For this reason, it is often used as a measure of vector similarity. If one of the vectors has a length of 1 (known as a "unit" vector), the dot product is the cosine of the angle between them, scaled by the length of the non-unit vector.

In the NEF, the activity of a neuron is proportional to this similarity measure (because the preferred direction vector is always a unit vector). If both

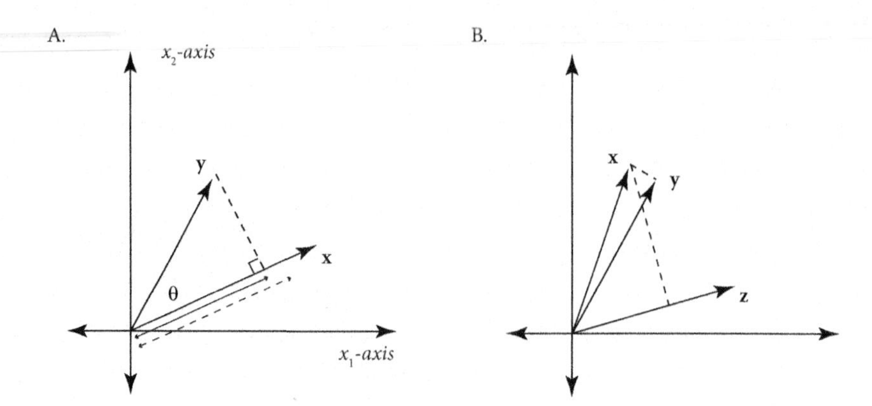

FIGURE A.3 Depiction of a dot product. **A**. A general dot product as a projection of one vector onto another. The solid double-ended line is the projection of **y** onto **x**, which is $\|\mathbf{y}\| \cos\theta$. The dashed double-ended line is the dot product (i.e., the projection scaled by $\|\mathbf{x}\|$). **B**. The dot product between unit vectors as a measure of similarity. Because all vectors have length 1, the dashed lines intersect the vectors at a length equal to the dot product. Hence, the dot product that captures **x** is much more similar to **y** than to **z**.

vectors are unit vectors, then the dot product is exactly proportional to the angle between them, and so a very natural measure of their "nearness" in the vector space, as shown in Figure A.3B. In the SPA this property is used to determine the similarity of concepts.

Notably, if two vectors are at 90° to one another, then their dot product will be zero, regardless of their lengths. Such vectors are known as "orthogonal." If the vectors are more than 90° apart, then their dot product becomes negative.

A.4 ■ Basis of a Vector Space

This section is not crucial for understanding the contents of the book, but I believe it gives helpful insight into how representation works in the NEF. Notice that when we express vectors as a set of numbers $[1, 2]$ and then graph this in a figure (as in Fig. A.1A), we are implicitly assuming that these numbers tell us how far to go to the right and how far to go up, respectively.

To be more clear about what the vector is, we should say what "right" and "up" mean. We can do this very naturally with more vectors. Specifically, "right" is the vector $[1,0]$ and "up" is the vector $[0,1]$ (these are often denoted **i** and **j**, respectively). So, in fact, when we write down any vector in this 2D vector space, we should always express it as a sum of these two vectors. That is, $[1,2]$ really means $1 \cdot [1,0] + 2 \cdot [0,1] = 1\mathbf{i} + 2\mathbf{j}$ (*see* Fig. A.4A).

It is somewhat cumbersome to do this, so we usually leave off the "right" and "up." However, by identifying these two vectors, we have identified a way

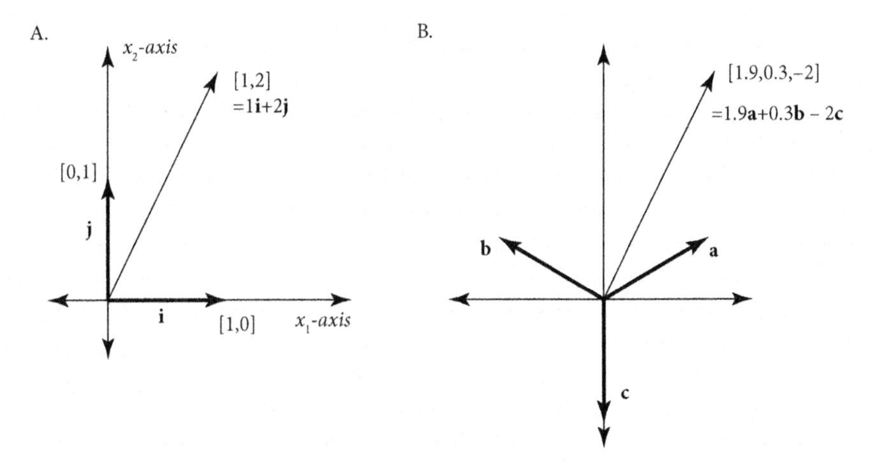

FIGURE A.4 The basis of a vector space. **A**. The standard Cartesian basis with the vectors [1,0] and [0,1] defining the space. The vector $\mathbf{x} = [1, 2]$ is shown. Thus, **i** and **j** form an orthonormal basis. **B**. An "overcomplete" basis with three unit vectors. The same object is plotted as in A but in a new basis, so that object is now $\mathbf{x} = [1.9, 0.3, -2]$ with respect to the new basis vectors **a**, **b**, and **c**.

of distinguishing vector spaces. Specifically, all vectors in the space can be written as a sum of those two vectors. If we change those two vectors, then to get to the same point we will need different numbers. For example, if instead of "right" we used "left," then [1,2] would become [−1,2]. Sets of vectors used to define a vector space in this manner are called the "basis" of that vector space.

All of the bases we usually run into are special, in that they tend to be orthogonal (notice, $\mathbf{i} \cdot \mathbf{j} = 0$). This is convenient because we can then specify a point by specifying movement in two *independent* directions. Moving along **i** does not move you along **j**, which would not be the case if the basis was not 90° apart. Nevertheless, we can specify all the points in a 2D space with a non-orthogonal basis as well. And, we can even add extra vectors (i.e., more than two), and specify a point as a sum of all of them (*see* Fig. A.4B). This kind of basis is called "overcomplete" because the numbers in the vector are redundant to some extent (e.g., it takes three numbers to specify a point in 2D space in Fig. A.4B).

Overcomplete bases are useful if there is significant uncertainty about the precise location of our vector. If we have many independently generated numbers to specify a low-dimensional point, then the uncertainty will often average out. This principle is used in all of the representations in the NEF– lots of noisy neurons with various preferred direction vectors are used to accurately represent a lower-dimensional space.

A.5 ■ Linear Transformations on Vectors

Often the point of defining a vector space is to capture relations between the vectors in that space. In a sense, we want to know how to "move" from one vector to another. The simplest such movements are "linear," meaning that they will always map straight lines (i.e., vectors that lie on a line) onto other straight lines (or zero).

Any linear mapping within a vector space can be defined using a *set* of vectors also in that space. Often, we need as many vectors in the set as there are dimensions in the space. To collect sets of vectors together, we use a matrix. For example, to group two 2D vectors we may define the matrix **M** as follows:

$$\mathbf{M} = [\mathbf{a};\mathbf{b}] = \begin{bmatrix} a_1 & a_2 \\ b_1 & b_2 \end{bmatrix} = \begin{bmatrix} m_{1,1} & m_{1,2} \\ m_{2,1} & m_{2,2} \end{bmatrix}.$$

Depending on which vectors we choose, we can define different linear transformations. Examples include scaling, rotating, shearing, and reflecting vectors (*see* Fig. A.5).

When we apply a linear mapping to a vector **x**, we can write this algebraically as

$$\mathbf{y} = \mathbf{M}\mathbf{x}, \text{or}$$

$$\begin{bmatrix} y_1 \\ y_2 \end{bmatrix} = \begin{bmatrix} m_{1,1} & m_{1,2} \\ m_{2,1} & m_{2,2} \end{bmatrix} \begin{bmatrix} x_1 \\ x_2 \end{bmatrix}$$

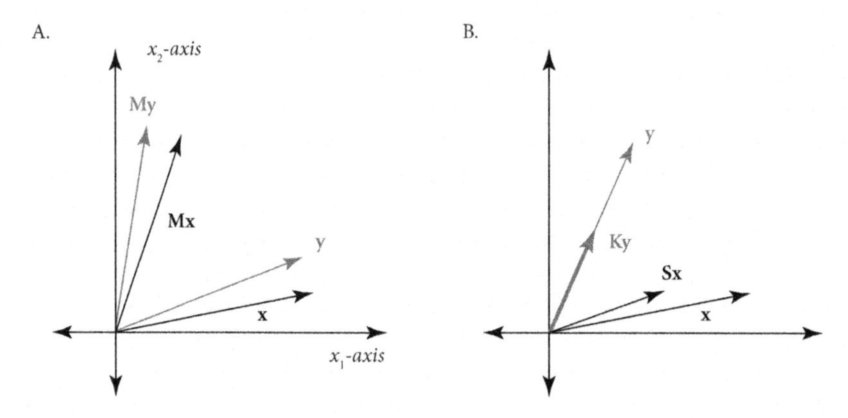

FIGURE A.5 Example linear transformations. **A.** Counterclockwise rotation by θ using $\mathbf{M} = \begin{bmatrix} \cos(\theta) & -\sin(\theta) \\ \sin(\theta) & \cos(\theta) \end{bmatrix}$. Rotations for $\theta = 60°$ are shown for two different vectors, **x** (black) and **y** (gray). **B.** Horizontal shear by a factor $s = -1$ using $\mathbf{S} = \begin{bmatrix} 1 & s \\ 0 & 1 \end{bmatrix}$ on **x** (black), and scaling by a factor $k = 0.5$ using $\mathbf{K} = \begin{bmatrix} k & 0 \\ 0 & k \end{bmatrix}$ on **y** (gray).

which means that y_1 is equal to the dot product between the first row of **M** and **x**, and y_2 is equal to the dot product between the second row of **M** and **x**. We can write this in summation notation as

$$y_j = \sum_{i=1}^{2} m_{ij} x_i,$$

which should be reminiscent of how neural networks are expressed for a weight matrix **M** connecting neurons in one layer **x** to another layer **y** (there is also a nonlinearity operating on **y** in most neural networks).

In short, all linear mappings in a vector space can be expressed as a series of dot products between vectors. Matrices are a useful way of collecting together sets of vectors that define a particular transformation. For example, holding **M** constant and changing **x** results in vectors **y** that always bear the same relation to **x**. Consequently, specific matrices are often identified with specific transformations. If **y** is in the same vector space as **x**, then we can apply **M** over and over to trace a path through the vector space. This is how simple dynamics are typically defined.

A.6 ■ Time Derivatives for Dynamics

The most common kind of derivatives I employ are time derivatives, because these are used to define dynamics. A time derivative specifies how the state of a system changes with each small change in time. The state of a system can be written as a vector **x** where each element of the vector is some measure characterizing the system, such as position (*see* Fig. A.6). The delta symbol Δ is often used to indicate a change (e.g., $\Delta t = t_2 - t_1$). So, we can write the time derivative of a system as

$$\frac{\Delta \mathbf{x}}{\Delta t} = \mathbf{M}\mathbf{x},$$

where **M** is specifying the transformation that occurs to the system state at every moment in time. To be more explicit, we can write this equation out, where $\Delta \mathbf{x} = \mathbf{x}(t + \Delta t) - \mathbf{x}(t)$, and the state vector is a function of time, t:

$$\mathbf{x}(t + \Delta t) - \mathbf{x}(t) = \mathbf{M}\mathbf{x}(t)\Delta t$$
$$\mathbf{x}(t + \Delta t) = \mathbf{x}(t) + \mathbf{M}\mathbf{x}(t)\Delta t,$$

which tells us exactly what the state at the next moment in time will be. This provides a rough sense of how specifying some linear transformation results in a particular series of states of a system if we apply that transformation over and over. Because time derivatives are so common, they are often written as just a dot–for example

$$\dot{\mathbf{x}} = \mathbf{M}\mathbf{x}.$$

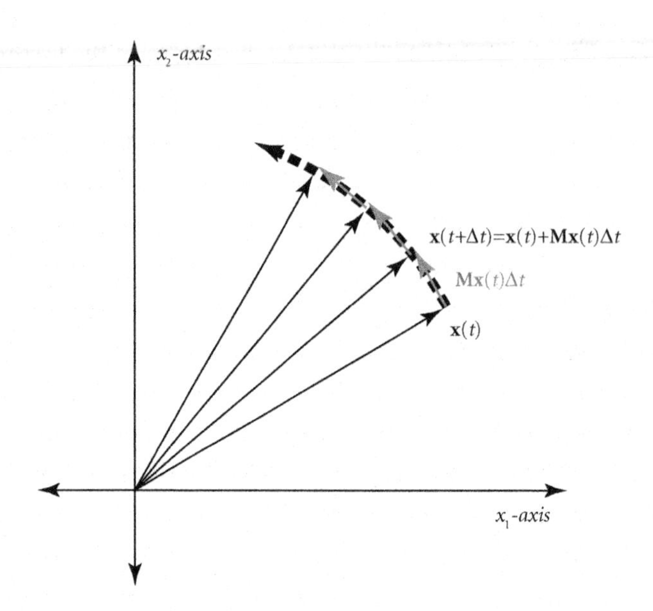

FIGURE A.6 Using time derivatives to specify a system's dynamics. The linear transformation in this case is a 90° rotation, $\mathbf{M} = \begin{bmatrix} 0 & -1 \\ 1 & 0 \end{bmatrix}$. Consequently, at each moment in time, the current state vector is rotated by 90° and scaled down by the time step Δt (gray). That vector is added to the current state to give the next state. This particular **M** results in a circular motion over time (dotted line).

As a simple example, suppose that **M** is a 90° rotation, like the 60° rotation shown in Figure A.5A. Then, if we plot each new state of **x** as we apply **M** and add it to the previous state, we would trace out a circular path as shown in Figure A.6. For different choices of **M**, we can characterize different paths through the vector space (also called the state space).

APPENDIX B. MATHEMATICAL DERIVATIONS FOR THE NEF

Detailed discussions and more in-depth derivations of each of these aspects of the NEF can be found in Eliasmith and Anderson (2003).

B.1 ■ Representation

B.1.1 Encoding

Consider a population of neurons whose activities $a_i(\mathbf{x})$ encode some D-dimensional vector, $\mathbf{x}$. These activities can be written

$$a_i(\mathbf{x}) = G_i\left[J_i(\mathbf{x})\right], \tag{B.1}$$

where G_i is the nonlinear function describing the neuron's response function, and $J_i(\mathbf{x})$ is the current entering the soma. The somatic current is defined by

$$J_i(\mathbf{x}) = \alpha_i \left\langle \mathbf{x}\mathbf{e}_i \right\rangle + J_i^{bias}, \tag{B.2}$$

where $\langle \cdot \rangle$ indicates a dot product, $J_i(\mathbf{x})$ is the current in the soma, α_i is a gain and conversion factor, $\mathbf{x}$ is the vector variable to be encoded, $\mathbf{e}_i$ is the D-dimensional encoding vector that picks out the "preferred stimulus" of the neuron (*see* Section 2.4), and J_i^{bias} is a bias current that accounts for background activity.

The nonlinearity G_i that describes the neuron's activity as a result of this current is determined by physiological properties of the neuron(s) being modeled. The most common model used in the book is the leaky integrate-and-fire (LIF) neuron. For descriptions of many other neural models, some of which are in Nengo, see Bower and Beeman (1998), Koch (1999), and Carnevale and Hines (2006).

The subthreshold evolution of the LIF neuron voltage is described by

$$\dot{V}(t) = -\frac{1}{\tau_{RC}}(V(t) - J(\mathbf{x})R), \tag{B.3}$$

where V is the voltage across the membrane, $J(\mathbf{x})$ is the input current, R is the passive membrane resistance, and τ_{RC} is the membrane time constant. When the membrane voltage crosses a threshold V_{thresh}, a spike is emitted, and the cell is reset to its resting state for a time period equal to the absolute refractory time constant τ_{ref}. The output activity of the cell is thus represented as a train of delta functions, placed at the times of threshold crossing (i.e., spikes) t_m as $a_i(\mathbf{x}) = \sum_m \delta(t - t_m)$.

The overall encoding equation is thus

$$\delta(t - t_{im}) = G_i\left[\alpha_i \langle \mathbf{x}\mathbf{e}_i \rangle + J_i^{bias}\right], \tag{B.4}$$

where $\delta(t)$ represents a neural spike, m indexes the spikes, i indexes the neurons, $G[]$ is the spiking neural model, α is the gain of the neuron, $\mathbf{x}$ is the variable being encoded, $\mathbf{e}$ is the encoding (or preferred direction) vector, and J^{bias} is the background current.

B.1.2 Decoding

To define the decoding, we need to determine the postsynaptic current (PSC) and the optimal decoding weight. A simple model of the PSC is

$$h(t) = e^{-t/\tau_{PSC}}, \tag{B.5}$$

where τ_{PSC} is the time constant of decay of the PSC. This varies with the type of neurotransmitter being used. Typical values are 5 ms for AMPA, 10 ms for GABA, and 100 ms for NMDA receptors.[1]

Given an input spike train $\delta(t - t_{im})$ generated from the encoding above, the "filtering" of the neural spikes by the PSC gives an "activity" of[2]

$$a_i(\mathbf{x}) = \sum_m h(t - t_{im}).$$

Mathematically, this equation is the result of *convolving* the spike train with the PSC filter.

[1] A survey of the variety of time constants, with supporting empirical evidence, can be found at `http://nengo.ca/build-a-brain/chapter2`.

[2] It is important to keep in mind that the PSC filtering and the weighting by a synaptic weight happen at the same time, not in sequence as I am describing here. In addition, the decoders here constitute only part of a neural connection weight.

To determine the optimal linear decoding weight to decode this activity (assuming some noise in transmission of the spikes), we need to minimize the error between the representation of the signal under noise and the original signal (see also Salinas & Abbott, 1994):

$$E = \int_R \left[\mathbf{x} - \sum_i^N (a_i(\mathbf{x}) + \eta_i)\mathbf{d}_i \right]^2 d\mathbf{x}d\eta \tag{B.6}$$

where N is the number of neurons, η is a random perturbation chosen from an i.i.d. Gaussian distribution with mean zero, $\mathbf{d}_i$ are the D-dimensional decoders to be determined, and R is the range over which the representation is to be optimized (e.g., a unit hypersphere). In matrix notation, the solution to this common kind of linear least-squares optimization is[3]

$$\mathbf{D} = \Gamma^{-1}\Upsilon,$$

where $\mathbf{D}$ is the $N \times D$ matrix of optimal decoders (one decoder $\mathbf{d}_i$ in each row of the matrix), Υ is the matrix where each row$\Upsilon_i = \int_R \mathbf{x}a(\mathbf{x})d\mathbf{x}$, and Γ is the correlation matrix of neural activities where each element $\Gamma_{ij} = \gamma_{ij} = \int_R a_i(\mathbf{x})a_j(\mathbf{x})d\mathbf{x}$ and $\Gamma_{ii} = \gamma_{ii} + \sigma^2$. The σ^2 is the variance of the noise from which the η_i are picked.

The overall decoding equation is thus

$$\hat{\mathbf{x}} = \sum_{i,m}^{N,M} h_i(t - t_{im})\mathbf{d}_i, \tag{B.7}$$

where N is the number of neurons, M is the number of spikes, i indexes the neurons, m indexes the spikes, $h(t)$ is the PSC of the neuron, $\hat{\mathbf{x}}$ is the estimate of the variable being represented, and $\mathbf{d}_i$ is the decoder for neuron i to estimate $\mathbf{x}$.

B.2 ■ Transformation

To define a transformational decoder, we follow the same procedure as in Section B.1.2 but minimize a slightly different error to determine the population decoders–namely,

$$E = \int_R \left[f(\mathbf{x}) - \sum_i^N (a_i(\mathbf{x}) + \eta_i)\mathbf{d}_i \right]^2 d\mathbf{x}d\eta,$$

[3] For a more detailed derivation, *see* any standard optimization text book or website–for example, `http://en.wikipedia.org/wiki/Linear_least_squares_(mathematics)`.

where we have simply substituted $f(\mathbf{x})$ for $\mathbf{x}$ in Equation B.6. This results in the related solution

$$\mathbf{D}^f = \Gamma^{-1}\Upsilon^f,$$

where $\Upsilon_i^f = \int_R f(\mathbf{x})a(\mathbf{x})d\mathbf{x}$, resulting in the matrix $\mathbf{D}^f$ where the transformation decoders $\mathbf{d}_i^f$ can be used to give an estimate of the original transformation $f(\mathbf{x})$ as

$$\hat{f}(\mathbf{x}) = \sum_i^N a_i(\mathbf{x})\mathbf{d}_i^f.$$

Thus, we have a decoding equation for transformations of spike trains as

$$\hat{f}(\mathbf{x}) = \sum_{i,m}^{N,M} h_i(t - t_{im})\mathbf{d}_i^f, \tag{B.8}$$

where N is the number of neurons, M is the number of spikes, i indexes the neurons, m indexes the spikes, $h(t)$ is the PSC of the neuron, $\hat{f}(\mathbf{x})$ is the estimate of the transformation being performed, $\mathbf{x}$ is the representation being transformed, and $\mathbf{d}_i^f$ is the decoder for neuron i to compute f.

B.3 ■ Dynamics

The equation describing Figure 2.12A is

$$\dot{\mathbf{x}}(t) = \mathbf{A}\mathbf{x}(t) + \mathbf{B}\mathbf{u}(t). \tag{B.9}$$

Notably, the input matrix $\mathbf{B}$ and the dynamics matrix $\mathbf{A}$ completely describe the dynamics of any linear time-invariant (LTI) system, given the state variables $\mathbf{x}(t)$ and the input $\mathbf{u}(t)$.

In this derivation, I employ the Laplace transform, which is convenient for analyzing dynamical systems. However, the relevant results summarized by Equations B.10 and B.11 can be applied directly to Equation B.9 in the time domain.

Taking the Laplace transform of Equation B.9 gives

$$\mathbf{X}(s) = H(s)\left[\mathbf{A}\mathbf{X}(s) + \mathbf{B}\mathbf{U}(s)\right],$$

where $H(s) = \frac{1}{s}$.

In the case of a neural system, the transfer function $H(s)$ is not $\frac{1}{s}$ but is determined by the intrinsic properties of the component cells. Because it is reasonable to assume that the dynamics of the synaptic PSC dominate the dynamics of the cellular response as a whole (Eliasmith & Anderson, 2003), it is reasonable to characterize the dynamics of neural populations based on their synaptic dynamics–that is using $h(t)$ from Equation B.5. The Laplace transform of this filter is

$$H'(s) = \frac{1}{1 + s\tau}.$$

Given the change in filters from $H(s)$ to $H'(s)$, we need to determine how to change $\mathbf{A}$ and $\mathbf{B}$ to preserve the dynamics defined in the original system (i.e., the one using $H(s)$). In other words, letting the neural dynamics be defined by $\mathbf{A}'$ and $\mathbf{B}'$, we need to determine the relation between matrices $\mathbf{A}$ and $\mathbf{A}'$ and matrices $\mathbf{B}$ and $\mathbf{B}'$ given the differences between $H(s)$ and $H'(s)$. To do so, we can solve for $s\mathbf{X}(s)$ in both cases and equate the resulting expressions:

$$\mathbf{X}(s) = \frac{1}{s}\left[\mathbf{AX}(s) + \mathbf{BU}(s)\right]$$

$$s\mathbf{X}(s) = \left[\mathbf{AX}(s) + \mathbf{BU}(s)\right]$$

and

$$\mathbf{X}(s) = \frac{1}{1+s\tau}\left[\mathbf{A}'\mathbf{X}(s) + \mathbf{B}'\mathbf{U}(s)\right]$$

$$s\mathbf{X}(s) = \frac{1}{\tau}\left(\mathbf{A}'\mathbf{X}(s) - \mathbf{X}(s) + \mathbf{BU}(s)\right)$$

so

$$\frac{1}{\tau}\left(\mathbf{A}'\mathbf{X}(s) - \mathbf{X}(s) + \mathbf{BU}(s)\right) = \left[\mathbf{AX}(s) + \mathbf{BU}(s)\right].$$

Rearranging and solving gives

$$\mathbf{A}' = \tau\mathbf{A} + \mathbf{I} \tag{B.10}$$

$$\mathbf{B}' = \tau\mathbf{B}, \tag{B.11}$$

where τ is the synaptic time constant of the neurons representing $\mathbf{x}$.

Notably, this procedure assumes nothing about $\mathbf{A}$ or $\mathbf{B}$. However, it does assume that the system is LTI. For many nonlinear and time-varying systems, however, similar derivations follow. For the time-varying case, it is possible to simply replace $\mathbf{A}$ with $\mathbf{A}(t)$ in the preceding derivation for LTI systems. For any nonlinear system that can be written in the form:

$$\mathbf{X}(s) = \frac{1}{s}F(\mathbf{X}(s), \mathbf{U}(s), s),$$

a similar derivation leads to

$$F'(\mathbf{X}(s), \mathbf{U}(s), s) = \tau F(\mathbf{X}(s), \mathbf{U}(s), s) + \mathbf{X}(s). \tag{B.12}$$

In short, the main difference between synaptic dynamics and the integration assumed by control theory is that there is an exponential "forgetting" of the current state with synaptic dynamics. As long as this difference is accounted for by τ (which determines the speed of forgetting) and $\mathbf{X}(s)$ (which is a "reminder" of the current state), any dynamical system can be implemented in a recurrent spiking network with PSC dynamics given by Equation B.5. Importantly, however, the implementation of a dynamical system may not be successful if the representation of the state space by the network is not sufficiently accurate, or the function F is difficult to compute. Typically, more such difficulties arise the less smooth the function is. More precisely, we can determine which functions a given set of neuron tuning curves are good at computing by decomposing the correlation matrix (*see* Eliasmith & Anderson, 2003, Chapter 7).

APPENDIX C. FURTHER DETAILS ON DEEP SEMANTIC MODELS

C.1 ■ The Perceptual Model

This model is based on that described in detail in Tang and Eliasmith (2010). The main purpose of that paper was to demonstrate a deep belief network (DBN) architecture that was robust to noise. The specific kind of DBN we employed was a hierarchical restricted Boltzmann machine (RBM; Hinton & Salakhutdinov, 2006a). Standard DBNs can fail terribly if the class of stimuli that are used for training is changed before recognition. For example, introducing a white border, an occluding block, or salt and pepper noise causes a network with a 1% error rate to have 66%, 34%, and 80% error rates, respectively. We addressed this issue, improving the robustness of the network, by including sparse connections between visual and hidden nodes and introducing a denoising RBM into the network.

An RBM with visible layer nodes $\mathbf{v}$ and hidden layer nodes $\mathbf{h}$ is defined by the energy function:

$$E(\mathbf{v},\mathbf{h};\theta) = -\mathbf{b}^T\mathbf{v} - \mathbf{c}^T\mathbf{h} - \mathbf{v}^T\mathbf{W}\mathbf{h},$$

where $\theta = [\mathbf{W},\mathbf{b},\mathbf{c}]$ are the model parameters. The probability distribution of the system can be written as:

$$\begin{aligned} p(\mathbf{v},\mathbf{h}) &= \frac{p^*(\mathbf{v},\mathbf{h})}{Z(\theta)} \\ &= \frac{exp^{E(\mathbf{v},\mathbf{h})}}{Z(\theta)} \end{aligned}$$

where $Z(\theta)$ is the normalization term $\sum_{\mathbf{v},\mathbf{h}} exp^{-E(\mathbf{v},\mathbf{h})}$.

In practice, contrastive divergence has been shown to be an effective way to train such a network (Hinton, 2002). Here we used a slight modification of this algorithm for the sparsely connected RBM:

$$\Delta W_{ij} \propto (\mathbb{E}_{data}[v_i h_j] - \mathbb{E}_{recon}[v_i h_j])\widetilde{W}_{ij},$$

where

$$\widetilde{W}_{ij} = \begin{cases} 1 & \text{if } v_i \text{ is in } h_j\text{'s receptive field} \\ 0 & \text{otherwise} \end{cases}$$

and $\mathbb{E}[\,\cdot\,]$ is the expectation with respect to the distribution after n steps of block Gibbs sampling.

As usual, to construct a DBN, several such RBM layers are trained independently and then stacked and fine-tuned using the up-down algorithm (Hinton et al., 2006). The resulting DBN is a statistical model of its input. The learned connection weights are the parameters, and the node transfer functions are sigmoids, which define the form of the model.

To implement this statistical model in the NEF, we employ the principles described in Chapter 2. Specifically, each layer of the DBM is taken to be a cortical layer. Each node in the DBM layer is taken to be one dimension of a high-dimensional vector represented and transformed by that cortical layer. So, for a given DBM layer $\mathbf{h}_1$, after Equation B.4 we have a cortical encoding in to neurons i:

$$\sum_m \delta(t - t_{im}) = G_i\left[\alpha_i \langle \mathbf{h}_1 \mathbf{e}_i \rangle + J_i^{bias}\right]$$

and after Equation B.7, the transformational decoding:

$$\hat{\mathbf{h}}_1 = \sum_{i,m} h(t - t_{im})\mathbf{d}_i^{sig},$$

where $\mathbf{d}_i^{sig}$ are the transformational decoders estimating the sigmoid function in the original network. Assuming a similar encoding into the next cortical neurons, indexed by j, allows us to compute the connection weights as follows:

$$\begin{aligned}
\sum_m \delta(t - t_{jm}) &= G_j\left[\alpha_j \left\langle \mathbf{h}_2 \mathbf{e}_j \right\rangle + J_j^{bias}\right] \\
&= G_j\left[\alpha_j \left\langle W_{12}\mathbf{h}_1 \mathbf{e}_j \right\rangle + J_j^{bias}\right] \\
&= G_j\left[\alpha_j \left\langle \left(W_{12} \sum_{i,m} h(t - t_{im})\mathbf{d}_i^{sig} \right) \mathbf{e}_j \right\rangle + J_j^{bias}\right] \\
&= G_j\left[\sum_i \omega_{ij} a_i + J_j^{bias}\right]
\end{aligned}$$

where $\omega_{ij} = \mathbf{e}_j \mathbf{W}_{12} \mathbf{d}_i^{sig}$ are the cortical connection weights between cortical layers one and two, $a_i = \sum_m h(t - t_{im})$ is the neural activity in the layer, and W_{12} are the parameters in the DBM between layers one and two. A similar process allows us to compute the connection weights between all layers of the cortical implementation of this DBM. It is this spiking cortical network that is used to generate the figures in Chapter 3.

A few comments are in order. First, the network employed here does not compress the representation as much as is typical in DBMs, because categorization is not the main concern. As discussed in Chapter 7, both categorization and representation of visual features are crucial for semantic pointers. Second, it may be possible to allow the learning itself to occur in a cortical network using a rule like that in Section 6.4. However, this would prove very computationally intensive, and so I have kept learning and neural implementation of the cortical model separate. Finally, running this model "backward" to perform "decompression," as mentioned in the main text, amounts to driving the top-layer inputs and allowing the transpose of the model parameters (i.e., $\mathbf{W}^T$) to determine the backward connection weights. This is essentially sampling from a generative statistical model by clamping the high-level priors (Hinton et al., 2006).

C.2 ■ The Motor Model

In this section, I present the derivation of a simplified hierarchical motor controller. It is simplified in the sense that the controlled model is linear, defined in terms of arm angles, it has only two links, and has only two hierarchical levels–none of these assumptions are made for the arm model presented in the main text. However, the core principles are the same, and the derivation is significantly shorter (for more in-depth discussions, *see* Liu & Todorov, 2009; Dewolf & Eliasmith, 2011).

We can begin by defining the lower level system state (i.e., the state of the arm itself) as joint angles and velocities $\mathbf{x} = [\theta_1, \theta_2, \dot{\theta}_1, \dot{\theta}_2]^T$ with system dynamics given by:

$$\dot{\mathbf{x}} = \mathbf{A}\mathbf{x} + \mathbf{B}\mathbf{u},$$

where $\mathbf{A} = \begin{bmatrix} \mathbf{0} & \mathbf{I} \\ \mathbf{0} & \mathbf{0} \end{bmatrix}$, $\mathbf{B} = \begin{bmatrix} \mathbf{0} \\ \mathbf{m}^{-1} \end{bmatrix}$, $\mathbf{u} = \begin{bmatrix} \tau_1 \\ \tau_2 \end{bmatrix}$ where $\mathbf{m} = \begin{bmatrix} m_1 & 0 \\ 0 & m_2 \end{bmatrix}$ (m_1 and m_2 are the mass of the first and second links), andτ_1 and τ_2 are the torques applied to the shoulder and elbow joints, respectively. These dynamics are an expression of Newtonian mechanics applied to this simple two-link arm model.

We can now define the high-level system separately, in Cartesian coordinates of the end of the arm (i.e., the hand). The system state will be

$\mathbf{y} = [p_x, p_y, \dot{p}_x, \dot{p}_y]^T$, relative to an absolute reference frame at the shoulder. The system dynamics are given by:

$$\dot{\mathbf{y}} = \mathbf{G}\mathbf{y} + \mathbf{F}\mathbf{v}, \tag{C.1}$$

where $\mathbf{G}$ and $\mathbf{F}$ are the same as $\mathbf{A}$ and $\mathbf{B}$, and $\mathbf{v} = \begin{bmatrix} f_x \\ f_y \end{bmatrix}$, which are the forces applied to the end effector.

Let the transformation between these two levels be

$$\mathbf{y} = h(\mathbf{x}). \tag{C.2}$$

We can then take the derivative of both sides, and apply the chain rule to give:

$$\dot{\mathbf{y}} = H(\mathbf{x})(\mathbf{A}\mathbf{x} + \mathbf{B}\mathbf{u}), \tag{C.3}$$

where $H(\mathbf{x}) = \partial h(\mathbf{x})/\partial \mathbf{x}$.

The mapping between the low- and high-level states can be found geometrically:

$$h(\mathbf{x}) = \begin{bmatrix} \cos(\theta_1)L_1 + \cos(\theta_1 + \theta_2)L_2 \\ \sin(\theta_1)L_1 + \sin(\theta_1 + \theta_2)L_2 \\ J_\theta(\mathbf{y})\dot{\theta} \end{bmatrix}$$

where L_1 and L_2 are the lengths of the upper and lower arm segments, and $J_\theta(\mathbf{y})$ is the Jacobian of the high-level state with respect to $\theta = [\theta_1, \theta_2]$.

Equating Equations C.3 and C.1 gives

$$H(\mathbf{x})\mathbf{A}\mathbf{x} + H(\mathbf{x})\mathbf{B}\mathbf{u} = \mathbf{F}\mathbf{y} + \mathbf{G}\mathbf{v}.$$

Because $\mathbf{F}\mathbf{y} = H(\mathbf{x})\mathbf{A}\mathbf{x}$ by definition, and because we allow the high-level controller to have online access to the low-level state $\mathbf{x}$, we have

$$H(\mathbf{x})\mathbf{B}\mathbf{u} = \mathbf{G}\mathbf{v}. \tag{C.4}$$

So, given a high-level control signal $\mathbf{v}$, we can generate the equivalent low-level control signal $\mathbf{u}$ by finding the minimal $\mathbf{u}$ such that Equation C.4 is satisfied. A method for efficiently generating the original optimal control signal in the end-effector space has been described in Dewolf (2010).

As mentioned in the introduction, the model described here is much simpler than that presented in the main text. In particular, that model includes additional levels in the hierarchy. We can repeat the above process of mapping dynamics between levels for additional levels as needed. Typically, this mapping will include a reduction in the dimensionality as we go up the hierarchy (e.g., the model presented in the text goes from six to three to two dimensions). In addition, the model in the text has (1) muscle dynamics modeled (using Hill's muscle model; Hill, 1938); (2) deals with configuration redundancy; and (3) has nonlinear dynamics. Both of these arm models can be downloaded from `http://nengo.ca/build-a-brain/chapter3`.

APPENDIX D. MATHEMATICAL DERIVATIONS FOR THE SEMANTIC POINTER ARCHITECTURE

D.1 ■ Binding and Unbinding Holographic Reduced Representations

The Semantic Pointer Architecture (SPA) uses circular convolution for binding and unbinding, as first proposed by Plate (1991). Circular convolution is a variant of the more common convolution operation. Convolution is widely used in linear system analysis and signal processing and has many properties in common with scalar and matrix multiplication (e.g., it is commutative, associative, distributive over addition, a zero vector and an identity vector both exist, and most vectors have exact inverses).

Convolution is useful because it can be used to compute the output of a linear system given any input signal. Given the impulse response filter $y(t)$ that defines a linear system and an input signal $x(t)$, the convolution is

$$z(t) = x(t) * y(t) = \int_{-\infty}^{\infty} x(\tau)y(t-\tau)\,d\tau.$$

It is standard to visualize convolution as flipping the filter around the ordinate (y-axis), and sliding it over the signal, computing the integral of the product of the overlap of the two functions at each position, giving $z(t)$. In the case where $y(t)$ and $x(t)$ are discrete vectors, the length of the result is equal to the sum of the lengths of the original vectors, minus one. So, if the vectors are the same size, then the result is approximately double the original length. This would lead to scaling problems if used for binding.

Circular convolution is a similar operation, but as the elements of $y(t)$ slide off of the end of $x(t)$, they are wrapped back onto the beginning of $x(t)$. Consequently, we can define the discrete circular convolution of two vectors as

$$\mathbf{z} = \mathbf{x} \circledast \mathbf{y}$$
$$z_j = \sum_{k=0}^{D-1} x_k y_{j-k},$$

where subscripts are modulo D (the length of the filter vector). The result of this form of convolution for two equal-length vectors is equal to their original length because of the wrapping.

A computationally efficient algorithm for computing the circular convolution of two vectors takes advantage of the Discrete Fourier Transform (DFT) and Inverse Discrete Fourier Transform (IDFT). In general, Fourier Transforms are closely related to convolution because the Fourier Transform of any convolution is a multiplication in the complementary domain. The same is true for circular convolution. So, the circular convolution of two finite-length vectors can be expressed in terms of DFT and IDFT as follows:

$$\mathbf{x} \circledast \mathbf{y} = IDFT\left(DFT\left(\mathbf{x}\right) \cdot DFT\left(\mathbf{y}\right)\right), \tag{D.1}$$

where "·" indicates element-wise multiplication.

In a neural circuit, we can take advantage of this identity by noting that $DFT(\mathbf{x})$ is a linear transformation of the vector $\mathbf{x}$ and can thus be written as a matrix multiplication, $DFT(\mathbf{x}) = \mathbf{C}_{DFT}\mathbf{x}$, where $\mathbf{C}_{DFT}$ is a $D \times D$ constant coefficient matrix that can be pre-computed for a given size of the vector space. The IDFT operation can similarly be written as a matrix multiplication $IDFT(\mathbf{x}) = \mathbf{C}_{DFT}^{-1}\mathbf{x} = \mathbf{C}_{IDFT}\mathbf{x}$. We can thus rewrite circular convolution as matrix multiplication–that is,

$$\mathbf{x} \circledast \mathbf{y} = \mathbf{C}_{IDFT}\left(\left(\mathbf{C}_{DFT}\mathbf{x}\right).\left(\mathbf{C}_{DFT}\mathbf{y}\right)\right), \tag{D.2}$$

where all of the matrices are constant for any values of $\mathbf{x}$ and $\mathbf{y}$.

The NEF can be used to compute arbitrary linear transformations of vectors, as well as point-wise vector products (see Section 3.8). Consequently, to compute the circular convolution $\mathbf{z} = \mathbf{x} \circledast \mathbf{y}$, we can combine these two techniques.

We must also be able to unbind vectors. To do so, we can invert the convolution operation by defining what is sometimes called a "correlation" operation ($\oplus$). More simply, we can think of this as standard matrix inversion–that is,

$$\mathbf{x} = \mathbf{z} \oplus \mathbf{y} = \mathbf{z} \circledast \mathbf{y}^{-1}.$$

As discussed by Plate (2003), although most vectors do have an exact inverse under convolution, it is numerically advantageous to use an approximate inverse (called the involution) of a vector in the unbinding process. This is because the exact inverse $\mathbf{y}^{-1}$ is unstable when elements of $DFT(\mathbf{y})$ are near zero.

Involution of a vector $\mathbf{x}$ is defined as $\mathbf{x}' = \left[x_0, x_{D-1}, x_{D-2}, \ldots, x_1\right]$. Notice that this is simply a flip of all the elements after the first element. Thus, it is a simple permutation of the original matrix. Unbinding then becomes

$$\mathbf{x} = \mathbf{z} \circledast \mathbf{y}'.$$

Notice that the involution is a linear transformation. That is, we can define a permutation matrix $\mathbf{S}$ such that $\mathbf{Sx} = \mathbf{x}'$. As a result, the neural circuit for circular convolution is easily modified to compute circular correlation by using the matrix $\mathbf{C}_S = \mathbf{C}_{DFT}\mathbf{S}$ in place of the second DFT matrix in the original circuit (Equation D.2). Specifically,

$$\mathbf{z} \circledast \mathbf{y}' = \mathbf{C}_{IDFT}\left((\mathbf{C}_{DFT}\mathbf{z}) . (\mathbf{C}_S\mathbf{y})\right),$$

where again all matrices are constant for any vectors that need unbinding.

D.2 ■ Learning High-Level Transformations

To model learning of different transformations in different contexts, we need to derive a biologically plausible learning rule that can infer these transformations. Neumann (2001) noted that to find some unknown transformation $\mathbf{T}$ between two vectors $\mathbf{A}$ and $\mathbf{B}$, we can solve

$$\mathbf{T} = circ\left(\sum_i^m \mathbf{B}_i \oplus \mathbf{B}_i\right)^{-1}\left(\sum_i^m \mathbf{B}_i \oplus \mathbf{A}_i\right), \tag{D.3}$$

where $circ(\,\cdot\,)$ is the circulant matrix, $\oplus$ is circular correlation, i indexes training examples, and m is the number of examples.

Noting $\mathbf{B}_i \oplus \mathbf{B}_i \approx 1$, Equation D.3 can be simplified to $\mathbf{T} = \frac{1}{m}\sum_i^m \mathbf{B}_i \oplus \mathbf{A}_i$. Thus, we can define a standard delta rule to determine how to update $\mathbf{T}$ as additional training examples are made available:

$$\mathbf{T}_{i+1} = \mathbf{T}_i - w_i\left(\mathbf{T}_i - \mathbf{B}_i \oplus \mathbf{A}_i\right), \tag{D.4}$$

where w_i is an adaptive learning rate inversely proportional to i. Or, we can allow w_i to be a constant, which would cause the effect of the training examples on the estimate to exponentially decay. Both approaches can be used successfully with the Raven's model described in Section 4.7.

Essentially, Equation D.4 updates the current estimate of the transformation vector $\mathbf{T}$ based on the difference between that transformation and what the current examples suggest it should be (i.e., $\mathbf{A}_i \oplus \mathbf{B}_i$). Intuitively, this means that we are estimating the transformation by trying to find the average transformation consistent with all of the available examples.

There are different means of implementing such a rule. For example, in Section 4.7, a working memory is used to retain a running estimate of **T** by updating the representations in the memory. In such a case, no neural connection weights are changed, as the activities of the neurons themselves represent **T**. In contrast, in Section 6.6, the hPES learning rule is used to learn the mapping from a context signal to desired transformation **T**. In this case, the neural connection weights between the context signal(s) and the population generating the relevant transformation vector(s) are updated. As such, the rule is indirectly implemented by using it to generate an error signal that trains the mapping. The rule could also be directly embedded into a set of connection weights by using the hPES rule on a circuit whose input was the vector to be transformed and constructing an error signal appropriately.

D.3 ■ Ordinal Serial Encoding Model

The equations that define the encoding and decoding of the OSE model for serial recall are as follows:

Encoding:

$$\mathbf{M}_i^{in} = \gamma \mathbf{M}_{i-1}^{in} + (\mathbf{P}_i \circledast \mathbf{I}_i) + \mathbf{I}_i$$

$$\mathbf{M}_i^{epis} = \rho \mathbf{M}_{i-1}^{epis} + (\mathbf{P}_i \circledast \mathbf{I}_i) + \mathbf{I}_i$$

$$\mathbf{M}_i^{OSE} = \mathbf{M}_i^{in} + \mathbf{M}_i^{epis}$$

Decoding:

$$\mathbf{I}_i = cleanup(\mathbf{M}^{OSE} \circledast \mathbf{P}_i')$$

where $\mathbf{M}^{in}$ is input working memory trace, γ is the rate of decay of the old memory trace, $\mathbf{M}^{epis}$ is the memory trace in the episodic memory, ρ is scaling factor related to primacy, M^{OSE} is the overall encode memory of the list, **P** is a position vector, **I** is an item vector, i indexes the associated item number, and *cleanup*() indicates the application of the clean-up memory to the semantic pointer inside the brackets.

D.4 ■ Spike-Timing Dependent Plasticity

As mentioned in the main text, the hPES rule is a combination of BCM (Bienenstock et al., 1982) and the spike-timing rule derived by MacNeil and Eliasmith (2011), which relates neural activity to the representation of vectors using the NEF methods. The BCM rule is

$$\Delta\omega_{ij} = \kappa a_i a_j (a_j - \theta), \tag{D.5}$$

where

$$\theta = E[a_j/c].$$

In these equations, ω is the connection weight, κ is the learning rate parameter, a is the activity of a neuron, i and j index the pre- and postsynaptic neurons, and θ is a variable threshold. The value of θ is determined as the expected value $E[\cdot]$ of neural activity over all possible input patterns divided by a constant, c, to account for scaling. In practice, θ is usually determined by tracking a slowly changing temporal average. Notably, the BCM is a purely rate-dependent rule, although there is good biological evidence for the effect accounted for by the adapting θ parameter. This parameter plays a homeostatic role in the rule.

The second element used by hPES is a Hebbian rule of the form

$$\Delta\omega_{ij} = \kappa\alpha_j \mathbf{e}_j \mathbf{E} a_i, \tag{D.6}$$

where ω is the connection weight, κ is the learning rate parameter, α is the gain of the cell, $\mathbf{e}$ is the encoding vector of the cell, $\mathbf{E}$ is an error term, a is the activity of the cell (i.e., PSC filtered spikes), and i and j index the pre- and postsynaptic cells, respectively. The gain and encoding vectors are described in more detail in Appendix B.1.1. The error term is an input to the cell that captures the error in the performance of the network. This term can be either a modulatory input, like dopamine, or another direct spiking input carrying the relevant error information. Thus it can be employed in either a reinforcement learning or supervised learning role.

Notably, the adaptive aspect of the BCM rule is the $a_j(a_j - \theta)$ term, which allows the neurons to be sensitive to their own activity. In essence this acts as a kind of unsupervised error, as it tells the neuron if it is firing more or less than usual. Substituting this error term into Equation D.6 and noticing that this is a scalar error (so $\mathbf{e}_j = 1$) gives a rule that can be implemented in a spiking network, but incorporates homeostasis:

$$\Delta\omega_{ij} = \kappa\alpha_j a_i a_j (a_j - \theta), \tag{D.7}$$

where all parameters are as before.

Combining the unsupervised (Equation D.7) and supervised (Equation D.6) rules additively results in the overall hPES rule

$$\Delta\omega_{ij} = \kappa_1 \alpha_j \mathbf{e}_j \mathbf{E} a_i + \kappa_2 \alpha_j a_i a_j (a_j - \theta) \tag{D.8}$$

$$= \alpha_j a_i \left[\kappa_1 \mathbf{e}_j \mathbf{E} + \kappa_2 a_j (a_j - \theta)\right]. \tag{D.9}$$

The relative contributions of the supervised and unsupervised elements of the rule are determined by the chosen learning rates κ_1 and κ_2, respectively. For example, the results in Figure 6.9 (timing and frequency effects of STDP) are generated with $\kappa_1 = 0$, whereas the results in Figure 6.10 (learning a 2D communication channel) are generated with $\kappa_2 = 0$. Overall, the rule provides a means of examining relative contributions of self-organized and directed learning in spiking neural networks, while remaining consistent with a broader set of neural data than any other rule of which I am aware.

D.5 ■ Number of Neurons for Representing Structure

The following calculations show the number of neurons required by the different architectures discussed. In the text I sometimes assume a simple vocabulary of 6,000 symbols, but here I assume a vocabulary of 60,000 symbols, which is approximately an adult-sized vocabulary (Crystal, 2003). The scaling properties of the SPA are discussed in Section 4.6, where I argue that a combination of binding, unbinding, and clean-up networks that handle adult-level representations can fit comfortably within 9 mm^2 of cortex.

For LISA (Hummel & Holyoak, 2003), we can divide the 60,000 symbols into 40,000 nouns and 20,000 relations, and assume all relations are of the form *relation(noun1, noun2)*. More complex relations (including higher-order relations) will require many more neurons, but I do not consider such relations here. So, for encoding first-order relations in LISA, we have:

$$\begin{aligned}\text{number of neural groups} &= \text{relations} \times \text{nouns} \times \text{nouns}\\ &= 20{,}000 \times 40{,}000 \times 40{,}000\\ &= 32{,}000\ 000\ 000\ 000.\end{aligned}$$

Each neural group requires multiple neurons to represent it, so I suppose 100 neurons per group:

$$\begin{aligned}\text{neurons} &= \text{groups} \times 100\\ &= 3{,}200\ 000\ 000\ 000\ 000.\end{aligned}$$

To determine the cortical area required, we assume ~20 million neurons per square centimeter (given 170,000 neurons per square mm calculated from Pakkenberg & Gundersen [1997], as discussed in Section 2.1).

$$\begin{aligned}\text{area} &= \text{neurons} / 20{,}000\ 000\\ &= 160{,}000\ 000\ \text{cm}^2.\end{aligned}$$

However, the average human adult has about 2,500 cm^2 of cortex (Peters & Jones, 1984). Another way of highlighting this problem is by noting that a similar calculation does not fit within cortex until we use only 1,500 words (500 verbs and 1,000 nouns).

For the Neural Blackboard Architecture (van der Velde & de Kamps, 2006), the number of neurons required is dominated by the connections between the symbols and the word assemblies. Each connection requires 8 neural groups; the authors suggest 100 assemblies per grammatical role (van der Velde & de Kamps, 2006). It is unclear how many grammatical roles are needed, so

I assume two to be conservative. I also assume that 100 neurons per neural group are needed to maintain accurate representation.

$$\begin{aligned}\text{groups} &= \text{words} \times \text{assemblies} \times 8\\ &= 60{,}000 \times 200 \times 8\\ &= 96{,}000\,000\\ \text{neurons} &= \text{groups} \times 100\\ &= 9{,}600\,000\,000\\ \text{area} &= \text{neurons} / 20{,}000\,000 = 480\ \text{cm}^2.\end{aligned}$$

This is about one-fifth of available cortex, which is too large for language areas (Ojemann et al., 1989)[1]. More critically, this estimate is only for representation, not transformation or processing.

To make a clear comparison to the SPA, recall that the number of dimensions required for similar structures (60,000 words, 8 possible bindings, 99% correct) is about 500 (*see* Fig. 4.7). If, by analogy to the assumptions in LISA and the NBA, we allow each dimension to be a population of 100 neurons, representing any such structure requires:

$$\begin{aligned}\text{neurons} &= \text{dimensions} \times 100 = 50{,}000\\ \text{area} &= \text{neurons}/170{,}000 \approx 1\ \text{mm}^2.\end{aligned}$$

This does not include the clean-up memory, which is crucial for decoding. As mentioned in the main text, this would add 3.5 mm^2 of cortical area for a total of approximately 5 mm^2 in the worst case (*see* Section 4.6).

For the ICS (Smolensky & Legendre, 2006a), scaling issues arise more directly with respect to encoding embedded structure, as discussed in Section 9.1.4. As a result, let me consider more interesting structure than *relation(noun1, noun2)*. Specifically, let's determine the number of neurons required for the sentence "Bill believes John loves Mary" or any similar structure. First, we allow the sentence to be represented the same as in the SPA:

$$\begin{aligned}\mathbf{S} =& \mathbf{relation} \otimes \mathbf{believes} + \mathbf{subject} \otimes \mathbf{Bill} + \mathbf{object} \otimes\\ &(\mathbf{relation} \otimes \mathbf{loves} + \mathbf{subject} \otimes \mathbf{John} + \mathbf{object} \otimes \mathbf{Mary}).\end{aligned}$$

[1] Specifically, this paper suggests that the total area of temporal and frontal cortex necessary for naming is typically 7 cm^2 or less. Wernicke's area is approximately 20 cm^2 (Lamb, 1999, p. 343), as is Broca's area (Dronkers et al., 2000). Consequently, adding in these areas accounts for only about 50 cm^2 of cortex. If we add to this supramarginal gyrus, angular gyrus, auditory cortex, motor cortex, and somatosensory cortex (which together comprise the full "implementation system"), then we have about 250 cm^2 of cortex (Dronkers et al., 2000). However, this large area is clearly responsible for far more than just structure representation in language.

Each vector is assumed to have 500 dimensions to represent any of 60,000 words, so the number of dimensions in the final structure **S** for the ICS is:

$$\text{dimensions} = 500 \times 500 \times 500 = 125{,}000\,000$$

$$\text{neurons} = \text{dimensions} \times 100 = 12{,}500\,000\,000$$

$$\text{area} = \text{neurons}/20{,}000\,000 = 625\ \text{cm}^2.$$

This is about a quarter the 2,500 cm^2 of available cortex, and again only for representation, not processing of the structure.

It has been suggested that ICS can use lower-dimensional vectors because recovery is guaranteed (Smolensky et al., 2010). However, this does not escape the exponentially poor scaling with depth of structure. If we take our lexical vectors to be structured themselves, as seems likely (*see* Section 4.8), ICS will scale poorly even with few dimensions per item. For example, with only eight dimensions per item, a two-level conceptual structure (e.g., $\mathbf{Mary} = \mathbf{isA} \otimes \mathbf{person} + \ldots = \mathbf{isA} \otimes (\mathbf{isA} \otimes \mathbf{mammal} + \ldots) + \ldots$) will scale worse than described above when used in the example sentence (as each item in that sentence will then be $8 \times 8 \times 8 = 512$ dimensions).

APPENDIX E. SEMANTIC POINTER ARCHITECTURE MODEL DETAILS

E.1 ■ Tower of Hanoi

This appendix includes two tables. Table E.1 lists all of the cortical elements of the Tower of Hanoi model. Table E.2 lists all of the rules that were used by those elements and the basal ganglia to solve the task.

TABLE E.1 *The Model Elements of the SPA Tower of Hanoi Model Found in Cortex.*

Name	Type	Description
state	buffer	Used to control the different stages of the problem-solving algorithm
focus	buffer	Stores the disk currently being attended to (D0, D1, D2, D3)
focuspeg	sensory	Automatically contains the location of the focus disk (A, B, C)
goal	buffer	Stores the disk we are trying to move (D0, D1, D2, D3)
goaltarget	buffer	Stores the location we want to move the goal disk to (A, B, C)
goalcurrent	sensory	Automatically contains the location of the goal disk (A, B, C)
goalfinal	sensory	Automatically contains the final desired location of the goal disk (A, B, C)

TABLE E.1 *(cont.)*

largest	sensory	Automatically contains the largest visible disk (D3)
mem1	memory	Stores an association between mem1 and mem2 in working memory
mem2	memory	Stores an association between mem1 and mem2 in working memory
request	memory	Indicates one element of a pair to attempt to recall from working memory
recall	memory	The vector associated with the currently requested vector
movedisk	motor	Tells the motor system which disk to move (D0, D1, D2, D3)
movepeg	motor	Tells the motor system where to move the disk to (A, B, C)
motor	sensory	Automatically contains DONE if the motor action is finished

The model also includes a basal ganglia and thalamus.

TABLE E.2 *A List of the 16 Rules Used in the Tower of Hanoi Simulation*

Rule name	Match (IF)	Execute (THEN)
LookDone	focus≠D0 + goal=focus + goalcurrent=goaltarget + state≠STORE	focus=goal⊛NEXT goal=goal⊛NEXT goaltarget=goalfinal
LookNotDone	focus≠D0 + goal=focus + goalcurrent≠goaltarget + state≠STORE	focus=goal⊛NEXT
InTheWay1	focus≠goal + focuspeg=goalcurrent + focuspeg≠goaltarget + state≠STORE	focus=goal⊛NEXT
InTheWay2	focus≠goal + focuspeg≠goalcurrent + focuspeg=goaltarget + state≠STORE	focus=goal⊛NEXT
NotInTheWay	focus≠goal + focuspeg≠goalcurrent + focuspeg≠goaltarget + focus≠D0	focus=focus⊛NEXT
MoveD0	focus=D0 + goal=D0 + goalcurrent≠goaltarget	movedisk=D0 movepeg=goaltarget

TABLE E.2 *(cont.)*

MoveGoal	focus=D0 + goal≠D0 + focuspeg≠goaltarget + goaltarget≠goalcurrent + focuspeg≠goalcurrent	movedisk=goal movepeg=goaltarget
MoveDone	motor=DONE + goal≠largest + state≠RECALL	state=RECALL goal=goal⊛$NEXT^{-1}$
MoveDone2	motor=DONE + goal=largest + state≠RECALL	focus=largest⊛$NEXT^{-1}$ goal=largest⊛$NEXT^{-1}$ goaltarget=goalfinal state=HANOI
Store	state=STORE + recall≠goaltarget	mem1=goal mem2=goaltarget request=goal
StoreDone	state=STORE + recall=goaltarget	state=FIND
FindFree1	state=FIND + focus≠goal + focuspeg=goalcurrent + focuspeg≠goaltarget	goaltarget=A+B+ C-focuspeg -goaltarget goal=focus state=HANOI
FindFree2	state=FIND + focus≠goal + focuspeg≠goalcurrent + focuspeg=goaltarget	goaltarget=A+B+ C-goalcurrent -goaltarget goal=focus state=HANOI
Recall	state=RECALL + recall≠(A+B+C)	request=goal
RecallDo	state=RECALL + recall=(A+B+C) + recall≠goalcurrent	state=HANOI focus=goal goaltarget=recall*4
RecallNext	state=RECALL + recall=(A+B+C) + recall=goalcurrent	goal=goal⊛$NEXT^{-1}$ request=goal

For the matching, equality signs indicate the result of a dot product, where "=" is summed and "≠" is subtracted. For the "execute" column, all statements refer to setting the left-hand model element to the value indicated on the right-hand side.

BIBLIOGRAPHY

Abeles, M. (1991). *Corticonics: Neural circuits of the cerebral cortex.* Cambridge, UK: Cambridge University Press.

Adolphs, R., Bechara, A., Tranel, D., Damasio, H., & Damasio, A. R. (1995). *Neuropsychological approaches to reasoning and decision-making.* Neurobiology of decision-making. New York: Springer Verlag.

Albin, R. L., Young, A. B., & Penney, J. B. (1989). The functional anatomy of basal ganglia disorders. *Trends in Neurosciences, 12,* 366–375.

Aldridge, J. W., Berridge, K., Herman, M., & Zimmer, L. (1993). Neuronal coding of serial order: Syntax of grooming in the neostriatum. *Psychological Science, 4*(6), 391–395.

Allport, D. A. (1985). Distributed memory, modular subsystems and dysphasia. In Newman, S.K. & Epstein, R. (Eds.), *Current Perspectives in Dysphasia* (pp. 207–244). Edinburgh: Churchill Livingstone.

Almor, A. & Sloman, S. A. (1996). Is deontic reasoning special? *Psychological Review, 103,* 374–380.

Altmann, E. M. & Trafton, J. G. (2002). Memory for goals: An activation-based model. *Cognitive Science, 26,* 39–83.

Amari, S. (1975). Homogeneous nets of neuron-like elements. *Biological Cybernetics, 17,* 211–220.

Amit, D. J. (1989). *Modeling brain function: The world of attractor neural networks.* New York Cambridge University Press.

Amit, D. J., Fusi, S., & Yakovlev, V. (1997). A paradigmatic working memory (attractor) cell in IT cortex. *Neural Computation, 9*(5), 1071–1092.

Ananthanarayanan, R. & Modha, D. S. (2007). Anatomy of a cortical simulator. In *Proceedings of the 2007 ACM/IEEE conference on Supercomputing - SC '07,* (p.1̃)., New York: ACM Press.

Andersen, R. A., Essick, G. K., & Siegel, R. M. (1985). The encoding of spatial location by posterior parietal neurons. *Science, 230*, 456–458.

Anderson, J. (1996). Act: A simple theory of complex cognition. *American psychologist, 51*(4), 355–365.

Anderson, J., John, B. E., Just, M., Carpenter, P. A., Kieras, D. E., & Meyer, D. E. (1995). Production system models of complex cognition. In Moore, J. D. & Lehman, J. F. (Eds.), *Proceedings of the Seventeenth Annual Conference of the Cognitive Science ...*, (p.9). New York, NY, Routledge.

Anderson, J. R. (1976). *Language, memory, and thought.* Hillsdale, NJ: Lawrence Erlbaum.

Anderson, J. R. (1983). *The architecture of cognition.* Cambridge, MA: Harvard University Press.

Anderson, J. R. (1990). *The adaptive character of thought.* New York, NY: Routledge.

Anderson, J. R. (2007). *How can the human mind occur in the physical universe?* New York, NY: Oxford University Press.

Anderson, J. R., Albert, M. V., & Fincham, J. M. (2005). Tracing problem solving in real time: fMRI analysis of the subject-paced Tower of Hanoi. *Journal of cognitive neuroscience, 17*(8), 1261–1274.

Anderson, J. R. & Betz, J. (2001). A hybrid model of categorization. *Psychonomic Bulletin and Review, 8*, 629–647.

Anderson, J. R., Bothell, D., Byrne, M. D., Douglass, S., Lebiere, C., & Qin, Y. (2004). An integrated theory of the mind. *Psychological Review, 111*(4), 1036–1060.

Anderson, J. R., Corbett, A., Koedinger, K. R., & Pelletier, R. (1995). Cognitive tutors: Lessons learned. *Journal of the Learning Sciences, 4*(2), 167–207.

Anderson, J. R., Kushmerick, N., & Lebiere, C. (1993). Tower of Hanoi and goal structures. In Anderson, J.R. (Ed.), *Rules of the Mind.* Mahwah, NJ: Erlbaum Associates.

Anderson, J. R. & Lebiere, C. (1998). *The atomic components of thought.* Mahwah, NJ: Erlbaum Associates.

Anderson, J. R. & Lebiere, C. (2003). The Newell Test for a theory of cognition. *The Behavioral and Brain Sciences, 26*(5), 587–601; discussion 601–648.

Anderson, M. (2010). Neural re-use as a fundamental organizational principle of the brain. *Behavioral and Brain Sciences, 33*(3), 245–266.

Askay, E., Baker, R., Seung, H. S., & Tank, D. (2000). Anatomy and discharge properties of pre-motor neurons in the goldfish medulla that have eye-position signals during fixations. *Journal of Neurophysiology, 84*, 1035–1049.

Aziz-Zadeh, L. & Damasio, A. (2008). Embodied semantics for actions: findings from functional brain imaging. *Journal of physiology, Paris, 102*(1-3), 35–39.

Baars, B. (1988). *A cognitive theory of consciousness.* New York: Cambridge University Press.

Baddeley, A. (1998). Recent developments in working memory. *Current opinion in neurobiology, 8*(2), 234–238.

Baddeley, A. (2003). Working memory: looking back and looking forward. *Nature reviews. Neuroscience, 4*(10), 829–839.

Ballard, D. H. (1991). Animate vision. *Artificial Intelligence, 48*, 57–86.

Barr, M. (2004). Visual objects in context. *Nature Reviews Neuroscience, 5*, 617–629.

Barsalou, L. W. (1999). Perceptual symbol systems. *Behavioral and Brain Sciences, 22*, 577–660.

Barsalou, L. W. (2003). Abstraction in perceptual symbol systems. *Philosophical Transactions of the Royal Society of London: Biological Sciences, 358*, 1177–1187.

Barsalou, L. W. (2009). Simulation, situated conceptualization, and prediction. *Philosophical Transactions of the Royal Society of London: Biological Sciences, 364*, 1281–1289.

Barto, A. G., Sutton, R. S., & Brouwer, P. S. (1981). Associative search network: A reinforcement learning associative memory. *Biological Cybernetics, 40*(3), 201–211.

Bayer, H. M. & Glimcher, P. W. (2005). Midbrain dopamine neurons encode a quantitative reward prediction error signal. *Neuron, 47*, 129–141.

Beal, M.J. (2003). Variational algorithms for approximate Bayesian inference. Physics. University of Cambridge. PhD Thesis.

Beal, M. J. & Ghahramani, Z. (2003). The variational Bayesian EM algorithm for incomplete data: with application to scoring graphical model structures. In Bernardo, J., Bayarri, M., Berger, J., Dawid, A., Heckerman, D., Smith, A., & West, M. (Eds.), *Bayesian statistics 7: proceedings of the seventh Valencia International Meeting, June 2-6, 2002*, (pp. 453). New York, NY: Oxford University Press.

Bechtel, W. (1988). *Philosophy of science: An introduction for cognitive science.* Hillsdale, NJ: Erlbaum.

Bechtel, W. (1998). Representations and cognitive explanations: Assessing the dynamicist challenge in cognitive science. *Cognitive Science, 22*, 295–318.

Bechtel, W. (2005). The challenge of characterizing operations in the mechanisms underlying behavior. *Journal of the Experimental Analysis of Behavior, 84*, 313–325.

Bechtel, W. (2008). *Mental mechanisms: Philosophical Perspectives on Cognitive Neuroscience.* New York, NY: Routledge.

Bechtel, W. & Abrahamsen, A. (2001). *Connectionism and the mind: Parallel processing, dynamics, and evolution in networks*, volume 2nd. Oxford: Blackwell.

Bechtel, W. & Graham, G. (1999). *A companion to cognitive science.* London: Blackwell.

Bechtel, W. & Richardson, R. C. (1993). *Discovering complexity: Decomposition and localization as strategies in scientific research.* Princeton, NJ: Princeton University Press.

Becker, S. (2005). A computational principle for hippocampal learning and neurogenesis. *Hippocampus, 15*(6), 722–738.

Beckmann, H. & Lauer, M. (1997). The human striatum in schizophrenia. II. Increased number of striatal neurons in schizophrenics. *Psychiatry research, 68*(2-3), 99–109.

Beiser, D. G. & Houk, J. C. (1998). Model of cortical-basal ganglionic processing: Encoding the serial order of sensory events. *Journal of Neurophysiology, 79*, 3168–3188.

Bengio, Y. & Lecun, Y. (2007). Scaling Learning Algorithms towards AI. In L. Bottou, O. Chapelle, D. DeCoste, & J. Weston (Eds.), *Large-Scale Kernel Machines*, number 1 (pp. 1–41). Cambridge, MA: MIT Press.

Bernstein, N. (1967). *The Coordination and Regulation of Movements.* New York: Pergamon Press.

Bi, G. Q. & Poo, M. M. (1998). Synaptic modifications in cultured hippocampal neurons: dependence on spike timing, synaptic strength, and postsynaptic cell type. *The Journal of neuroscience : the official journal of the Society for Neuroscience, 18*(24), 10, 464–472.

Bialek, W. & Rieke, F. (1992). Reliability and information transmission in spiking neurons. *Trends in Neuroscience, 15*(11), 428–434.

Bialek, W. & Zee, A. (1990). Coding and computation with neural spike trains. *Journal of Statistical Physics, 59*, 103–115.

Bienenstock, E., Cooper, L. N., & Munro, P. (1982). On the development of neuron selectivity: Orientation specificity and binocular interaction in visual cortex. *Journal of Neuroscience, 2*, 32–48.

Bingham, E. & Mannila, H. (2001). Random projection in dimensionality reduction. In *Proceedings of the seventh ACM SIGKDD international conference on knowledge discovery and data mining - KDD '01*, (pp. 245–250), New York, New York: ACM Press.

Block, N. (1986). Advertisement for a semantics for psychology, chapter X, (pp. 615–678). *Midwest Studies in Philosophy.* Minneapolis, MN: University of Minnesota Press.

Blumberg, M. (2011). Dynamic field theory homepage. http://www.uiowa.edu/delta-center/research/dft/index.html Accessed Jan. 2013.

Bobier, B. (2011). *The attentional routing circuit: A model of selective visuospatial attention.* PhD, Waterloo, ON: University of Waterloo.

Bogacz, R., Usher, M., Zhang, J., & McClelland, J. L. (2007). Extending a biologically inspired model of choice: multi-alternatives, nonlinearity and value-based multidimensional choice. *Philosophical transactions of the Royal Society of London. Series B, Biological sciences, 362*(1485), 1655–1670.

Bors, D. (2003). The effect of practice on Raven's Advanced Progressive Matrices. *Learning and Individual Differences, 13*(4), 291–312.

Bors, D. & Stokes, T. (1998). Raven's Advanced Progressive Matrices: Norms for First-Year University Students and the Development of a Short Form. *Educational and Psychological Measurement, 58*(3), 382–398.

Botvinick, M. M. & Plaut, D. C. (2006). Short-Term Memory for Serial Order: A Recurrent Neural Network Model. *Psychological Review, 113*(2), 201–233.

Bower, J. M. & Beeman, D. (1998). *The book of GENESIS: Exploring realistic neural models with the GEneral NEural SImulation System.* New York, NY: Springer Verlag.

Bowers, J. S. (2009). On the Biological Plausibility of Grandmother Cells: Implications for Neural Network Theories in Psychology and Neuroscience. *Psychological Review, 116*(1), 220–251.

Brady, M. (2009). Speech as a Problem of Motor Control in Robotics. In Taatgen, N. & van Rijn, H. (Eds.), *Proceedings of the Thirty-First Annual Conference of the Cognitive Science Society*, Austin, TX.(pp. 2558–2563).

Braitenburg, V. & Shuz, A. (1998). *Cortex: Statistics and geometry of neuronal connectivity.* New York, Springer.

Brogan, W. L. (1990). *Modern Control Theory (3rd Edition).* Englewood Cliffs, NJ: Prentice Hall.

Brooks, R. (1991). Intelligence without representation. *Artificial Intelligence, 47*, 139–159.

Bruner, J. S., Goodnow, J. J., & Austin, G. A. (1956). *A study of thinking.* New York: Wiley.

Buckner, R. L. & Wheeler, M. E. (2001). The cognitive neuroscience of remembering. *Nature Reviews Neuroscience, 2*, 626–634.

Budiu, R. & Anderson, J. R. (2004). Interpretation-Based Processing: A Unified Theory of Semantic Sentence Processing. *Cognitive Science, 28*, 1–44.

Buffalo, E. A., Fries, P., Landman, R., Liang, H., & Desimone, R. (2010). A backward progression of attentional effects in the ventral stream. *Proceedings of the National Academy of Sciences of the United States of America, 107*(1), 361–365.

Bunge, S. A. (2006). A Brain-Based Account of the Development of Rule Use in Childhood. *Current Directions in Psychological Science, 15*(3), 118–121.

Burger, P., Mehlb, E., Cameron, P. L., Maycoxa, P. R., Baumert, M., Friedrich, L., et al (1989). Synaptic vesicles immunoisolated from rat cerebral cortex contain high levels of glutamate. *Neuron, 3*(6), 715–720.

Busemeyer, J. R. & Townsend, J. T. (1993). Decision field theory: A dynamic-cognitive approach to decision making in an uncertain environment. *Psychological review, 100*(3), 432–459.

Calabresi, P., Picconi, B., Tozzi, A., & Di Filippo, M. (2007). Dopamine-mediated regulation of corticostriatal synaptic plasticity. *Trends in neurosciences, 30*(5), 211–219.

Canessa, N., Gorini, A., Cappa, S. F., Piattelli-Palmarini, M., Danna, M., Fazio, F., et al. (2005). The effect of social content on deductive reasoning: an fMRI study. *Human brain mapping, 26*(1), 30–43.

Caporale, N. & Dan, Y. (2008). Spike timing-dependent plasticity: a Hebbian learning rule. *Annual review of neuroscience, 31*, 25–46.

Carnap, R. (1931). Psychology in the language of physics. *Erkenntnis, 2*, 107-142.

Carnevale, N. & Hines, M. (2006). *The NEURON Book.* Cambridge, UK: Cambridge University Press.

Carpenter, P., Just, M., & Shell, P. (1990). What one intelligence test measures: a theoretical account of the processing in the Raven Progressive Matrices Test. *Psychological Review, 97*(3), 404–431.

Chaaban, I. & Scheessele, M. R. (2007). Human Performance on the USPS Database. Technical report, South Bend, IN: Indiana University South Bend.

Cheng, P. W. & Holyoak, K. J. (1985). Pragmatic reasoning schemas. *Cognitive psychology, 17*, 391–416.

Chomsky, N. (1959). A review of B. F. Skinner's Verbal Behavior. *Language, 35*(1), 26–58.

Choo, X. (2010). *The Ordinal Serial Encoding Model: Serial Memory in Spiking Neurons.* Masters, Waterloo, ON: University of Waterloo.

Churchland, M. M., Cunningham, J. P., Kaufman, M. T., Ryu, S. I., & Shenoy, K. V. (2010). Cortical Preparatory Activity: Representation of Movement or First Cog in a Dynamical Machine? *Neuron, 68*(3), 387–400.

Churchland, P. (1979). *Scientific realism and the plasticity of mind.* Cambridge, UK: Cambridge University Press.

Churchland, P. & Churchland, P. (1981). Functionalism, qualia and intentionality. *Philosophical Topics, 12*, 121–145.

Churchland, P. S., Ramachandran, V. S., & Sejnowski, T. J. (1994). *A critique of pure vision.* Large-scale neuronal theories of the brain. Cambridge, MA: MIT Press.

Clark, A. (1997). *Being there: Putting brain, body and world together again.* Cambridge, MA: MIT Press.

Clark, A. & Chalmers, D. (2002). *The extended mind.* Philosophy of mind: Classical and contemporary readings. New York, NY: Oxford University Press.

Colby, C. L., Duhamel, J. R., & Goldberg, M. E. (1993). Ventral intraparietal area of the macaque: anatomic location and visual response properties. *Journal of Neurophysiology, 69*(3), 902–914.

Collins, A. & Quillian, M. (1969). Retrieval time from semantic memory. *Journal of Verbal Learning and Verbal Behavior, 8*(2), 240–247.

Conklin, J. & Eliasmith, C. (2005). An attractor network model of path integration in the rat. *Journal of Computational Neuroscience, 18*, 183–203.

Cooper, R. (2002). *Modelling High-Level Cognitive Processes.* Hillsdale, N.J.: Lawrence Erlbaum Associates.

Cordes, S., Gelman, R., Gallistel, C. R., & Whalen, J. (2001). Variability signatures distinguish verbal from nonverbal counting for both large and small numbers. *Psychonomic Bulletin & Review, 8*(4), 698–707.

Cosmides, L. (1989). The logic of social exchange: Has natural selection shaped how humans reason? Studies with the Wason selection task. *Cognition, 31*, 187–276.

Cowan, N. (2001). The magical number 4 in short-term memory: A reconsideration of mental storage capacity. *Behavioral and Brain Sciences, 24*, 87–185.

Cox, J. R. & Griggs, R. A. (1982). The effects of experience on performance in Wason's selection task. *Memory & cognition, 10*(5), 496–502.

Craik, F. I. M. & Lockhart, R. S. (1972). Levels of processing: A framework for memory research. *Journal of Verbal Learning and Verbal Behavior, 11,* 671–684.

Craver, C. (2007). *Explaining the brain.* Oxford, UK: Oxford University Press.

Crystal, D. (2003). *Cambridge encyclopedia of the english language.* Cambridge, UK: Cambridge University Press.

Cummins, R. (1989). *Meaning and mental representation.* Cambridge, MA: MIT Press.

Damasio, A. R. (1989). Time-locked multiregional retroactivation: a systems-level proposal for the neural substrates of recall and recognition. *Cognition, 33,* 25–62.

d'Avila Garcez, A. S., Lamb, L. C., & Gabbay., D. M. (2008). *Neural-Symbolic Cognitive Reasoning. Cognitive Technologies.* New York, NY: Springer.

de Garis, H., Shuo, C., Goertzel, B., & Ruiting, L. (2010). A world survey of artificial brain projects, Part I: Large-scale brain simulations. *Neurocomputing, 74*(1–3), 3–29.

Deerwester, S., Dumais, S. T., Furnas, G. W., Landauer, T. K., & Harshman, R. (1990). Indexing By Latent Semantic Analysis. *Journal of the American Society For Information Science, 41,* 391–407.

Deerwester, S., Furnas, G. W., Landauer, T. K., Harshman, R., & Dumais, S. T. (1990). Indexing by latent semantic analysis. *Journal of the American Society For Information Science, 41*(6), 391–407.

DeLong, M. R. (1990). Primate models of movement disorders of basal ganglia origin. *Trends in Neurosciences, 13*(7), 281–285.

Dempster, A., Laird, N., & Rubin, D. (1977). Maximum Likelihood from Incomplete Data via the EM Algorithm. *Journal of the Royal Statistical Society Series B, 39*(1), 1–38.

Dennett, D. & Viger, C. (1999). Sort-of symbols? *Behavioral and Brain Sciences, 22*(4), 613.

D'Esposito, M., Postle, B. R., Ballard, D., & Lease, J. (1999). Maintenance versus manipulation of information held in working memory: an event-related fMRI study. *Brain and cognition, 41*(1), 66–86.

Dewolf, T. (2010). *NOCH: A framework for biologically plausible models of neural motor control.* Masters thesis, Waterloo, ON: University of Waterloo.

Dewolf, T. & Eliasmith, C. (2010). NOCH: A framework for biologically plausible models of neural motor control. In *Neural Control of Movement 20th Annual Conference,* Naples, FL: University of Waterloo.

Dewolf, T. & Eliasmith, C. (2011). The neural optimal control hierarchy for motor control. *Neural Engineering, 8*(6), 21.

Dorf, R. C. & Bishop, R. H. (2004). *Modern Control Systems (10th Edition).* Englewood Cliffs, NJ: Prentice Hall.

Dosher, B. A. (1999). Item Interference and Time Delays in Working Memory: Immediate Serial Recall. *International Journal of Psychology, 34*(5–6), 276–284.

Doumas, L. A. A., Hummel, J. E., & Sandhofer, C. M. (2008). A theory of the discovery and predication of relational concepts. *Psychological Review, 115*, 1–43.

Dretske, F. (1981). *Knowledge and the flow of information.* Cambridge, MA: MIT Press.

Dretske, F. (1994). *If you can't make one, you don't know how it works*, chapter XIX, (pp. 615–678). Midwest Studies in Philosophy. Minneapolis, MN: University of Minnesota Press.

Dronkers, N., Pinker, S., & Damasio, A. (2000). Language and the aphasias. In Kandel, E., Schwartz, J., & Jessell, T. (Eds.), *Principles in Neural Science* (4th ed.). (pp. 1169–1187). New York: McGraw-Hill.

Edelman, S. & Breen, E. (1999). On the virtues of going all the way. *Behavioral and Brain Sciences, 22*(4), 614.

Eliasmith, C. (1996). The third contender: a critical examination of the dynamicist theory of cognition. *Philosophical Psychology, 9*(4), 441–463.

Eliasmith, C. (1997). Computation and dynamical models of mind. *Minds and Machines, 7*, 531–541.

Eliasmith, C. (1998). Commentary: Dynamical models and van Gelder's dynamicism: Two different things. *Behavioral and Brain Sciences, 21*, 616–665.

Eliasmith, C. (2000). *How neurons mean: A neurocomputational theory of representational content.* PhD thesis, Washington University in St. Louis, Department of Philosophy, St. Louis, MO.

Eliasmith, C. (2003). Moving beyond metaphors: Understanding the mind for what it is. *Journal of Philosophy, 100*, 493–520.

Eliasmith, C. (2004). Learning context sensitive logical inference in a neurobiological simulation. In S. Gayler & R. Levy (Eds.), *Compositional Connectionism in Cognitive Science* (pp. 17–20). Palo Alto, CA: AAAI Press.

Eliasmith, C. (2005a). A unified approach to building and controlling spiking attractor networks. *Neural computation, 17*(6), 1276–1314.

Eliasmith, C. (2005b). Cognition with neurons: A large-scale, biologically realistic model of the Wason task. In Bara, G., Barsalou, L., & Bucciarelli, M. (Eds.), *Proceedings of the 27th Annual Meeting of the Cognitive Science Society*, Cognitive Science Society, Austin, TX, (pp. 624–630). Cognitive Science Society, Austin, TX.

Eliasmith, C. (2006). Neurosemantics and categories. *Handbook of Categorization in Cognitive Science.* Amsterdam: Elsevier.

Eliasmith, C. (2007). How to build a brain: From function to implementation. *Synthese, 153*, 373–388.

Eliasmith, C. (2009a). Dynamics, control, and cognition. In P. Robbins & M. Aydede (Eds.), *Cambridge Handbook of Situated Cognition.* New York, NY: Cambbridge University Press.

Eliasmith, C. (2009b). How we ought to understand computation in the brain. *Studies in History and Philosophy of Science, 41*, 313–320.

Eliasmith, C. & Anderson, C. H. (2001). Beyond bumps: Spiking networks that store sets of functions. *Neurocomputing, 38*, 581–586.

Eliasmith, C. & Anderson, C. H. (2003). *Neural engineering: Computation, representation and dynamics in neurobiological systems.* Cambridge, MA: MIT Press.

Eliasmith, C., Stewart T. C., Choo X., Bekolay T., DeWolf T., Tang Y., et al. (2012). A large-scale model of the functioning brain. *Science.* 338(6111), 1202–1205.

Eliasmith, C. & Thagard, P. (1997). Particles, waves and explanatory coherence. *British Journal of the Philosophy of Science, 48,* 1–19.

Engel, A. K., Fries, P., & Singer, W. (2001). Dynamic predictions: oscillations and synchrony in top-down processing. *Nature reviews.Neuroscience, 2*(10), 704–716.

Evans, G. (1982). *Varieties of reference.* New York: Oxford University Press.

Faubel, C. & Sch, G. (2010). Learning Objects on the Fly: Object Recognition for the Here and Now. In *International Joint Conference on Neural Networks.* Orlando, FL: IEEE Press.

Feeney, A. & Handley, S. J. (2000). The suppression of Q card selections: Evidence for deductive inference in Wason's selection task. *Quarterly Journal of Experimental Psychology, 53*(4), 1224–1242.

Feldman, D. E. (2009). Synaptic mechanisms for plasticity in neocortex. *Annual review of neuroscience, 32,* 33–55.

Felleman, D. J. & Van Essen, D. C. (1991a). Distributed hierarchical processing in primate visual cortex. *Cerebral Cortex, 1,* 1–47.

Felleman, D. J. & Van Essen, D. C. (1991b). Distributed Hierarchical Processing in the Primate Cerebral Cortex. *Cerebral Cortex, 1*(1), 1–47.

Felleman, D. J., Xiao, Y., & McClendon, E. (1997). Modular organization of occipito-temporal pathways: Cortical connections between visual area 4 and visual area 2 and posterior inferotemporal ventral area in macaque monkeys. *Journal of Neuroscience, 17*(9), 3185–3200.

Fillmore, C. (1975). An alternative to checklist theories of meaning. In *First Annual Meeting of the Berkeley Linguistics Society,* (pp. 123–131). Berkeley Linguistics Society, Berkeley, CA.

Fischer, B. J. (2005). *A model of the computations leading to a representation of auditory space in the midbrain of the barn owl.* PhD, Washington University in St. Louis. St. Louis, MO.

Fischer, B. J., Peña, J. L., & Konishi, M. (2007). Emergence of multiplicative auditory responses in the midbrain of the barn owl. *Journal of neurophysiology, 98*(3), 1181–1193.

Fishbein, J. M. (2008). *Integrating Structure and Meaning: Using Holographic Reduced Representations to Improve Automatic Text Classification.* Masters, Waterloo, ON: University of Waterloo.

Fleetwood, M. & Byrne, M. (2006). Modeling the Visual Search of Displays: A Revised ACT-R Model of Icon Search Based on Eye-Tracking Data. *Human-Computer Interaction, 21*(2), 153–197.

Fodor (1974). Special sciences (or: The disunity of science as a working hypothesis). *Synthese, 28*(2), 97.

Fodor, J. (1975). *The language of thought.* New York: Crowell.

Fodor, J. (1981). *Representations.* Cambridge, MA: MIT Press.

Fodor, J. (1987). *Psychosemantics.* Cambridge, MA: MIT Press.

Fodor, J. (1995). West coast fuzzy: Why we don't know how brains work (review of Paul Churchland's The engine of reason, the seat of the soul). *The Times Literary Supplement,* (August). London, UK.

Fodor, J. (1998). *Concepts: Where cognitive science went wrong.* New York: Oxford University Press.

Fodor, J. & McLaughlin, B. (1990). Connectionism and the problem of systematicity: Why Smolensky's solution doesn't work. *Cognition, 35,* 183–204.

Fodor, J. & Pylyshyn, Z. (1988). Connectionism and cognitive architecture: A critical analysis. *Cognition, 28,* 3–71.

Forbes, A. R. (1964). An item analysis of the advanced matrices. *British Journal of Educational Psychology, 34,* 1–14.

Frank, M. J. & O'Reilly, R. C. (2006). A mechanistic account of striatal dopamine function in human cognition: psychopharmacological studies with cabergoline and haloperidol. *Behavioral neuroscience, 120*(3), 497–517.

Fries, P. (2009). Neuronal gamma-band synchronization as a fundamental process in cortical computation. *Annual review of neuroscience, 32,* 209–224.

Friston, K. (2010). The free-energy principle: a unified brain theory? *Nature reviews. Neuroscience, 11*(2), 127–138.

Fuster, J. M. (2000). Executive frontal functions. *Experimental brain research, 133*(1), 66–70.

Gayler, R. W. (1998). Multiplicative binding, representation operators and analogy. In Holyoak, K., Gentner, D., & Kokinov, B. (Eds.), *Advances in analogy research: Integration of theory and data from the cognitive, computational, and neural sciences.* Sofia, BG, NBU Press.

Gayler, R. W. (2003). Vector Symbolic Architectures answer Jackendoff's challenges for cognitive neuroscience. In Slezak, P. (Ed.), *ICCS/ASCS International Conference on Cognitive Science,* (pp. 133–138). Sydney, Australia: University of New South Wales.

Gentner, D. (1983). Structure mapping: A theoretical framework for analogy. *Cognitive Science, 7,* 155–170.

Georgopoulos, A. P., Kalasaka, J. F., Crutcher, M. D., Caminiti, R., & Massey, J. T. (1984). The representation of movement direction in the motor cortex: Single cell and population studies. In Edelman, G.M. Gail, W.E. & Cowan, W. M. (Eds.), *Dynamic aspects of neocortical function.* Neurosciences Research Foundation. publisher is Neurosciences Research Foundation, location: New York, NY.

Georgopoulos, A. P., Schwartz, A. B., & Kettner, R. E. (1986). Neuronal population coding of movement direction. *Science, 243*(1416-1419).

Georgopoulos, A. P., Taira, M., & Lukashin, A. (1993). Cognitive neurophysiology of the motor cortex. *Science, 260,* 47–52.

Gershman, S., Cohen, J., & Niv, Y. (2010). Learning to selectively attend. In *Proceedings of the 32nd annual conference of the cognitive science society*, (pp. 1270–1275). Cognitive Science Society, Austin, TX.

Gibbs Jr., R. W. (2006). *Embodiment and Cognitive Science. Cambridge.* Cambridge: Cambridge University Press: Cambridge, UK.

Gibson, J. J. (1977). The Theory of Affordances. In Shaw, R. & Bransford, J. (Eds.), *Perceiving, Acting, and Knowing.* Hillsdale, NJ: Lawrence Erlbaum Associates.

Gisiger, T. & Boukadoum, M. (2011). Mechanisms Gating the Flow of Information in the Cortex: What They Might Look Like and What Their Uses may be. *Frontiers in Computational Neuroscience, 5*(January), 1–15.

Glaser, W. R. (1992). Picture naming. *Cognition, 42*, 61–105.

Goel, V. (2005). *Cognitive Neuroscience of Deductive Reasoning.* Cambridge Handbook of Thinking & Reasoning. Cambridge, UK: Cambridge University Press.

Gonchar, Y. & Burkhalter, A. (1999). Connectivity of GABAergic calretinin-immunoreactive neurons in rat primary visual cortex. *Cerebral Cortex, 9*(7), 683–696.

Goodale, M. A. & Milner, A. D. (1992). Separate pathways for perception and action. *Trends in Neuroscience, 15*, 20–25.

Gray, J. R., Chabris, C. F., & Braver, T. S. (2003). Neural mechanisms of general fluid intelligence. *Nature Neuroscience, 6*(3), 316–322.

Griffiths, T. L. & Tenenbaum, J. B. (2006). Optimal predictions in everyday cognition. *Psychological Science, 17*, 767–773.

Gunzelmann, G., Richard Moore Jr., L., Salvucci, D. D., & Gluck, K. A. (2011). Sleep loss and driver performance: Quantitative predictions with zero free parameters. *Cognitive Systems Research, 12*(2), 154–163.

Gupta, A., Wang, Y., & Markram, H. (2000). Organizing Principles for a Diversity of GABAergic Interneurons and Synapses in the Neocortex. *Science, 287*, 273–278.

Gurney, K., Prescott, T., & Redgrave, P. (2001). A computational model of action selection in the basal ganglia. *Biological Cybernetics, 84*, 401–423.

Hadley, R. F. (2009). The problem of rapid variable creation. *Neural computation, 21*(2), 510–532.

Halsband, U., Ito, N., Tanji, J., & Freund, H.-J. (1993). The role of premotor cortex and the supplementary motor area in the temporal control of movement in man. *Brain, 116*(1), 243–266.

Hardie, J. & Spruston, N. (2009). Synaptic depolarization is more effective than back-propagating action potentials during induction of associative long-term potentiation in hippocampal pyramidal neurons. *The Journal of neuroscience : the official journal of the Society for Neuroscience, 29*(10), 3233–3241.

Harman, G. (1982). Conceptual role semantics. *Notre Dame Journal of Formal Logic, 23*, 242–256.

Harnad, S. (1990). The symbol grounding problem. *Physica D, 42*, 335–346.

Hasselmo, M. (2011). *How We Remember: Brain Mechanisms of Episodic Memory.* Cambridge, MA: MIT Press.

Haugeland, J. (1993). Mind embedded and embodied. In *Mind and Cognition: An International Symposium*, Taipei, Taiwan: Academia Sinica.

Hayes-Roth, B. & Hayes-Roth, F. (1979). A cognitive model of planning. *Cognitive Science, 3*, 275–310.

Hazy, T. E., Frank, M. J., & O'reilly, R. C. (2007). Towards an executive without a homunculus: computational models of the prefrontal cortex/basal ganglia system. *Philosophical transactions of the Royal Society of London. Series B, Biological sciences, 362*(1485), 1601–1613.

Hebb, D. O. (1949). *The organization of behavior.* New York: Wiley.

Heekeren, H. R., Marrett, S., & Ungerleider, L. G. (2008). The neural systems that mediate human perceptual decision making. *Nature reviews. Neuroscience, 9*(6), 467–479.

Hellwig, B. (2000). A quantitative analysis of the local connectivity between pyramidal neurons in layers 2/3 of the rat visual cortex. *Biological Cybernetics, 82*(2), 111–121.

Hempel, C. G. (1966). *Philosophy of natural science.* Englewood Cliffs, N J: Prentice-Hall.

Henson, R. (1998). Short-term memory for serial order: The start-end model. *Cognitive psychology, 36*, 73–137.

Henson, R., Noriss, D., Page, M., & Baddeley, A. (1996). Unchained memory: Error patterns rule out chaining models of immediate serial recall. *The quarterly journal of experimental psychology A, 49*(1), 80–115.

Herd, S. A., Banich, M. T., & O'Reilly, R. C. (2006). Neural mechanisms of cognitive control: an integrative model of stroop task performance and FMRI data. *Journal of cognitive neuroscience, 18*(1), 22–32.

Hier, D., Yoon, W., Mohr, J., Price, T., & Wolf, P. (1994). Gender and aphasia in the stroke data bank. *Brain and Language, 47*, 155–167.

Hill, A. V. (1938). The Heat of Shortening and the Dynamic Constants of Muscle. *Proceedings of the Royal Society of London. Series B - Biological Sciences, 126*(843), 136–195.

Hinton, G. (1990). Connectionist learning procedures, (pp. 185–234). Machine learning: Paradigms and methods. Cambridge, MA: MIT Press.

Hinton, G. (2010). Where do features come from? In *Outstanding questions in cognitive science: A symposium honoring ten years of the David E. Rumelhart prize in cognitive science.* Austin, TX: Cognitive Science Society.

Hinton, G. E. (2002). Training Products of Experts by Minimizing Contrastive Divergence. *Neural Computation, 14*(8), 1771–1800.

Hinton, G. E. (2007). Learning multiple layers of representation. *Trends in Cognitive Science, 11*, 428–434.

Hinton, G. E. & Anderson, J. R. (1981). *Parallel models of associative memory.* Hillsdale, NJ: Erlbaum.

Hinton, G. E., Osindero, S., & Teh, Y.-W. (2006). A fast learning algorithm for deep belief nets. *Neural computation, 18*(7), 1527–1554.

Hinton, G. E. & Salakhutdinov, R. R. (2006a). Reducing the dimensionality of data with neural networks. *Science, 313*(5786), 504–507.

Hinton, G. E. & Salakhutdinov, R. R. (2006b). Reducing the dimensionality of data with neural networks. *Science, 313*(5786), 504–507.

Hochstein, E. (2011). *Intentionality as methodology.* PhD thesis, Waterloo, ON: University of Waterloo.

Hollerman, J. R. & Schultz, W. (1998). Dopamine neurons report an error in the temporal prediction of reward during learning. *Nature neuroscience, 1*(4), 304–309.

Holmgren, C., Harkany, T., Svennenfors, B., & Zilberter, Y. (2003). Pyramidal cell communication within local networks in layer 2/3 of rat neocortex. *The Journal of Physiology, 551*(Pt 1), 139–153.

Holyoak, K. & Thagard, P. (1995). *Mental leaps: Analogy in creative thought.* Cambridge, MA: MIT Press.

Hoshi, E. (2006). Functional specialization within the dorsolateral prefrontal cortex: a review of anatomical and physiological studies of non-human primates. *Neuroscience Research, 54*(2), 73–84.

Hummel, J. E., Burns, B., & Holyoak, K. J. (1994). *Analogical mapping by dynamic binding: Preliminary investigations*, chapter 2. Advances in connectionist and neural computation theory: Analogical connections. Norwood, NJ: Ablex.

Hummel, J. E. & Holyoak, K. J. (1997). Distributed representations of structure: a theory of analogical access and mapping. *Psychology Review, 104*(3), 427–466.

Hummel, J. E. & Holyoak, K. J. (2003). A symbolic-connectionist theory of relational inference and generalization. *Psychological review, 110*(2), 220–264.

Hung, C. P., Kreiman, G., Poggio, T., & DiCarlo, J. J. (2005). Fast readout of object identity from macaque inferior temporal cortex. *Science, 310*(5749), 863–866.

Hurwitz, M. (2010). *Dynamic judgments of spatial extent: Behavioural, neural and computational studies.* PhD, Waterloo, ON: University of Wateroo.

Hussain, F. (2010). *Cognitive Modelling of Attentional Networks: Efficiencies, Interactions, Impairments and Development.* DPhil, Sussex, UK: University of Sussex.

Hutchins, E. (1995). *Cognition in the Wild.* Cambridge, MA: MIT Press.

Hurzook, A (2012). "A mechanistic model of motion processing in the early visual system" MASc Thesis, Systems Design Engineering, University of Waterloo.

Ikezu, T. & Gendelman, H. (2008). *Neuroimmune Pharmacology.* New York, NY: Springer.

Izhikevich, E. M. & Edelman, G. M. (2008). Large-scale model of mammalian thalamocortical systems. *Proceedings of the National Academy of Sciences of the United States of America, 105*(9), 3593–3598.

Jackendoff, R. (2002). *Foundations of language: Brain, meaning, grammar, evolution.* New York, NY: Oxford University Press.

Jahnke, J. C. (1968). Delayed recall and the serial-position effect of short-term memory. *Journal of Experimental Psychology, 76*(4), 618.

James Bergstra, Bengio, Y., Lamblin, P., Desjardins, G., & Louradour, J. (2010). Image classification with complex cell neural networks. Front. Neurosci. Computational and Systems Neuroscience 2010 In *Computational and systems neuroscience (COSYNE)*, Salt Lake City, UT.

Jancke, D., Erlhagen, W., Dinse, H., Akhavan, A., Steinhage, A., Schöner, G., et al. (1999). Population representation of retinal position in cat primary visual cortex: interaction and dynamics. *Journal of Neuroscience, 19*(20), 9016–9028.

Jilk, D., Lebiere, C., O'Reilly, R., & Anderson, J. (2008). SAL: an explicitly pluralistic cognitive architecture. *Journal of Experimental & Theoretical Artificial Intelligence, 20*(3), 197–218.

Johnson-Laird, P. N. (1983). *Mental models: Towards a cognitive science of language, inference, and consciousness.* Cambridge, MA: Harvard Press.

Johnson-Laird, P. N., Legrenzi, P., Legrenzi, S., & Legrenzi, M. (1972). Reasoning and a sense of reality. *British Journal of Psychology, 63,* 395–400.

Jones, R. M., Laird, J. E., Nielsen, P. E., Coulter, K. J., Kenny, P., & Koss, F. V. (1999). Automated Intelligent Pilots for Combat Flight Simulation. *AI Magazine, 20*(1), 27–41.

Kalisch, R., Korenfeld, E., Stephan, K. E., Weiskopf, N., Seymour, B., & Dolan, R. J. (2006). Context-dependent human extinction memory is mediated by a ventromedial prefrontal and hippocampal network. *Journal of Neuroscience, 26*(37), 9503–9511.

Kan, I. P., Barsalou, L. W., Solomon, K. O., Minor, J. K., & Thompson-Schill, S. L. (2003). Role of mental imagery in a property verification task: fMRI evidence for perceptual representations of conceptual knowledge. *Cognitive Neuropsychology, 20,* 525–540.

Kandel, E., Schwartz, J. H., & Jessell, T. M. (1991). *Principles of neural science.* New York: McGraw Hill.

Kane, M. J. & Engle, R. W. (2002). The role of prefrontal cortex in working-memory capacity, executive attention, and general fluid intelligence: An individual-differences perspective. *Psychonomic Bulletin & Review, 9*(4), 637–671.

Kanerva, P. (1994). *The spatter code for encoding concepts at many levels,* chapter 1, (pp. 226–229). Proceedings of the International Conference on Artificial Neural Networks. Sorrento, Italy: Springer-Verlag.

Kaplan, G. B. & Gzelis, C. (2001). Hopfield networks for solving Tower of Hanoi problems. *ARI: An Interdisciplinary Journal of Physical and Engineering Sciences, 52,* 23.

Kawato, M. (1995). Cerebellum and motor control. In M. Arbib (Ed.), *The handbook of brain theory and neural networks.* Cambridge, MA: MIT Press.

Khan, M., Lester, D., Plana, L., Rast, A., Jin, X., Painkras, E., et al. (2008). *SpiNNaker: Mapping neural networks onto a massively-parallel chip multiprocessor.* IEEE. International Joint Conference on Neural Networks. 2849–2856.

Kieras, D. E. & Meyer, D. E. (1997). An overview of the EPIC architecture for cognition and performance with application to human-computer interaction. *Human-Computer Interaction, 4*(12), 391–438.

Kim, H., Sul, J. H., Huh, N., Lee, D., & Jung, M. W. (2009). Role of striatum in updating values of chosen actions. *The Journal of neuroscience : the official journal of the Society for Neuroscience, 29*(47), 14, 701–712.

Kim, R., Alterman, R., Kelly, P. J., Fazzini, E., Eidelberg, D., Beric, A., et al. (1997). Efficacy of bilateral pallidotomy. *Neurosurgical FOCUS, 2*(3), E10.

Kirkwood, A., Rioult, M. C., & Bear, M. F. (1996). Experience-dependent modification of synaptic plasticity in visual cortex. *Nature, 381*(6582), 526–528.

Kitcher, P. (1993). *The advancement of science.* Oxford: Oxford University Press.

Klahr, D., Chase, W. G., & Lovelace, E. A. (1983). Structure and process in alphabetic retrieval. *Journal of Experimental Psychology, 9*(3), 462–477.

Knill, D. C. & Pouget, A. (2004). The Bayesian brain: the role of uncertainty in neural coding and computation. *Trends in Neuroscience, 27*(12), 712–719.

Knoblauch, A. (2011). Neural associative memory with optimal Bayesian learning. *Neural computation, 23*(6), 1393–1451.

Koch, C. (1999). *Biophysics of computation: Information processing in single neurons.* New York: Oxford University Press.

Kosslyn, S. M., Ganis, G., & Thompson, W. L. (2000). Neural foundations of imagery. *Nature Reviews Neuroscience, 2,* 635–642.

Koulakov, A. A., Raghavachari, S., Kepecs, A., & Lisman, J. E. (2002). Model for a robust neural integrator. *Nature neuroscience, 5*(8), 775–782.

Krajbich, I. & Rangel, A. (2011). Multialternative drift-diffusion model predicts the relationship between visual fixations and choice in value-based decisions. *Proceedings of the National Academy of Sciences of the United States of America, 108*(33), 13, 852–857.

Kreiman, G., Koch, C., & Fried, I. (2000). Category-specific visual responses of single neurons in the human medial temporal lobe. *Nature Neuroscience, 3,* 946–953.

Kringelbach, M. L. (2005). The human orbitofrontal cortex: linking reward to hedonic experience. *Nature reviews. Neuroscience, 6*(9), 691–702.

Krueger, L. E. (1989). Reconciling Fechner and Stevens: Toward a unified psychophysical law. *Behavioral and Brain Sciences, 12,* 251–267.

Kuindersma, S., Grupen, R., & Barto, A. (2011). Learning Dynamic Arm Motions for Postural Recovery. In *Proceedings of the 11th IEEE-RAS International Conference on Humanoid Robots,* Bled, Slovenia. IEEE.

Kulic, D., Ott, C., Lee, D., Ishikawa, J., & Nakamura, Y. (2011). Incremental learning of full body motion primitives and their sequencing through human motion observation. *The International Journal of Robotics Research.*

Kuo, D. & Eliasmith, C. (2005). Integrating behavioral and neural data in a model of zebrafish network interaction. *Biological Cybernetics, 93*(3), 178–187.

Lakoff, G. (1987). *Women, fire, and dangerous things.* Chicago: University of Chicago Press.

Lamb, S. M. (1999). *Pathways of the brain: the neurocognitive basis of language.* 4. Current issues in linguistic theory. John Benjamins Publishing Company. Amsterdam, NL.

Landauer, T. (1962). Rate of implicit speech. *Perceptual and Motor Skills, 1,* 646.

Landauer, T. (1986). How much do people remember? some estimates of the quantity of learned information in long-term memory. *Cognitive Science, 10*(4), 477–493.

Landauer, T. & Dumais, S. T. (1997). A solution to plato's problem: The latent semantic analysis theory of acquisition, induction and representation of knowledge. *Psychological Review, 104*(2), 211–240.

Landers, D. M. & Feltz, D. L. (2007). The Effects of Mental Practice on Motor Skill Learning and Performance: A Meta-Analysis. In D. E. Smith & M. Bar-Eli (Eds.), *Essential Readings in Sport and Exercise Psychology*. Champaign, IL: Human Kinetics.

Langacker, R. W. (1986). An introduction to cognitive grammar. *Cognitive Science, 10*, 1–40.

Langley, P., Laird, J. E., & Rogers, S. (2009). Cognitive architectures: Research issues and challenges. *Cognitive Systems Research, 10*(2), 141–160.

Lass, Y. & Abeles, M. (1975). Transmission of information by the axon. I: Noise and memroy in the myelinated nerve fiber of the frog. *Biological Cybernetics, 19*, 61–67.

Laubach, M., Caetano, M. S., Liu, B., Smith, N. J., Narayanan, N. S., & Eliasmith, C. (2010). Neural circuits for persistent activity in medial prefrontal cortex. In *Society for Neuroscience Abstracts*, (pp. 200.18). Washington, DC: Society for Neuroscience.

Lebiere, C. & Anderson, J. R. (1993). A connectionist implementation of the ACT-R production system. In *Fifteenth Annual Conference of the Cognitive Science Society*, (pp. 635–640), Hillsdale, NJ: Erlbaum.

Lee, H., Ekanadham, C., & Ng, A. (2007). Sparse deep belief net model for visual area V2. *Advances in neural information processing systems, 20*, 1–8.

Lee, J. & Maunsell, J. (2010). The effect of attention on neuronal responses to high and low contrast stimuli. *Journal of neurophysiology, 104*(2), 960–971.

Lee, W. (1984). Neuromotor synergies as a basis for coordinated intentional action. *Journal of Motor Behavior, 16*, 135–170.

Legendre, G., Miyata, Y., & Smolensky, P. (1994). *Principles for an integrated connectionist/symbolic theory of higher cogntion*. Hillsdale, NJ: Lawrence Erlbaum Associates.

Lerman, M., Danckert, J., & Eliasmith, C. (2005). A computational model of the effects of prism adaptation in spatial neglect. In *Society for Neuroscience*, (pp. 287.15), Washington, DC: Society for Neuroscience.

Lewis, F. L. (1992). *Applied optimal control and estimation*. New York: Prentice-Hall.

Lieberman, P. (2006). *Toward an evolutionary biology of language*. Cambridge, MA: The Belknap Press of Harvard University Press.

Lieberman, P. (2007). The Evolution of Human Speech: Its Anatomical and Neural Bases. *Current Anthropology, 48*(1), 39–66

Liepa, P. (1977). Models of content addressable distributed associative memory. Technical report, University of Toronto.

Lipinski, J., Spencer, J. P., Samuelson, L. K., & Schöner, G. (2006). Spam-ling: A dynamical model of spatial working memory and spatial language. In *Proceedings of the 28th Annual Conference of the Cognitive Science Society*, (pp. 768–773), Vancouver, Canada: Cognitive Science Society.

Litt, A., Eliasmith, C., & Thagard, P. (2008). Neural affective decision theory: Choices, brains, and emotions. *Cognitive Systems Research, 9*, 252–273.

Liu, B., Caetano, M., Narayanan, N., Eliasmith, C., & Laubach, M. (2011). A neuronal mechanism for linking actions to outcomes in the medial prefrontal cortex. In *Computational and Systems Neuroscience 2011*. Computational and Systems Neuroscience (CoSyNe) 2011, Salt Lake City, UT, USA.

Liu, D. & Todorov, E. (2009). Hierarchical optimal control of a 7-dof arm model. In *IEEE Symp. on Adaptive Dynamic Programming and Reinforcement Learning*, (pp. 50–57). Orlando, FL: IEEE

Liu, Y. & Jagadeesh, B. (2008). Neural selectivity in anterior inferotemporal cortex for morphed photographic images during behavioral classification or fixation. *Journal of Neurophysiology, 100*(2), 966–982.

Logothetis, N. K. & Wandell, B. A. (2004). Interpreting the BOLD Signal. *Annual review of physiology*, 66, 735-69.

Luenberger, D. (1992). A double look at duality. *IEEE Transactions on Automatic Control, 37*(10), 1474–1482.

Lund, J. S., Yoshioka, T., & Levitt, J. B. (1993). Comparison of intrinsic connectivity in different areas of macaque monkey cerebral cortex. *Cerebral cortex, 3*(2), 148–162.

Ma, L., Steinberg, J. L., Hasan, K. M., Narayana, P. A., Kramer, L. A., & Moeller, F. G. (2011). Working memory load modulation of parieto-frontal connections: Evidence from dynamic causal modeling. *Human brain mapping*. 33(8), 1850-1867.

Machamer, P., Darden, L., & Craver, C. (2000). Thinking about mechanisms. *Philosophy of Science, 67*, 1–25.

Machens, C. K., Romo, R., & Brody, C. D. (2010). Functional, but not anatomical, separation of "what" and "when" in prefrontal cortex. *The Journal of Neuroscience, 30*(1), 350–360.

Machery, E. (2009). *Doing Without Concepts*. New York: Oxford University Press.

MacNeil, D. & Eliasmith, C. (2011). Fine-Tuning and the Stability of Recurrent Neural Networks. *PLoS ONE, 6*(9), e22885.

Maia, T. V. (2009). Reinforcement learning, conditioning, and the brain: Successes and challenges. *Cognitive, affective & behavioral neuroscience, 9*(4), 343–364.

Maratsos, M. & Kuczaj, S. (1976). Is "Not N't"? A Study in Syntactic Generalization. Technical report, Education Resources Information Centre, Standford University Dept. of Linguistics. Stanford, CA: Stanford University.

Marcus, G. F. (2001). *The algebraic mind*. Cambridge, MA: MIT Press.

Markram, H. (2006). The blue brain project. *Nature reviews. Neuroscience, 7*(2), 153–160.

Markram, H., Lübke, J., Frotscher, M., & Sakmann, B. (1997). Regulation of synaptic efficacy by coincidence of postsynaptic APs and EPSPs. *Science, 275*(5297), 213–215.

Marr, D. (1982). *Vision*. San Francisco, CA: Freeman.

Martin, A. (2007). The representation of object concepts in the brain. *Annual Review of Psychology, 58*, 25–45.

McCormick, D. A., Connors, B. W., Lighthall, J. W., & Prince, D. A. (1985). Comparative electrophysiology of pyramidal and sparsely spiny stellate neuron. *Journal of Neurophysiology, 54*, 782–806.

McKay, T. J. (1999). *Reasons, Explanations, and Decisions: Guidelines for Critical Thinking.* Stamford, CT: Wadsworth Publishing.

McNorgan, C., Reid, J., & McRae, K. (2011). Integrating conceptual knowledge within and across representational modalities. *Cognition, 118*(2), 211–233.

Mehta, A. D. (2000). Intermodal Selective Attention in Monkeys. I: Distribution and Timing of Effects across Visual Areas. *Cerebral Cortex, 10*(4), 343–358.

Miikkulainen, R. (1991). Trace feature map: A model of episodic associative memory. *Biological Cybernetics.* 66(3), 273-282.

Miller, G. (1956). The magical number seven, plus or minus two: Some limits on our capacity for processing information. *Psychological review, 63*, 81–97.

Miller, J., Jacobs, G. A., & Theunissen, F. (1991). Representation of sensory information in the cricket cercal sensory system. I: Response properties of the primary interneurons. *Journal of Neurophysiology, 66*, 1680–1703.

Miller, P., Brody, C. D., Romo, R., & Wang, X. J. (2003). A Recurrent Network Model of Somatosensory Parametric Working Memory in the Prefrontal Cortex. *Cerebral Cortex, 13*, 1208–1218.

Millikan, R. G. (1984). *Language, thought and other biological categories.* Cambridge, MA: MIT Press.

Mink, J. W. (1996). The basal ganglia: Focused selection and inhibition of competing motor programs. *Progress in Neurobiology, 50*, 381–425.

Miyake, A. & Shah, P. (Eds.). (1999). *Models of Working Memory: Mechanisms of Active Maintenance and Executive Control.* New York: Cambridge University Press.

Moran, D. W. & Schwartz, A. B. (1999). Motor cortical representation of speed and direction during reaching. *Journal of Neurophysiology, 82*, 2676–2692.

Mumford, D. (1996). Pattern theory: A unifying perspective. In D. C. Knill & W. Richards (Eds.), *Perception as Bayesian inference.* Cambridge, UK: Cambridge University Press.

Murdock, B. B. (1983). A distributed memory model for serial-order information. *Psychological review, 90*(4), 316–338.

Murdock, B. B. (1993). Todam2: A model for the storage and retrieval of item, associative and serial-order information. *Psychological review, 100*(2), 183–203.

Murphy, G. L. (2002). *The Big Book of Concepts.* Cambridge MA: MIT Press.

Nambu, A., Tokuno, H., & Takada, M. (2002). Functional significance of cortico-subthalamo-pallidal 'hyperdirect' pathway. *Neuroscience Research, 43*, 111–117.

Neumann, J. (2001). PhD Thesis. Holistic Processing of Hierarchical Structures in Connectionist Networks. Edinburgh, UK: University of Edinburgh.

Newell, A. (1980). Physical symbol systems. *Cognitive Science, 4*(2), 135–183.

Newell, A. (1990). *Unified theories of cognition.* Cambridge, MA: Harvard University Press.

Newell, A., Shaw, C., & Simon, H. (1958). Elements of a theory of human problem solving. *Psychological review, 65*, 151–166.

Newell, A. & Simon, H. (1963). GPS: A Program that simulates human thought. In Feldman, E.A. & Feigenbaum, J. (Eds.), *Computers and Thought.* New York: McGraw-Hill.

Newell, A. & Simon, H. A. (1976). *GPS, a program that simulates human thought.* Computers and thought. New York: McGraw-Hill.

Niklasson, L. F. & van Gelder, T. (1994). On being systematically connectionist. *Mind and Language, 9*, 288–302.

Nise, N. S. (2007). *Control Systems Engineering, 5th Edition.* Hoboken, NJ: Wiley.

Norman, D. A. (1986). Reflections on Cognition and parallel distributed processing. In McClelland, J. L. & Rumelhart, D. E. (Eds.), *Parallel distributed processing: Explorations in the microstructure of cognition. Vol. 2: Psychological and biological models.* Cambridge, MA: MIT Press/Bradford.

Oaksford, M. & Chater, N. (1994). A rational analysis of the selection task as optimal data selection. *Psychological review, 101*(4), 608–631.

Oaksford, M. & Chater, N. (1996). Rational explanation of the selection task. *Psychological Review, 103*(2), 381–391.

Oaksford, M. & Chater, N. (Eds.). (1998). *Rational models of cognition.* New York, NY: Oxford University Press.

Ojemann, G., Ojemann, J., Lettich, E., & Berger, M. (1989). Cortical language localization in left, dominant hemisphere. An electrical stimulation mapping investigation in 117 patients. *Journal Of Neurosurgery, 71*(3), 316–326.

Olshausen, B. A., Anderson, C. H., & Essen, D. C. V. (1993). A neurobiological model of visual attention and invariant pattern recognition based on dynamic routing of information. *Journal of Neuroscience, 13*(11), 4700–4719.

Olshausen, B. A. & Field, D. J. (1996). Emergence of simple-cell receptive field properties by learning a sparse code for natural images. *Nature, 381*, 607–609.

Oppenheim, P. & Putnam, H. (1958). Unity of science as a working hypothesis. In Feigl,H. Maxwell, G. & Scriven, M.(Eds.), *Minnesoia Studies in the Philosophy of Science* (pp. 3–36). Minneapolis, MN: University of Minnesota Press.

O'Reilly, R. C. & Frank, M. J. (2006). Making working memory work: a computational model of learning in the prefrontal cortex and basal ganglia. *Neural Computation, 18*(2), 283–328.

O'Reilly, R. C., Frank, M. J., Hazy, T. E., & Watz, B. (2007). PVLV: the primary value and learned value Pavlovian learning algorithm. *Behavioral Neuroscience, 121*(1), 31–49.

O'Reilly, R. C., Herd, S. A., & Pauli, W. M. (2010). Computational models of cognitive control. *Current Opinion in Neurobiology, 20*(2), 257–261.

O'Reilly, R. C. & Munakata, Y. (2000). *Computational Explorations in Cognitive Neuroscience: Understanding the Mind by Simulating the Brain* (1 ed.). Cambridge, MA: The MIT Press.

Osherson, D. N. & Smith, E. E. (1981). On the adequacy of prototype theory as a theory of concepts. *Cognition, 9*(1), 935-958.

Owen, A. M. (2004). Working memory: imaging the magic number four. *Current Biology, 14*(14), R573–R574.

Oztop, E., Kawato, M., & Arbib, M. (2006). Mirror neurons and imitation : A computationally guided review. *Neural Networks, 19*, 254–271.

Paaß, G., Kindermann, J., & Leopold, E. (2004). Learning prototype ontologies by hierachical latent semantic analysis. In *Lernen, Wissensentdeckung und Adaptivität*, (pp. 193–205). ECAI-2004.

Page, M. & Norris, D. (1998). The primacy model: A new model of immediate serial recall. *Psychological Review, 105*(4), 761–781.

Paivio, A. (1971). *Imagery and verbal processes.* Holt, New York, NY: Rinehart and Winston.

Paivio, A. (1986). *Mental representations: A dual coding approach.* New York: Oxford University Press.

Pakkenberg, B. & Gundersen, H. J. (1997). Neocortical neuron number in humans: effect of sex and age. *The Journal of Comparative Neurology, 384*(2), 312–320.

Parisien, C., Anderson, C. H., & Eliasmith, C. (2008). Solving the problem of negative synaptic weights in cortical models. *Neural Computation, 20*, 1473–1494.

Parisien, C. & Thagard, P. (2008). Robosemantics: How Stanley the Volkswagen represents the world. *Minds and Machines, 18*, 169–178.

Parks, R. W. & Cardoso, J. (1997). Parallel distributed processing and executive functioning: Tower of Hanoi neural network model in healthy controls and left frontal lobe patients. *International Journal of Neuroscience, 89*, 217.

Parsons, L. & Osherson, D. (2001). New evidence for distinct right and left brain systems for deductive versus probabilistic reasoning. *Cerebral Cortex, 11*, 954–965.

Parsons, L., Osherson, D., & Martinez, M. (1999). Distinct neural mechanisms for propositional logic and probabilistic reasoning. In *Proceedings of the Psychonomic Society Meeting*, Green Valley, AZ: Psychonomic Society, (pp.61–62).

Perfetti, B., Saggino, A., Ferretti, A., Caulo, M., Romani, G. L., & Onofrj, M. (2009). Differential patterns of cortical activation as a function of fluid reasoning complexity. *Human Brain Mapping, 30*(2), 497–510.

Pesaran, B., Pezaris, J. S., Sahani, M., Mitra, P. P., & Andersen, R. A. (2002). Temporal structure in neuronal activity during working memory in macaque parietal cortex. *Nature Neuroscience, 5*(8), 805–811.

Peters, A. & Jones, E. G. (1984). *Cerebral Cortex*, volume 1. New York: Plenum Press.

Petersen, S., Robinson, D., & Keys, W. (1985). Pulvinar nuclei of the behaving rhesus monkey: visual responses and their modulation. *Journal of Neurophysiology, 54*(4), 867.

Petersen, S., Robinson, D., & Morris, J. (1987). Contributions of the pulvinar to visual spatial attention. *Neuropsychologia, 25*, 97–105.

Pew, R. W. & Mavor, A. S. (Eds.). (1998). *Modeling human and organizational behavior: Application to military simulations.* Washington, DC: National Academy Press.

Pfister, J.-P. & Gerstner, W. (2006). Triplets of spikes in a model of spike timing-dependent plasticity. *Journal of Neuroscience, 26*(38), 9673–9682.

Plate, T. A. (1991). Holographic reduced representations: Convolution algebra for compositional distributed representations. In Mylopoulos, J. (Ed.), *Proceedings of the 12th International Joint Conference on Artificial Intelligence.* New York, NY: Morgan Kaufamann.

Plate, T. A. (1994). Distributed representations and nested compositional structure. PhD Thesis, Computer Science, Dept of Computer Science, Toronto, ON: University of Toronto.

Plate, T. A. (2003). *Holographic reduced representations.* Stanford, CA: CSLI Publication.

Plaut, D. C. & Shallice, T. (1994). Word Reading in Damaged Connectionist Networks: Computational and Neuropsychological Implications. In Mammone, R. (Ed.), *Artificial Neural Networks for Speech and Vision* (pp. 294–323). London: Chapman & Hall.

Poirazi, P., Brannon, T., & Mel, B. W. (2003). Pyramidal Neuron as Two-Layer Neural Network. *Neuron, 37*(6), 989–999.

Pollack, J. (1990). Recursive distributed representations. *Artificial Intelligence, 46*, 77–105.

Pollack, J. B. (1988). Recursive auto-associative memory: devising compositional distributed representations. In *Proceedings of the 10th Annual Conference of the Cognitive Science Society.*

Polsky, A., Mel, B. W., & Schiller, J. (2004). Computational subunits in thin dendrites of pyramidal cells. *Nature neuroscience, 7*(6), 621–627.

Popper, K. (1959). *The logic of scientific discovery.* London: Hutchinson.

Port, R. & van Gelder, T. (1995). *Mind as motion: Explorations in the dynamics of cognition.* Cambridge, MA: MIT Press.

Postman, L. & Phillips, L. (1965). Short-term temporal changes in free recall. *The Quarterly Journal of Experimental Psychology, 17*(2), 132–138.

Prabhakaran, V., Smith, J., Desmond, J., Glover, G., & Gabrieli, E. (1997). Neural substrates of fluid reasoning: an fMRI study of neocortical activation during performance of the Raven's Progressive Matrices Test. *Cognitive Psychology, 33*, 43–63.

Prinz, J. (2002). *Furnishing the Mind: Concepts and the Perceptual Basis.* Cambridge, UK: MIT Press.

Pulvermüller, F. (1999). Words in the brain's language. *Behavioral and Brain Sciences, 22*, 253–336.

Putnam, H. (1975). *Philosophy and our mental life*, chapter 2, (pp. 291–303). Mind, language and reality: Philosophical papers. Cambridge, UK: Cambridge University Press.

Qi, X., Katsuki, F., Meyer, T., Rawley, J., Zhou, X., Douglas, K., (2010). Frontiers: Comparison of neural activity related to working memory in primate dorsolateral prefrontal and posterior parietal cortex. *Frontiers in Systems Neuroscience, 4*, 12.

Quine, W. V. O. & Ullian, J. (1970). *The web of belief.* New York: Random House.

Quirk, G. J. & Beer, J. S. (2006). Prefrontal involvement in the regulation of emotion: convergence of rat and human studies. *Current Opinion in Neurobiology, 16*(6), 723–727.

Quiroga, R. Q., Reddy, L., Kreiman, G., Koch, C., & Fried, I. (2005). Invariant visual representation by single neurons in the human brain. *Nature, 435*(7045), 1102–1107.

Quyen, M. L. V., Foucher, J., Lachaux, J.-P., Rodriguez, E., Lutz, A., Martinerie, J., et al. (2001). Comparison of Hilbert transform and wavelet methods for the analysis of neuronal synchrony. *Journal of Neuroscience Methods, 111*(2), 83–98.

Ramscar, M. & Gitcho, N. (2007). Developmental change and the nature of learning in childhood. *Trends in cognitive sciences, 11*(7), 274–279.

Ranzato, M., Boureau, Y., & Lecun, Y. (2007). Sparse feature learning for deep belief networks. *Advances in Neural Information Processing Systems.*

Rasmussen, D. (2010). *A neural modelling approach to investigating general intelligence.* Masters thesis, Waterloo, ON: University of Waterloo.

Rasmussen, D. & Eliasmith, C. (2011a). A neural model of rule generation in inductive reasoning. *Topics in Cognitive Science, 3*(1), 140–153.

Rasmussen, D. & Eliasmith, C. (2011b). A spiking neural model that accounts for human performance and cognitive decline on Raven's Advanced Progressive Matrices. *submitted.*

Raven, J. (1962). *Advanced Progressive Matrices (Sets I and II).* London: Lewis.

Raven, J., Raven, J., & Court, J. (2004). *Manual for Raven's Progressive Matrices and Vocabulary Scales.* San Antonio, TX: Harcourt Assessment.

Redgrave, P., Prescott, T., & Gurney, K. (1999). The basal ganglia: a vertebrate solution to the selection problem? *Neuroscience, 86*, 353–387.

Regan, J. K. O. & Noë, A. (2001). A sensorimotor account of vision and visual consciousness. *Behavioral and Brain Sciences, 24*(5), 939–1031.

Reitman, J. (1974). Without surreptitious rehearsal, information in short-term memory decay. *Journal of Verbal Learning and Verbal Behavior, 13*, 365–377.

Reynolds, J. H., Chelazzi, L., & Desimone, R. (1999). Competitive mechanisms sub- serve attention in macaque areas V2 and V4. *Journal of Neuroscience, 19*, 1736–1753.

Rieke, F., Warland, D., de Ruyter van Steveninick, R., & Bialek, W. (1997). *Spikes: Exploring the neural code.* Cambridge, MA: MIT Press.

Rinella, K., Bringsjord, S., & Yang, Y. (2001). *Efficacious logic instruction: People are not irremediably poor deductive reasoners*, (pp. 851–856). Proceedings of the 23rd Annual Conference of the Cognitive Science Society. Mahwah, NJ: Lawrence Erlbaum Associates.

Ringach, D. L. (2002). Spatial Structure and Symmetry of Simple-Cell Receptive Fields in Macaque Primary Visual Cortex. *J Neurophysiol, 88*(1), 455–463.

Ritter, D. A., Bhatt, D. H., & Fetcho, J. R. (2001). In Vivo Imaging of Zebrafish Reveals Differences in the Spinal Networks for Escape and Swimming Movements. *Journal of Neuroscience, 21*(22), 8956–8965.

Ritter, F. E., Kukreja, U., & St. Amant, R. (2007). Including a model of visual processing with a cognitive architecture to model a simple teleoperation task. *Journal of Cognitive Engineering and Decision Making, 1*(2), 121–147.

Roberts, S. & Pashler, H. (2000). How persuasive is a good fit? A comment on theory testing. *Psychological Review, 107*(2), 358–367.

Rockel, A. J., Hiorns, R. W., & Powell, T. P. S. (1980). The basic uniformity in structure of the neocortex. *Brain, 103*(2), 221–244.

Rogers, T. T. & McClelland, J. L. (2004). *Semantic cognition: a parallel distributed processing approach.* Cambridge, MA: MIT Press.

Roitman, J. D. & Shadlen, M. N. (2002). Response of neurons in the lateral intraparietal area during a combined visual discrimination reaction time task. *The Journal of neuroscience : the official journal of the Society for Neuroscience, 22*(21), 9475–9489.

Romo, R., Brody, C. D., Hernández, A., & Lemus, L. (1999). Neuronal correlates of parametric working memory in the prefrontal cortex. *Nature, 399*, 470–473.

Romo, R., Brody, C. D., Hernandez, A., Lemus, L., & Hernández, A. (1999). Neuronal correlates of parametric working memory in the prefrontal cortex. *Nature, 399*, 470–473.

Rosenberg, A. (1994). *Instrumental Biology, or the Disunity of Science.* Chicago: University of Chicago Press.

Rosenbloom, P. S., Laird, J. E., & Newell, A. (1993). *The Soar papers: Research on integrated intelligence.* Cambridge, MA: MIT Press.

Rosenblueth, A., Wiener, N., & Bigelow, J. (1943). Behavior, purpose, and teleology. *Philosophy of Science, 10*, 18–24.

Rumelhart, D. E. & McClelland, J. L. (1986a). *Parallel distributed processing: Explorations in the microstructure of cognition.* Number 1. Cambridge MA: MIT Press/Bradford Books.

Rumelhart, D. E. & McClelland, J. L. (1986b). *Parallel distributed processing: Explorations in the microstructure of cognition.* Number 1. Cambridge MA: MIT Press/Bradford Books.

Rundus, D. (1971). Analysis of rehearsal processes in free recall. *Journal of experimental psychology, 89*(1), 63–77.

Ryan, L. & Clark, K. (1991). The role of the subthalamic nucleous in the response of globus pallidus neurons to stimulation of the prelimbic and agranular frontal cortices in rats. *Experimental Brain Research, 86*(3), 641–651.

Salinas, E. & Abbott, L. F. (1994). Vector reconstruction from firing rates. *Journal of Computational Neuroscience, 1*, 89–107.

Salinas, E. & Abbott, L. F. (1997). Invariant Visual Responses From Attentional Gain Fields. *J Neurophysiol, 77*(6), 3267–3272.

Salvucci, D. (2006). Modeling driver behavior in a cognitive architecture. *Human Factors, 48*, 368–380.

Scheier, C. & Pfeifer, R. (1995). Classification as Sensory-Motor Coordination–A Case Study . Chacon (Eds.) Advances in Artificial Life: Proceedings of the Third European Conference on Artifical Life, New York: Springer Verlag, 657–667.

Schiller, J., Major, G., Koester, H. J., & Schiller, Y. (2000). NMDA spikes in basal dendrites of cortical pyramidal neurons. *Nature, 404*(6775), 285–289.

Schöner, G. (2008). Dynamical systems approaches to cognition. *Cambridge Handbook of Computational Cognitive Modeling,* 101–126.

Schöner, G. & Thelen, E. (2006). Using dynamic field theory to rethink infant habituation. *Psychological Review, 113,* 273–299.

Schultheis, H. (2009). Computational and Explanatory Power of Cognitive Architectures: The Case of ACT-R. In Howes, A., Peebles, D., & Cooper, R. P. (Eds.), *International Conference on Cognitive Modeling,* (pp. 384–389).

Schultz, W., Dayan, P., & Montague, P. R. (1997). A neural substrate of prediction and reward. *Science, 275,* 1593–1599.

Schutte, A. R., Spencer, J. P., & Schöner, G. (2003). Testing the dynamic field theory: Working memory for locations becomes more spatially precise over development. *Child Development,* (74), 1393–1417.

Schwartz, A. B. (1994). Direct cortical representation of drawing. *Science, 265,* 540–542.

Searle, J. (1984). *Minds, Brains and Science: The 1984 Reith Lectures.* Cambridge, MA: Harvard University Press.

Seger, C. A. & Cincotta, C. M. (2006). Dynamics of frontal, striatal, and hippocampal systems during rule learning. *Cerebral cortex, 16*(11), 1546–1555.

Seung, H. S. (1996). How the brain keeps the eyes still. *Proceedings of the National Academy of Sciences USA, 93,* 13, 339–344.

Shadlen, M. & Newsome, W. (1996). Motion perception: seeing and deciding. *Proceedings of the National Academy of Science, 93,* 628–633.

Shadlen, M. N. & Movshon, J. A. (1999). Synchrony Unbound: A Critical Evaluation of the Temporal Binding Hypothesis. *Neuron, 24*(1), 67–77.

Shadlen, N. N. & Newsome, W. T. (2001). Neural basis of a perceptual decision in the parietal cortex (area lip) of the rhesus monkey. *Journal of Neurophysiology, 86,* 1916–1936.

Shastri, L. & Ajjanagadde, V. (1993). From simple associations to systematic reasoning: A connectionist representation of rules, variables, and dynamic bindings. *Behavioral and Brain Sciences, 16,* 417–494.

Silver, R., Boahen, K., Grillner, S., Kopell, N., & Olsen, K. L. (2007). Neurotech for neuroscience: Unifying concepts, organizing principles, and emerging tools. *Journal of Neuroscience, 27*(44), 807–819.

Simmons, W. K., Hamann, S. B., Harenski, C. L., Hu, X. P., & Barsalou, L. W. (2008). fMRI evidence for word association and situated simulation in conceptual processing. *Journal of Physiology, Paris, 102,* 106–119.

Simon, H. A. (1975). The functional equivalence of problem solving skills. *Cognitive Psychology, 7*(2), 268–288.

Simon, H. A. (1996). *Models of my life.* Cambridge, MA: MIT Press.

Singh, R. & Eliasmith, C. (2006). Higher-dimensional neurons explain the tuning and dynamics of working memory cells. *Journal of Neuroscience, 26*, 3667–3678.

Smith, E. E. (1989). *Concepts and induction*, (pp. 501–526). Foundations of cognitive science. Cambridge, MA: MIT Press.

Smolensky, P. (1986). Information processing in dynamical systems: Foundations of harmony theory. In Rumelhart, D. E. & McClelland, J. L. (Eds.), *Parallel Distributed Processing: Explorations in the Microstructure of Cognition. Volume 1: Foundations.* (pp. 194–281). Cambridge, MA: MIT Press.

Smolensky, P. (1988). On the proper treatment of connectionism. *Behavioral and Brain Sciences, 11*(1), 1–23.

Smolensky, P. (1990). Tensor product variable binding and the representation of symbolic structures in connectionist systems. *Artificial Intelligence, 46*, 159–217.

Smolensky, P., Goldrick, M., & Mathis, D. (2010). Optimization and Quantization in Gradient Symbol Systems: A Framework for Integrating the Continuous and the Discrete in Cognition. *Rutgers Optimality Archive*, 1–36.

Smolensky, P. & Legendre, G. (2006a). *The Harmonic Mind: From Neural Computation to Optimality-Theoretic Grammar Volume 1: Cognitive Architecture.* Cambridge MA: MIT Press.

Smolensky, P. & Legendre, G. (2006b). *The Harmonic Mind: From Neural Computation to Optimality-Theoretic Grammar Volume 2: Linguistic and Philosophical Implication.* Cambridge MA: MIT Press.

Solomon, K. O. & Barsalou, L. W. (2004). Perceptual simulation in property verification. *Memory and Cognition, 32*, 244–259.

Song, S., Sjostrom, P. J., Reigl, M., Nelson, S., & Chklovskii, D. B. (2005). Highly nonrandom features of synaptic connectivity in local cortical circuits. *PLoS biology, 3*(3), e68.

Sperber, D., Cara, E., & Girotto, R. (1995). Relevance theory explains the selection task. *Cognition, 57*, 31–95.

Spruston, N., Jonas, P., & Sakmann, B. (1995). Dendritic glutamate receptor channel in rat hippocampal CA3 and CA1 pyramidal neurons. *Journal of Physiology, 482*, 325–352.

Squire, L. (1992). Memory and the hippocampus: a synthesis from findings with rats, monkeys, and humans. *Psychological Review, 99*, 195–231.

Stepniewska, I. (2004). The pulvinar complex. In Kaas, J. & Collins, C. (Eds.), *The primate visual system.* Cambridge, MA: CRC Press LLC.

Stevens, C. F. & Wang, Y. (1994). Changes in reliability of synaptic function as a mechanism for plasticity. *Nature, 371*, 704–707.

Stewart, T., Bekolay, T., & Eliasmith, C. (2012). Learning to select actions with spiking neurons in the basal ganglia. *Frontiers in Decision Neuroscience, 6*.

Stewart, T. & Eliasmith, C. (2011a). Compositionality and biologically plausible models. *Oxford Handbook of Compositionality.* New York, NY: Oxford University Press.

Stewart, T. & Eliasmith, C. (2011b). Neural Cognitive Modelling: A Biologically Constrained Spiking Neuron Model of the Tower of Hanoi Task. In Carlson, L., Hölscher, C., & Shipley, T. (Eds.), *Proceedings of the 33rd Annual Conference of the Cognitive Science Society*, Austin, TX. Cognitive Science Society.

Stewart, T., Tang, Y., & Eliasmith, C. (2010). A biologically realistic cleanup memory: Autoassociation in spiking neurons. *Cognitive Systems Research.* 12(2), 84-92.

Stewart, T. C., Bekolay, T., & Eliasmith, C. (2011). Neural representations of compositional structures: Representing and manipulating vector spaces with spiking neurons. *Connection Science.* 3(2), 145-153.

Stewart, T. C., Choo, X., & Eliasmith, C. (2010). Dynamic Behaviour of a Spiking Model of Action Selection in the Basal Ganglia. In Salvucci, D. D. & Gunzelmann, G. (Eds.), *10th International Conference on Cognitive Modeling.* Philadelphia, PA: Drexel University.

Stewart, T. C., F-X., C., & C., E. (2010). Symbolic Reasoning in Spiking Neurons: A Model of the Cortex/Basal Ganglia/Thalamus Loop. In *32nd Annual Meeting of the Cognitive Science Society.* Austin, TX: Cognitive Science Society.

Sun, R. (2006). The CLARION cognitive architecture: Extending cognitive modeling to social simulation. In Sun, R. (Ed.), *Cognition and Multi-Agent Interaction.* Cambridge, UK: Cambridge University Press.

Taatgen, N. & Anderson, J. (2010). The past, present, and future of cognitive architectures. *Topics In Cognitive Science, 2,* 693–704.

Taatgen, N. & Anderson, J. R. (2002). Why do children learn to say "broke"? A model of learning the past tense wihout feedback. *Cognition, 86,* 123–155.

Tanaka, K. (1993). Neuronal mechanisms of object recognition. *Science, 262*(5134), 685–688.

Tang, Y. (2010). *Robust Visual Recognition Using Multilayer Generative Neural Networks.* Masters of mathematics, University of Waterloo.

Tang, Y. & Eliasmith, C. (2010). Deep networks for robust visual recognition. In Fürnkranz, J. & Joachims, T. (Eds.), *Proceedings of the 27th International Conference on Machine Learning.* ICML, Omnipress: University of Waterloo.

Taylor, G., Fergus, R., LeCun, Y., & Bregler, C. (2010). Convolutional Learning of Spatio-temporal Features. In *Proc. European Conference on Computer Vision (ECCV).* Lecture Notes in Computer Science, vol. 6311-6316. *Springer, Berlin.*

Taylor, G., Sigal, L., Fleet, D., & Hinton, G. E. (2010). Dynamic binary latent variable models for 3D human pose tracking. In *IEEE Conference on Computer Vision and Pattern Recognition.* IEEE, Orlando, FL.

Tenenbaum, J. B. & Griffiths, T. L. (2001). Generalization, similarity, and Bayesian inference. *Behavioral and Brain Sciences, 24,* 629–641.

Tenenbaum, J. B., Kemp, C., Griffiths, T. L., & Goodman, N. D. (2011). How to grow a mind: statistics, structure, and abstraction. *Science, 331*(6022), 1279–85.

Thagard, P. (2011). Cognitive architectures. In Frankish, K. & Ramsay, W. (Eds.), *The Cambridge handbook of cognitive science.* Cambridge: Cambridge University Press.

Thagard, P. (2012). *The cognitive science of science: Explanation, discovery, and conceptual change.* Cambridge, MA: MIT Press.

Thagard, P. & Litt, A. (2008). Models of scientific explanation. In Sun, R. (Ed.), *The Cambridge handbook of computational psychology* (pp. 549–564). Cambridge, UK: Cambridge University Press.

Thelen, E., Schöner, G., Scheier, C., & Smith., L. (2001). The dynamics of embodiment: A field theory of infant perseverative reaching. *Brain and Behavioral Sciences,* (24), 1–33.

Todorov, E. (2000). Direct cortical control of muscle activation in voluntary arm movements: A model. *Nature Neuroscience, 3,* 391–398.

Todorov, E. (2004). On the role of primary motor cortex in arm movement control. In Latash, M. & Levin, M. (Eds.), *In Progress in Motor Control III* chapter 6, (pp. 125–166). Chicago, IL: Human Kinetics.

Todorov, E. (2007). *Optimal control theory.* In *Bayesian Brain: Probabilistic Approaches to Neural Coding,* Doya K at al (eds), chap 12, pp 269-298, Cambridge, MA: MIT Press.

Todorov, E. (2009). Parallels between sensory and motor information processing. In Gazzaniga, M. S. (Ed.), *The cognitive neurosciences.* Cambridge, MA: MIT Press.

Tovée, M. & Rolls, E. (1992). The functional nature of neuronal oscillations. *Trends in Neuroscience, 15,* 387.

Treue, S. & Martinez-Trujillo, J. (1999). Feature-based attention influences motion processing gain in macaque visual cortex. *Nature, 399,* 575–579.

Tripp, B. P. & Eliasmith, C. (2007). Neural populations can induce reliable postsynaptic currents without observable spike rate changes or precise spike timing. *Cerebral Cortex, 17*(8), 1830–1840.

Turing, A. M. (1950). Computing machinery and intelligence. *Mind, 59,* 433–460.

Turner, R. S. & Desmurget, M. (2010). Basal ganglia contributions to motor control: a vigorous tutor. *Current opinion in neurobiology, 20*(6), 704–716.

Turrigiano, G. G. & Nelson, S. B. (2004). Homeostatic plasticity in the developing nervous system. *Nature reviews. Neuroscience, 5*(2), 97–107.

Ulanovsky, N., Las, L., Farkas, D., & Nelken, I. (2004). Multiple time scales of adaptation in auditory cortex neurons. *The Journal of Neuroscience: the Official Journal of the Society for Neuroscience, 24*(46), 10 440–453.

Usher, M. & McClelland, J. L. (2001). On the time course of perceptual choice: The leaky competing accumulator model. *Psychological Review, 108,* 550–592.

Van Der Meer, M. A. A., Johnson, A., Schmitzer-Torbert, N. C., & Redish, A. D. (2010). Triple Dissociation of Information Processing in Dorsal Striatum, Ventral Striatum, and Hippocampus on a Learned Spatial Decision Task. *Neuron, 67*(1), 25–32.

van der Velde, F. & de Kamps, M. (2006). Neural blackboard architectures of combinatorial structures in cognition. *Behavioral and Brain Sciences*, (29), 37–108.

Van Essen, D., Olshausen, B., Anderson, C., & Gallant, J. (1991). Pattern recognition, attention, and information bottlenecks in the primate visual system. In *Proceedings of SPIE Conf. on Visual Information Processing: From Neurons to Chips, 1473*, 17.

van Gelder, T. (1998). The dynamical hypothesis in cognitive science. *Behavioral and Brain Sciences, 21*(5), 615–665.

van Gelder, T. & Gelder, T. V. (1995). What might cognition be if not computation? *Journal of Philosophy, 92*(7), 345–381.

van Gelder, T. & Port, R. (1995). *It's about time: An overview of the dynamical approach to cognition.* Mind as motion: Explorations in the dynamics of cognition. Cambridge, MA: MIT Press.

Varela, J. A., Sen, K., Gibson, J., Fost, J., Abbott, L. F., & Nelson, S. B. (1997). A Quantitative Description of Short-Term Plasticity at Excitatory Synapses in Layer 2/3 of Rat Primary Visual Cortex. *Journal of Neuroscience, 17*(20), 7926–7940.

Verguts, T. & De Boeck, P. (2002). The induction of solution rules in Ravens Progressive Matrices Test. *European Journal of Cognitive Psychology, 14*, 521–547.

Vigneau, F., Caissie, A., & Bors, D. (2006). Eye-movement analysis demonstrates strategic influences on intelligence. *Intelligence, 34*(3), 261–272.

von der Malsburg, C. (1981). The correlation theory of brain function.

Vrabie, D. & Lewis, F. (2009). Neural network approach to continuous-time direct adaptive optimal control for partially unknown nonlinear systems. *Neural Networks, 22*(3), 237–246.

Wang, H., Fan, J., & Yang, Y. (2004). Toward a Multilevel Analysis of Human Attentional Networks. In Forbus, K., Genter, D., & Regier, T. (Eds.), *Proceedings of 26th Annual conference of the cognitive science society*, (pp. 1428–1433)., Mahwah, NJ: Lawrence Erlbaum.

Warden, M. R. & Miller, E. K. (2007). The representation of multiple objects in prefrontal neuronal delay activity. *Cerebral Cortex , 17*(suppl_1), i41–i50.

Wason, P. C. (1966). *Reasoning.* New horizons in psychology. Harmondsworth: Penguin, UK.

Weinrich, M., Wise, S. P., & Mauritz, K. H. (1984). A neurophysiological study of the premotor cortex in the rhesus monkey. *Brain: a Journal of Neurology, 107 (2)*, 385–414.

Weiskopf, D. A. (2010). Embodied cognition and linguistic comprehension. *Studies in History and Philosophy of Science, 41*, 294–304.

West, L. J., Pierce, C. M., & Thomas, W. (1962). Lysergic Acide Diethylamides: Its effects on a male asiatic elephant. *Science, 138*, 1100–1103.

White, E. (1989). *Cortical circuits.* Boston, MA: Birkhauser.

Whitlock, J. R., Heynen, A. J., Shuler, M. G., & Bear, M. F. (2006). Learning induces long-term potentiation in the hippocampus. *Science (New York, N.Y.)*, *313*(5790), 1093–1097.

Wolfrum, P. & von der Malsburg, C. (2007). What is the optimal architecture for visual information routing? *Neural Computation*, *19*(12), 3293–3309.

Wolpert, D. M. & Kawato, M. (1998). Multiple paired forward and inverse models for motor control. *Neural Networks*, *11*, 1317–1329.

Womelsdorf, T., Anton-Erxleben, K., Pieper, F., & Treue, S. (2006). Dynamic shifts of visual receptive fields in cortical area MT by spatial attention. *Nature Neuroscience*, *9*(9), 1156–1160.

Womelsdorf, T., Anton-Erxleben, K., & Treue, S. (2008). Receptive field shift and shrinkage in macaque middle temporal area through attentional gain modulation. *The Journal of neuroscience: the official journal of the Society for Neuroscience*, *28*(36), 8934–8944.

Zipser, D., Kehoe, B., Littlewort, G., & Fuster, J. (1993). A spiking network model of short-term active memory. *Journal of Neuroscience*, *13*(8), 3406–3420.

Zucker, R. S. (1973). Changes in the statistics of transmitter release during facilitation. *Journal of Physiology*, *229*(3), 787–810.

INDEX

CPSIA information can be obtained at www.ICGtesting.com
Printed in the USA
BVOW09s1516040316

438863BV00015B/12/P

9 780190 262129